AVAILABLE

The programs described in this book are available for your IBM PC (and most compatibles) on a 1.2 Mb high density 5¼″ floppy disk.

Order the Program Disk today priced £19.95 including VAT (UK)/$35.90 (elsewhere) from your computer store, bookseller, or by using the order form below.

G. Beer and J.O. Watson: Introduction to Finite and Boundary Element Methods for Engineers — Program Disk.

Please send me .... copies of the **G. Beer and J.O. Watson: Introduction to Finite and Boundary Element Methods for Engineers** — Program Disk at £19.95 including VAT/$35.90 each.

IBM PC 0 471 93820 3

POSTAGE AND HANDLING FREE FOR CASH WITH ORDER OR PAYMENT BY CREDIT CARD.

☐ Remittance enclosed .... Allow approx. 14 days for delivery

☐ Please charge this to my credit card (All orders subject to credit approval)

Delete as necessary: — AMERICAN EXPRESS, DINERS CLUB, BARCLAYCARD/VISA, ACCESS

CARD NUMBER ☐☐☐☐☐☐☐☐☐☐☐☐☐☐☐☐ Expiration date ....

☐ Please send me an invoice for prepayment. A small postage and handling charge will be made.

Software purchased for professional purposes is generally recognized as tax deductable.

☐ Please keep me informed of new books in my subject area which is ....

NAME ....

ADDRESS ....

....

OFFICIAL ORDER No. .... SIGNATURE

If you have any queries please contact:

Helen Ramsey
John Wiley & Sons Limited
Baffins Lane
Chichester
West Sussex
PO19 1UD
England

Affix stamp here

Customer Service Department
John Wiley & Sons Limited
Shripney Road
Bognor Regis
West Sussex
PO22 9SA
England

Introduction to Finite and Boundary Element Methods for Engineers

Introduction to Finite and Boundary Element Methods for Engineers

G. Beer
Division of Geomechanics
Commonwealth Scientific and Industrial Research Organisation
Australia

J. O. Watson
Department of Mining Engineering
University of New South Wales
Australia

JOHN WILEY & SONS
Chichester · New York · Brisbane · Toronto · Singapore

Other Wiley Editorial Offices

John Wiley & Sons, Inc., 605 Third Avenue,
New York, NY 10158-0012, USA

Jacaranda Wiley Ltd, G.P.O. Box 859, Brisbane,
Queensland 4001, Australia

John Wiley & Sons (Canada) Ltd, 22 Worcester Road,
Rexdale, Ontario M9W 1L1, Canada

John Wiley & Sons (SEA) Pte Ltd, 37 Jalan Pemimpin #05-04,
Block B, Union Industrial Building, Singapore 2057

British Library Cataloguing in Publication Data

A catalogue record for this book is available from the British Library

ISBN 0 471 92813 5

Typeset in 10/12pt Computer Modern Roman from authors' disks by
Text Processing Department, John Wiley & Sons Ltd, Chichester
Printed and bound in Great Britain by Biddles Ltd, Guildford, Surrey

Contents

Preface

The last three decades have seen a great increase in the power of digital computers. This development has facilitated the use of numerical methods for the design of objects made by man ranging from aircraft to bridges, from machine components to tunnels. Two of these numerical methods are the finite element method (FEM) and the boundary element method (BEM). Whereas the former is now firmly established, the latter is a more recent development which offers advantages in the analysis of infinite regions and stress concentration problems. The combination of the methods dates from about 1977 when Professor O. C. Zienkiewicz delivered his lecture entitled 'Marriage a la mode—finite elements and boundary integrals'. The ideas presented in that lecture have given impetus to combining the methods, and motivation for writing this book.

As one of us has a background in finite elements and the other in boundary elements, we hope that we present an objective view. We do not find it useful to consider that FEM and BEM belong to a common class of residual methods, or to regard BEM as just a special case of FEM in which global instead of local shape functions are used. Instead, we recognise the differences between the methods and discuss them in detail. By reference to simple examples we point out the shortcomings as well as the advantages of each method, and do not attempt to 'sell' one method or the other.

The emphasis of the book is on physical understanding and practical implementation of the methods. In the presentation of boundary element methods we make extensive use of matrix notation to render the theory and implementation as easy as possible to understand. Also, we begin the discussion at the root of the BEM which is the method devised by Trefftz for solving boundary value problems.

Although our background is solid mechanics, we discuss both flow problems (Laplace equation) and elasticity. However, the reader may detect some bias towards elasticity and in particular geomechanics, the area in which we are both working at present. To emphasise the practical, applied nature of this book we include many examples to demonstrate how the methods work. We use simple examples which have analytical solutions to show advantages and disadvantages of each method. Large scale industrial applications are also included to demonstrate that the methods are not only of academic interest. In most chapters, exercises are provided to enable the reader to test knowledge acquired.

Two computer programs, one FEM and the other BEM, are available on floppy disc so that readers may be in a position to apply the methods to their own problems. The programs, written in FORTRAN, can be run on a PC and are described in some detail in appendices; they may be used as a starting point by those wishing to develop their own codes.

The text is intended to be an introduction and many advanced topics are not discussed. Its purpose is to provide a firm basis of understanding from which readers may venture into their own special applications or progress to publications in which such applications are discussed. The book is also intended to serve as a reference, and for BEM includes a comprehensive list of two and three dimensional fundamental solutions presented in a concise and clear manner. It is hoped that the book will contribute to the unified use of both methods, resulting in more efficient numerical modelling of complex engineering problems.

G. Beer
J. O. Watson

Acknowledgements

To my father, Professor Herman Beer, who encouraged and inspired my career in Engineering. This is for the book he should have written.

To Professor O. C. Zienkiewicz for the initial inspiration he gave me, through his books, lectures and discussions, to work in the field of Finite Elements. To Professor John Meek, who supervised my PhD thesis and shared my enthusiasm in Finite Elements. To the people at T.D.V., Austria, who provided an atmosphere for pioneering work in FEM programming and industrial applications in the 1970s. A few examples of this early involvement in Finite Elements are included in this book.

To my co-author for helping my understanding of Boundary Elements and the many stimulating discussions. To the people of Mining Research at Mount Isa Mines who supported research in coupling FEM and BEM.

To Professor G. Swoboda, who was the first to apply the results of the research into coupling to practical problems in tunnelling.

To the chiefs of the Commonwealth Scientific and Industrial Research Organisation of Australia (CSIRO), Division of Geomechanics, Dr Barry Brady and Professor Bruce Hobbs, who supported the writing of this book. To Dell Dunn of the Long Pocket Laboratories who typed the first chapters of the book. Very special thanks are due to Brett Poulson who patiently and without complaint prepared most of the manuscript using TEX. To Dan Sullivan who looked after the drafting of the figures. To Irma Laredo for doing such a good job with the artwork used in Chapter 1. Thanks are also due to the many colleagues for their discussions, in particular to Professors H. Mang and A. Ingraffea and to Dr M. Fama and R. Yang, who checked the manuscript.

To Dr John Napier of the Chamber of Mines in South Africa (COMRO) who supplied the fundamental solutions for a displacement discontinuity in Appendix C.

Last but not least to my wife Sylvia for the support and help.

G. Beer

Thanks are due to the numerous individuals who have variously helped, encouraged or enabled me to progress through the educational system, gain practical experience of the development of numerical methods and engineering applications software, and press on with the development of boundary element methods: my parents, my excellent mentors at school and university, and many of those with whom I have since worked.

I thank in particular my supervisor of research at the University of Southampton, Hugh Tottenham, who introduced me to boundary integral equations; R. B. Moon and Walter Brown at Computation Research and Development, London who initiated me in the art of software development; Jean-Claude Lachat at Centre Technique des Industries Mécaniques, Senlis, France; Professor T. A. Cruse; Professor E. T. Brown who was instrumental in securing for me a Research Fellowship at Imperial College; and Professor F. F. Roxborough of the University of New South Wales.

Thanks are due to Lloyds Register of Shipping, London and to BP Research, Sunbury, UK for permission to present three dimensional elastic analyses by the boundary element method.

I join with my co-author in thanking those at CSIRO who prepared the manuscript and drawings with such meticulous care. Finally I thank my co-author for his initiative in developing a consistent and readable notation for both finite and boundary elements, and my wife Claire for her forbearance whilst I have been working on my part of the book.

J. O. Watson

1

Introduction to Finite and Boundary Element Methods

A journey of a million miles starts with the first step

Confucius

1.1 BACKGROUND

Numerical methods of stress analysis have been with us for some time now. They are routinely used by mechanical, civil and mining engineers in computer aided design (CAD) of mechanical components, structures and excavations.

CAD has played an important role in the design of aircraft and ship structures replacing to a large extent expensive prototype testing in wind tunnels or water tanks. Numerical modelling has also helped mining engineers to predict rock mass behaviour due to the excavation of ore materials. This has led to improvements in the economy and safety of mines. Numerical modelling has also recently entered such conservative areas as tunnel design. It is very probable modern engineers have been exposed to numerical modelling during their career.

Example 1. In Figure 1.1 is shown a ferro-concrete shell structure. The design question is what shape of the shell is optimal, i.e. is safe under various loadings including cyclonic wind conditions. A series of *finite element* analyses of different shapes gives the answer.

Example 2. In Figure 1.2 are shown mining excavations in a rock mass which contains a fault (to understand Figure 1.2 the reader should note that what is shown are surfaces of the fault and of the excavations, as if the rock

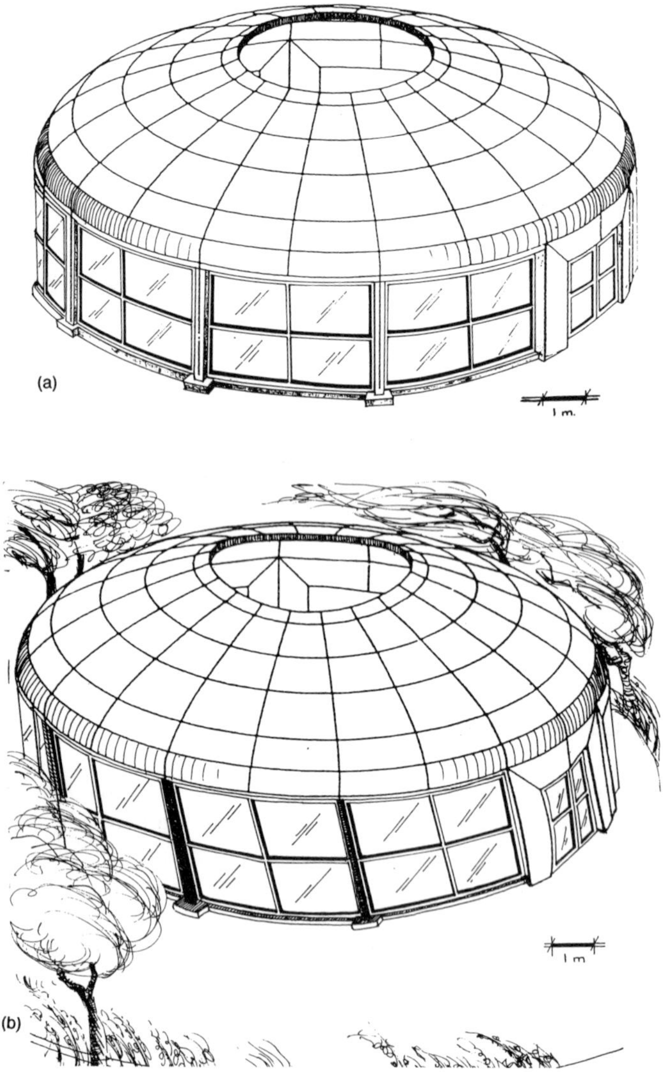

Figure 1.1 Design of a ferro-concrete roof. (a) subdivision into finite elements, and (b) displaced shape due to cyclonic wind. The analysis helped find the optimal shape of the shell

mass itself were transparent). Previous mining activity produced a sudden substantial movement on the fault and concern was expressed regarding further excavation. A *boundary element* analysis revealed that an extension of an existing excavation would have produced only minor additional movement of the fault.

This book is about the two most important numerical modelling techniques: the *finite element* method (FEM) and the *boundary element* method (BEM). They are discussed together because it is the authors' opinion that they complement each other and that the engineer should be given a choice. Indeed, we believe that software in which the methods are combined will find wide application in the future. As will become apparent in this book each of the methods is suited to particular problems, and for a given problem one is generally more efficient while the other is less efficient.

Basically the *finite element* methods require the subdivision of the region into small elements, the response characteristics of each of which are described in a simplified way. In the limit for infinitesimally small elements the solution of the problem will be exact. The *boundary element* method on the other hand only requires the subdivision of the boundary of the region because the solution of the problem to be solved is taken to be a combination of exact solutions inside the region. The latter method works extremely well for problems with a high ratio of volume to surface area but not so well for those with a low ratio. The opposite is the case for the finite element method. This is why in the first example the FEM was chosen whereas for the second the BEM was more efficient. In some cases a combination of the two methods is required to analyse the problem efficiently, and as will be shown in this book it is possible to use a combination of finite and boundary elements to solve a problem.

A list of text books on both the finite and boundary element methods is given at the end of this chapter. There are now over 40 text books on the FEM compared with less than ten books on the BEM, emphasising the fact that the BEM is a relatively new numerical method. In most of the text books on the BEM tensor notation is used, and most practising engineers find this notation hard to follow.

There are no reasons why tensor notation must be used except tradition and the fact that fundamental solutions can be presented in a concise way in that notation. The basic principles can be explained very well in matrix notation, and the authors believe that matrix notation enhances clarity of presentation as fewer subscripts are used. For this reason tradition is broken with here in that no tensor notation is used in explaining the BEM.

1.2 COMPUTATIONAL STEPS

In both the finite element and the boundary element methods basic computational steps are followed as broadly defined here. Details of the computational algorithms will be discussed in the relevant chapters.

The basic steps of the *finite element method* are as follows.

Step 1: Discretise the region. The first step is to divide the problem domain into a number of finite subdomains (finite elements) each of simple

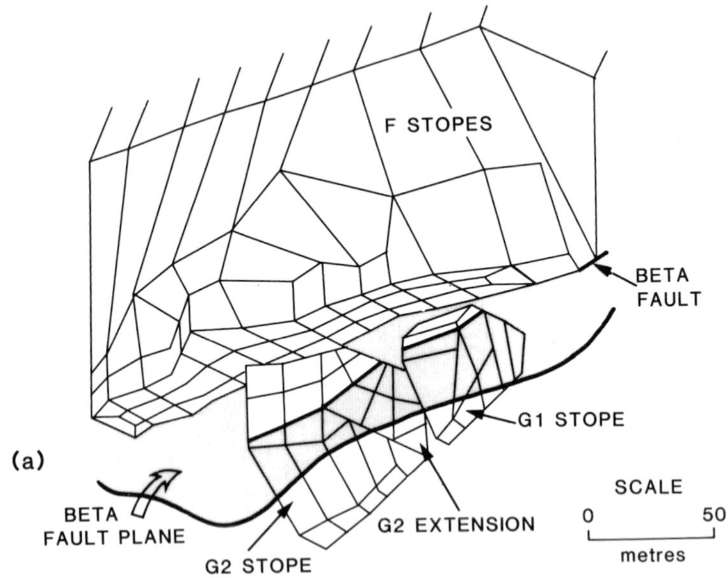

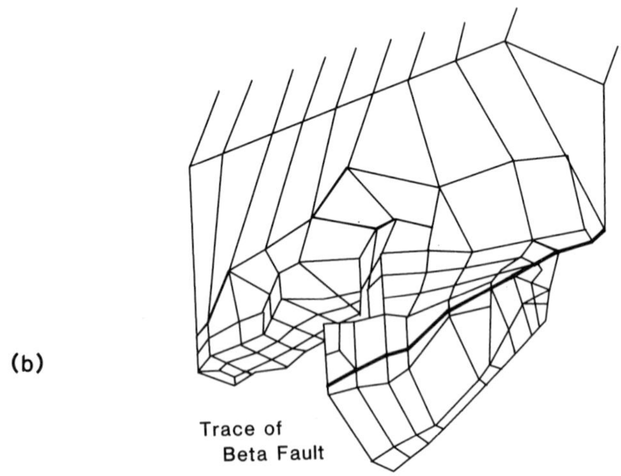

Figure 1.2 Mining excavations: prediction of slip on major fault (a) (b) Boundary Element meshes and (c) (plate) contours of slip before and after mining step 5

geometry. An element has a number of nodal points the locations in space of which are given by coordinates relative to a set of global axes. The shape of each element is defined in terms of these coordinates by interpolation or shape functions. This is also known as solid modelling and is not only a feature of finite element analysis. In Chapter 2 solid modelling is discussed in more detail and various forms of element are described.

Step 2: Assume a variation of the unknown inside each finite element. An interpolation function is proposed for the variation of the unknown (e.g. displacement, temperature) inside each element in terms of values at the nodal points. These interpolation functions in many cases will be the same as the shape functions used to describe the element shape which are discussed in Chapter 2.

Step 3: Find element response matrices. For each finite element, coefficient matrices which describe the response characteristics of the element are determined. In solid mechanics applications for example we compute a matrix of stiffness coefficients. Most of the methods used for determining the element coefficient matrices are based on the minimisation of total potential energy. In the limit as the number of elements used to model the domain tends to infinity, this is equivalent to solving the differential equation exactly.

In order to determine the response matrix the material behaviour has to be defined and this is discussed in Chapter 3. In Chapters 4 and 5 response matrices for solid elements, plate and shell elements and joint elements are derived.

Step 4: Assemble the element matrices to obtain the response matrix of the problem domain. To find the response matrix of the system to be modelled we must combine the response matrices of the individual elements to form a matrix equation expressing the behaviour of the entire solution region.

Step 5: Solve the system of equations. Before the assembled system of equations can be solved boundary conditions have to be specified. In most problems the number of equations is large, thus special solution techniques have to be employed to solve the system efficiently taking advantage of properties such as symmetry and sparseness of the matrix of equation coefficients. After solution the values of the dependent variable at all the nodes of the domain are known.

Step 6: Determine additional quantities as required. Using the nodal values and interpolation functions, we can determine additional parameters such as strain, stress and heat flux inside each element.

The basic steps for *boundary element methods* are as follows.

Step 1: Discretise the boundary. By contrast with the finite element method only the boundary of the problem domain need be discretised. The boundary elements are analogous to finite elements except in that their dimension is reduced by one. For plane problems one-dimensional (line) elements are used, whereas for the discretisation of a three-dimensional domain surface elements are required. These elements are discussed together with finite elements in Chapter 2.

Step 2: Assume a variation of the unknown inside each boundary element. This is the same as in the finite element method except in that only boundary values are interpolated.

Step 3: Use a fundamental solution of the governing differential equation. A fundamental solution is used which satisfies the governing differential equation exactly. In most cases this is a solution due to point loads or sources in an infinite domain. For elasticity problems the Kelvin solution is applied.

Step 4: Calculate element contributions to the coefficient matrices. The system of equations is, as in the finite element method, obtained by first computing element contributions. The element contributions are obtained by integrating over the boundary elements. Details are given in Chapters 6 and 7.

Step 5: Assemble element contributions. The next step is to combine all element contributions to form the response coefficient matrices for the problem to be solved. Here we already incorporate given boundary values such as known traction and displacement values over boundary elements.

Step 6: Solve system of equations. As in the FEM the system of equations is now solved. In the indirect methods the unknowns are fictitious parameters from which the boundary values and values at interior points can be determined. In the direct method the primary unknown values are obtained directly.

Step 7: Compute values inside the domain. By contrast with the FEM the solution of the boundary element equations give only unknowns (e.g. displacements, surface tractions) at the boundary or surface of the problem domain. If displacement or stress values are required at specific points inside the domain they can be calculated from those boundary values. This is one of the distinct features of the boundary element method in that output may be selectively obtained after the analysis has been completed. Also, because the fundamental solutions are exact solutions of the governing differential equations the values obtained will exactly satisfy those equations. The only error introduced is in the approximation of the boundary values.

1.3 NOTATION

Good notation is difficult. This is especially true for a book which incorporates two different numerical methods. Nearly all the books on the FEM use the matrix notation whereas all books on the BEM use tensor notation. Traditionally, engineers are more familiar with the matrix notation, and mathematicans with the tensor notation. Consider the system of equations written in longhand

$$\begin{aligned} k_{xx}u_x + k_{xy}u_y + k_{xz}u_z &= f_x \\ k_{yx}u_x + k_{yy}u_y + k_{yz}u_z &= f_y \\ k_{zx}u_x + k_{zy}u_y + k_{zz}u_z &= f_z \end{aligned}$$

In matrix notation this becomes

$$\mathbf{Ku} = \mathbf{f}$$

where

$$\mathbf{K} = \begin{bmatrix} k_{xx} & k_{xy} & k_{xz} \\ k_{yx} & k_{yy} & k_{yz} \\ k_{zx} & k_{zy} & k_{zz} \end{bmatrix}$$

$$\mathbf{u} = \begin{Bmatrix} u_x \\ u_y \\ u_z \end{Bmatrix}$$

$$\mathbf{f} = \begin{Bmatrix} f_x \\ f_y \\ f_z \end{Bmatrix}$$

In indicial (tensor) notation this is written as

$$k_{ij}u_j = f_i \qquad \begin{aligned} i &= 1,2,3 \\ j &= 1,2,3 \end{aligned}$$

It can be seen that the matrix notation uses no indices in this case but requires the definition of the matrices.

In our notation we distinguish between *vectors* or first-order tensors, *pseudo-vectors* or one-dimensional matrices and *multidimensional matrices.*

Vectors are denoted by lower case bold letters and may have two or three components. For example the displacement components in x, y and z directions at an arbitrary point are given by

$$\mathbf{u} = \begin{Bmatrix} u_x \\ u_y \\ u_z \end{Bmatrix}$$

The displacements at the jth node of element e are given by $\mathbf{u}_j^e$ and the displacements at a node n of the mesh are given by $\mathbf{u}_n$.

Pseudo-vectors are bold lower case Greek letters or bold letters which are enclosed by curly brackets and are used to store values in a one-dimensional matrix. For example the stress *pseudo-vector* is given by

$$\boldsymbol{\sigma} = \begin{Bmatrix} \sigma_x \\ \sigma_y \\ \sigma_z \\ \tau_{xy} \\ \tau_{yz} \\ \tau_{xz} \end{Bmatrix}$$

whereas the vector containing the displacement components at all the nodes of the mesh is

$$\{\mathbf{u}\} = \begin{Bmatrix} \mathbf{u}_1 \\ \mathbf{u}_2 \\ \vdots \\ \mathbf{u}_n \end{Bmatrix}$$

Two-dimensional matrices are denoted by bold upper case letters sometimes enclosed in square brackets. For example a differential operator matrix for strain is given by

$$\mathbf{B} = \begin{bmatrix} \frac{\partial}{\partial x} & 0 \\ 0 & \frac{\partial}{\partial y} \\ \frac{\partial}{\partial y} & \frac{\partial}{\partial x} \end{bmatrix}$$

whereas an element stiffness matrix is given by

$$\mathbf{K}^e = \begin{bmatrix} \mathbf{K}^e_{11} & \mathbf{K}^e_{12} & \mathbf{K}^e_{13} \\ \mathbf{K}^e_{21} & \mathbf{K}^e_{22} & \mathbf{K}^e_{23} \\ \mathbf{K}^e_{31} & \mathbf{K}^e_{32} & \mathbf{K}^e_{33} \end{bmatrix}$$

where

$$\mathbf{K}^e_{ij} = \begin{bmatrix} K_{xx} & K_{xy} & K_{xz} \\ K_{yx} & K_{yy} & K_{yz} \\ K_{zx} & K_{zy} & K_{zz} \end{bmatrix}^e_{i,j}$$

are the coefficients linking the ith and jth node of element e.

1.4 CONCLUDING REMARKS

This book is an introduction to two important numerical methods, the finite and the boundary element method. It is intended for practising engineers; the emphasis is on physical interpretation and lengthy mathematical derivations are consigned to the Appendices. The many practical applications and examples shown should enable the engineer to assess the potential of the methods and their applicability to specific design problems. The emphasis of the book is on implementation and use of the methods and two FORTRAN programs which operate on PCs are included with this text. These serve not only to demonstrate the implementation of the methods described in the book, but also to help readers who are contemplating writing their own code.

In the first two chapters are shown procedures used in both methods to describe the shape of the problem domain or its boundaries, and material properties. Finite element methods are then introduced. In Chapter 5 special finite elements for plate and shell problems and materials containing joints are presented. Boundary element methods are introduced in Chapter 6, and in Chapter 7 the procedures necessary for the implementation of these methods on the computer are described in some detail.

Methods for combining both methods are discussed in Chapter 8. An introduction to non-linear problems is presented in Chapter 9 where the treatment of material non-linear behaviour is discussed in detail. The next chapter deals with the important question of mesh design, error analysis and automatic mesh refinement procedures.

The book closes with a discussion of pre- and post-processing using computer graphics.

The book is intended as an introduction and it is hoped that the reader, having gained some familiarity with both methods, will graduate to other text books and monographs which deal with wider applications such as dynamics and large displacement theory.

1.5 TEXT BOOKS ON THE FINITE ELEMENT METHOD

Zienkiewicz, O. C. (1967) *The Finite Element Method in Engineering Science*, (4th edn 1989) McGraw-Hill, London.

Desai, C., and Abel, J. (1971) *Introduction to the Finite Element Method*, Van Nostrand-Reinhold, New York.

Oden, J. T. (1972) *Finite Elements of Nonlinear Continua*, McGraw-Hill, New York.

Martin, H. C., and Carey, G. F. (1973) *Introduction to Finite Element Analysis*, McGraw-Hill, New York.

Norrie, D. H., and de Vries, G. (1973) *The Finite Element Method*, Academic Press, New York.

Robinson, J. (1973) *Integrated Theory of Finite Element Methods*, John Wiley, London.

Cook, R. D. (1974) *Concepts and Applications for Finite Element Analysis*, (2nd edn 1981) John Wiley, New York.

Gallagher, R. H. (1975) *Finite Element Analysis: Fundamentals*, Prentice-Hall, Englewood Cliffs, New Jersey

Bowes, W. H., and Russel, L.T. (1975) *Stress Analysis by the Finite Element Method for Practising Engineers*, Lexington Books, Lexington, Mass.

Fenner, R. T. (1975) *Finite Element Methods for Engineers*, Macmillan, London.

Rockey, K. C. (1975) *The Finite Element Method: A Basic Introduction*, John Wiley, New York.

Wachspress, E. L. (1975) *A Rational Finite Element Basis*, Academic Press, New York.

Connor, J. J., and Brebbia, C. A. (1976) *Finite Element Techniques for Fluid Flow*, Newnes-Butterworths, London.

Bathe, K. J., and Wilson, E. L. (1976) *Numerical Methods in Finite Element Analysis*, Prentice-Hall, Englewood Cliffs, New Jersey.

Odgen, J. T., and Reddy, J. N. (1976) *An Introduction to the Mathematical Theory of Finite Elements*, John Wiley, New York.

Segerlind, L. J. (1976) *Applied Finite Element Analysis*, John Wiley, New York.

Hinton, E., and Owen, D. R. J. (1977) *Finite Element Programming*, Academic Press, New York.

Mitchell, A. R., and Wait, R. (1977) *The Finite Element Method in Partial Differential Equations*, John Wiley, New York.

Tong, T., and Rossettos, J. N. (1977) *Finite Element Method: Basic Technique and Implementation*, MIT Press, Cambridge, Mass.

Chung, T. J. (1978) *Finite Element Analysis in Fluid Dynamics*, McGraw-Hill, New York.

Norrie, D. H., and de Vries, G. (1978) *An Introduction to Finite Element Analysis*, 2nd edn, Academic Press, New York.

Desai, C. S. (1979) *Elementary Finite Element Method*, Prentice-Hall, Englewood Cliffs, New Jersey.

Cheung, Y. K. and Yeo, M. F. (1979) *A Practical Introduction to Finite Element Analysis*, Pitman, London.

Rockey, K. C., Evans, H. R., Griffiths, D. W. and Nethercott, D. A. (1980) *The Finite Element Method—A Basic Introduction*, 2nd edn, Halsted Press, New York.

Irons, B., and Ahmad, S. (1980) *Techniques of Finite Elements*, Halsted Press, New York.

Hinton, H., and Owen, D. R. J. (1980) *A Simple Guide to Finite Elements*, Pineridge Press, Swansea.

Davies, A. J. (1980) *The Finite Element Method: A First Approach*, Clarendon Press, Oxford.

Oden, J. T., Carey, G. F., and Becker, E. B. (1981) *Finite Elements*, Prentice-Hall, Englewood Cliffs, New Jersey.

Akin, J. E. (1981) *Applications and Implementation of Finite Element Methods*, Academic Press, New York.

Owen, D. R. J., and Hinton, E. (1981) *Finite Elements in Plasticity- Theory and Practice*, Pineridge Press, Swansea.

Chari, M. V. K. (1981) *Finite Elements in Electrical and Magnetic Field Problems*, John Wiley, New York.

Smith, I. (1982) *Programming the Finite Element Method with Applications in Geomechanics*, John Wiley, Chichester.

Bathe, K. J. (1982) *Finite Element Procedures in Engineering Analysis*, Prentice-Hall, Englewood Cliffs, New Jersey.

Heubner, K. H., and Thornton, F. A. (1982) *The Finite Element Method for Engineers*, John Wiley, Chichester.

Irons, B., and Shrive, N. (1983) *Finite Element Primer*, Ellis Horwood, Chichester.

Zienkiewicz, O. C., and Morgan, K. (1983) *Finite Elements and Approximations*, John Wiley, Chichester.
Livesley, R. K. (1983) *Finite Elements—An Introduction to Engineers*, Cambridge University Press, Cambridge, U.K.
Dhatt, G., and Touzot, G. (1984) *The Finite Element Method Displayed*, John Wiley, Chichester.
White, R. E. (1985) *An Introduction to the Finite Element Method with Applications to Nonlinear Problems*, John Wiley, Chichester.

1.6 TEXT BOOKS ON THE BOUNDARY ELEMENT METHOD

Jaswon, M. A. and Symm, G. T. (1977) *Integral equation methods in potential theory and elastostatics.* Academic Press, New York.
Brebbia, C. A. (1978) *The Boundary Element Method for Engineers*, Pentech Press, London.
Brebbia, C. A., and Walker, S. (1980) *Boundary Element Techniques in Engineering*, Newnes-Butterworths, London.
Banerjee, P. K., and Butterfield, R. (1981) *Boundary Element Methods in Engineering Science*, McGraw-Hill, London.
Mukherjee, S. (1982) *Boundary Element Methods for Creep and Fracture*, Applied Science, London.
Crouch, S. L., and Starfield, A. M. (1983) *Boundary Element Methods in Solid Mechanics*, George Allen & Unwin, London.
Brebbia, C. A., Telles, J. C. F., and Wrobel, L. C. (1983) *Boundary Element Techniques*, Springer-Verlag, Berlin.
Liggett, J. A., and Liu, P. L. F. (1983) *The Boundary Integral Equation Method for Porous Media Flow*, George Allen & Unwin, London.

2

Solid Modelling—Finite and Boundary Elements

Nature is indifferent
towards the difficulties it
causes to a mathematician

Fourier

2.1 INTRODUCTION

One of the fundamental requirements for numerical modelling is a description of the problem, its boundaries, boundary conditions and material properties, in a mathematical way. The exact definition of the complex shape of a particular problem to be analysed requires the specification of the location (relative to the origin of a set of axes) of an infinite number of points on the surface. In order to be able to model such a problem with a reasonable amount of input data, only a limited number of points may be defined and the shape between the points approximated by functions. This is known as solid modelling[1] and is a technique which is not only restricted to the finite and boundary element methods. Solid modelling is being used for example to describe the shape of car bodies in mechanical engineering and ore bodies in mining, for the purpose of generating displays on computer graphics terminals. Thus, for example, a new form of car body can be visualised in perspective from various angles even before a scale model is built. Alternatively the location and grade of ore bodies can be displayed and optimal excavation strategies for mining it can be planned.

In the following we will discuss one-, two- and three-dimensional elements, as defined by the number of intrinsic (element) coordinate directions.

An element can exist in any cartesian space of dimension equal to or greater than its own dimension.

Thus in this chapter we consider *discretisation* methods used in the boundary and finite element methods.

2.2 ONE-DIMENSIONAL ELEMENTS

For the description of the boundary of plane sections of a problem domain or alternatively for the discretisation of one-dimensional structural components (cables, etc) one-dimensional elements are used. In the simplest case we have linear elements which connect two nodes i and j, the positions of which are defined by cartesian coordinates as shown in Figure 2.1. For each element we now define an element or intrinsic coordinate ξ which follows the direction of the element, and equals zero at the centre and ± 1 at the ends (Figure 2.2).

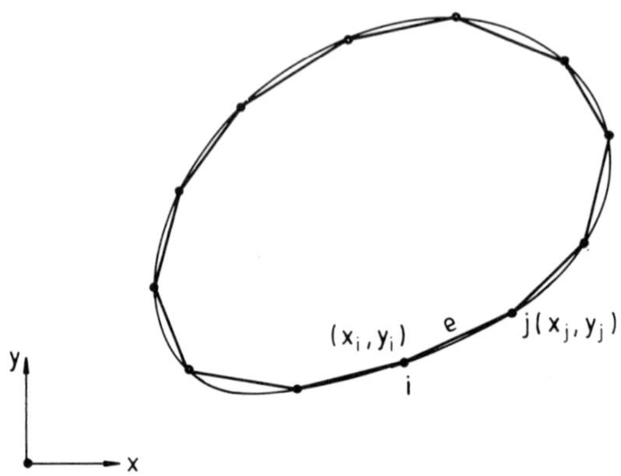

Figure 2.1 Plane domain. Boundary approximated by linear elements.

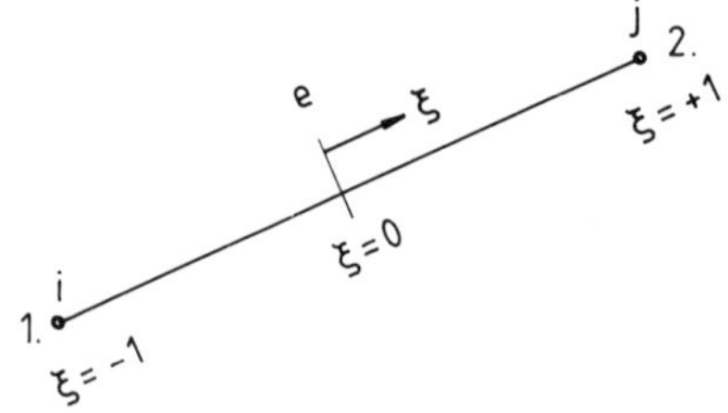

Figure 2.2 Linear element with local coordinate system.

It can easily be verified that the coordinates of a point on element e with the intrinsic coordinate ξ are given by

$$x(\xi) = \frac{x_i + x_j}{2} + \frac{x_j - x_i}{2}\xi$$
$$y(\xi) = \frac{y_i + y_j}{2} + \frac{y_j - y_i}{2}\xi \qquad (2.1)$$

This equation can be checked by substituting $\xi = -1$ and $\xi = +1$ to obtain the coordinates of nodes i and j.

It is now convenient to substitute for the global coordinates:

$$\left.\begin{aligned} x_i &= x_1{}^e \\ y_i &= y_1{}^e \end{aligned}\right\} * \text{ first node of element } e$$

$$\left.\begin{aligned} x_j &= x_2{}^e \\ y_j &= y_2{}^e \end{aligned}\right\} \text{ second node of element } e$$

In this way we establish a link between local and global numbering of nodes:
the first node has global node number i.
the second node has global node number j.

The global numbers of the nodes which belong to the element are referred to as '*element incidences*' or '*element connectivity*'. From now on we will work in the local numbering system and use the element incidences to extract coordinates from the global values.

We can rewrite equation (2.1) as

$$x(\xi) = \frac{1}{2}(1-\xi)x_1^e + \frac{1}{2}(1+\xi)x_2^e$$
$$y(\xi) = \frac{1}{2}(1-\xi)y_1^e + \frac{1}{2}(1+\xi)y_2^e \qquad (2.2)$$

or in abbreviated form

$$x = \sum_{n=1}^{2} N_n(\xi)x_n{}^e$$
$$y = \sum_{n=1}^{2} N_n(\xi)y_n{}^e \qquad (2.3)$$

Or still shorter and more general,

$$\begin{Bmatrix} x \\ y \end{Bmatrix} = \sum_{n=1}^{L} N_n(\xi) \begin{Bmatrix} x_n{}^e \\ y_n{}^e \end{Bmatrix} \qquad (2.4)$$

where L is the number of element nodes and N_n are element 'shape' functions. Equation (2.4) can be written in matrix notation.

$$\mathbf{x} = \sum N_n(\xi)\mathbf{x}_n^e \tag{2.5}$$

where $\mathbf{x}$ is a vector containing coordinates of a point on element e and $\mathbf{x}_n^e$ is a vector of coordinates of the nth node of element e. Note that $\mathbf{x}$ can have up to three components (x, y, z).

For the two-node element just derived the shape functions are (Figure 2.3).

$$\begin{aligned} N_1 &= \frac{1}{2}(1-\xi) \\ N_2 &= \frac{1}{2}(1+\xi) \end{aligned} \tag{2.6}$$

or in short

$$N_n = \frac{1}{2}(1+\xi_n\xi) \tag{2.7}$$

where

$$\begin{aligned} n = 1 \quad & \xi_n = -1 \\ n = 2 \quad & \xi_n = 1 \end{aligned} \tag{2.8}$$

Complicated shapes can accurately be described by a smaller number of elements with three nodes and quadratic shape functions (Figure 2.4). The coordinate ξ now follows the element shape, i.e. is curvilinear and the third

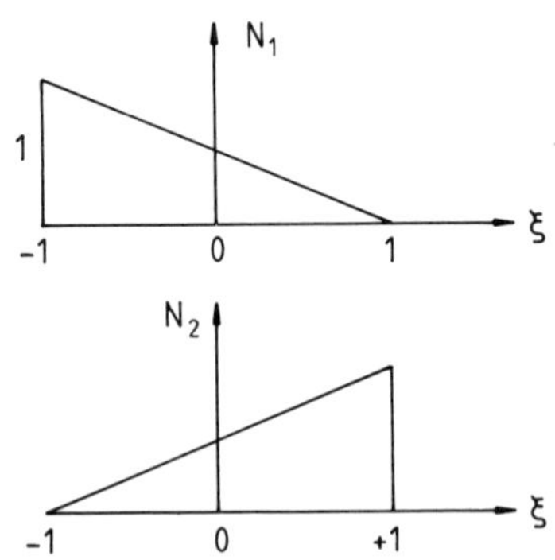

Figure 2.3 Linear shape functions.

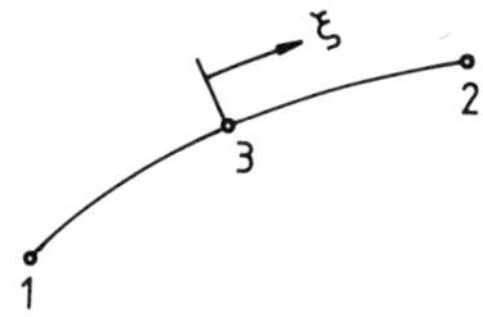

Figure 2.4 Quadratic element.

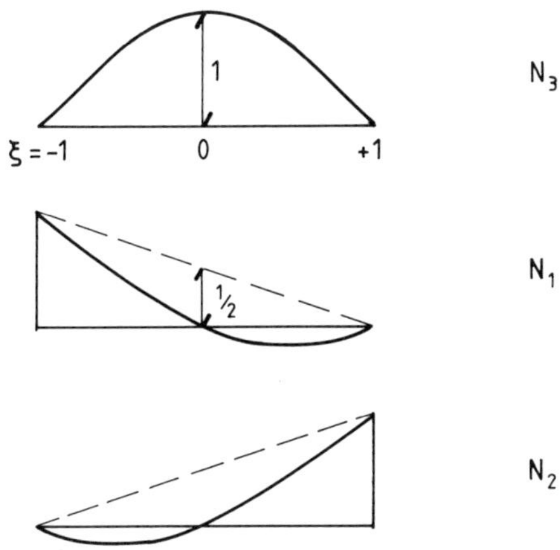

Figure 2.5 Quadratic shape functions.

node is placed at $\xi = 0$. The shape function associated with the midside node is a parabola which has unit value at the third node and zero value at the other nodes; that is,

$$N_3(\xi) = 1 - \xi^2 \tag{2.9}$$

The end node shape functions can be obtained by subtracting half of the centre node function from each of the linear shape functions (Figure 2.5):

$$N_n = \frac{1}{2}(1 + \xi_n \xi) - \frac{1}{2}N_3 \qquad gn = 1, 2 \tag{2.10}$$

The shape functions presented here have so far not been derived mathematically but written down intuitively. Shape functions derived this way have been called *Serendipity*† *functions*[2]. It can be seen that the shape functions derived so far have the following properties:

† Serendipity: the faculty of making desirable but unsought discoveries.

$$
\begin{aligned}
&\text{(a) } N_n(\xi_n) &&= 1 \\
&\text{(b) } N_n(\xi_i) &&= 0 \quad \text{when } i \neq n \\
&\text{(c) } \Sigma N_n(\xi) &&= 1
\end{aligned}
\tag{2.11}
$$

These are indeed the properties all the serendipity shape functions must have.

The mathematical derivation of functions which satisfy conditions (2.11) is possible using Lagrange polynomials[3]. For the parabolic elements the Lagrange shape functions are defined as

$$L_i(\xi) = A_{i1}A_{i2}A_{i3} \tag{2.12}$$

where

$$
\begin{aligned}
A_{i\ell} &= \frac{(\xi - \xi_\ell)}{(\xi_i - \xi_\ell)} && \text{if } i \neq \ell \\
A_{i\ell} &= 1 && \text{if } i = \ell
\end{aligned}
\tag{2.13}
$$

The reader can verify that the Lagrange and serendipity shape functions are identical for one-dimensional elements. We shall, however, use Lagrange polynomials to construct shape functions for two- and three-dimensional elements which differ from the serendipity shape functions.

The element shapes discussed so far allow discretisations which are continuous in the sense that no 'jumps' occur at interfaces between elements. This is also known as C^0 continuity. The continuity of slope (C^1 continuity) is not, however, guaranteed and the discretised surface may have 'kinks'.

If a smoother surface discretisation is needed the Hermitian polynomial shape functions may be used. In this case the element shape is specified not only by the nodal coordinates but also by the tangents at the nodes. In Figure 2.6 is shown an element with two nodes, the *Hermitian cubic* element[4]. There are two parameters for each node (coordinates and tangent), so a total of four shape functions is needed to describe the element geometry. The cartesian coordinates of a point on the element with intrinsic coordinate ξ are given by

$$\mathbf{x} = \sum_{n=1}^{2} H_n \mathbf{x}_n^e + \sum_{n=1}^{2} \widehat{H}_n \widehat{\mathbf{x}}_n^e \tag{2.14}$$

where $\widehat{\mathbf{x}}_n^e$ is defined as

$$\widehat{\mathbf{x}}_n^e = \left\{ \begin{matrix} \frac{\mathrm{d}x}{\mathrm{d}\xi} \\ \frac{\mathrm{d}y}{\mathrm{d}\xi} \end{matrix} \right\}_n \tag{2.15}$$

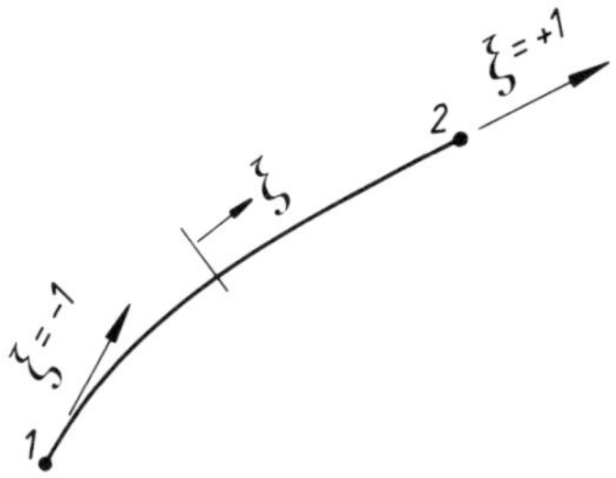

Figure 2.6 Hermitian cubic element.

The shape functions H_n and $\widehat{H}_n$ are given by

$$H_1(\xi) = \frac{1}{4}(1-\xi)^2(2+\xi)$$

$$H_2(\xi) = \frac{1}{4}(1+\xi)^2(2-\xi)$$

$$\widehat{H}_1(\xi) = \frac{1}{4}(1-\xi)^2(1+\xi)$$

$$\widehat{H}_2(\xi) = -\frac{1}{4}(1+\xi)^2(1-\xi)$$

These shape functions are shown in Figure 2.7 and the reader may recognise them as deflected shapes of a beam to which unit displacements or rotations have been applied at the ends. It can be seen that the shape functions are of order 3, hence the name Hermite cubic.

An example of how the various element forms discussed here can be applied is given in Figure 2.8, in which is shown the discretisation of a circular boundary using linear, parabolic and Hermite cubic elements. It can be seen that fewer of the more sophisticated elements are required to describe the boundary accurately.

2.3 TWO-DIMENSIONAL ELEMENTS

For the description of the boundary of three-dimensional problems or the domain of plane problems, two-dimensional elements are used. Their derivation is analogous to that of the one-dimensional elements described previously except in that two intrinsic coordinates (ξ, η) are used as shown in Figure 2.9. The coordinates of a point with the intrinsic coordinates (ξ, η) are obtained by

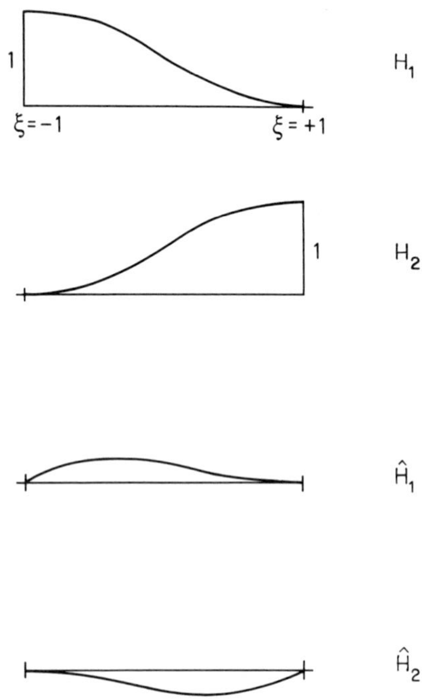

Figure 2.7 Hermite cubic shape functions.

$$\mathbf{x} = \sum_{n=1}^{L} N_n(\xi, \eta)\mathbf{x}_n^e \tag{2.16}$$

For the quadilateral element in Figure 2.10 bilinear shape functions are used:

$$N_n = \frac{1}{2}(1 + \xi_n \xi)\frac{1}{2}(1 + \eta_n \eta) \tag{2.17}$$

The shape function N_1 is shown in Figure 2.10. It describes a curved surface consisting of straight lines in the ξ, η directions. The surface, also called a hyper-surface, has been a widely used shape for concrete shells because the formwork is simple to construct.

Again we can derive a higher order element by adding midside nodes on the element sides. A quadratic element is shown in Figure 2.11 and the following local node numbering adopted.

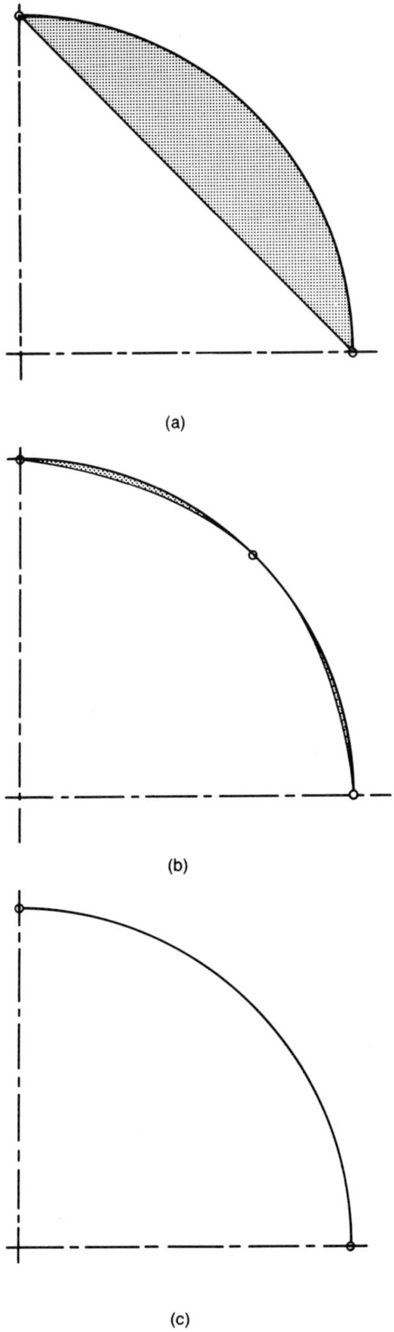

Figure 2.8 Discretisation of $\frac{1}{4}$ circle with linear, parabolic and Hermite cubic elements. Shaded area shows error.

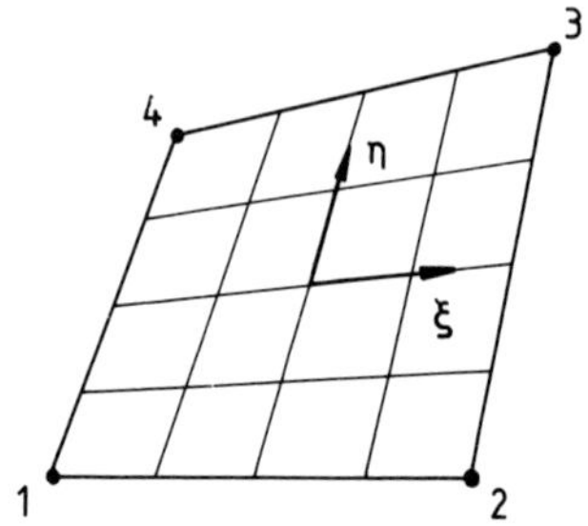

Figure 2.9 Quadrilateral element.

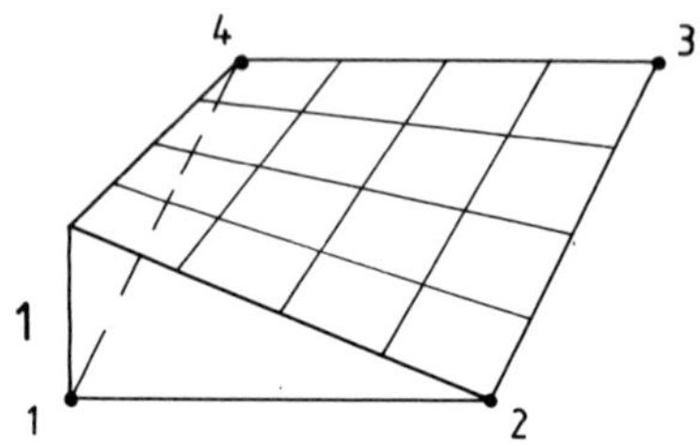

Figure 2.10 Bilinear shape function N_1

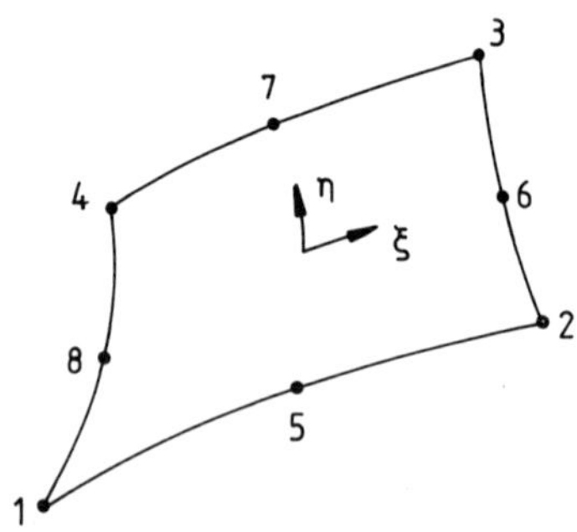

Figure 2.11 Quadratic serendipity element

n	ξ_n	η_n
1	−1.	−1.
2	1.	−1.
3	1.	1.
4	−1.	1.
5	0.	−1.
6	1.	0.
7	0.	1.
8	−1.	0.

(2.18)

The shape functions for the midside nodes are given by

$$N_n = \frac{1}{2}(1 - \xi^2)(1 + \eta_n \eta) \quad \text{for } n = 5,7$$

$$N_n = \frac{1}{2}(1 - \eta^2)(1 + \xi_n \xi) \quad \text{for } n = 6,8 \tag{2.19}$$

The corner node functions are constructed in a similar way to that for the one-dimensional element (Figure 2.12):

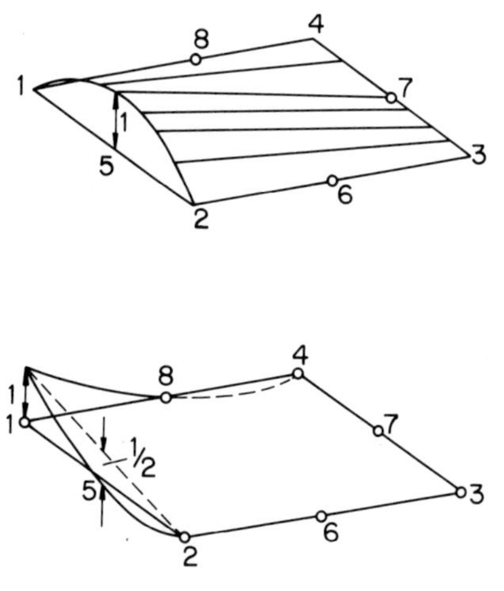

Figure 2.12 Shape functions for midside and corner nodes.

$$
\begin{aligned}
N_1 &= \frac{1}{4}(1-\xi)(1-\eta) - \frac{1}{2}N_5 - \frac{1}{2}N_8 \\
N_2 &= \frac{1}{4}(1+\xi)(1-\eta) - \frac{1}{2}N_5 - \frac{1}{2}N_6 \\
N_3 &= \frac{1}{4}(1+\xi)(1+\eta) - \frac{1}{2}N_6 - \frac{1}{2}N_7 \\
N_4 &\equiv \frac{1}{4}(1-\xi)(1+\eta) - \frac{1}{2}N_7 - \frac{1}{2}N_8
\end{aligned}
\tag{2.20}
$$

By writing down the shape functions in this manner it is possible to derive elements with variable numbers of nodes by deleting appropriate terms. For example for an element with no midside node 5 a linear shape is assumed between nodes 1 and 2 and the shape functions are obtained by simply setting $N_5 = 0$.

Hermitian polynomials may also be expanded to two dimensions in a similar way.

If the element shape functions for the quadratic element are derived from Lagrange polynomials then there is an additional node at the centre of the element (Figure 2.13). The shape functions are given by

$$L_{n(i,j)} = A_{i1}A_{i2}A_{i3}B_{j1}B_{j2}B_{j3} \tag{2.21}$$

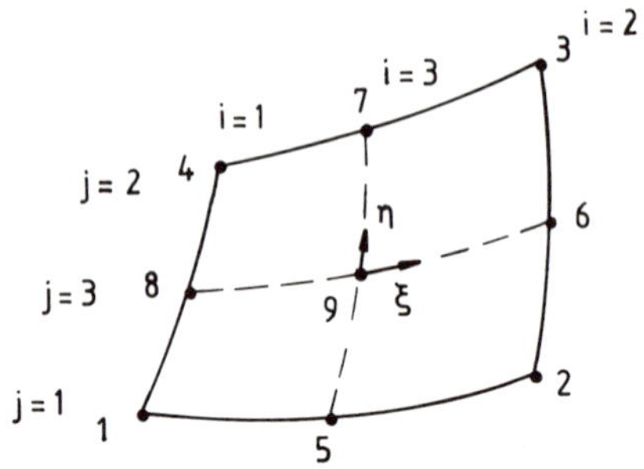

Figure 2.13 Quadratic Lagrange element

where $A_{i,\ell}$ is as defined in equation (2.13) and

$$
\begin{aligned}
B_{j\ell} &= \frac{(\eta - \eta_\ell)}{(\eta_j - \eta_\ell)} \qquad \text{if } j \neq \ell \\
B_{j\ell} &= 1 \qquad\qquad\quad \text{if } j = \ell
\end{aligned}
\tag{2.22}
$$

where i and j are the column and row numbers of the nodes. This numbering is defined in Figure 2.13. The nodes are given by

$$\begin{array}{lll} n(1,1) = 1 & n(2,1) = 2 & n(3,1) = 5 \\ n(1,2) = 4 & n(2,2) = 3 & n(3,2) = 7 \\ n(1,3) = 8 & n(2,3) = 6 & n(3,3) = 9 \end{array} \tag{2.23}$$

The serendipity and Lagrange shape functions are compared in Figure 2.14. The Lagrange element has an additional 'bubble mode' and is therefore able to describe complicated shapes more accurately.

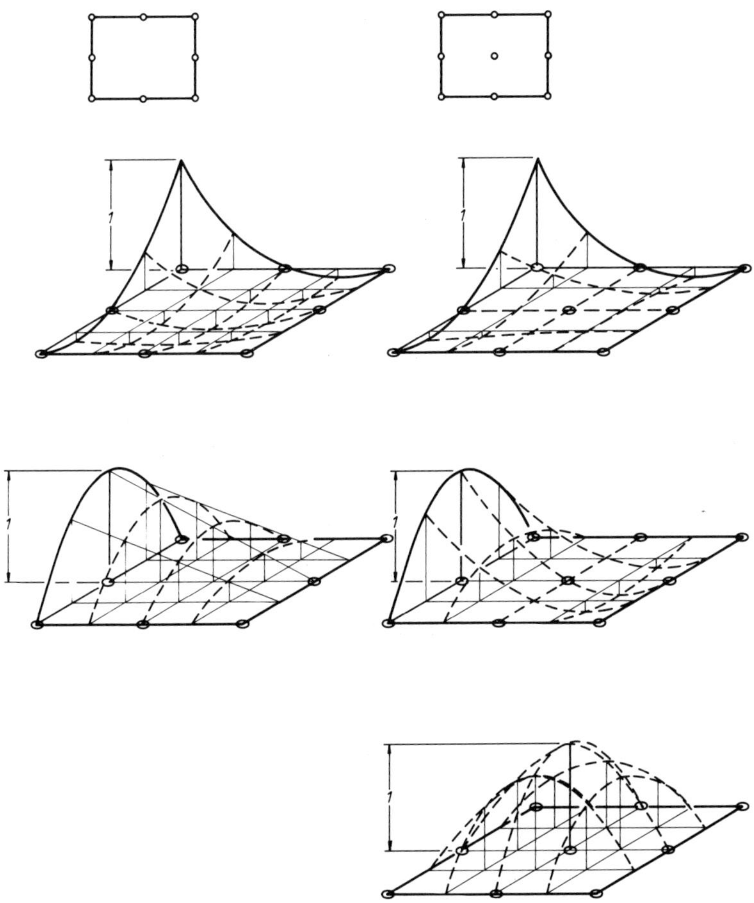

Figure 2.14 Serendipity and Lagrange shape functions.

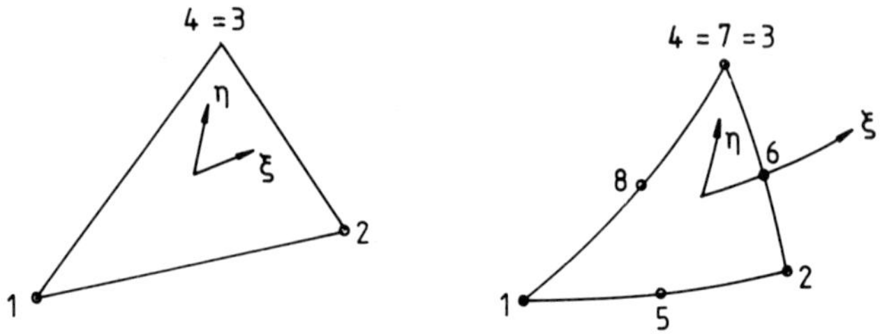

Figure 2.15 Linear and parabolic degenerate elements.

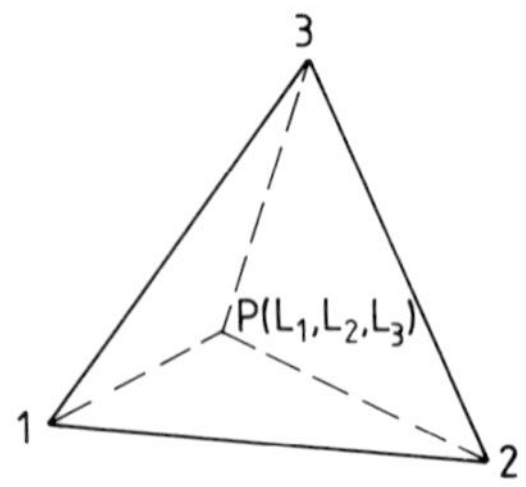

Figure 2.16 Triangular linear element.

Triangular elements can be formed from quadilateral elements by assigning to two or three nodes the same global node number and coordinates. Such 'degenerate' elements are shown in Figure 2.15.

Alternatively triangular elements can be derived using area coordinates[5] instead of the curvilinear ones. A convenient set of coordinates L_1, L_2, L_3 is defined as (Figure 2.16).

$$L_1 = \frac{\text{Area P23}}{\text{Area 123}}$$
$$L_2 = \frac{\text{Area P13}}{\text{Area 123}} \tag{2.24}$$
$$L_3 = 1 - L_1 - L_2$$

The coordinates of point P for the linear element are given by

$$\mathbf{x} = \sum_{n=1}^{3} L_n \mathbf{x}_n^e \tag{2.25}$$

It is also possible to derive quadratic and higher order element shape functions using area coordinates.

2.4 THREE-DIMENSIONAL ELEMENTS

For the description of a solid mass of material, three-dimensional elements are used. Three intrinsic coordinates are defined as shown in Figure 2.17

$$\mathbf{x} = \sum_{n=1}^{L} N_n(\xi, \eta, \zeta)\mathbf{x}_n^e \tag{2.26}$$

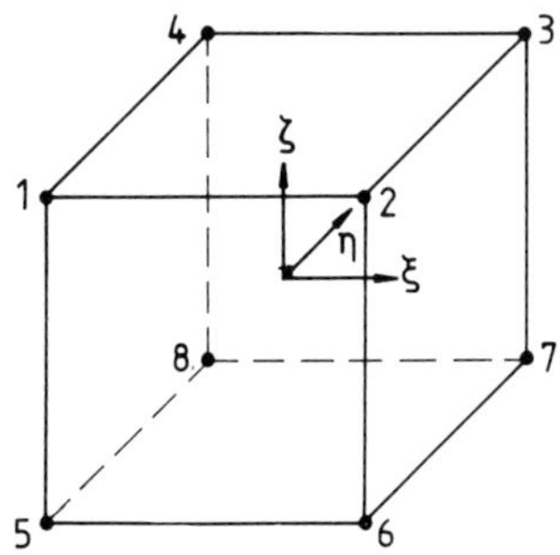

Figure 2.17 Solid 'brick' element with linear shape functions.

For the hexahedral element in Figure 2.17 the trilinear shape functions are simply a generalisation of (2.17):

$$N_n = \frac{1}{2}(1+\xi_n\xi)\frac{1}{2}(1+\eta_n\eta)\frac{1}{2}(1+\zeta_n\zeta) \tag{2.27}$$

where the local node numbering is as follows:

n	ξ_n	η_n	ζ_n	
1	−1	−1	1	
2	1	−1	1	
3	1	1	1	
4	−1	1	1	(2.28)
5	−1	−1	−1	
6	1	−1	−1	
7	1	1	−1	
8	−1	1	−1	

Again we can derive a higher order (quadratic) element by adding midside nodes as shown in Figure 2.18.

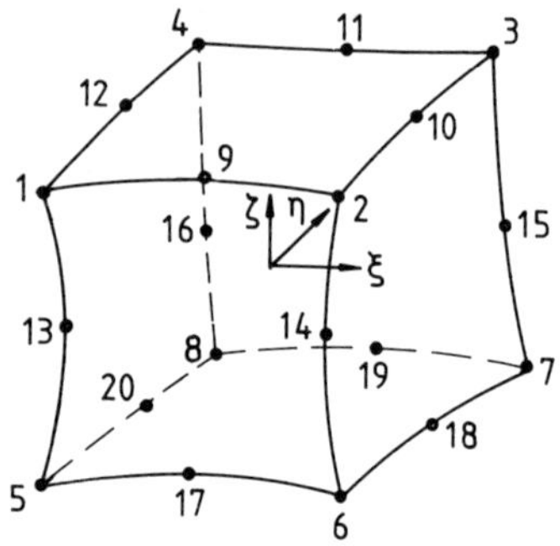

Figure 2.18 Solid 'brick' element with parabolic shape function (Serendipity family).

The shape functions for midside nodes are given by

$$\begin{aligned} N_n &= \frac{1}{4}(1-\xi^2)(1+\eta_n\eta)(1+\zeta_n\zeta) \\ N_n &= \frac{1}{4}(1-\eta^2)(1+\xi_n\xi)(1+\zeta_n\zeta) \\ N_n &= \frac{1}{4}(1-\zeta^2)(1+\xi_n\xi)(1+\eta_n\eta) \end{aligned} \tag{2.29}$$

for nodes in the planes $\xi = 0$, $\eta = 0$ and $\zeta = 0$, respectively.

The shape functions for the corner nodes are constructed in a way analogous to the two-dimensional case, for example

$$N_1 = \frac{1}{2}(1-\xi)\frac{1}{2}(1-\eta)\frac{1}{2}(1+\zeta) - \frac{1}{2}N_9 - \frac{1}{2}N_{12} - \frac{1}{2}N_{13} \tag{2.30}$$

Again an element can be derived from Lagrange polynomials in a similar way as before. This element has 27 nodes and the shape functions are given by (Figure 2.19)

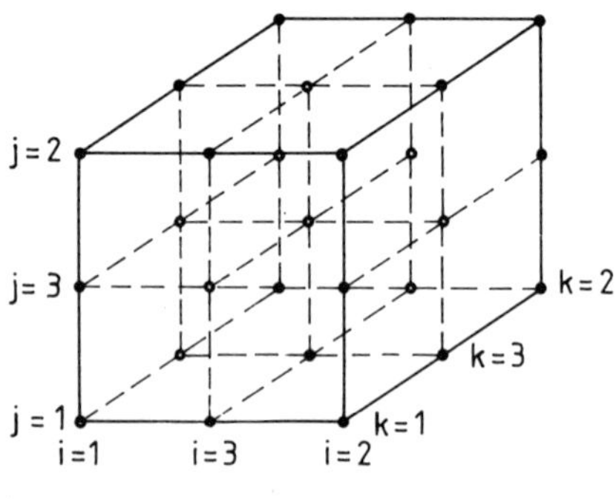

Figure 2.19 Three-dimensional Lagrange element.

$$L_{n(i,j,k)} = A_{i1}A_{i2}A_{i3}B_{j1}B_{j2}B_{j3}C_{k1}C_{k2}C_{k3} \tag{2.31}$$

where A and B are as defined previously and

$$\begin{aligned} C_{k\ell} &= \frac{(\zeta - \zeta_\ell)}{(\zeta_k - \zeta_\ell)} \quad &\text{for } k \neq \ell \\ C_{k\ell} &= 1 \quad &\text{for } k = \ell \end{aligned} \tag{2.32}$$

As with the two-dimensional elements, various special element shapes can be constructed by degeneration, that is (as shown for example in Figure 2.20), by assigning to several nodes the same global node number and coordinates.

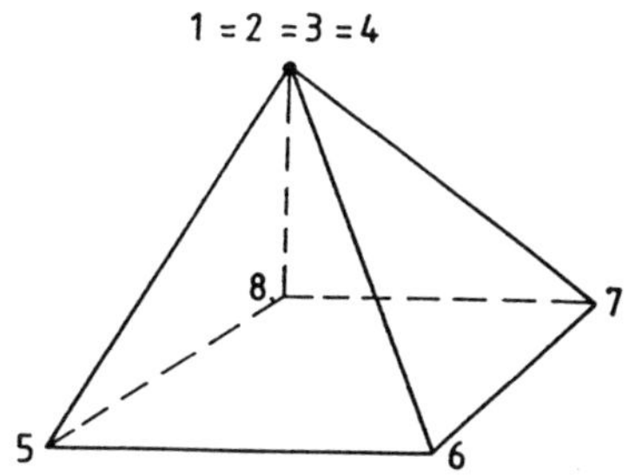

Figure 2.20 Degenerate three-dimensional element.

2.5 ELEMENTS OF INFINITE EXTENT

It is sometimes necessary to describe a domain of infinite extent. Examples are found in geomechanics in which the rock or soil mass is assumed to be infinitely large. To describe the geometry of an element of infinite extent in one intrinsic coordinate direction, say ξ, we may use special shape functions which tend to infinity as ξ tends to $+1$. For the one-dimensional element shown in Figure 2.21 the coordinate transformation

$$\mathbf{x} = \sum_{n=1}^{2} N_n^{\infty}(\xi)\mathbf{x}_n^e \tag{2.33}$$

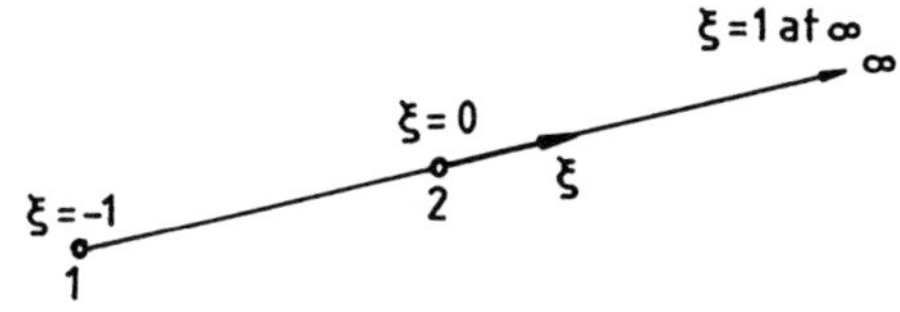

Figure 2.21 One-dimensional infinite element.

gives infinite cartesian coordinates at $\xi = 1$ if the shape functions N_n^∞ are taken to vary as follows. The element is divided into a finite portion $(-1 < \xi < 0)$ and an infinite portion $(0 < \xi < 1)$, and for the finite portion a linear relationship is adopted:

$$\left.\begin{aligned} N_1^\infty &= -\xi \\ N_2^\infty &= \xi + 1 \end{aligned}\right\} \quad \xi < 0 \qquad (2.34a)$$

For the infinite portion we write

$$\left.\begin{aligned} N_1^\infty &= -\xi/(1-\xi) \\ N_2^\infty &= \xi/(1-\xi) + 1 \end{aligned}\right\} \quad \xi > 0 \qquad (2.34b)$$

The concept can be extended to two dimensions. The geometry of the two-dimensional element shown in Figure 2.22 for example is described by

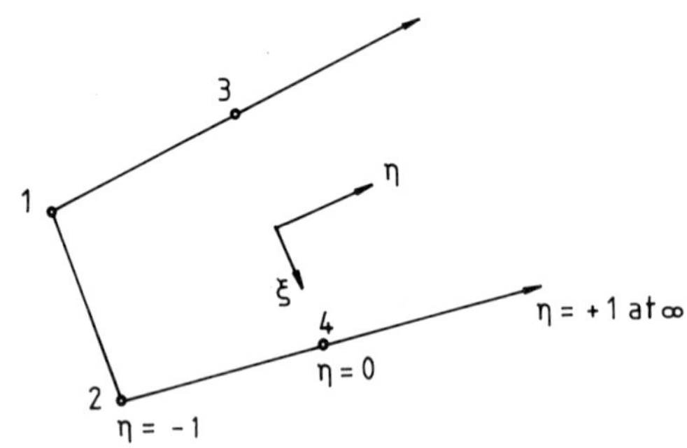

Figure 2.22 Two-dimensional infinite element.

$$\mathbf{x} = \sum_{n=1}^{4} N_n(\xi) N_n^\infty(\eta) \mathbf{x}_n^e \qquad (2.35)$$

where $N_n(\xi)$ is a serendipity shape function as discussed previously and N_n^∞ is defined by

$$N_n^\infty = \eta_o + 1 + \eta_n \qquad (2.35a)$$

where

$$\eta_o = (2\eta_n + 1)\eta \qquad \eta \leq 0$$

and

$$\eta_o = (2\eta_n + 1)\eta/(1-\eta) \qquad \eta > 0$$

where η_n is the intrinsic coordinate of node n, i.e. $\eta_n = -1$ for $n =$ 1,2 and $\eta_m = 0$ for $n =$ 3,4.

Similarly the geometry of the three dimensional infinite element is described by (Figure 2.23)

$$\mathbf{x} = \sum_{n=1}^{8} N_n(\xi,\eta) N_n^{\infty}(\zeta)\mathbf{x}_n^e \tag{2.36}$$

with $N_n^{\infty}(\zeta)$ defined as

$$N_n^{\infty}(\zeta) = \zeta_o + \zeta_n + 1$$

where

$$\zeta_o = (2\zeta_n + 1)\zeta \qquad \text{for} \quad \zeta < 0$$

and

$$\zeta_o = (2\zeta_n + 1)\zeta/(1-\zeta) \qquad \text{for} \quad \zeta > 0$$

where ζ_n is the intrinsic coordinate of node n, i.e. $\zeta_n = -1$ for $n =$ 1,2,3,4 and $\zeta_n = 0$ for $n =$ 5,6,7,8.

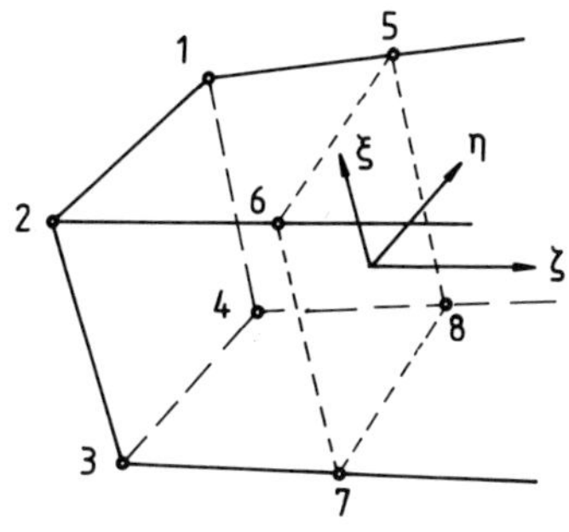

Figure 2.23 Three-dimensional infinite element.

2.6 DESCRIPTION OF PHYSICAL QUANTITIES

In addition to defining the shape of the solid to be modelled it is sometimes required to specify the variation of physical quantities either on the boundaries or within the solid. These can for example be displacements or temperature on the surface or inside the solid. Here we can use exactly the same method as that used for the modelling of the geometry. The physical quantities are specified at nodal points of the element and interpolation functions are used to calculate values between the nodes. The value at a point inside an element e can therefore be written as

$$q = \sum N_n q_n^e \tag{2.37}$$

where q_n^e is the value of the function at the nth node of element e and N_n are interpolation functions identical with the shape functions discussed previously.

If for a particular element the same functions are used for the element shape and for the interpolations of physical quantities inside the element then the element is called 'isoparametric' (i.e. same number of parameters).

An example of an isoparametric boundary element is shown in Figure 2.24, in which the shape and the variation of q are described using the same shape functions:

$$\begin{aligned} &\text{Geometry:} \quad \mathbf{x} = \sum N_n(\xi)\mathbf{x}_n^e \\ &\text{Physical quantity:} \quad q(\xi) = \sum N_n(\xi) q_n^e \end{aligned} \tag{2.38}$$

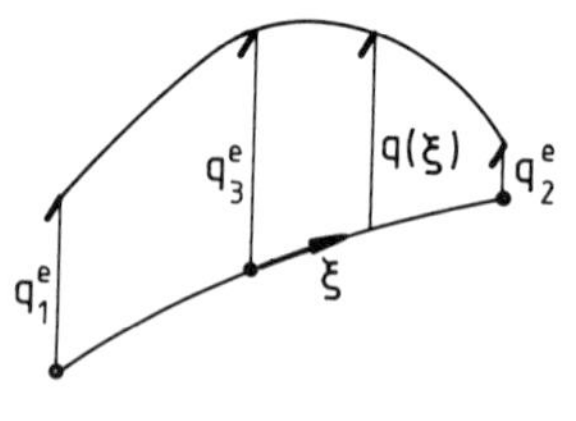

Figure 2.24 Variation of q along quadratic boundary element.

The variation of physical quantities on the surface of two-dimensional elements or inside plane elements can be described (Figure 2.25) in a similar way:

$$q(\xi, \eta) = \sum N_n(\xi, \eta) q_n^e \tag{2.39}$$

By extending the range of intrinsic coordinates to include ζ the variation of q inside a solid element can be described:

$$q(\xi,\eta,\zeta) = \sum N_n(\xi,\eta,\zeta) q_n^e \tag{2.40}$$

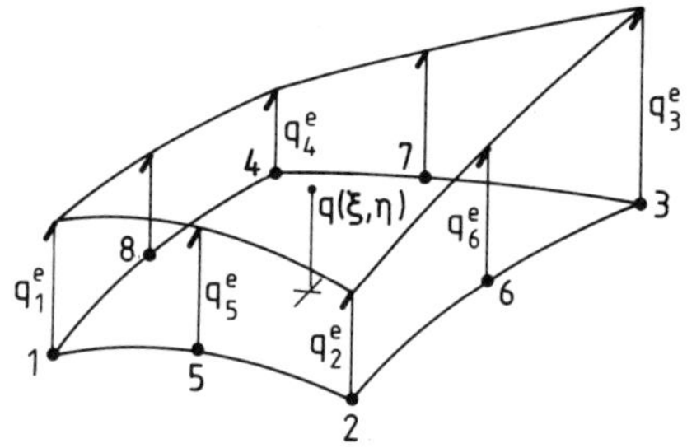

Figure 2.25 Variation of q over quadratic surface element.

Note that q may be a scalar or a vector.

The physical quantities are defined for each element separately, so they may be discontinuous at nodes shared by two elements: they may be either C^0 discontinuous i.e. have 'jumps', or C^1 discontinuous i.e. have 'kinks' as shown in Figure 2.26. If serendipity or Lagrange shape functions are used, C^0 continuity can be enforced between elements by specifying the same function value for each element at a shared node. C^1 continuity can only be specified if Hermitian cubic shape functions are used and the tangential derivative of the function is the same for each element at the shared node.

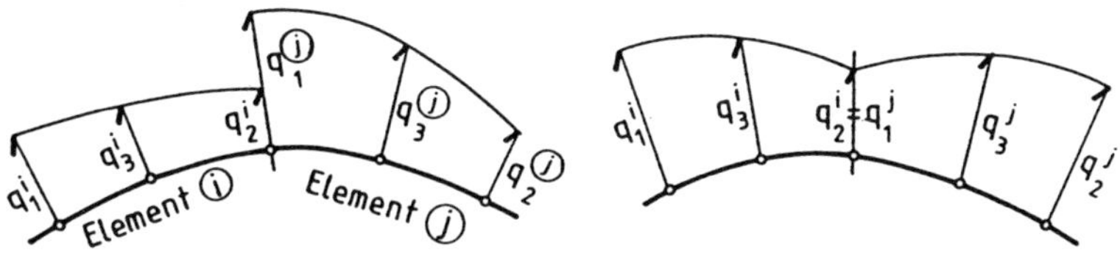

Figure 2.26 Distributions with 'jumps' and 'kinks'.

2.7 COORDINATE TRANSFORMATION

When defining the coordinates of a node it might be convenient to refer to a local cartesian coordinate system and compute the global coordinates afterwards.

A local coordinate system is defined by the location of its origin, $\mathbf{x}_0$, and the directions of its axes. The global co-ordinates of a point defined by local coordinates $\mathbf{x}'$ are given by

$$\mathbf{x} = \mathbf{x}_0 + \mathbf{T}_g \mathbf{x}' \tag{2.41}$$

For two-dimensional problems the geometric transformation matrix is

$$\mathbf{T}_g = \begin{bmatrix} \cos\alpha & -\sin\alpha \\ \sin\alpha & \cos\alpha \end{bmatrix} \tag{2.42}$$

where α is the angle from the x-axis to the x'-axis.

For three-dimensional problems we refer to unit vectors in the local coordinate directions, i.e. $\mathbf{v}_1$ along x', $\mathbf{v}_2$ along y' and $\mathbf{v}_3$ along z', and hence

$$\mathbf{T}_g = \begin{bmatrix} v_{1x} & v_{2x} & v_{3x} \\ v_{1y} & v_{2y} & v_{3y} \\ v_{1z} & v_{2z} & v_{3z} \end{bmatrix} \tag{2.43}$$

The reader may verify that the inverse relation is obtained by transposing the matrix $\mathbf{T}_g$:

$$\mathbf{T}_g^{-1} = \mathbf{T}_g^T \tag{2.44}$$

The local coordinates $\mathbf{x}'$ are given in terms of global coordinates $\mathbf{x}$ by

$$\mathbf{x}' = \mathbf{T}_g^T(\mathbf{x} - \mathbf{x}_0) \tag{2.45}$$

2.8 DIFFERENTIAL GEOMETRY

In many cases it is necessary to work out the direction normal to a line or surface. A typical application is in solid modelling where the reflection of light may be displayed according to the orientation of the surface as defined by the direction of the outward normal. Also the implementation of shell elements and boundary elements requires the computation of directions normal and tangential to the surface.

The best way to determine these directions is by using vector algebra.

Consider a one-dimensional element (Figure 2.27); the vector in the direction of ξ is

$$\mathbf{V}_\xi = \frac{\partial}{\partial \xi}\mathbf{x} \tag{2.46}$$

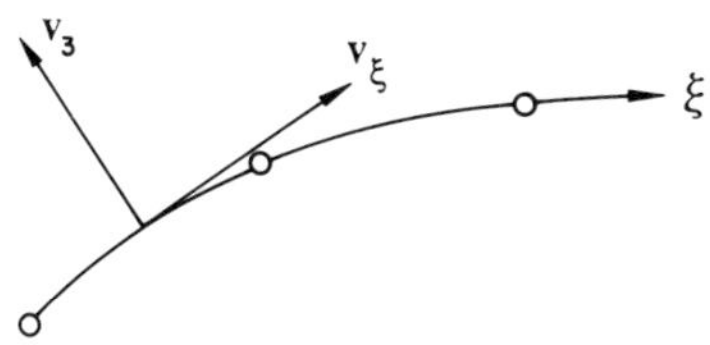

Figure 2.27 Normal and tangential vectors for one-dimensional element.

where $\mathbf{x}$ may have two or three cartesian components. By equation (2.5),

$$\frac{\partial}{\partial \xi}\mathbf{x} = \sum \frac{\partial N_n}{\partial \xi}\mathbf{x}_n^e \tag{2.47}$$

A vector normal to the line element, $\mathbf{V}_3$, may then be computed by taking the cross-product of $\mathbf{V}_\xi$ with another unit vector (for example in the z-direction):

$$\mathbf{V}_3 = \mathbf{V}_\xi \times \mathbf{v}_z \tag{2.48}$$

For a three-dimensional problem

$$\mathbf{V}_3 = \begin{Bmatrix} V_{3x} \\ V_{3y} \\ V_{3z} \end{Bmatrix} = \begin{Bmatrix} \frac{\mathrm{d}x}{\mathrm{d}\xi} \\ \frac{\mathrm{d}y}{\mathrm{d}\xi} \\ \frac{\mathrm{d}z}{\mathrm{d}\xi} \end{Bmatrix} \times \begin{Bmatrix} 0 \\ 0 \\ 1 \end{Bmatrix} = \begin{Bmatrix} \frac{\mathrm{d}y}{\mathrm{d}\xi} \\ -\frac{\mathrm{d}x}{\mathrm{d}\xi} \\ 0 \end{Bmatrix} \tag{2.49}$$

The length of the vector $\mathbf{V}_3$ is equal to

$$|\,\mathbf{V}_3\,| = \left(V_{3x}^2 + V_{3y}^2 + V_{3z}^2\right)^{1/2}$$

and therefore the unit vector in the direction normal to a line element is given by

$$\mathbf{v}_3 = \mathbf{V}_3 \frac{1}{|\,V_3\,|} \tag{2.50}$$

For three-dimensional problems the computation breaks down if $\mathbf{V}_\xi$ is exactly in the z-direction, i.e. if $\frac{\mathrm{d}x}{\mathrm{d}\xi} = \frac{\mathrm{d}y}{\mathrm{d}\xi} = 0$. In this case the cross-product with a unit vector in the direction of the x-axis may be taken.

For two-dimensional surface elements (Figure 2.28) there are two tangential vectors, $\mathbf{V}_\xi$ in the ξ-direction and

$$\mathbf{V}_\eta = \frac{\mathrm{d}}{\mathrm{d}\eta}\mathbf{x} \tag{2.51}$$

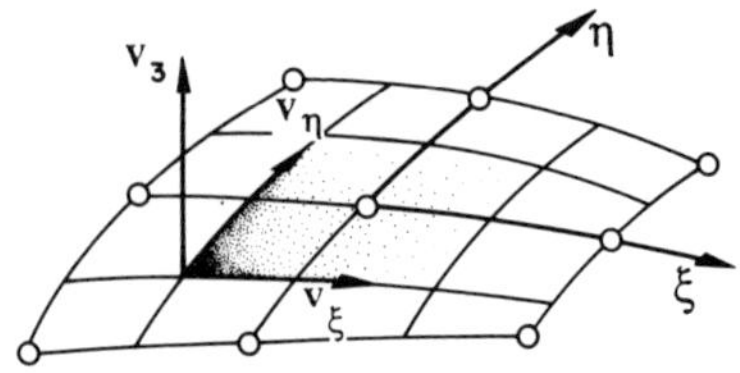

Figure 2.28 Normal and tangential vectors for two-dimensional (surface) element.

in the η-direction where

$$\frac{\partial}{\partial\eta}\mathbf{x} = \sum \frac{\partial N_n}{\partial\eta}\mathbf{x}_n^e \tag{2.52}$$

The vector normal to the surface may be computed by taking the cross-product of $\mathbf{V}_\xi$ and $\mathbf{V}_\eta$:

$$\mathbf{V}_3 = \mathbf{V}_\xi \times \mathbf{V}_\eta \tag{2.53}$$

that is

$$\mathbf{V}_3 = \begin{Bmatrix} \frac{dx}{d\xi} \\ \frac{dy}{d\xi} \\ \frac{dz}{d\xi} \end{Bmatrix} \times \begin{Bmatrix} \frac{dx}{d\eta} \\ \frac{dy}{d\eta} \\ \frac{dz}{d\eta} \end{Bmatrix} = \begin{Bmatrix} \frac{dy}{d\xi}\frac{dz}{d\eta} - \frac{dy}{d\eta}\frac{dz}{d\xi} \\ \frac{dx}{d\eta}\frac{dz}{d\xi} - \frac{dx}{d\xi}\frac{dz}{d\eta} \\ \frac{dx}{d\xi}\frac{dy}{d\eta} - \frac{dx}{d\eta}\frac{dy}{d\xi} \end{Bmatrix} \tag{2.54}$$

The unit normal vector $\mathbf{v}_3$ is obtained as indicated previously by normalisation.

To determine orthogonal directions tangential to the surface we quite arbitrarily choose the tangential direction 1, $\mathbf{v}_1$, to be defined by the cross-product of the vector $\mathbf{v}_3$ and the unit vector in the y-direction, $\mathbf{v}_y$:

$$\mathbf{v}_1 = \mathbf{v}_3 \times \mathbf{v}_y \tag{2.55}$$

To construct an orthogonal system of axes it follows that the tangential direction 2, $\mathbf{v}_2$ is obtained by

$$\mathbf{v}_2 = \mathbf{v}_1 \times \mathbf{v}_3 \tag{2.56}$$

Again a special case might exist in which $\mathbf{v}_3$ is exactly in the y-direction. In this case the vector $\mathbf{v}_1$ is obtained by taking the cross product with a vector in the x-direction.

If the element is in two-dimensional cartesian space then $\frac{dz}{d\xi} = \frac{dz}{d\eta} = 0$

and

$$\mathbf{V}_3 = \begin{Bmatrix} 0 \\ 0 \\ \frac{\mathrm{d}x}{\mathrm{d}\xi}\frac{\mathrm{d}y}{\mathrm{d}\eta} - \frac{\mathrm{d}x}{\mathrm{d}\eta}\frac{\mathrm{d}y}{\mathrm{d}\xi} \end{Bmatrix} \tag{2.57}$$

2.9 INTEGRATION OVER ELEMENTS

The main reason for selecting a range of +1 to −1 for the intrinsic coordinates is to enable the use of numerical (Gauss quadrature) integration over the elements. In the discussion of both finite and boundary element methods we will find that integration plays an important role. The functions to be integrated over elements are often quite complex and therefore require numerical treatment. One prerequisite for numerical integration is an expression for rate of change of length, area or volume with respect to intrinsic coordinates. For example, for a one-dimensional element the total element length is given by

$$S^e = \int_{-1}^{+1} |\, J \,|\, \mathrm{d}\xi \tag{2.58}$$

where $|\, J \,| = \frac{\mathrm{d}s}{\mathrm{d}\xi}$ is equal to the length of the vector $\mathbf{V}_3$ normal to the element (equation (2.49)).

Similarly the area of a plane element is given by

$$A^e = \int_{-1}^{+1}\int_{-1}^{+1} |\, J \,|\, \mathrm{d}\xi \mathrm{d}\eta \tag{2.59}$$

where $|\, J \,| \equiv \partial(x,y)/\partial(\xi,\eta)$ (known as the Jacobian) represents the area of a quadrilateral of dimension $\mathrm{d}\xi\mathrm{d}\eta$ divided by the product $\mathrm{d}\zeta\mathrm{d}\eta$. This is equal to the length of $\mathbf{V}_3$ in equation (2.57).

2.10 DERIVATIVES OF SHAPE FUNCTIONS

In later chapters we will be required to compute the derivatives of shape functions with respect to intrinsic and global coordinates. For the general three-dimensional case the local derivatives (in ξ, η, ζ-directions) of the shape functions which are expressed in intrinsic ξ, η, ζ coordinates are, according to the chain rule,

$$\frac{\partial N_j}{\partial \xi} = \frac{\partial N_j}{\partial x}\frac{\partial x}{\partial \xi} + \frac{\partial N_j}{\partial y}\frac{\partial y}{\partial \xi} + \frac{\partial N_j}{\partial z}\frac{\partial z}{\partial \xi} \tag{2.60}$$

etc.

The relationship between local and global derivatives is therefore given by

$$\begin{Bmatrix} \dfrac{\partial N_j}{\partial \xi} \\[2mm] \dfrac{\partial N_j}{\partial \eta} \\[2mm] \dfrac{\partial N_j}{\partial \zeta} \end{Bmatrix} = \begin{bmatrix} \dfrac{\partial x}{\partial \xi} & \dfrac{\partial y}{\partial \xi} & \dfrac{\partial z}{\partial \xi} \\[2mm] \dfrac{\partial x}{\partial \eta} & \dfrac{\partial y}{\partial \eta} & \dfrac{\partial z}{\partial \eta} \\[2mm] \dfrac{\partial x}{\partial \zeta} & \dfrac{\partial y}{\partial \zeta} & \dfrac{\partial z}{\partial \zeta} \end{bmatrix} \begin{Bmatrix} \dfrac{\partial N_j}{\partial x} \\[2mm] \dfrac{\partial N_j}{\partial y} \\[2mm] \dfrac{\partial N_j}{\partial z} \end{Bmatrix} \tag{2.61}$$

or $\frac{\partial N_j}{\partial \boldsymbol{\xi}} = \mathbf{J}\frac{\partial N_j}{\partial \mathbf{x}}$ where $\mathbf{J}$ is the Jacobian matrix. The inverse relationship is

$$\frac{\partial N_j}{\partial \mathbf{x}} = \mathbf{J}^{-1}\frac{\partial N_j}{\partial \boldsymbol{\xi}} \tag{2.62}$$

For the two-dimensional (plane) case the Jacobian matrix is obtained by deleting the last row and column in equation (2.62):

$$\begin{Bmatrix} \dfrac{\partial N_j}{\partial \xi} \\[2mm] \dfrac{\partial N_j}{\partial \eta} \end{Bmatrix} = \begin{bmatrix} \dfrac{\partial x}{\partial \xi} & \dfrac{\partial y}{\partial \xi} \\[2mm] \dfrac{\partial x}{\partial \eta} & \dfrac{\partial y}{\partial \eta} \end{bmatrix} \begin{Bmatrix} \dfrac{\partial N_j}{\partial x} \\[2mm] \dfrac{\partial N_j}{\partial y} \end{Bmatrix} \tag{2.63}$$

Note that the determinant of the Jacobian matrix (the Jacobian) is

$$|\,J\,| = \frac{\partial x}{\partial \xi}\frac{\partial y}{\partial \eta} - \frac{\partial y}{\partial \xi}\frac{\partial x}{\partial \eta} = |\,\mathbf{V}_3\,| \tag{2.64}$$

2.11 CONCLUDING REMARKS

In this chapter we have discussed techniques for describing the problem geometry in a mathematical way. These techniques can be used to display graphically geometrical shapes and boundaries on a computer graphics workstation. In addition they allow the computation of certain quantities such as area or volume. The interpolation techniques used for the modelling of geometry can also be used for the description of the variation of physical quantities such as traction and displacement as in the boundary and finite element methods. Solid modelling techniques are essential to both numerical methods.

Two practical applications of solid modelling are given here. In Figure 2.29 is shown a computer display of a hull of a ship. This model will not only allow the designer to view the hull even before a prototype is built but also allow the

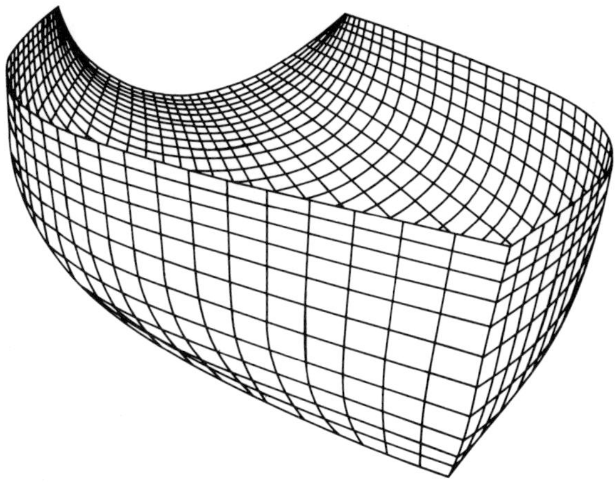

Figure 2.29 Computer graphical display of ship's hull.

design of the most efficient shape using a boundary element hydrodynamic code. A further example is shown in Figure 2.30 in which the surfaces of mining excavations in a rock mass are displayed. The model of the excavations can

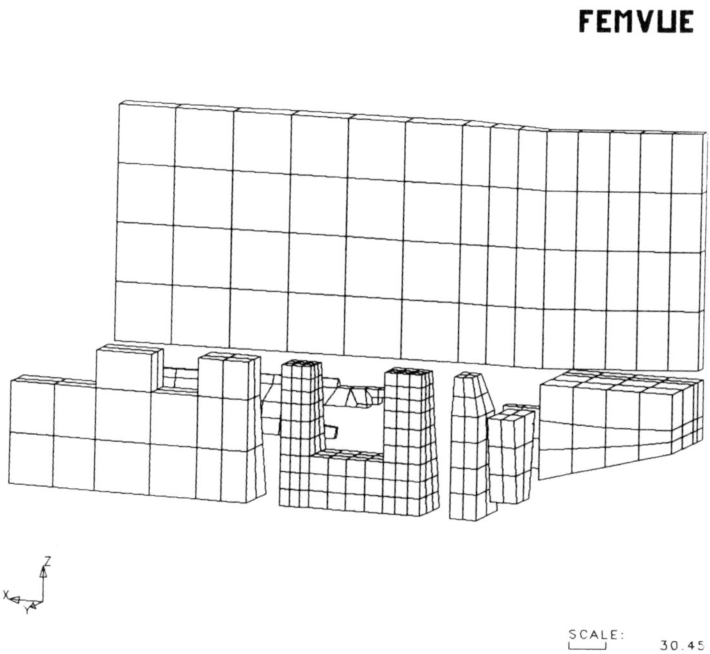

Figure 2.30 Computer display of mining excavations.

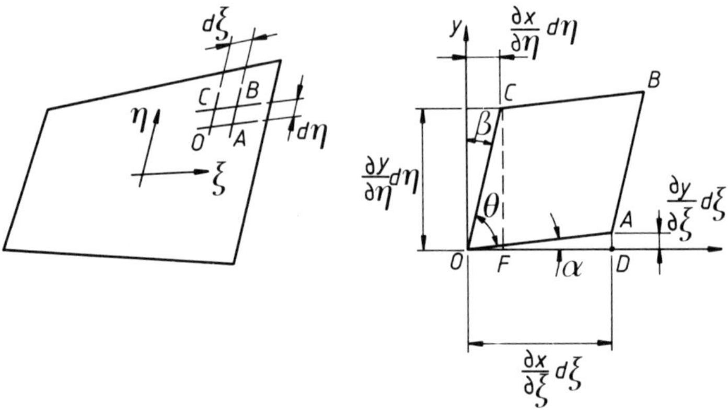

Figure 2.31 Physical interpretation of the Jacobian.

be used to compute excavation volume and hence the amount of ore extracted or may serve as input to a boundary element analysis to determine stress concentrations around the openings.

2.12 EXERCISES

2.1 Using equation (2.4) plot the element shape for a quadratic element the nodal coordinates of which are as follows.

n	x	y
1	0	1.0
2	1.0	0
3	0.7071	0.7071

2.2 Determine the maximum and mean deviation of the shape of this element from a quarter circle with radius 1 and origin at (0,0).

2.3 Referring to a standard mathematical text, work out the inverse of the Jacobian matrix (equation (2.61)) explicitly using Cramer's rule. Note that during the inversion process components of the vector normal to the element surface ($\mathbf{V}_3$) can be obtained.

2.4 Work out derivatives with respect to intrinsic coordinates for all the shape functions presented in Figures 2.2, 2.3, 2.4 and 2.5.

2.5 For the bilinear element shown in Figure 2.11 plot the variation along a diagonal line from nodes 1 to 3. What type of curve do you get?

2.6 On a two-dimensional example verify that the inverse of the geometric transformation matrix ($\mathbf{T}_g^{-1}$) is equal to the transform ($\mathbf{T}_g^T$).

2.7 Using geometrical considerations only derive the Jacobian for a two-dimensional element in two-dimensional space. Use Figure 2.31 as a guide.

2.13 REFERENCES

1. Encarnacao, J., and Schlechtendahl, J. (1983) *Computer Aided Design.* Springer Verlag, Berlin.
2. Irons, B. M. (1966) Engineering applications of numerical integration in stiffness method, *J.A.I.A.A.*, **14**, 2035–7.
3. Zienkiewicz, O. C., Irons, B. M., Ergatoudis, J. G., Ahmad, S., and Scott, F. C. (1969) Iso-parametric and associate element families for two- and three-dimensional analysis. *Finite Element Methods in Stress Analysis.* (eds I. Holland and K. Bell), Tapir Press, Norway, Chap. 14.
4. Watson, J. O. (1986) Hermitian cubic and singular elements for plane strain. *Developments in Boundary Element Methods - 4*, Elsevier Applied Science, London, Chap. 1.
5. Fraejis de Veubeke, B. (1965) Displacement and equilibrium models in the finite element method. *Stress Analysis*, (eds O.C. Zienkiewicz and G.S. Holister), John Wiley, Chichester, Chap. 9.
6. Argyris, J. H. (1965) Triangular elements with linearly varying strain for the matrix displacement method, *J. Roy. Aero. Soc. Tech. Note*, **69**, 711–713.
7. Beer, G., and Meek, J. L. (1981) 'Infinite domain' elements, *Int. J. Num. Meth. Eng.* **17**, 43–52.
8. Zienkiewicz, O. C., Emson, E., and Bettess, P. (1983) A novel boundary infinite element, *Int. J. Num.Meth. Eng.*, **19**, 393–404.
9. Irons, B. M., and Shrive, N. G. (1984) Jacobian determinant in isoparametric elements, *Eng Comput.*, **1**, 91–92.

3

Material Modelling

If you can measure what you are speaking about, and express it in numbers, you know something about it.

Lord Kelvin

3.1 INTRODUCTION

In addition to specifying the geometry of the problem, it is necessary to describe the physical response of the material in a mathematical way. This is done by defining the response characteristics of an infinitesimally small portion of the solid. For example a relationship may exist between heat flow in the material and the temperature gradient or between the displacement per unit length and stress.

The constants in such relationships are characteristic values or properties of the material.

We distinguish between materials the properties of which are independent of direction, known as isotropic materials, and materials the properties of which are dependent upon direction, known as anisotropic materials. For some materials the response may be the same at any location of the solid. These materials are called *homogeneous*. On the other hand material response varies from point to point for *inhomogeneous* materials.

Throughout this book (except in the discussion of joint elements) we take the problem domain to be a continuum. In other words, the domain is assumed to be infinitely subdivisible. Most materials in nature do not satisfy this condition. For example a rock mass is usually transected by numerous small joints and faults. It is uneconomic to model each of these joints individually. Materials of this nature may, however, be treated as an *equivalent continuum* and the effects of discontinuities 'smeared' over the domain. For example the effect of a joint set in a rock mass may be to increase the permeability or reduce the strength in a particular direction. Similarly a reinforced concrete structure may be treated as a homogeneous material with *equivalent elastic properties* which represent the combined effect of concrete and reinforcement.

We distinguish between linear and nonlinear material response. For a non-linear response the material characteristics may be dependent on the temperature or stress level, for example.

3.2 HEAT CONDUCTION AND SEEPAGE

Heat conduction in solids and flow in porous media (seepage) are diffusion problems and can be treated concurrently because they are governed by the same differential equation (Laplace in the case of an isotropic material).

Heat conduction in a solid is governed by the Fourier law. The steady state heat flow $\mathbf{q}$ per unit area is related to temperature u by (Figure 3.1)

$$\mathbf{q} = -\widehat{\mathbf{D}}\nabla u \tag{3.1}$$

where

$$\mathbf{q} = \begin{Bmatrix} q_x \\ q_y \\ q_z \end{Bmatrix} \tag{3.2}$$

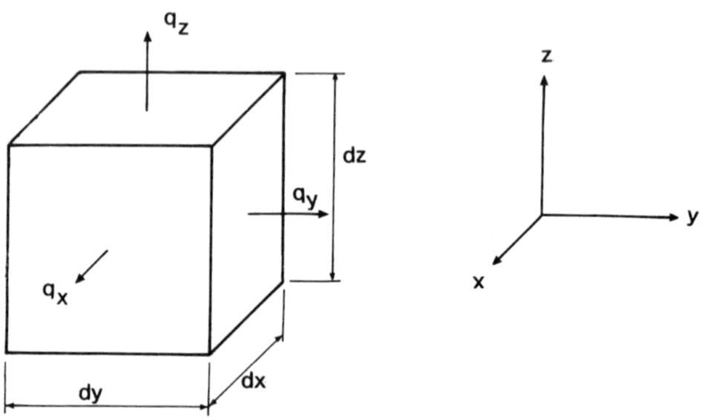

Figure 3.1 Heat flow in infinitesimal cube.

For an anisotropic material the principal directions of which coincide with the directions x, y and z, the conductivity matrix $\mathbf{D}$ is given by

$$\mathbf{D} = \begin{bmatrix} k_x & 0 & 0 \\ 0 & k_y & 0 \\ 0 & 0 & k_z \end{bmatrix} \tag{3.3}$$

where k_x, k_y, k_z are conductivity constants in the x, y, z directions.

The differential operator ∇ is defined as follows:

$$\nabla = \begin{Bmatrix} \frac{\partial}{\partial x} \\ \frac{\partial}{\partial y} \\ \frac{\partial}{\partial z} \end{Bmatrix} \tag{3.4}$$

For seepage problems $\mathbf{q}$ is defined as the fluid flow per unit area and u as potential.

3.3 SOLID MECHANICS—STRESS AND STRAIN

In solid mechanics applications a relationship between stress and strain must be established. Stress is defined as force per unit area.

Stresses are forces per unit area inside the solid. They can be visualised by cutting the solid on a particular plane (i.e. z–y, x–z, y–z plane). In Figure 3.2 are shown the definitions of these stresses for an infinitesimally small cube. There are three normal stress components σ_x, σ_y, σ_z and three shear stress components τ_{xy}, τ_{xz}, τ_{yz}.

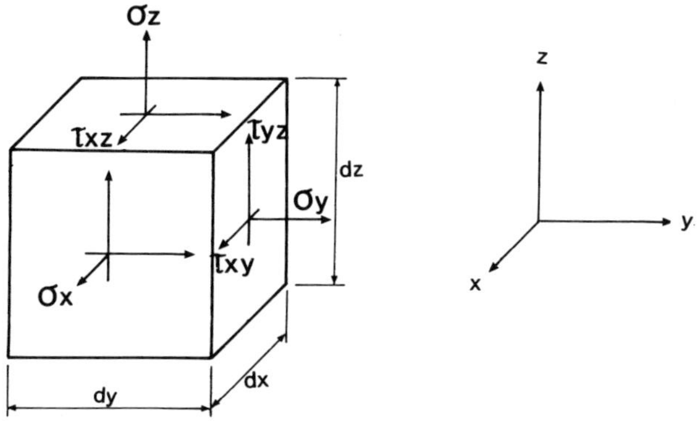

Figure 3.2 Stresses acting on infinitesimal cube.

To facilitate the derivations of response matrices later on the stresses are put into a pseudo-vector as follows:

$$\boldsymbol{\sigma} = \begin{Bmatrix} \sigma_x \\ \sigma_y \\ \sigma_z \\ \tau_{xy} \\ \tau_{yz} \\ \tau_{xz} \end{Bmatrix} \tag{3.5}$$

In some problems of solid mechanics certain stresses are zero or can be assumed to be negligible. Special cases are as follows.

Uniaxial states of stress such as may exist in members of a truss, in which all stresses except the normal stress in the axial direction of a member are zero. In this case $\boldsymbol{\sigma}$ is a scalar.

Plane stress, as in thin plates subject to in-plane loading, in which all stresses associated with the z direction are zero, i.e. $\sigma_z = \tau_{xz} = \tau_{yz} = 0$ and $\boldsymbol{\sigma}$ is defined as

$$\boldsymbol{\sigma} = \begin{Bmatrix} \sigma_x \\ \sigma_y \\ \tau_{xy} \end{Bmatrix} \tag{3.6}$$

Infinitesimal strains are defined in terms of displacement components in the x, y, z directions (u_x, u_y, u_z) by

$$\begin{aligned}
\epsilon_x &= \frac{\partial u_x}{\partial x} \\
\epsilon_y &= \frac{\partial u_y}{\partial y} \\
\epsilon_z &= \frac{\partial u_z}{\partial z} \\
\gamma_{xy} &= \frac{\partial u_x}{\partial y} + \frac{\partial u_y}{\partial x} \\
\gamma_{yz} &= \frac{\partial u_y}{\partial z} + \frac{\partial u_z}{\partial y} \\
\gamma_{xz} &= \frac{\partial u_x}{\partial z} + \frac{\partial u_z}{\partial x}
\end{aligned} \tag{3.7}$$

Again these can be put into a pseudo-vector as

$$\boldsymbol{\epsilon} = \begin{Bmatrix} \epsilon_x \\ \epsilon_y \\ \epsilon_z \\ \gamma_{xy} \\ \gamma_{yz} \\ \gamma_{xz} \end{Bmatrix} \tag{3.8}$$

In some circumstances simplifications can be made and certain strain components taken to be zero.

A state of *plane strain* can be assumed if the solid extends a long distance in the z-direction, the loading is uniform in this direction, and $u_z = 0$ everywhere. We then have $\epsilon_z = \gamma_{xz} = \gamma_{yz} = 0$.

A condition of axisymmetry can be assumed when the solid and loading are axially symmetric. In this case the strains (and stresses) are redefined as shown in Figure 3.3:

$$\begin{aligned} \epsilon_z &= \frac{\partial u_z}{\partial z} \\ \epsilon_r &= \frac{\partial u_r}{\partial r} \\ \epsilon_\theta &= \frac{u_r}{r} \\ \gamma_{rz} &= \frac{\partial u_r}{\partial z} + \frac{\partial u_z}{\partial r} \end{aligned} \tag{3.9}$$

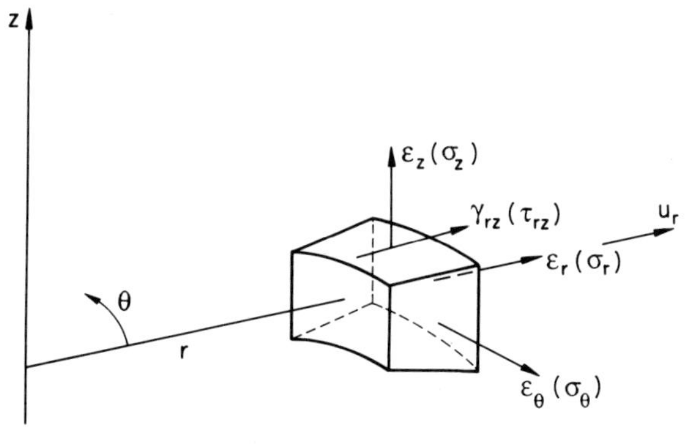

Figure 3.3 Axisymmetric solid and definition of stresses and strains.

Another special case is a state of complete plane strain[2,3], in which derivatives in the z direction of all displacements are taken to be zero, but u_z may be non-zero. This gives

$$
\begin{aligned}
\epsilon_x &= \frac{\partial u_x}{\partial x} \\
\epsilon_y &= \frac{\partial u_y}{\partial y} \\
\epsilon_z &= 0 \\
\gamma_{xy} &= \frac{\partial u_y}{\partial x} + \frac{\partial u_x}{\partial y} \\
\gamma_{zy} &= \frac{\partial u_z}{\partial y} \\
\gamma_{zx} &= \frac{\partial u_z}{\partial x}
\end{aligned}
\tag{3.10}
$$

Complete plane strain can be split into the plane strain case already discussed and an antiplane strain or St Venant torsion component for which $\epsilon_x = \epsilon_y = \epsilon_z = \gamma_{xy} = 0$ and

$$
\begin{aligned}
\gamma_{zy} &= \frac{\partial u_z}{\partial y} \\
\gamma_{zx} &= \frac{\partial u_z}{\partial x}
\end{aligned}
\tag{3.11}
$$

In complete plane strain it is possible to have shear strains and stresses acting in the z-direction.

Sometimes it is necessary to compute the magnitudes of stress or strain in directions which do not coincide with the global axes. In this case a *transformation* of stress or strain is necessary.

The transformation of local stresses $\boldsymbol{\sigma}'$ acting on planes in the material parallel with the x', y' and z' axes to global stresses $\boldsymbol{\sigma}$ acting on cuts parallel with the x, y, z axes can be written as

$$
\boldsymbol{\sigma} = \mathbf{T}_\sigma \boldsymbol{\sigma}' \tag{3.12}
$$

For the two-dimensional case in which the local axes are defined by a rotation about the z-axis, $\mathbf{T}_\sigma$ is obtained by the two transformations shown in Figure 3.4:

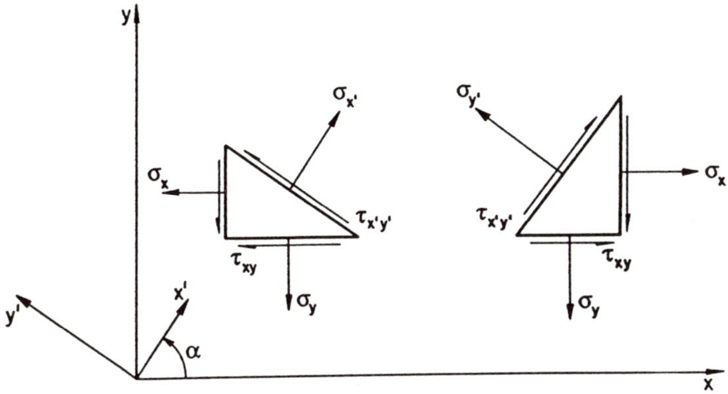

Figure 3.4 Transformation of stress in two dimensions.

$$\mathbf{T}_\sigma = \begin{bmatrix} \cos^2\alpha & \sin^2\alpha & -a\sin\alpha\cos\alpha \\ \sin^2\alpha & \cos^2\alpha & a\sin\alpha\cos\alpha \\ b\cos\alpha\sin\alpha & -b\cos\alpha\sin\alpha & \cos^2\alpha - \sin^2\alpha \end{bmatrix} \tag{3.13}$$

where the factors a and b are taken to be 2 and 1, respectively.

For the stress transformation in three-dimensional space it is convenient to refer to the components of unit vectors in the directions of the local axes, Figure 3.5. For example we denote by

$$\mathbf{v}_1 = \begin{Bmatrix} v_{1x} \\ v_{1y} \\ v_{1z} \end{Bmatrix} \tag{3.14}$$

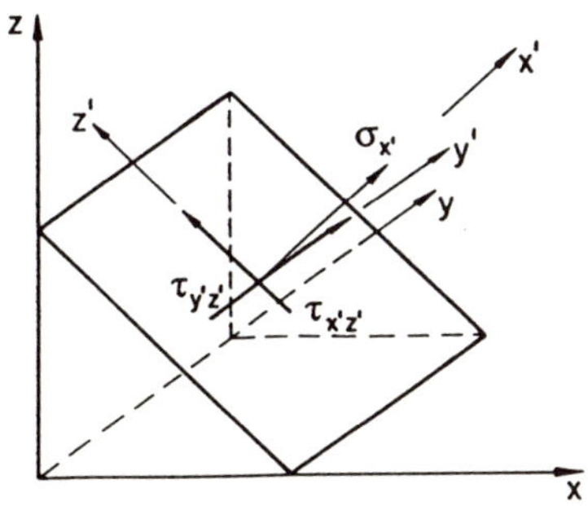

Figure 3.5 Transformation of stress in three dimensions.

the unit vector in the direction of the x' - axis. Similarly, $\mathbf{v}_2$ and $\mathbf{v}_3$ are unit vectors along the y'- and z'-axes. In terms of these vector components the matrix $\mathbf{T}_\sigma$ is written as

$$
\mathbf{T}_\sigma = \begin{bmatrix} \mathbf{T}_{\sigma 11} & \mathbf{T}_{\sigma 12} \\ \mathbf{T}_{\sigma 21} & \mathbf{T}_{\sigma 22} \end{bmatrix}
$$

$$
\mathbf{T}_{\sigma 11} = \begin{bmatrix} v_{1x}^2 & v_{2x}^2 & v_{3x}^2 \\ v_{1y}^2 & v_{2y}^2 & v_{3y}^2 \\ v_{1z}^2 & v_{2z}^2 & v_{3z}^2 \end{bmatrix}
$$

$$
\mathbf{T}_{\sigma 21} = \begin{bmatrix} av_{1y}v_{1x} & av_{2y}v_{2x} & av_{3y}v_{3x} \\ av_{1y}v_{1z} & av_{2y}v_{2z} & av_{3y}v_{3z} \\ av_{1x}v_{1z} & av_{2x}v_{2z} & av_{3x}v_{3z} \end{bmatrix}
$$

$$
\mathbf{T}_{\sigma 21} = \begin{bmatrix} bv_{1x}v_{2x} & bv_{2x}v_{3x} & bv_{1x}v_{3x} \\ bv_{1y}v_{2y} & bv_{2y}v_{3y} & bv_{1y}v_{3y} \\ bv_{1z}v_{2z} & bv_{2z}v_{3z} & bv_{1z}v_{3z} \end{bmatrix} \tag{3.15}
$$

$$
\mathbf{T}_{\sigma 22} = \begin{bmatrix} v_{1x}v_{2y} + v_{1y}v_{2x} & v_{2x}v_{3y} + v_{2y}v_{3x} & v_{1x}v_{3y} + v_{1y}v_{3x} \\ v_{1y}v_{2z} + v_{1z}v_{2y} & v_{2y}v_{3z} + v_{2z}v_{3y} & v_{1y}v_{3z} + v_{1z}v_{3y} \\ v_{1x}v_{2z} + v_{1z}v_{2x} & v_{2x}v_{3z} + v_{2z}v_{3x} & v_{1x}v_{3z} + v_{1z}v_{3x} \end{bmatrix}
$$

where the factor a is equal to 2, and b is equal to 1.

The transformation of strains is similar to that for stresses, i.e.

$$
\epsilon = \mathbf{T}_\epsilon \epsilon' \tag{3.16}
$$

where $\mathbf{T}_\epsilon$ is equal to $\mathbf{T}_\sigma$ as defined by equations (3.15) or (3.13) with the factors a and b equal to 1 and 2, respectively.

There is one transformation of the stresses which results in zero local shear stresses. The transformed normal stresses are in this case called *principal stresses.*

3.4 ELASTIC MATERIAL

In the general three dimensional case the material response is governed by Hooke's law[1]. For an isotropic material this is

$$
\begin{aligned}
\epsilon_x &= \frac{1}{E}\left[\sigma_x - \nu(\sigma_y + \sigma_z)\right] \\
\epsilon_y &= \frac{1}{E}\left[\sigma_y - \nu(\sigma_x + \sigma_z)\right] \\
\epsilon_z &= \frac{1}{E}\left[\sigma_z - \nu(\sigma_x + \sigma_y)\right] \\
\gamma_{xy} &= \frac{1}{G}\tau_{xy}, \quad \gamma_{yz} = \frac{1}{G}\tau_{yz}, \quad \gamma_{xz} = \frac{1}{G}\tau_{xz}
\end{aligned} \tag{3.17}
$$

where E is the modulus of elasticity, ν Poisson's ratio and G the shear modulus, given by

$$G = \frac{E}{2(1+\nu)} \tag{3.18}$$

Equation (3.17) can be conveniently written in matrix form

$$\epsilon = \mathbf{C}\boldsymbol{\sigma} \tag{3.19}$$

where the matrix $\mathbf{C}$ is defined as

$$\mathbf{C} = \frac{1}{E}\begin{bmatrix} 1 & -\nu & -\nu & 0 & 0 & 0 \\ -\nu & 1 & -\nu & 0 & 0 & 0 \\ -\nu & -\nu & 1 & 0 & 0 & 0 \\ 0 & 0 & 0 & \frac{E}{G} & 0 & 0 \\ 0 & 0 & 0 & 0 & \frac{E}{G} & 0 \\ 0 & 0 & 0 & 0 & 0 & \frac{E}{G} \end{bmatrix} \tag{3.20}$$

The inverse relationship can be derived by solving (3.19) for the stresses and is

$$\begin{aligned} \sigma_x &= C_1\left[\epsilon_x + C_2(\epsilon_y + \epsilon_z)\right] \\ \sigma_y &= C_1\left[\epsilon_y + C_2(\epsilon_x + \epsilon_z)\right] \\ \sigma_z &= C_1\left[\epsilon_z + C_2(\epsilon_x + \epsilon_y)\right] \\ \tau_{xy} &= G\gamma_{xy} \\ \tau_{yz} &= G\gamma_{yz} \\ \tau_{xz} &= G\gamma_{xz} \end{aligned} \tag{3.21}$$

where

$$\begin{aligned} C_1 &= E\frac{(1-\nu)}{(1+\nu)(1-2\nu)} \\ C_2 &= \nu/(1-\nu) \end{aligned} \tag{3.22}$$

The above equations can again be written in matrix form

$$\boldsymbol{\sigma} = \mathbf{D}\epsilon \tag{3.23}$$

where the elasticity matrix $\mathbf{D}$ is defined as

$$\mathbf{D} = C_1 \begin{bmatrix} 1 & C_2 & C_2 & 0 & 0 & 0 \\ C_2 & 1 & C_2 & 0 & 0 & 0 \\ C_2 & C_2 & 1 & 0 & 0 & 0 \\ 0 & 0 & 0 & G/C_1 & 0 & 0 \\ 0 & 0 & 0 & 0 & G/C_1 & 0 \\ 0 & 0 & 0 & 0 & 0 & G/C_1 \end{bmatrix} \tag{3.24}$$

Special cases of the elasticity matrix can be derived by setting the appropriate strain or stress values to zero.

For a general anisotropic material 21 elastic constants are required. In most cases these are difficult or impossible to determine. A special case of an anisotropic material is the stratified material. Examples are laminated materials and bedded rocks, Figure 3.6. These materials are characterised by the fact that they have different values of modulus of elasticity and shear modulus in two orthogonal directions. Assuming, for convenience, that direction $1(E_1,\, G_1)$ is associated with the x–y plane and direction $2(E_2,\, G_2)$ is orthogonal to it the $\mathbf{D}$ matrix is given by [5]

$$\mathbf{D} = \begin{bmatrix} C_1' & C_2' & C_3' & 0 & 0 & 0 \\ & C_1' & C_3' & 0 & 0 & 0 \\ \text{symmetric} & & C_4' & 0 & 0 & 0 \\ & & & G_1 & 0 & 0 \\ & & & & G_2 & 0 \\ & & & & & G_2 \end{bmatrix} \tag{3.25}$$

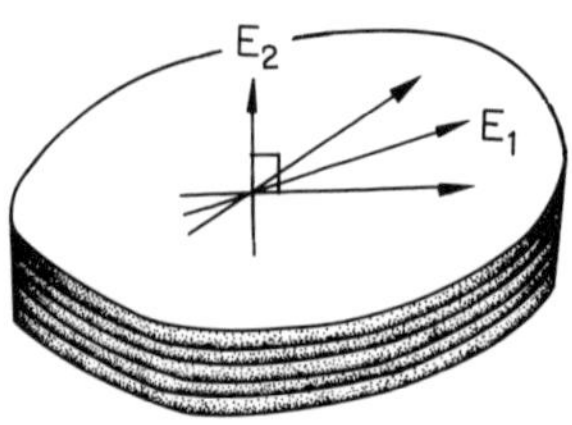

Figure 3.6 Laminate material.

where

$$
\begin{aligned}
C_1' &= n\left(1-\nu_2^2\right)C \\
C_2' &= n\left(\nu_1+n\nu_2^2\right)C \\
C_3' &= n\nu_2\left(1+\nu_1\right)C \\
C_4' &= \left(1+\nu_1^2\right)C \\
C &= \frac{E_2}{(1-\nu_1)\left(1-\nu_1-2n\nu_2^2\right)} \\
n &= \frac{E_1}{E_2} \quad \text{and} \quad G_1 = \frac{E_1}{2(1+\nu_1)}
\end{aligned}
\tag{3.26}
$$

In some cases initial strains or stresses not caused by the applied loading may be present. In geomechanics for example the virgin stress (i.e. the stress in the rock or soil mass before excavation) plays an important role. Alternatively there may exist thermal strains caused by temperature changes in the materials. Equation (3.23) can be generalised to include these effects:

$$\boldsymbol{\sigma} = \mathbf{D}(\epsilon - \epsilon_0) + \boldsymbol{\sigma}_0 \tag{3.27}$$

where ϵ_0 and $\boldsymbol{\sigma}_0$ are initial strains, and initial stresses, respectively.

3.5 INELASTIC MATERIAL

Once a certain stress level has been reached then the stress–strain relationship becomes non-linear and no longer obeys Hooke's law. In addition, if the loading is subsequently removed the strains do not revert to zero as they do for elastic material. The strains present after the loading has been removed are called plastic strains and the stress level at which plastic strains start to occur is the yield stress.

Let us consider the simple one-dimensional example shown in Figure 3.7, a bar subjected to tensile stress at its ends. The stress–strain curve shows that there is a linear stress–strain relationship until a yield stress σ_Y is reached. After this the strain increases without any increase in stress (as elastic–perfectly plastic material). In other words when the stress level

$$\sigma = \sigma_Y \tag{3.28}$$

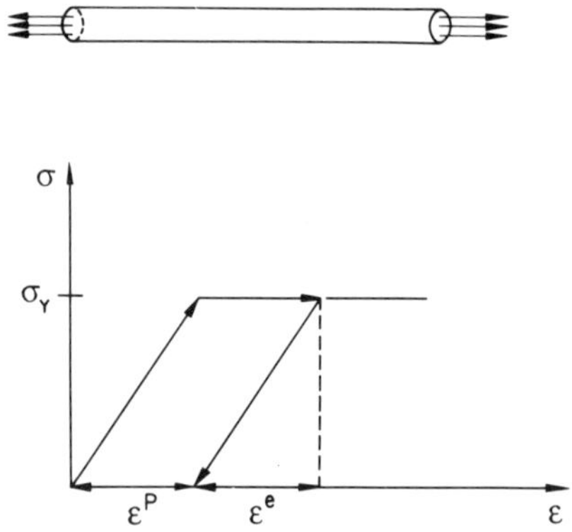

Figure 3.7 Bar subjected to tension.

has been reached then in addition to elastic strains ϵ^{e}, there are also plastic strains, ϵ^{p}. The total strain is written as

$$\epsilon = \epsilon^{\text{e}} + \epsilon^{\text{p}} \tag{3.29}$$

For this simple one-dimensional example ϵ^{p} is solely dependent on how much we allow the ends of the bar to move apart. Equation (3.28) can be rewritten as

$$F(\sigma) = \sigma - \sigma_Y = 0 \tag{3.30}$$

$F(\sigma)$ is known as a *yield function.* The above can be extended to multidimensional stress states, for which we write that yielding occurs if

$$F(\boldsymbol{\sigma}) = 0 \tag{3.31}$$

and that the total strain consists of elastic and plastic components

$$\epsilon = \epsilon^{\text{e}} + \epsilon^{\text{p}} \tag{3.32}$$

3.5.1 Yield functions—isotropic material

The yield function F is generally a function of all six stress components, and is normally expressed in terms of principal stresses or stress invariants. Stress invariants are characteristic values of the stress field which do not depend on the directions of the current system of axes, i.e. they do not change (are invariant) when the axes are rotated. The three invariants are

$$J_1 = \sigma_x + \sigma_y + \sigma_z$$

$$J_2 = \frac{1}{2}\left[S_x^2 + S_y^2 + S_z^2\right] + \tau_{xy}^2 + \tau_{yz}^2 + \tau_{zx}^2 \tag{3.33}$$

$$J_3 = S_x S_y S_z + \tau_{xy}\tau_{yz}\tau_{zx} - S_x\tau_{yz}^2 - S_y\tau_{zx}^2 - S_z\tau_{xy}^2$$

where the deviatoric stresses are defined as

$$S_x = \sigma_x - \frac{J_1}{3}$$

$$S_y = \sigma_y - \frac{J_1}{3}$$

$$S_z = \sigma_z - \frac{J_1}{3}$$

In terms of principal stresses (σ_1, σ_2, σ_3) the stress invariants are

$$J_1 = \sigma_1 + \sigma_2 + \sigma_3$$

$$J_2 = \sigma_1\sigma_2 + \sigma_2\sigma_3 + \sigma_3\sigma_1 \tag{3.34}$$

$$J_3 = \sigma_1\sigma_2\sigma_3$$

The following are some of the more commonly used yield functions.

von Mises for a ductile material such as steel:

$$F(\boldsymbol{\sigma}) = \sqrt{3J_2} - \sigma_Y \tag{3.35}$$

A *no-tension* yield function is applicable to brittle materials such as certain type of rocks:

$$F = \sigma_{\max} \tag{3.36}$$

Here $\sigma_{\max}$ is the maximum principal stress (tension positive convention).

The *Mohr–Coulomb* yield function is applicable to materials such as soil and rock:

$$F = \frac{J_1}{3}\sin\phi + J_2^{1/2}\cos\theta - \frac{J_2^{1/2}}{\sqrt{3}}\sin\phi\sin\theta - c\cos\phi \tag{3.37}$$

where the 'lode angle' θ is given by

$$\theta = \frac{1}{3}\sin^{-1}\left(-\frac{3\sqrt{3}}{2}\frac{J_3}{J_2^{3/2}}\right) \quad \text{where} -\frac{\pi}{6} < \theta < \frac{\pi}{6}$$

If the friction angle ϕ is equal to zero this reverts to the well known *Tresca* yield condition.

The *Drucker–Prager* yield function is a modification of Mohr–Coulomb and is often used in preference to it because it is simpler to program:

$$F = \alpha J_1 + J_2^{1/2} - K \tag{3.38}$$

where

$$\alpha = \frac{2\sin\phi}{\sqrt{3}(3-\sin\phi)}$$

$$K = \frac{6c\cos\phi}{\sqrt{3}(3-\sin\phi)}$$

For $\phi = 0$ (no friction) the *von Mises* yield condition is obtained.

The above yield conditions can be expressed as functions of the principal stresses and plotted in principal stress space as three-dimensional surfaces which are shown in Figure 3.8.

There are a number of more sophisticated yield functions which model the behaviour of various materials and the reader is referred to the relevant literature on the subject.[6]

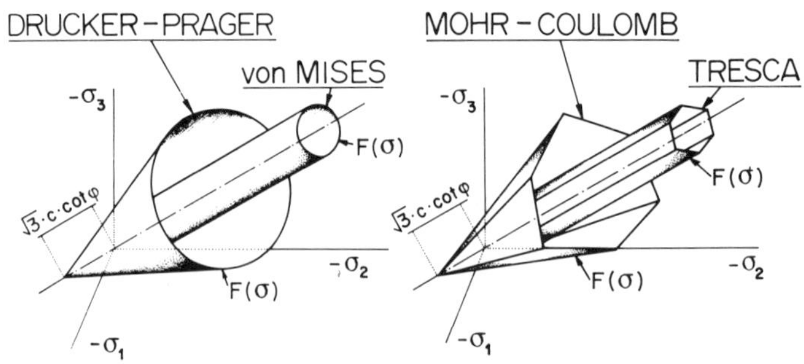

Figure 3.8 Isotropic yield conditions plotted as surfaces in principal stress space.

3.5.2 Yield conditions—anisotropic material

For some materials the elastoplastic behaviour is dependent upon direction and we can no longer work with stress invariants. For example a rock mass may be transected by fractures and joints which constitute planes of weakness on which plastic strains may develop.

Here we may use a 'multi-laminate' model. For the ith plane of weakness two yield conditions may be written (Figure 3.9):

$$F_1^{(i)} = |\,\tau^{(i)}\,| - c^{(i)} + \sigma_n^{(i)} \tan \phi^{(i)} \tag{3.39}$$

$$F_2^{(i)} = \sigma_n^{(i)} \tag{3.40}$$

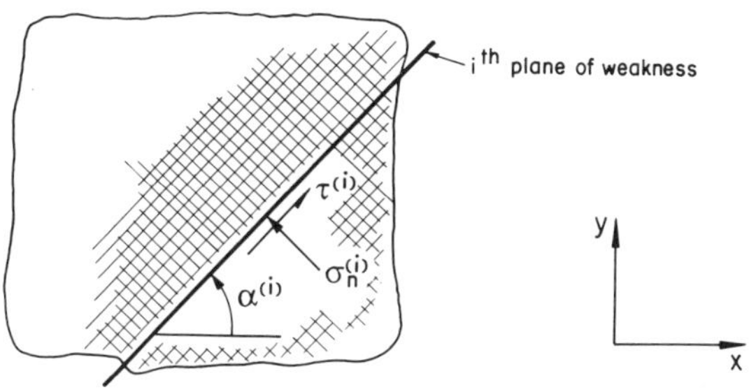

Figure 3.9 'Planes of weakness' in rock mass.

F_1 is the Mohr–Coulomb and F_2 a no-tension yield condition. $c^{(i)}$ and $\phi^{(i)}$ are the cohesion and angle of friction of the ith joint set and the stresses $\sigma_n^{(i)}$ and $\tau^{(i)}$ can be obtained from the global stress components by means of the transformation

$$\begin{Bmatrix} \sigma_n \\ — \\ \tau \end{Bmatrix}^{(i)} = \left(\mathbf{T}_\sigma^{(i)}\right)^T \boldsymbol{\sigma} \tag{3.41}$$

where $\mathbf{T}_\sigma^{(i)}$ is the stress transformation matrix for the ith plane of weakness.

3.5.3 Material behaviour at yield

Once a material has yielded the relationship between stresses and strains is no longer linear and only an incremental relationship between stress and strain can be written. For an infinitesimal increment of stress the strain increment is

$$\mathrm{d}\boldsymbol{\epsilon} = \mathrm{d}\boldsymbol{\epsilon}^{\mathrm{e}} + \mathrm{d}\boldsymbol{\epsilon}^{\mathrm{p}} \tag{3.42}$$

The plastic strain increment is given by

$$\mathrm{d}\boldsymbol{\epsilon}^{\mathrm{p}} = \mathrm{d}\epsilon^{\mathrm{p}} \frac{\partial Q}{\partial \boldsymbol{\sigma}} \tag{3.43}$$

where $\mathrm{d}\epsilon^{\mathrm{p}}$ is an equivalent plastic strain increment, which is basically a measure of the amount of plastic straining that has occurred. $\mathrm{d}\epsilon^{\mathrm{p}}$ (a scalar) is sometimes referred in the literature as $\mathrm{d}\lambda$ (load factor). Our definition is thus more general.

Q is a plastic potential and

$$\frac{\partial Q}{\partial \boldsymbol{\sigma}} = \left\{ \begin{array}{c} \frac{\partial Q}{\partial \sigma_x} \\ \frac{\partial Q}{\partial \sigma_y} \\ \frac{\partial Q}{\partial \tau_{xy}} \end{array} \right\} \tag{3.44}$$

for the plane stress case.

The plastic potential is dependent on the type of material modelled. For example if $Q \equiv F$ (yield function) then the elasto-plastic behaviour is called *associative* otherwise it is *non-associative*.

To illustrate this, let us consider an anisotropic material with one plane of weakness. For this plane we may write the associative yield criterion and flow-rule as

$$F \equiv Q = |\,\tau\,| - c + \sigma_n \tan\phi \tag{3.45}$$

The increment of plastic strain is given by

$$\mathrm{d}\boldsymbol{\epsilon}^{\mathrm{p}} = \left\{ \begin{array}{c} \mathrm{d}\gamma \\ \mathrm{d}\epsilon_n \end{array} \right\}_{\mathrm{p}} = \left\{ \begin{array}{c} \frac{\partial Q}{\partial \tau} \\ \frac{\partial Q}{\partial \sigma_n} \end{array} \right\} \mathrm{d}\epsilon^{\mathrm{p}} \tag{3.46}$$

$$\mathrm{d}\boldsymbol{\epsilon}^{\mathrm{p}} = \left\{ \begin{array}{c} \mathrm{sgn}(\tau) \\ \tan\phi \end{array} \right\} \mathrm{d}\epsilon^{\mathrm{p}} \tag{3.47}$$

This means that each plastic strain increment $\mathrm{d}\epsilon^{\mathrm{p}}$ will produce a strain ($\mathrm{d}\epsilon^{\mathrm{p}} \tan\phi$) in the direction normal to the plane of weakness, i.e. the inelastic shearing produces a dilation. This can be visualised by supposing that the

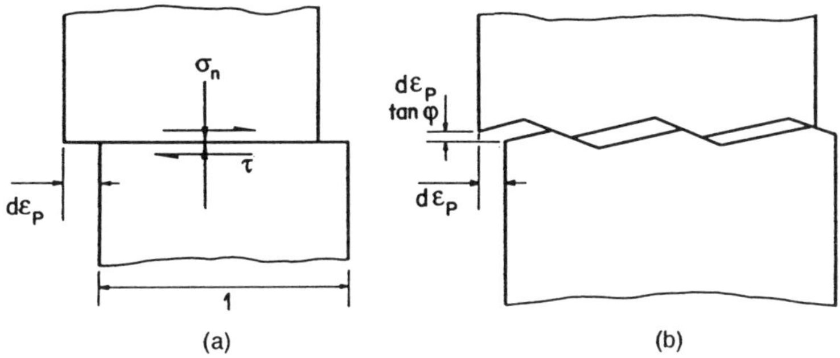

Figure 3.10 Smooth and rough joint.

plane of weakness is a rough joint as shown in Figure 3.10b. If the joint were to be perfectly smooth (Figure 3.10a) then a non-associative flow rule with

$$Q = |\tau| \tag{3.48}$$

would be adopted, giving zero dilation, because $\frac{\partial Q}{\partial \sigma_n}$ is zero.

The incremental relationship between total strain and stress at yield is

$$\mathrm{d}\boldsymbol{\epsilon} = \mathbf{C}\mathrm{d}\boldsymbol{\sigma} + \frac{\partial Q}{\partial \boldsymbol{\sigma}}\mathrm{d}\epsilon^{\mathrm{p}} \tag{3.49}$$

where the unknown $\mathrm{d}\epsilon^{\mathrm{p}}$ can be determined from the condition that we must always have $F \leq 0$, i.e. F must not become positive. As we shall see later this restriction may be relaxed if time dependent plasticity is considered. For the present case, $\mathrm{d}F$ must be less than or equal to zero if $F = 0$.

Taking $\mathrm{d}F = 0$ and expanding $\mathrm{d}F$ we can write

$$\mathrm{d}F = \frac{\partial F}{\partial \sigma_x}\mathrm{d}\sigma_x + \frac{\partial F}{\partial \sigma_y}\mathrm{d}\sigma_y + \ldots = 0 \tag{3.50}$$

or

$$\mathrm{d}F = \left(\frac{\partial F}{\partial \boldsymbol{\sigma}}\right)^{\mathrm{T}} \mathrm{d}\boldsymbol{\sigma} = 0 \tag{3.51}$$

Equation (3.49) and (3.51) may be written as a single matrix equation:

$$\left\{ \begin{array}{c} \mathrm{d}\boldsymbol{\epsilon} \\ 0 \end{array} \right\} = \left[\begin{array}{cc} \mathbf{C} & \frac{\partial Q}{\partial \boldsymbol{\sigma}} \\ \left(\frac{\partial F}{\partial \boldsymbol{\sigma}}\right)^{\mathrm{T}} & 0 \end{array} \right] \left\{ \begin{array}{c} \mathrm{d}\boldsymbol{\sigma} \\ \mathrm{d}\epsilon^{\mathrm{p}} \end{array} \right\} \tag{3.52}$$

It is possible to eliminate $d\epsilon^p$ from (3.52) and obtain an incremental relationship between stress and strain at yield

$$d\boldsymbol{\sigma} = \mathbf{D}_{ep} d\boldsymbol{\epsilon} \tag{3.53}$$

where
$$\mathbf{D}_{ep} = \mathbf{D} - \mathbf{D}\left(\frac{\partial Q}{\partial \boldsymbol{\sigma}}\right)\left(\frac{\partial F}{\partial \boldsymbol{\sigma}}\right)^{T}\mathbf{D}\frac{1}{C_1}$$

and
$$C_1 = \left(\frac{\partial F}{\partial \sigma}\right)^{T}\mathbf{D}\left(\frac{\partial Q}{\partial \boldsymbol{\sigma}}\right)$$

When the material has reached the yield point, the incremental relationship between stress and strain (3.53) is used instead of the relationship (3.23). Non-linear solution algorithms are discussed in Chapter 9.

3.5.4 Time-dependent plasticity, creep

In the plasticity theory just outlined it is assumed that plastic strains occur instantaneously and that there is no time dependence. We know, however, from observation that in some materials inelastic strains take time to develop fully.

The concept of viscoplasticity allows the yield function F to be greater than zero during the development of inelastic strain and then to approach the condition $F = 0$ asymptotically with time. Considering now the time effect we write an equation for the plastic strain rate instead of the strain increment.

The (scalar) equivalent plastic strain rate can be written quite generally as

$$\dot{\epsilon}^p = \frac{1}{\eta}\phi(F) \tag{3.54}$$

where η is a material parameter (viscosity) and $\phi(F)$ is a function of the yield function which is zero if $F < 0$. For $F > 0$ various forms of ϕ may be proposed. The most widely used are the power law

$$\phi(F) = F^n \tag{3.55}$$

where n is an additional parameter (n = 1–3 for most materials), and the exponential law

$$\phi(F) = e^F - 1 \tag{3.56}$$

The parameters and type of function ϕ to be used for a particular material can be determined by laboratory tests or by back analysis[9]. The equivalent plastic strain increment (a scalar) which develops in a time increment Δt is given by

$$\Delta\epsilon^{\mathrm{p}} = \int_{\Delta t} \dot{\epsilon}^{\mathrm{p}} \mathrm{d}t \tag{3.57}$$

If small time steps are taken in the analysis the plastic strain increment between time t_0 and $t_0 + \Delta t$ can be approximated by

$$\Delta\epsilon^{\mathrm{p}} = \dot{\epsilon}^{\mathrm{p}}_{t0}\Delta t = \frac{\Delta t}{\eta}\phi(F) \tag{3.58}$$

when $\dot{\epsilon}^{\mathrm{p}}_{t0}$ is the equivalent plastic strain rate at time t_0.

The total actual strain increment (a vector) during time step Δt is given by

$$\Delta\boldsymbol{\epsilon} = \Delta\epsilon^{\mathrm{e}} + \frac{\partial Q}{\partial \boldsymbol{\sigma}}\Delta\epsilon^{\mathrm{p}} \tag{3.59}$$

The time step Δt cannot be chosen freely but has to satisfy certain stability criteria to ensure an asymptotic approach to $F = 0$ and prevent oscillations and divergence. How this *time stability limit* is determined is discussed in detail in Reference 8. The time stability limit is a function of the viscosity η. Swoboda *et al*[9] show applications of viscoplastic analysis in tunnelling and how the viscosity η and the exponent n in (3.55) are determined for geomaterials.

If the material parameter η is not known then an analysis can be made with a 'fictitious' time step of Δt equal to the stability limit and a number of time steps applied until $F(\sigma) = 0$ everywhere.

3.6 CLOSURE

In this chapter we have discussed some models which can be used to describe the response of materials which are subjected to a temperature, flow or stress field.

The list of material models discussed here is by no means complete. A great number of such models exists and it is not appropriate to cover all of them in an introductory test. Material models can include such effects as strain hardening and softening, the effect of pore pressure and fluid flow through the material, and visco-elasticity. The reader is referred to References 10 and 11 for a more detailed discussion of elasto-plastic behaviour.

3.7 EXERCISES

3.1 Modify the elasticity matrix **D** (equation (3.25)) for a laminated material in plane strain.

3.2 Work out the **D** matrix for a laminated material where the direction 2 is inclined to the z-axis.

3.3 Write a FORTRAN subroutine which computes the stress invariants J_1, J_2 and J_3 for a particular stress state.

3.4 Write a subroutine which transforms two-dimensional global stress components into stresses in local directions inclined an angle of α to the x-axis.

3.5 Using the above subroutine or otherwise check that J_1, J_2 and J_3 are invariant, i.e. have the same value computed from the global and rotated stresses.

3.6 Show how $\boldsymbol{\epsilon}^{\mathrm{p}}$ can be eliminated from equation (3.52) to give equation (3.53).

3.8 REFERENCES

1. Timoshenko, S., and Goodier, J. N. (1951) *Theory of Elasticity*, 2nd Edn, McGraw-Hill, New York.
2. Brady, B. H. G. (1979) A direct formulation of the boundary element method of stress analysis for complete plane strain, *Int. J. Rock Mech. Min. Sci.*, **16**, 235–244.
3. Zienkiewicz, O. C., and Pande, G. (1977) Multilaminate models for rock, *Int. J. Num. Anal. Meth. Geomech.*, **1**.
4. Zienkiewicz, O. C., Taylor, R. L., and Pande, G. N. (1978) Quasi - plane strain in the analysis of geological problems, *Computer Methods in Tunnel Design*, Institution of Civil Engineers, London.
5. Lekhnitzkii, S. G. (1963) *Theory of Elasticity of an Anisotropic Elastic Body* (Translation from Russian), Holden Day, U.S.A.
6. Desai, C. S. (Ed.) (1976) *Numerical Methods in Geomechanics*, Vol. 1–3, Am. Soc. Civil Eng, Special Publication.
7. Desai, C. S. (Ed.) (1981) *Implementation of Computer Procedures and Stress-Strain Laws in Geotechnical Engineering*, Proc., Acorn Press.
8. Zienkiewicz, O. C., and Cormeau, I. C. (1974) Visco-plasticity and creep in elastic solids—a unified numerical solution approach, *Int. J. Num. Meth. Eng.*, **8**, 821–45.
9. Swoboda, G., Mertz, W., and Beer, G. (1987). Rheological analysis of tunnel excavations by means of coupled finite element (FEM)—boundary element (BEM) analysis. *Int. J. Num. Anal. Meth. Geomech.*, **11**, 159–129.
10. Zienkiewicz, O. C. (1977) *The Finite Element Method* 3rd Edn, McGraw-Hill, New York.
11. Hinton, E., and Owen, D. R. J. (1980) *Finite Elements in Plasticity—Theory and Practice*, Pineridge Press, Swansea.

4

Finite Elements of a Continuum

It is the mark of an educated man to look for precision in each class of things just so far as the nature of the subject admits.

Aristotle

4.1 INTRODUCTION

In the previous chapters the prerequisites for a finite or boundary element analysis, namely the modelling of the solid domain and the material, were discussed.

Here we shall introduce the first method of analysis, the finite element method. As discussed previously this method requires a subdivision of the problem domain into subdomains known as finite elements. For each of these elements a matrix of coefficients which describes the response of the element is obtained.

Thus, the first step in any finite element analysis is to obtain a matrix of response coefficients for each individual element (Figure 4.1 (left)). The next step is to assemble the individual element contributions to form a response matrix for the problem domain (Figure 4.1 (right)). The system of equations is solved subject to the given boundary conditions and loading. By using the interpolation (shape) functions the results at any point inside the domain can be computed.

In the following we will first describe how the matrix of response coefficients is computed for both elasticity and potential problems. To introduce the basic principles, first one-dimensional problems are discussed. The basic theory is then extended to multidimensional problems.

Next we discuss how the element coefficient matrices can be assembled to form the structure response matrix.

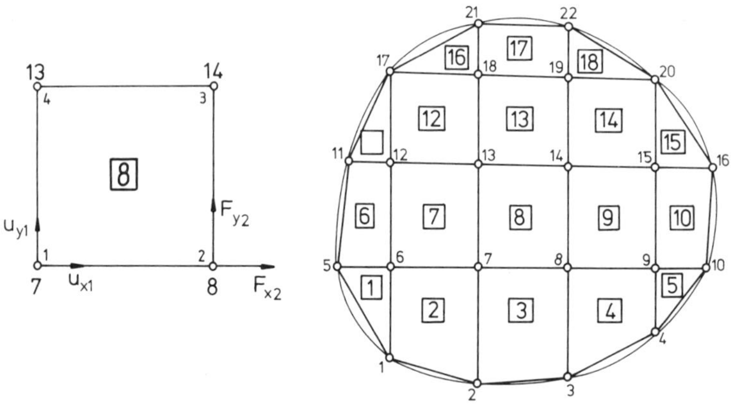

Figure 4.1 Individual element and assembly of elements in two dimensions

A fairly detailed description is then given how the system of equations can be solved subject to given boundary conditions and loading. Finally, postprocessing procedures are discussed for obtaining results inside elements.

4.2 DERIVATION OF ELEMENT RESPONSE MATRICES

4.2.1 Elasticity problems—one-dimensional elements

To illustrate the basic steps in the derivation of the element response matrix consider the weightless bar (cross sectional area of A) in Figure 4.2 which is fixed at one end and subjected to a force at the other end.

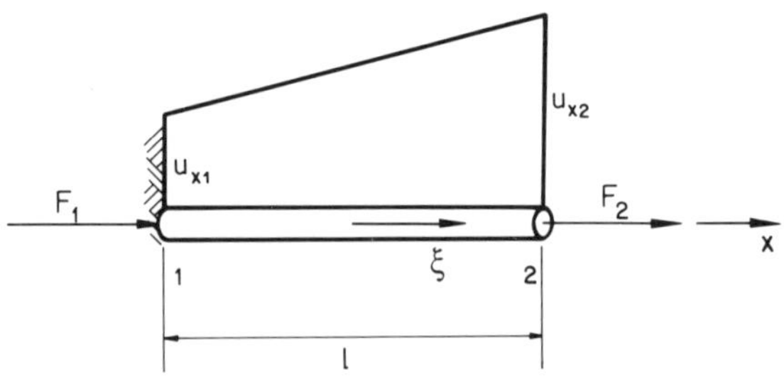

Figure 4.2 A bar subject to forces

The first step is the description of the element geometry. The coordinate x of a point with intrinsic coordinate ξ is given by

$$x(\xi) = N_1 x_1 + N_2 x_2 \tag{4.1}$$

The actual displacement $\hat{u}_x$ in the bar is approximated by

$$u_x = N_1 u_{x1} + N_2 u_{x2} \tag{4.2}$$

where u_{x1} and u_{x2} are displacements at nodes 1 and 2.

The strain in the bar is given by

$$\epsilon = \frac{\mathrm{d}u_x}{\mathrm{d}x} = \frac{\partial N_1}{\partial x} u_{x1} + \frac{\partial N_2}{\partial x} u_{x2} \tag{4.3}$$

or in matrix form

$$\epsilon = \mathbf{B}^e \mathbf{u}^e \tag{4.4}$$

where the element strain–displacement matrix is given by

$$\mathbf{B}^e = \left[\frac{\partial N_1}{\partial x}, \frac{\partial N_2}{\partial x} \right] \tag{4.5}$$

and the pseudo-vector of element displacements is

$$\mathbf{u}^e = \begin{Bmatrix} u_{x1} \\ u_{x2} \end{Bmatrix} \tag{4.6}$$

Since the shape functions are expressed in terms of the intrinsic coordinate ξ we have by the chain rule

$$\frac{\partial N_n}{\partial \xi} = \frac{\partial N_n}{\partial x} \frac{\mathrm{d}x}{\mathrm{d}\xi} \tag{4.7}$$

or

$$\frac{\partial N_n}{\partial x} = \frac{\partial N_n}{\partial \xi} \left(\frac{\mathrm{d}x}{\mathrm{d}\xi} \right)^{-1} \tag{4.8}$$

For a linear element the shape functions are given by

$$\begin{aligned} N_1 &= \frac{1}{2}(1 - \xi) \\ N_2 &= \frac{1}{2}(1 + \xi) \end{aligned} \tag{4.9}$$

and hence

$$\frac{\partial N_1}{\partial \xi} = -\frac{1}{2} \qquad \frac{\partial N_2}{\partial \xi} = \frac{1}{2} \tag{4.10}$$

and from equation (4.1)

$$\frac{\partial x}{\partial \xi} = \frac{\partial N_1}{\partial \xi} x_1 + \frac{\partial N_2}{\partial \xi} x_2 = \frac{1}{2}(x_2 - x_1) = \frac{l}{2} \tag{4.11}$$

The $\mathbf{B}^e$ matrix is therefore given by

$$\mathbf{B}^e = \left[-\frac{1}{l}, \frac{1}{l} \right] \tag{4.12}$$

As we know, the strain in the bar is constant and the stress is given by

$$\sigma = E\epsilon = E\mathbf{B}^e\mathbf{u}^e \tag{4.13}$$

where E is the Young's modulus.

Assume that the forces at the ends of the bar are F_1 and F_2 then for equilibrium we must have

$$\begin{aligned} F_1 &= -\sigma A \\ F_2 &= \sigma A \end{aligned} \tag{4.14}$$

where A is the cross sectional area. Written in matrix form this becomes

$$\begin{aligned} \mathbf{F}^e &= EA \begin{bmatrix} -\mathbf{B}^e \\ \mathbf{B}^e \end{bmatrix} \mathbf{u}^e \\ &= \mathbf{K}^e\mathbf{u}^e \end{aligned} \tag{4.15}$$

where the stiffness matrix of the bar is given by

$$\mathbf{K}^e = \frac{EA}{l} \begin{bmatrix} 1 & -1 \\ -1 & 1 \end{bmatrix} \tag{4.16}$$

Let us look at an alternative way of deriving the element stiffness matrix. Instead of ensuring equilibrium at the ends of the bar only, we endeavour to satisfy the equilibrium condition along the total length of the bar.

Another way of stating the equilibrium condition is to write down the total potential energy in the bar and minimise it with respect to the nodal displacements. A proof that this will satisfy equilibrium is given in reference 1.

If we take the zero energy state to be the state of zero displacements then the strain energy in the bar is given by

$$U = \frac{A}{2} \int_{x=0}^{l} \epsilon\sigma \, \mathrm{d}x \tag{4.17}$$

which in our case is simply

$$U = \frac{lA}{2} \epsilon E \epsilon = \frac{lA}{2} (\mathbf{B}^e \mathbf{u}^e)^{\mathrm{T}} E \mathbf{B}^e \mathbf{u}^e \tag{4.18}$$

Using the relationship between transpose matrices this becomes

$$U = \frac{lA}{2} (\mathbf{u}^e)^{\mathrm{T}} \mathbf{B}^{e\mathrm{T}} E \mathbf{B}^e \mathbf{u}^e \tag{4.19}$$

The potential energy of the external forces is given by

$$W = -(\mathbf{u}^e)^{\mathrm{T}} \mathbf{F}^e \tag{4.20}$$

and the total potential energy is

$$\pi = \frac{lA}{2} (\mathbf{u}^e)^{\mathrm{T}} \mathbf{B}^{e\mathrm{T}} E \mathbf{B}^e \mathbf{u}^e - (\mathbf{u}^e)^{\mathrm{T}} \mathbf{F}^e \tag{4.21}$$

The condition that π must be a minimum can be written using variational calculus as

$$\delta\pi = \frac{\partial \pi}{\partial \mathbf{u}^e} \delta \mathbf{u}^e = 0 \tag{4.22}$$

Since we exclude here $\delta\mathbf{u}^e = 0$ it follows that

$$\frac{\partial \pi}{\partial \mathbf{u}^e} = \mathbf{0} \tag{4.23}$$

Differentiating equation (4.21) using the formula for the differentiation of a product gives

$$\frac{\partial \pi}{\partial \mathbf{u}^e} = \left(\frac{l}{2} A \mathbf{B}^{e\mathrm{T}} E \mathbf{B}^e + \frac{l}{2} A \left(\mathbf{B}^{e\mathrm{T}} E \mathbf{B}^e \right)^{\mathrm{T}} \right) \mathbf{u}^e - \mathbf{F}^e \tag{4.24}$$

or since E is a scalar

$$\frac{\partial \pi}{\partial \mathbf{u}^e} = lA \mathbf{B}^{e\mathrm{T}} E \mathbf{B}^e \mathbf{u}^e - \mathbf{F}^e \tag{4.25}$$

and setting this to zero gives

$$\mathbf{K}^e\mathbf{u}^e - \mathbf{F}^e = \mathbf{0} \tag{4.26}$$

where

$$\begin{aligned}\mathbf{K}^e &= lA\mathbf{B}^{e\mathrm{T}}E\mathbf{B}^e \\ &= \frac{EA}{l}\begin{bmatrix} 1 & -1 \\ -1 & 1 \end{bmatrix}\end{aligned} \tag{4.27}$$

which is exactly the same result as previously obtained.

This is not surprising since the linear element satisfies equilibrium at the ends as well as inside the bar exactly.

4.2.2 Potential problems—one-dimensional elements

Turning to potential problems, consider the one-dimensional heat flow in Figure 4.3 where the temperature does not change in the y direction.

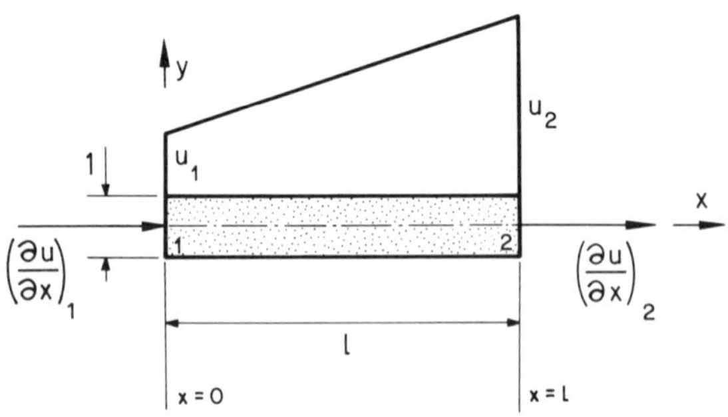

Figure 4.3 One-dimensional heat flow example

Consider a small element of size 1.dx. The condition that the outflow of heat from that element $q + \frac{\mathrm{d}q}{\mathrm{d}x}\,\mathrm{d}x$ must be equal to the inflow of heat q plus the heat generated inside the element, $Q\,\mathrm{d}x$, can be stated as

$$q + \frac{\mathrm{d}q}{\mathrm{d}x}\mathrm{d}x = q + Q\,\mathrm{d}x \tag{4.28}$$

or

$$\frac{\mathrm{d}q}{\mathrm{d}x} = Q \tag{4.29}$$

From Chapter 3 we know that the relationship between heat flow q and temperature $\hat{u}$ is

$$q = -k\frac{\mathrm{d}\hat{u}}{\mathrm{d}x} \tag{4.30}$$

where k is the thermal conductivity. Substitution of equation (4.30) into equation (4.29) gives

$$k\frac{\mathrm{d}^2\hat{u}}{\mathrm{d}x^2} + Q = 0 \tag{4.31}$$

For this type of problem we will use a completely different approach. Instead of using the terms equilibrium and minimum potential energy we use the fact that the differential equation has to be satisfied everywhere in the continuum.

We know that equation (4.31) is satisfied everywhere in the continuum, however if we have an approximation u to $\hat{u}$, of the type

$$u = \sum N_j u_j \tag{4.32}$$

then in general

$$k\frac{\mathrm{d}^2 u}{\mathrm{d}x^2} + Q = R_\Omega \neq 0 \tag{4.33}$$

Here R_Ω is called a residual and is a function of the position inside the domain Ω.

In an attempt to reduce this residual over the whole domain we state that an appropriate number of integrals of the error over the domain weighted in different ways is zero. This is written as

$$\int_\Omega W_l \left(k\frac{\mathrm{d}^2 u}{\mathrm{d}x^2} + Q\right) \mathrm{d}\Omega = 0 \tag{4.34}$$

where $W_l (l = 1, 2, 3, \ldots, M)$ is a set of different, spatially distributed *weighting functions*. This method is called a *weighted residual* method.

The requirement for convergence is that the approximate solution u approaches the exact solution $\hat{u}$ as M approaches infinity. This can, of course, only be true if R_Ω approaches zero at all points inside the domain Ω.

There are various choices we can make regarding the weighting function W_l. One choice is

$$W_l = N_l \tag{4.35}$$

i.e. that the shape functions and weighting functions are identical. This is also known as the *Galerkin* method.

Let us now solve the heat conduction problem in Figure 4.3 assuming that the internal heat generation Q is zero. As with the previous demonstration example we use only one linear finite element.

The weighted residual statement is

$$\int_{\Omega} N_l \left(k \frac{\mathrm{d}^2 u}{\mathrm{d}x^2} \right) \, \mathrm{d}x = 0 \tag{4.36}$$

We note that a second-order derivative occurs in the integral which means a continuity of first derivatives of the trial function is required. If we are using linear or parabolic shape functions this requirement is not satisfied. To relax this requirement we apply integration by parts

$$\int_{x=x_0}^{x=x_l} \phi \, \mathrm{d}\psi = - \int_{x=x_0}^{x=x_l} \psi \, \mathrm{d}\phi + \phi\psi\big|_{x=x_l} - \phi\psi\big|_{x=x_0} \tag{4.37}$$

Applying this to equation (4.36) we obtain (substituting $\phi = N_l$ and $\psi = k\frac{\partial u}{\partial x}$)

$$- \int_0^l k \left(\frac{\mathrm{d}N_l}{\mathrm{d}x} \cdot \frac{\mathrm{d}u}{\mathrm{d}x} \right) \mathrm{d}x + kN_l \frac{\mathrm{d}u}{\mathrm{d}x}\Big|_{x=l} - kN_l \frac{\mathrm{d}u}{\mathrm{d}x}\Big|_{x=0} = 0 \tag{4.38}$$

Now only C° continuity of u and N_l is required which is satisfied by piecewise linear functions.

Substitution of

$$u = \sum_{j=1}^{2} N_j u_j^e \tag{4.39}$$

gives

$$\mathbf{K}^e \mathbf{u}^e = \mathbf{F}^e \tag{4.40}$$

where

$$\mathbf{K}^e = kl \begin{bmatrix} \left(\frac{\mathrm{d}N_1}{\mathrm{d}x}\right)^2 & \frac{\mathrm{d}N_1}{\mathrm{d}x}\frac{\mathrm{d}N_2}{\mathrm{d}x} \\ \frac{\mathrm{d}N_2}{\mathrm{d}x}\frac{\mathrm{d}N_1}{\mathrm{d}x} & \left(\frac{\mathrm{d}N_2}{\mathrm{d}x}\right)^2 \end{bmatrix} \tag{4.41}$$

or after substitution of equation (4.8)

$$\mathbf{K}^e = \frac{k}{l} \begin{bmatrix} 1 & -1 \\ -1 & 1 \end{bmatrix} \tag{4.42}$$

$\mathbf{F}^e$ is a vector which contains the heat inflow at point 1 and outflow at point 2.

4.2.3 Elasticity problems—two-dimensional elements

To derive the stiffness matrix for a plane element (Figure 4.4) we express the strain $\boldsymbol{\epsilon}$ inside the element in terms of nodal displacements by

$$\boldsymbol{\epsilon} = \mathbf{B}^e \mathbf{u}^e \tag{4.43}$$

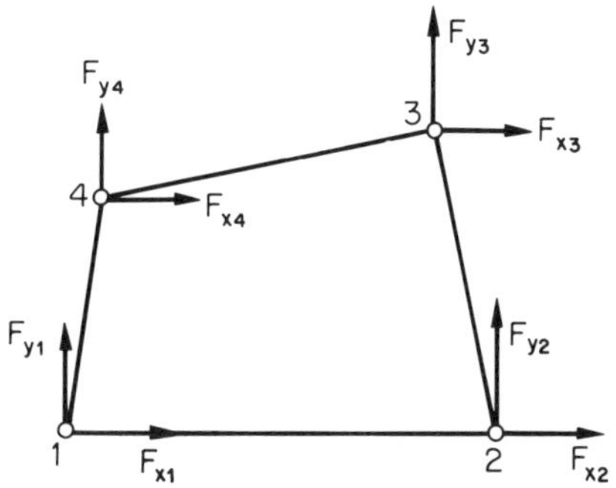

Figure 4.4 Plane quadrilateral element

The strain pseudo-vector $\boldsymbol{\epsilon}$ is given by

$$\boldsymbol{\epsilon} = \begin{Bmatrix} \epsilon_x \\ \epsilon_y \\ \gamma_{xy} \end{Bmatrix} = \begin{Bmatrix} \dfrac{\partial u_x}{\partial x} \\ \dfrac{\partial u_y}{\partial y} \\ \dfrac{\partial u_y}{\partial x} + \dfrac{\partial u_x}{\partial y} \end{Bmatrix} \tag{4.44}$$

where

$$\frac{\partial \mathbf{u}}{\partial x} = \sum \frac{\partial N_n}{\partial x} \mathbf{u}_n^e \quad \text{etc} \tag{4.45}$$

For an element with four nodes the matrix $\mathbf{B}^e$ is therefore defined as

$$\mathbf{B}^e = \begin{bmatrix} \dfrac{\partial N_1}{\partial x} & 0 & \dfrac{\partial N_2}{\partial x} & 0 & \dfrac{\partial N_3}{\partial x} & 0 & \dfrac{\partial N_4}{\partial x} & 0 \\ 0 & \dfrac{\partial N_1}{\partial y} & 0 & \dfrac{\partial N_2}{\partial y} & 0 & \dfrac{\partial N_3}{\partial y} & 0 & \dfrac{\partial N_4}{\partial y} \\ \dfrac{\partial N_1}{\partial y} & \dfrac{\partial N_1}{\partial x} & \dfrac{\partial N_2}{\partial y} & \dfrac{\partial N_2}{\partial x} & \dfrac{\partial N_3}{\partial y} & \dfrac{\partial N_3}{\partial x} & \dfrac{\partial N_4}{\partial y} & \dfrac{\partial N_4}{\partial x} \end{bmatrix} \tag{4.46}$$

The global derivatives of N are computed as explained in Section 2.10. The vector $\mathbf{u}^e$ contains the displacements at all nodes of the element

$$\mathbf{u}^e = \begin{Bmatrix} \mathbf{u}_1^e \\ \mathbf{u}_2^e \\ \mathbf{u}_3^e \\ \mathbf{u}_4^e \end{Bmatrix} \tag{4.47}$$

The equilibrium condition is stated using the minimum potential energy principle as explained previously.

The strain energy is given by

$$U = \frac{1}{2}\int_V \boldsymbol{\epsilon}^{\mathrm{T}} \boldsymbol{\sigma} \,\mathrm{d}V \tag{4.48}$$

which is after substitution of equation (4.43)

$$U = \frac{1}{2}\int_V (\mathbf{B}^e\mathbf{u}^e)^{\mathrm{T}} \mathbf{D}\mathbf{B}^e\mathbf{u}^e \,\mathrm{d}V \tag{4.49}$$

The potential energy of the nodal point forces is

$$W = -(\mathbf{u}^e)^{\mathrm{T}}\mathbf{F}^e \tag{4.50}$$

Using the relationship between transpose matrices $(\mathbf{B}^e\mathbf{u}^e)^{\mathrm{T}} = \mathbf{u}^{e\mathrm{T}}\mathbf{B}^{e\mathrm{T}}$ and considering that the vector $\mathbf{u}^e$ containing nodal values of displacement can be taken out of the integral, we can write

$$\pi = (\mathbf{u}^e)^{\mathrm{T}}\frac{1}{2}\left(\int_V \mathbf{B}^{e\mathrm{T}}\mathbf{D}\mathbf{B}^e \,\mathrm{d}V\right)\mathbf{u}^e - (\mathbf{u}^e)^{\mathrm{T}}\mathbf{F}^e \tag{4.51}$$

Taking the derivative with respect to the nodal displacements we have

$$\frac{\partial \pi}{\partial \mathbf{u}^e} = \left(\frac{1}{2}\int_V \mathbf{B}^{e\mathrm{T}}\mathbf{D}\mathbf{B}^e \,\mathrm{d}V + \frac{1}{2}\int_V (\mathbf{B}^{e\mathrm{T}}\mathbf{D}\mathbf{B}^e)^{\mathrm{T}} \,\mathrm{d}V\right)\mathbf{u}^e - \mathbf{F}^e = \mathbf{0} \tag{4.52}$$

or

$$\mathbf{K}^e\mathbf{u}^e - \mathbf{F}^e = \mathbf{0} \tag{4.53}$$

where, provided $\mathbf{D}$ is symmetric, the stiffness matrix (which is also symmetric) is given by

$$\mathbf{K}^e = \int_V \mathbf{B}^{e\mathrm{T}}\mathbf{D}\mathbf{B}^e \,\mathrm{d}V \tag{4.54}$$

4.2.4 Potential problem—two-dimensional elements

For two-dimensional steady state potential problems we have from the requirement of the conservation of heat

$$Q\,\mathrm{d}x\,\mathrm{d}y = \left(q_x + \frac{\partial q_x}{\partial x}\,\mathrm{d}x - q_x\right)\,\mathrm{d}y + \left(q_y + \frac{\partial q_y}{\partial y}\,\mathrm{d}y - q_y\right)\,\mathrm{d}x \tag{4.55}$$

or

$$\frac{\partial q_x}{\partial x} + \frac{\partial q_y}{\partial y} - Q = 0 \tag{4.56}$$

Introducing the Fourier law:

$$\begin{aligned} q_x &= -k_x \frac{\partial \hat{u}}{\partial x} \\ q_y &= -k_y \frac{\partial \hat{u}}{\partial y} \end{aligned} \tag{4.57}$$

where $\hat{u}$ is the exact temperature distribution, we obtain

$$\frac{\partial}{\partial x}\left(k_x \frac{\partial \hat{u}}{\partial x}\right) + \frac{\partial}{\partial y}\left(k_y \frac{\partial \hat{u}}{\partial y}\right) + Q = 0 \tag{4.58}$$

Using the Galerkin method we obtain (setting $Q = 0$)

$$\int_\Omega N_l \left[\frac{\partial}{\partial x}\left(k_x \frac{\partial u}{\partial x}\right) + \frac{\partial}{\partial y}\left(k_y \frac{\partial u}{\partial y}\right)\right]\,\mathrm{d}x\,\mathrm{d}y = 0 \tag{4.59}$$

Again we seek to relax the condition of continuity of shape functions. Here we use Green's lemma (see reference 5, Appendix 3) which states that for suitably differentiable functions ϕ and ψ we can write

$$\int_\Omega \phi \frac{\partial \psi}{\partial x}\,\mathrm{d}x\,\mathrm{d}y = -\int_\Omega \frac{\partial \phi}{\partial x}\psi\,\mathrm{d}x\,\mathrm{d}y + \int_\Gamma \phi\psi\, n_x\,\mathrm{d}s \tag{4.60}$$

and

$$\int_\Omega \phi \frac{\partial \psi}{\partial y}\,\mathrm{d}x\,\mathrm{d}y = -\int_\Omega \frac{\partial \phi}{\partial y}\psi\,\mathrm{d}x\,\mathrm{d}y + \int_\Gamma \phi\psi\, n_y\,\mathrm{d}s \tag{4.61}$$

where n_x, n_y are components of the unit vector normal to the boundary.

Using the fact that

$$n_x \frac{\partial u}{\partial x} + n_y \frac{\partial u}{\partial y} = \frac{\partial u}{\partial n}$$

and Green's theorem we can transform equation (4.59) into

$$-\int_{\Omega}\left(\frac{\partial N_l}{\partial x}k_x\frac{\partial u}{\partial x}+\frac{\partial N_l}{\partial y}k_y\frac{\partial u}{\partial y}\right)\,\mathrm{d}x\,\mathrm{d}y+\int_{\Gamma}\left[k_x\frac{\partial u}{\partial x}n_x+k_y\frac{\partial u}{\partial y}n_y\right]N_l\,\mathrm{d}s=0 \tag{4.62}$$

Substituting $u=\sum N_j\,u_j^e$ we obtain

$$\mathbf{K}^e\mathbf{u}^e=\mathbf{F}^e \tag{4.63}$$

where

$$\mathbf{K}^e=\int_V \mathbf{B}^{e\mathrm{T}}\mathbf{D}\mathbf{B}^e\,\mathrm{d}V \tag{4.64}$$

and (keeping the same notation as for elasticity problems) $\mathbf{B}^e$ is now defined differently as

$$\mathbf{B}^e=\begin{bmatrix}\frac{\partial N_1}{\partial x}, & \frac{\partial N_2}{\partial x}, & \frac{\partial N_3}{\partial x}, & \frac{\partial N_4}{\partial x}\\ \frac{\partial N_1}{\partial y}, & \frac{\partial N_2}{\partial y}, & \frac{\partial N_3}{\partial y}, & \frac{\partial N_4}{\partial y}\end{bmatrix} \tag{4.65}$$

and the matrix analogous to the elasticity matrix is given by

$$\mathbf{D}=\begin{bmatrix}k_x & 0\\ 0 & k_y\end{bmatrix}$$

The pseudo-vector $\mathbf{F}^e$ contains values of heat flow at the node points of elements.

4.2.5 Numerical integration

The integration for the stiffness matrix is best carried out in the intrinsic coordinate system because the limits of integration are then simple. For two-dimensional elements the expression for the stiffness matrix is

$$\mathbf{K}^e=\int_V \mathbf{B}^{e\mathrm{T}}\mathbf{D}\mathbf{B}^e\,\mathrm{d}V=\int_{-1}^{+1}\int_{-1}^{+1}\mathbf{B}^{e\mathrm{T}}\mathbf{D}\mathbf{B}^e J\,\mathrm{d}\xi\,\mathrm{d}\eta \tag{4.66}$$

Analytical integration, although cumbersome, is possible for linear elements. Numerical integration is recommended for these elements and essential for higher order elements. In numerical integration techniques the integral is approximated by the sum of values of the integrand at a number of points multiplied by factors known as weights.

$$I = \int_{-1}^{+1} f(\xi)\,\mathrm{d}\xi \approx \sum_{l=1}^{L} W_l\, f(\xi_l) \tag{4.67}$$

In the above, W_l are weights and ξ_l are intrinsic coordinates of integration (sampling) points. The well known *trapezoidal rule*[2] for example gives

$$\int_{-1}^{+1} f(\xi)\,\mathrm{d}\xi \approx f(-1) + f(1) \tag{4.68}$$

the function being determined at two points only.

A more accurate integration with fewer points can be carried out using *Gauss Quadrature*[3], in which we suppose that $f(\xi)$ can be approximated by a polynomial expression of the form.

$$f(\xi) \approx a_0 + a_1\xi + a_2\xi^2 + \ldots + a_p\xi^p \tag{4.69}$$

where the coefficients a_i are adjusted to give a best fit to $f(\xi)$. For the numerical integration we determine the locations of sampling points ξ_l and values of weights W_l from the condition that this polynomial is to be exactly integrated.

The number of sampling points L required to integrate exactly a polynomial is obviously dependent on the degree of the polynomial p, and we tabulate below the relationship between L and p for Gauss quadrature.

No. of Sampling Points L	**Degree of Polynomial** p
1	1 (linear)
2	3 (cubic)
3	5 (quintic)

The location of the sampling (Gauss) points and corresponding weights are given in reference 4 and for up to three integration points they are:

L	ξ_l	W_l
1	0	2.
2	+0.577 35, −0.577 35	1., 1.
3	0.774 59, 0, −0.774 59	0.555 55, 0.888 88, 0.555 55

Two-dimensional integration requires a double sum (Figure 4.5):

$$\int_{-1}^{+1}\int_{-1}^{+1} f(\xi,\eta)\,\mathrm{d}\xi\,\mathrm{d}\eta \approx \sum_{l=1}^{L}\sum_{m=1}^{M} W_l W_m\, f(\xi_l,\eta_m) \tag{4.70}$$

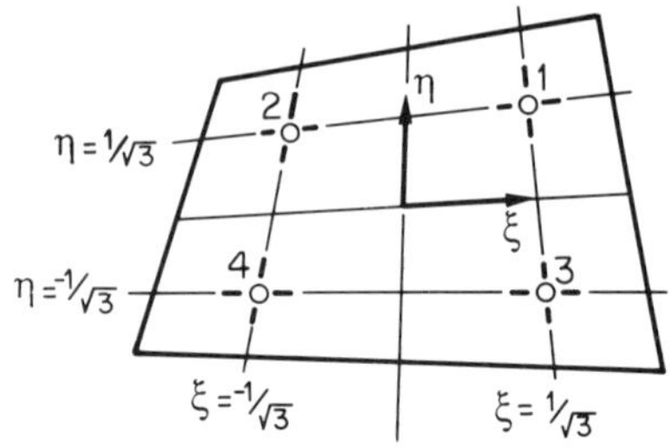

Figure 4.5 Gauss integration points for quadrilateral element (2 × 2 integration)

Similarly the integration formula for three dimensions is given by

$$\int_{-1}^{+1}\int_{-1}^{+1}\int_{-1}^{+1} f(\xi,\eta,\zeta)\,\mathrm{d}\xi\,\mathrm{d}\eta\,\mathrm{d}\zeta = \sum_{l=1}^{L}\sum_{m=1}^{M}\sum_{n=1}^{N} W_l W_m W_n f(\xi_l,\eta_m,\zeta_n) \tag{4.71}$$

Since the order of integration depends on the degree of polynomial to be integrated, different integration formulae are used depending on the type of shape function. Consider a linear element of dimensions 2×2 the sides of which are parallel with the x and y axes. Then $\frac{\partial N_i}{\partial x}$ equals $\frac{\partial N_i}{\partial \xi}$ is independent of ξ and varies linearly with respect to η. Similarly, $\frac{\partial N_i}{\partial y}$ equals $\frac{\partial N_i}{\partial \eta}$, is independent of η and varies linearly with respect to ξ. The integrand in equation (4.66) contains products of derivatives of the shape functions and hence is quadratic or of order 2. A 2×2 Gauss integration is exact for polynomials up to order 3, so it will give the exact value of the integral for the element just discussed.

For the same element geometry and a parabolic shape function we find—by the same reasoning as before—that the integrand is of order 4 and hence a 3×3 integration will be exact.

It is difficult to determine the number of integration points for elements of general shape. Obviously the integration will no longer be exact for distorted elements and the error will increase with element distortion.

As it turns out it is not necessary to integrate expression (4.54) to a very high degree of accuracy. In fact is has been found that exactly integrated elements are always too stiff and experience has shown that an integration order of less than that derived above gives the best results. This experience is supported by theoretical considerations[5].

The basic steps in computing the stiffness matrix are quite general and can be used for any type of isoparametric element, whether linear or parabolic, plane or three dimensional. This is found to be a particularly useful feature when writing general purpose finite element programs. For example the basic steps in the computation of the response matrix are:

FOR each integration point (coordinates ξ_l, η_m, ζ_n) **DO**
  Determine the Jacobian matrix $\mathbf{J}$ and the Jacobian J.
  Determine the global derivatives $\frac{\partial N_1}{\partial x}$, etc.
  Assemble the $\mathbf{B}^e$ matrix.
  Calculate $\mathbf{B}^{e\mathrm{T}}\mathbf{D}\mathbf{B}^e J \cdot W_l \cdot W_m \cdot W_n$
  Add to other contributions
END DO

4.3 DERIVATION OF CONTINUUM RESPONSE–ASSEMBLY

So far we have derived matrices which describe the response of an individual element. The response of the problem domain can be obtained by assembling the element response matrices, that is by adding the contributions of individual elements.

Consider the example in Figure 4.6 consisting of two elements of length $l/2$ with different Young's moduli E_1 and E_2.

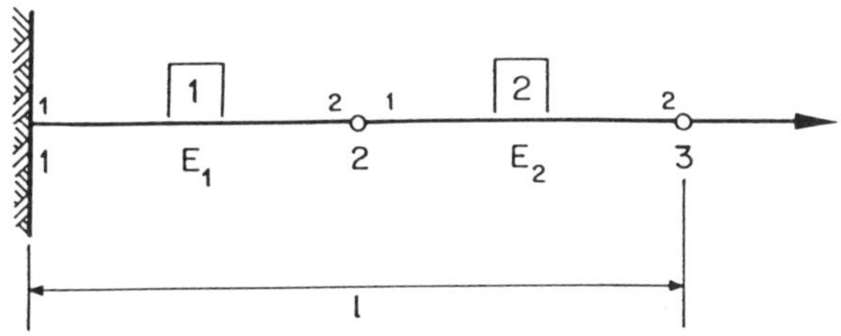

Figure 4.6 Assembly of two elements

The element response matrices (as expressed in a local numbering system) are given by

$$\begin{Bmatrix} F_{x1}^1 \\ F_{x2}^1 \end{Bmatrix} = \frac{2E_1A}{l}\begin{bmatrix} 1 & -1 \\ -1 & 1 \end{bmatrix}\begin{Bmatrix} u_{x1}^1 \\ u_{x2}^1 \end{Bmatrix}$$

$$\begin{Bmatrix} F_{x1}^2 \\ F_{x2}^2 \end{Bmatrix} = \frac{2E_2A}{l}\begin{bmatrix} 1 & -1 \\ -1 & 1 \end{bmatrix}\begin{Bmatrix} u_{x1}^2 \\ u_{x2}^2 \end{Bmatrix} \tag{4.72}$$

To be able to construct the response matrix of the assembly of elements we have to establish a global numbering scheme for the nodes. According to Figure 4.6 we have

$$\begin{aligned} u_{x1} &= u^1_{x1} \\ u_{x2} &= u^1_{x2} = u^2_{x1} \\ u_{x3} &= u^2_{x2} \end{aligned} \tag{4.73}$$

Equilibrium at the nodes stipulates that, for a set of externally applied loads F_{x1}, F_{x2},F_{x3}.

$$\begin{aligned} -F_{x1} + F^1_{x1} &= 0 \\ -F_{x2} + F^1_{x2} + F^2_{x1} &= 0 \\ -F_{x3} + F^2_{x2} &= 0 \end{aligned} \tag{4.74}$$

Substituting equations (4.72) and using the equivalence of local and global displacement components we obtain for the global response

$$-\begin{Bmatrix} F_{x1} \\ F_{x2} \\ F_{x3} \end{Bmatrix} + \frac{2A}{l}\begin{bmatrix} E_1 & -E_1 & 0 \\ -E_1 & E_1 + E_2 & -E_2 \\ 0 & -E_2 & E_2 \end{bmatrix}\begin{Bmatrix} u_{x1} \\ u_{x2} \\ u_{x3} \end{Bmatrix} = \mathbf{0} \tag{4.75}$$

Consider now the two-dimensional example shown in Figure 4.7 in which the problem domain is modelled by two four-node quadrilateral elements. In Figure 4.7, in addition to the global node numbering, the local (element) numbering (for which the element stiffness matrices have been derived) is shown. The element incidences or connectivities are as follows:

	1.	2.	3.	4.
Element e_1	1	2	5	4
Element e_2	2	3	6	5

(4.76)

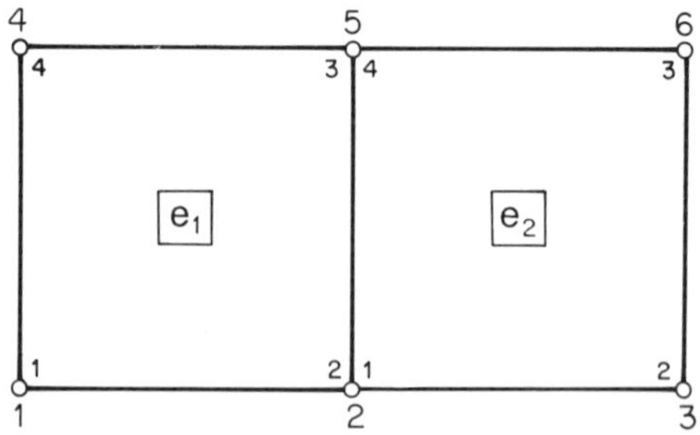

Figure 4.7 Assembly of two quadrilateral elements

The stiffness matrix of an individual element can be written as

$$\mathbf{K}^e = \begin{matrix} & 1. & 2. & 3. & 4. \\ 1. & \mathbf{k}_{11} & \mathbf{k}_{12} & \mathbf{k}_{13} & \mathbf{k}_{14} \\ 2. & \mathbf{k}_{21} & \mathbf{k}_{22} & \mathbf{k}_{23} & \mathbf{k}_{24} \\ 3. & \mathbf{k}_{31} & \mathbf{k}_{32} & \mathbf{k}_{33} & \mathbf{k}_{34} \\ 4. & \mathbf{k}_{41} & \mathbf{k}_{42} & \mathbf{k}_{43} & \mathbf{k}_{44} \end{matrix}^{\,e} \tag{4.77}$$

where $\mathbf{k}_{ij}$ are either submatrices or scalar response coefficients. For example for two-dimensional problems of elasticity $\mathbf{k}_{ij}$ is a submatrix

$$\mathbf{k}_{ij} = \begin{bmatrix} k_{xx} & k_{xy} \\ k_{yx} & k_{yy} \end{bmatrix}_{ij} \tag{4.78}$$

whereas for potential problems it is a scalar.

Before we can add the individual element stiffness matrices we have to express each one in the global node numbering system. This can be visualised as 'scattering' the stiffness matrix as shown below:

$$\begin{matrix} \text{Node} = & 1 & 2 & 3 & 4 & 5 & 6 \end{matrix}$$

$$\mathbf{K}^{e_1} = \begin{matrix} 1 \\ 2 \\ 3 \\ 4 \\ 5 \\ 6 \end{matrix} \begin{bmatrix} \mathbf{k}_{11} & \mathbf{k}_{12} & 0 & \mathbf{k}_{14} & \mathbf{k}_{13} & 0 \\ \mathbf{k}_{21} & \mathbf{k}_{22} & 0 & \mathbf{k}_{24} & \mathbf{k}_{23} & 0 \\ 0 & 0 & 0 & 0 & 0 & 0 \\ \mathbf{k}_{41} & \mathbf{k}_{42} & 0 & \mathbf{k}_{44} & \mathbf{k}_{43} & 0 \\ \mathbf{k}_{31} & \mathbf{k}_{32} & 0 & \mathbf{k}_{34} & \mathbf{k}_{33} & 0 \\ 0 & 0 & 0 & 0 & 0 & 0 \end{bmatrix}^{e_1} \tag{4.79}$$

$$\begin{matrix} \text{Node} = & 1 & 2 & 3 & 4 & 5 & 6 \end{matrix}$$

$$\mathbf{K}^{e_2} = \begin{matrix} 1 \\ 2 \\ 3 \\ 4 \\ 5 \\ 6 \end{matrix} \begin{bmatrix} 0 & 0 & 0 & 0 & 0 & 0 \\ 0 & \mathbf{k}_{11} & \mathbf{k}_{12} & 0 & \mathbf{k}_{14} & \mathbf{k}_{13} \\ 0 & \mathbf{k}_{21} & \mathbf{k}_{22} & 0 & \mathbf{k}_{24} & \mathbf{k}_{23} \\ 0 & 0 & 0 & 0 & 0 & 0 \\ 0 & \mathbf{k}_{41} & \mathbf{k}_{42} & 0 & \mathbf{k}_{44} & \mathbf{k}_{43} \\ 0 & \mathbf{k}_{31} & \mathbf{k}_{32} & 0 & \mathbf{k}_{34} & \mathbf{k}_{33} \end{bmatrix}^{e_2} \tag{4.80}$$

Note that the 'scattered stiffness matrix' for element 1 has no stiffness coefficients for nodes 3 and 6 since that element has no connection with these nodes.

The response matrix of the problem domain is now obtained by writing

$$[\mathbf{K}] = \mathbf{K}^{e_1} + \mathbf{K}^{e_2} \tag{4.81}$$

Note that actual addition of stiffness coefficients only occurs at nodes 2 and 5, that is the two nodes which are common to both elements.

This procedure to assemble the structure stiffness matrix is for demonstration purposes only. An expansion of the element stiffness matrix before assembly is not necessary as submatrices can be directly assembled into $[\mathbf{K}]$ by reference to the element connectivity. The procedure is as follows. First the structure stiffness matrix is set to zero

$$[\mathbf{K}] = \mathbf{0} \tag{4.82}$$

Then the stiffness submatrix of each element is added into $\mathbf{K}$ at the locations corresponding to the global node numbers ('destinations'). For example if the element connectivity is

$$\begin{array}{lcccc} & 1 & 2 & 3 & 4 \\ \text{Element } e & i & j & k & l \end{array} \tag{4.83}$$

then the destination of $\mathbf{k}_{12}^e$ is the ith column and jth row of submatrices in $[\mathbf{K}]$.

After assembly we have a matrix of coefficients which relates the displacements $\{\mathbf{u}\}$ to the equivalent nodal forces $\{\mathbf{F}\}$ for all nodes of the problem domain

$$[\mathbf{K}]\{\mathbf{u}\} = \{\mathbf{F}\} \tag{4.84}$$

Since the element response matrices are symmetric the assembled response matrix must also be symmetric.

4.4 BOUNDARY CONDITIONS AND APPLIED LOADS

The system of equations (4.84) is singular and can only be solved if boundary conditions are specified. Basically there are two types of boundary conditions.

(1) Prescribed value of unknown at a nodal point.
In potential problems we may have at node n

$$u_n = \bar{u}$$

This is known as a *Dirichlet* boundary condition.
In elasticity problems we have

$$u_{xn} = \bar{u}_x, \; u_{yn} = \bar{u}_y \qquad \text{etc}$$

or

$$u_{xn} = 0 \qquad \text{etc}$$

This is known as a prescribed displacement or support boundary condition. For finite elasticity problems we know that a minimum of such boundary conditions is necessary to prevent rigid body displacements.

(2) Prescribed value of flux or traction on the boundary.

In potential problems we have for a prescribed flow normal to the boundary (for an isotropic material):

$$k\frac{\partial u}{\partial n} = q \qquad \text{on boundary } S_q$$

This is known as a *Neumann* boundary condition.

Alternatively for elasticity problems we have a prescribed value of traction on the boundary.

Finally we have to account for heat generation inside the domain or body forces. These are specified per unit volume.

4.4.1 Body forces

To explain how loading and boundary conditions are considered when solving problems in elasticity let us return to the problem of the weightless bar consisting of two elements. Let us assume that a constant body force b_x (per unit volume) acts in the x direction. Then for one element in equation (4.17) in addition to the strain energy we have

$$U_Q = -A\int u_x b_x \,\mathrm{d}x \tag{4.85}$$

where the integration is taken over the length of the element.

After substitution of the shape functions this becomes

$$U_Q = -Ab_x\int (N_1 u_{x1} + N_2 u_{x2})\,\mathrm{d}x \tag{4.86}$$

Taking the variation of the potential energy as in equation (4.24) we obtain for element e

$$\mathbf{K}^e\mathbf{u}^e - \mathbf{F}^e - \mathbf{F}_b^e = \mathbf{0} \tag{4.87}$$

where

$$\mathbf{F}_b^e = Ab_x\begin{Bmatrix} \int N_1\,\mathrm{d}x \\ \int N_2\,\mathrm{d}x \end{Bmatrix} = Ab_x\frac{l}{2}\begin{Bmatrix} \frac{1}{2} \\ \frac{1}{2} \end{Bmatrix} \tag{4.88}$$

Taking $E_2 = E_1 = E$, $A = l = 1$ and $F_{x1} = F_{x2} = F_{x3} = 0$ we obtain the following system of equations:

$$E\begin{bmatrix} 2 & -2 & 0 \\ -2 & 4 & -2 \\ 0 & -2 & 2 \end{bmatrix}\begin{Bmatrix} u_{x1} \\ u_{x2} \\ u_{x3} \end{Bmatrix} = b_x\begin{Bmatrix} \frac{1}{4} \\ \frac{1}{4}+\frac{1}{4} \\ \frac{1}{4} \end{Bmatrix} \tag{4.89}$$

Assuming that the bar is fixed on one end, then $u_{x1} = 0$ and we can delete the first row and the first column of the system of equations. This leads to

$$\begin{aligned} 4u_{x2} - 2u_{x3} &= \frac{b_x}{2E} \\ -2u_{x2} + 2u_{x3} &= \frac{b_x}{4E} \end{aligned} \tag{4.90}$$

Using Gauss elimination we replace the second equation with that equation plus 1/2 of the first equation.

This gives us the solution

$$u_{x3} = \frac{b_x}{2E} \tag{4.91}$$

and

$$u_{x2} = \frac{3b_x}{8E} \tag{4.92}$$

4.4.2 Tractions

Let us consider the element shown in Figure 4.8 which is subjected to distributed tractions on one side. On the boundaries S_t, tractions $\mathbf{t}$ are prescribed. The potential energy of these tractions is given by

$$W = -\int_{S_t} \mathbf{u}^{\mathrm{T}}\mathbf{t}\, \mathrm{d}s \tag{4.93}$$

where

$$\mathbf{t} = \begin{Bmatrix} t_x \\ t_y \end{Bmatrix} \tag{4.94}$$

and $\mathrm{d}s$ indicates integration with respect to arc length.

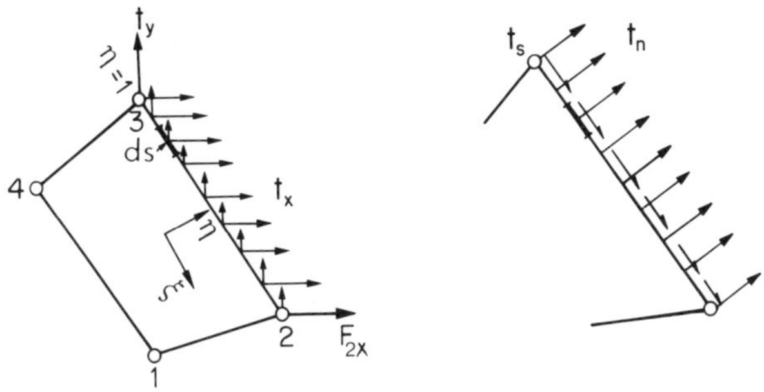

Figure 4.8 Distributed loading on boundary

Substitution of $\mathbf{u} = \sum N_j \mathbf{u}_j^e$ gives

$$W = -(\mathbf{u}^e)^{\mathrm{T}} \int_{S_t} \mathbf{N}^{e\mathrm{T}} \mathbf{t} \, \mathrm{d}S \tag{4.95}$$

where

$$\mathbf{N}^e = \begin{bmatrix} N_1 & 0 & N_2 & 0 & N_3 & 0 & N_4 & 0 \\ 0 & N_1 & 0 & N_2 & 0 & N_3 & 0 & N_4 \end{bmatrix}$$

Taking a variation of the total potential energy as explained in Section 4.2.3 we obtain

$$\mathbf{K}^e \mathbf{u}^e - \mathbf{F}^e - \mathbf{F}_t^e = \mathbf{0}$$

where

$$\mathbf{F}_t^e = \int_{S_t} \mathbf{N}^{e\mathrm{T}} \mathbf{t} \, \mathrm{d}s \tag{4.96}$$

Note that for loading on the element side $\eta = 1, N_4 = N_1 = 0$ and so there is no contribution to forces at nodes 4 and 1.

Sometimes it is convenient to specify the components t_n and t_s normal and tangential to the boundary. In this case the corresponding global traction components are given by

$$\mathbf{t} = \mathbf{T}_{\mathrm{g}} \mathbf{t}' \tag{4.97}$$

where $\mathbf{T}_{\mathrm{g}}$ is the geometric transformation matrix (Section 2.7) and

$$\mathbf{t}' = \begin{Bmatrix} t_n \\ t_{\mathrm{s}} \end{Bmatrix} \tag{4.98}$$

4.4.3 Boundary flux

For potential problems the analogue to tractions is flux on the boundary. On boundaries $\mathbf{S}_q$ where flux is prescribed we have for isotropic material

$$-\left(k\frac{\partial u}{\partial n}+\bar{q}\right)=0 \qquad \text{on } S_q$$

We now simply include in the residual statement equation (4.59) the term

$$-\int_{S_q}\left(k\frac{\partial u}{\partial n}+\bar{q}\right)N_l \,\mathrm{d}S \tag{4.99}$$

After integration by parts, equation (4.59) and this term become:

$$\begin{aligned}-\int_{\Omega}k\left(\frac{\partial N_l}{\partial x}k\frac{\partial u}{\partial x}+\frac{\partial N_l}{\partial y}k\frac{\partial u}{\partial y}\right)\,\mathrm{d}x\,\mathrm{d}y+\int_{S_q}k\frac{\partial u}{\partial n}N_l\,\mathrm{d}s\\ -\int_{S_q}\left(k\frac{\partial u}{\partial n}+q\right)N_l\,\mathrm{d}s=0\end{aligned} \tag{4.100}$$

which gives for the element right hand side

$$\mathbf{F}_q^e=\left\{\begin{matrix}\int N_1 q\,\mathrm{d}s\\ \cdot\\ \cdot\\ \cdot\\ \int N_n q\,\mathrm{d}s\end{matrix}\right\}=\int_{S_q}\{\mathbf{N}\}^e q\,\mathrm{d}s \tag{4.101}$$

where

$$\{\mathbf{N}\}^e=\left\{\begin{matrix}N_1\\ \cdot\\ \cdot\\ \cdot\\ N_n\end{matrix}\right\}$$

4.4.4 Initial stress and strain

For elasticity problems stresses and strains which are not caused by loading may be present in the solid. These are called initial stresses and strains.

Initial strains $\boldsymbol{\epsilon}_0$ can for example be caused by a temperature change ΔT in the solid; for plane strain,

$$\boldsymbol{\epsilon}_0 = \begin{Bmatrix} \alpha\Delta T \\ \alpha\Delta T \\ 0 \end{Bmatrix} \tag{4.102}$$

where α is the coefficient of thermal expansion.

Initial stresses $\boldsymbol{\sigma}_0$ will be present for example in materials such as rock and soil due to the process of geological formation (tectonic stresses), and due to overburden.

Let us suppose that an element is subjected to initial strain $\boldsymbol{\epsilon}_0$ (for example strain caused by a temperature increase) but we prevent the element from deforming (i.e. apply zero displacements at all the nodes). There is now an additional strain energy term due to the fact that we force the element to remain in its original shape:

$$W = -\int_V \boldsymbol{\sigma}^{\mathrm{T}} \boldsymbol{\epsilon}_0 \, \mathrm{d}V \tag{4.103}$$

Substitution of equation (4.43) gives

$$W = -\int (\mathbf{D}\mathbf{B}^e \mathbf{u}^e)^{\mathrm{T}} \boldsymbol{\epsilon}_0 \, \mathrm{d}V \tag{4.104}$$

and this gives

$$\mathbf{K}^e \mathbf{u}^e - \mathbf{F}^e - \mathbf{F}^e_{\epsilon} = \mathbf{0}$$

where

$$\mathbf{F}^e_{\epsilon} = \int_V \mathbf{B}^{e\mathrm{T}} \mathbf{D} \boldsymbol{\epsilon}_0 \, \mathrm{d}V \tag{4.105}$$

If initial stresses are present there will be no additional energy created unless material is removed (see Figure 4.9). If this is the case the energy term is given by

$$W = \int_V \boldsymbol{\epsilon}^{\mathrm{T}} \boldsymbol{\sigma}_0 \, \mathrm{d}V \tag{4.106}$$

which after substitution of equation (4.43) becomes

$$W = \int_V (\mathbf{B}^e \mathbf{u}^e)^{\mathrm{T}} \boldsymbol{\sigma}_0 \, \mathrm{d}V \tag{4.107}$$

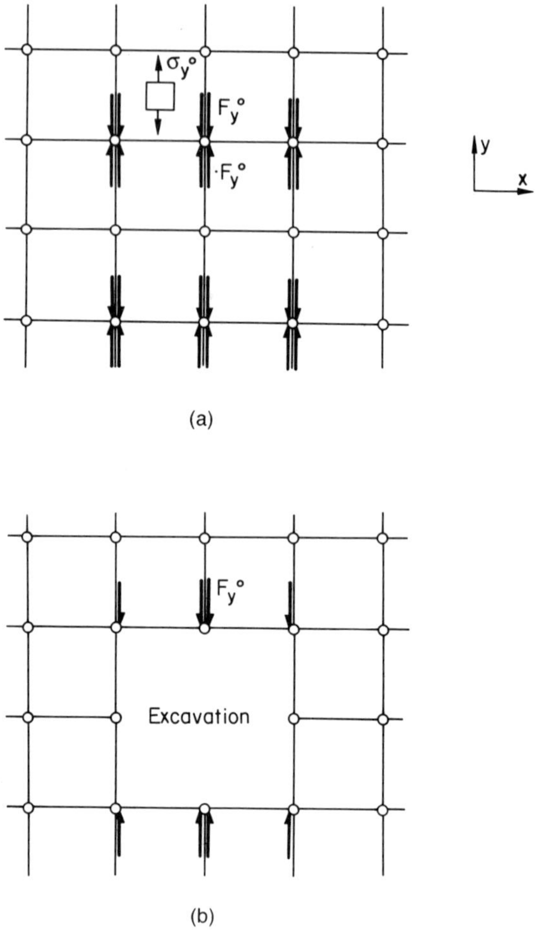

Figure 4.9 Modelling of excavation

and this gives

$$\mathbf{K}^e \mathbf{u}^e - \mathbf{F}^e + \mathbf{F}^e_\sigma = \mathbf{0}$$

where

$$\mathbf{F}^e_\sigma = \int_V \mathbf{B}^{e\mathrm{T}} \boldsymbol{\sigma}_0 \, \mathrm{d}V \tag{4.108}$$

4.5 SOLUTION

After assembling all the element response matrices we obtain a set of simultaneous equations

$$[\mathbf{K}]\{\mathbf{u}\} = \{\mathbf{F}\}$$

where $\{\mathbf{F}\}$ may include all the forces discussed above.

The solution of the system of equations is usually done by Gauss elimination. This involves two steps:

STEP 1: Reduction of equations

Here we introduce zeros below the diagonal for n=1, 2, ..., N. For example to introduce a zero in the nth column of ith equation.

$$\begin{array}{ll} n\text{th equation} & k_{nn}X_n + \ldots + k_{nj}X_j \ldots = F_n \\ i\text{th equation} & k_{in}X_n + \ldots + k_{ij}X_j \ldots \ = F_i \end{array} \tag{4.109}$$

We subtract k_{in}/k_{nn} times equation n from equation i to eliminate the coefficient k_{in}

$$\begin{aligned} k_{nn}X_n + \ldots + k_{nj}X_j \ldots \qquad\qquad & = F_n \\ 0 + \ldots + \left(k_{ij} - \frac{k_{in}k_{nj}}{k_{nn}}\right) X_j \ldots & = F_i - \frac{k_{in}F_n}{k_{nn}} \end{aligned} \tag{4.110}$$

The recursive formulae are for the left hand side

$$k_{ij}^* = k_{ij} - \frac{k_{in}k_{nj}}{k_{nn}} \tag{4.111}$$

and

$$F_i^* = F_i - \frac{k_{in}F_n}{k_{nn}} \tag{4.112}$$

for the right-hand side.

STEP 2: Backsubstitution

Here we obtain the results by computing the unknown from the last equation (which now involves one unknown only) and by substituting the known values of X_n as we go up from the last to the first equation. The recursion formula is given by

$$X_n = \frac{-1}{k_{nn}} \left(\sum_{i=n+1}^{N} k_{ni}X_i - F_n \right) \tag{4.113}$$

where N is the total number of equations.

One of the main tasks of a finite element analysis is the reduction of equations.

A typical finite element mesh may have a large number of elements and many degrees of freedom leading to substantial requirements of computer storage and execution time. For example we may have 10 000 degrees of freedom. To store the coefficient matrix would require one hundred million storage locations and the number of calculations needed to solve the equations would be in the order of billions—a formidable task for even the largest and fastest computers.

To make the solution of equations efficient we have to exploit certain properties of the coefficient matrix.

(1) The matrix is symmetric.

(2) It is sparsely populated i.e. contains a large number of zeros.

The crux of the problem is to store as little information as possible and to avoid multiplication with zero as much as possible when solving the system of equations.

To explain this consider the example of Figure 4.10a with three elements and one degree of freedom per node. Let us number the degrees of freedom

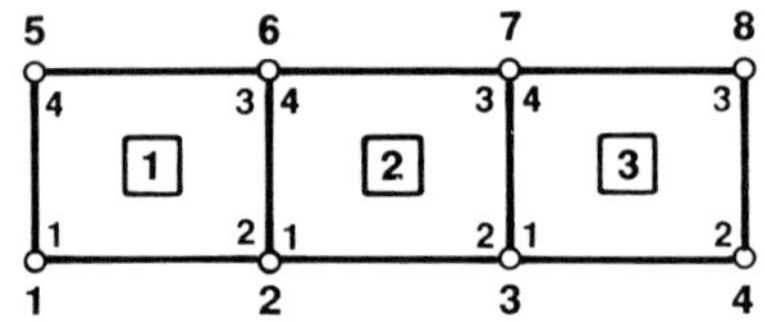

1	2	3	4	5	6	7	8	
k^1_{11}	k^1_{12}	0	0	k^1_{14}	k^1_{13}	0	0	1
k^1_{21}	$k^1_{22}+k^2_{11}$	k^2_{12}	0	k^1_{24}	$k^1_{23}+k^2_{14}$	k^2_{13}	0	2
0	k^2_{21}	$k^2_{22}+k^3_{11}$	k^3_{12}	0	k^2_{24}	$k^2_{23}+k^3_{14}$	k^3_{13}	3
0	0	k^3_{21}	k^3_{22}	0	0	k^3_{24}	k^3_{23}	4
k^1_{41}	k^1_{42}	0	0	k^1_{44}	k^1_{43}	0	0	5
k^1_{31}	$k^1_{32}+k^1_{41}$	k^2_{42}	0	k^1_{34}	$k^1_{33}+k^2_{44}$	k^2_{43}	0	6
0	k^2_{31}	$k^2_{32}+k^3_{41}$	k^3_{42}	0	k^2_{34}	$k^2_{33}+k^3_{44}$	k^3_{43}	7
0	0	k^3_{31}	k^3_{32}	0	0	k^3_{34}	k^3_{33}	8

(a)

Figure 4.10 Assembled global response matrix for different node numbering schemes (superscript denotes element number)

from left to right (same as the element numbering). The assembled response matrix based on the local node numbering shown, according to Section 4.3, is also shown in Figure 4.10a.

Following the task, to store the minimum information required, we find that we only need to store the lower portion of the matrix as all the coefficients in the upper portions can be obtained from symmetry.

Furthermore we need only to store non-zero coefficients. This is not as easy as it may seem because the relative position of a coefficient in the matrix must not be disturbed.

One scheme would be to store only up to the last non-zero coefficient i.e. the shaded area in Figure 4.10a. However we would not gain much in such a scheme because there would still be a number of 'embedded' zeros.

Let us now renumber the nodes of the mesh as shown in Figure 4.10b. We immediately see that we have made the storage much more efficient as we have reduced the number of zeros to be stored. As a consequence we have also reduced the width of the shaded area, the *bandwidth*.

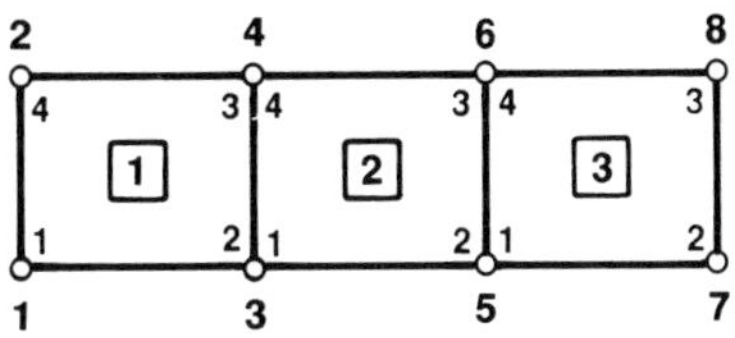

1	2	3	4	5	6	7	8	
k^1_{11}								1
k^1_{41}	k^1_{44}							2
k^1_{21}	k^1_{24}	$k^1_{22} + k^2_{11}$						3
k^1_{31}	k^1_{34}	$k^1_{32} + k^2_{41}$	$k^1_{33} + k^2_{44}$					4
0	0	k^2_{21}	k^2_{24}	$k^2_{22} + k^3_{11}$				5
0	0	k^2_{31}	k^2_{34}	$k^2_{32} + k^3_{41}$	$k^2_{33} + k^3_{44}$			6
0	0	0	0	k^3_{21}	k^3_{24}	k^3_{22}		7
0	0	0	0	k^3_{31}	k^3_{34}	k^3_{32}	k^3_{33}	8

(b)

Figure 4.10 (*cont.*)

We may now store the global coefficient matrix as follows:

$$[\hat{\mathbf{K}}] = \overset{p \rightarrow}{\begin{bmatrix} k_{11} & k_{21} & k_{31} & k_{41} \\ k_{22} & k_{32} & k_{42} & 0 \\ k_{33} & k_{43} & k_{53} & k_{63} \\ k_{44} & k_{54} & k_{64} & 0 \\ k_{55} & k_{65} & k_{75} & k_{85} \\ k_{66} & k_{76} & k_{86} & 0 \\ k_{77} & k_{87} & 0 & 0 \\ k_{88} & 0 & 0 & 0 \end{bmatrix}} m \downarrow \tag{4.114}$$

where k_{ij} are now the assembled coefficients i.e. $k_{33} = k^1_{22} + k^2_{11}$ etc.

This is a very compact storage. Exploiting the symmetry of the system saves us about 44% and storing the non zero band saves us another 6%. Whilst the 6% does not seem significant for this small problem, for large systems of equations encountered in practice there could be as much benefit as 90% for storing the band only.

Whilst storing the coefficients in such a way is very efficient there is some housekeeping involved now. Consider the Gauss elimination of the coefficient matrix in Figure 4.10b. The recursive formula for elimination of the nth variable is

$$k^*_{ij} = k_{ij} - k_{nj}\frac{k_{ni}}{k_{nn}} \tag{4.115}$$

When we address the coefficients in the band matrix $[\hat{\mathbf{K}}]$ then

$$k_{ij} = k_{ji} = \hat{k}_{mp} \tag{4.116}$$

where

$$m = \min(i, j)$$
$$p = \max(i, j) - m + 1$$

Obviously when either k_{nj} or k_{ni} is zero then there is no change to k_{ij} and therefore the recursion formula (4.115) is only applied within the band.

It should have become clear to the reader by now that the efficiency of the '*sliced band*' technique depends on how the user of a program numbers the nodes of the mesh. Obviously the mesh in Figure 4.10b will have a more efficient numbering system. Whilst it is fairly easy to choose the right system for this simple example it is very difficult for a complex three-dimensional object. In fact with CAD packages and graphical preprocessors we do away largely with numbers and it is uneconomical to attempt to renumber all the nodes of the mesh to reduce the bandwidth.

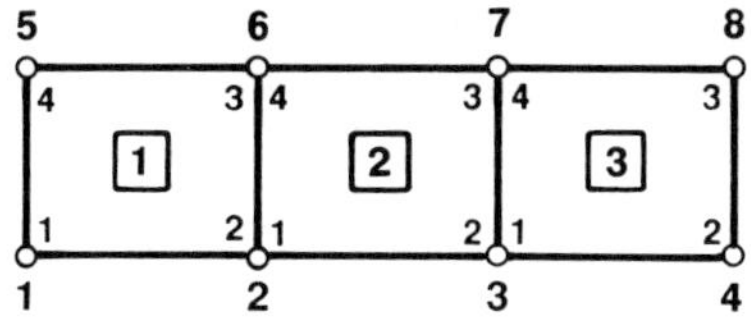

NODE	1	2	6	5			
DESTINATION	1	2	3	4			
	k^1_{11}	k^1_{21}	k^1_{31}	k^1_{41}	1	1	ELEMENT 1
	k^1_{12}	k^1_{22}	k^1_{32}	k^1_{42}	2	2	
	k^1_{13}	k^1_{23}	k^1_{33}	k^1_{43}	3	6	
	k^1_{14}	k^1_{24}	k^1_{34}	k^1_{44}	4	5	

NODE	(3)	2	6	(7)			
DESTINATION	1	2	3	4			
	k^2_{22}	k^2_{21}	k^2_{24}	k^2_{23}	1	(3)	ELEMENT 2
	k^2_{12}	$k^2_{11} + k^{1*}_{22}$	$k^2_{14} + k^{1*}_{32}$	k^2_{13}	2	2	
	k^2_{42}	$k^2_{41} + k^{1*}_{23}$	$k^2_{44} + k^{1*}_{33}$	k^2_{43}	3	6	
	k^2_{32}	k^2_{31}	k^2_{34}	k^2_{33}	4	(7)	

Figure 4.11 Storage of coefficients in the frontal solution scheme

The crux of the problem is in the way we store the equations or unknowns. So far we have stored the equations in such a way that the column and row number correspond to the node number.

Let us look at an alternative scheme whereby we store the equations in the same sequence as we assemble them. Consider Figure 4.11. Each node has a 'destination' which gives its position in the coefficient matrix. As we assemble the first element, nodes 1 to 4, we assign destinations 1 to 4 to nodes 1, 2, 6, 5 and so on. Now we see that it does not matter in which way we number the nodes, the 'profile' of the matrix will stay the same.

Let us now look at further savings that we can make in precious core storage. After the assembly of element 1 we find that equations 1 and 4 (corresponding to nodes 1 and 5) are in their final form i.e. will not have any coefficients added as we assemble element 2 and subsequent elements. This means that in equation (4.115) coefficients k_{nj}, k_{ni} and k_{nn} are in their final form.

We can now 'eliminate' these equations i.e. make $k_{in} (i \neq n)$ zero for all other equations because it does not matter that the k_{ij} are not completely summed as it makes no difference if the assembly occurs before or after the 'starring' operation.

The idea of a frontal solver is to interleave assembly and equation solution. One advantage of such a scheme becomes immediately obvious. Once equations 1 and 4 have been eliminated they will not be used any more in the reduction of the coefficient matrix yet occupy precious storage space. However since they will be used later on for back-substitution we have to write these equations to backup storage (such as disk) and retrieve them later. By doing this we make space for new equations. As we assemble element 2 the destinations will change as we attempt to reuse the space for equations 1 and 4 (Figure 4.11). After assembly of element 2 we see that we can eliminate equations 2 and 3, which are now fully summed, leaving space for the last two equations.

We can see that this ingenious scheme, introduced in 1970 by Bruce Irons, gives us an enormous saving in core storage because we only have to store the front matrix. (The number of equations in the front matrix is also known as the front width.) Because of symmetry we only have to store the shaded area i.e. only 10 coefficients versus 32 for the sliced band technique. However we now have the overhead of bookkeeping i.e. keeping track of the destinations.

A vector of destinations is established for the degrees of freedom of each element. For example the destination vector for element 1 in Figure 4.11 is 1, 2, 6, 5 and for element 2 is 3, 2, 6, 7. The computation of the destination vectors for all elements is conveniently done before the actual solution (Prefront program). Here the allocation of destinations may be even more optimised by looking at the 'life span' of equations. In such schemes we put the equations which have the shortest life span at the end so that they can be eliminated first leading to an immediate reduction in the size of the front matrix[7].

We now see that although the numbering of nodes is irrelevant the order in which we assemble the elements is. It is left as an exercise to the reader to verify this by numbering the elements 1, 3, 2 and repeating the assembly/elimination process just explained. This numbering scheme will increase the front width from 4 to 8.

Another advantage of the frontal solution is that, using the element destination vector, results are obtained element by element in the local numbering system and this is convenient for the computation of element stress and fluxes.

4.5.1 Computation of stresses and fluxes

Once the system of equations has been solved the displacements at all the nodes are known.

Stresses (and fluxes) at any point inside an element can be obtained from

$$\boldsymbol{\sigma} = \mathbf{D}\mathbf{B}^e\mathbf{u}^e \tag{4.117}$$

Note that the vector $\mathbf{u}^e$ containing element displacements in local numbering must first be established by 'gathering' global values with reference to the element connectivity vector. If a frontal solver is used then the element results are gathered during back substitution using the element destination vector.

The stresses are computed independently for each element and we find that in most cases we obtain different results for each element at an interelement boundary. This means that there may be a lack of equilibrium locally at points on interelement boundaries. This arises because in the computation of the response matrices, equilibrium is only enforced in an integral sense.

We find indeed that the stresses computed at a node to which elements connect have n different values where n is the number of elements adjacent to the node. Usually a unique value at the node is computed by averaging these values.

Regarding these stresses however, an interesting discovery has been made. The Gauss points are the points where the stresses are most accurate. To explain this let us consider the stress distribution in Figure 4.12 which has been obtained from a rather coarse mesh, and note that the predicted stress distribution is not very accurate. Apart from the mismatch of values at the interelement boundaries the distribution, assumed here to be quadratic within each element, does not match the actual solution very well.

However it can be shown that a straight line connecting the computed values of stress at the Gauss points of the two-point formula gives a much better approximation. It can also be shown that the straight line through the two Gauss points is in this case a least squares approximation to the computed (parabolic) distribution.

The fact that the Gauss points are optimal sampling points for stresses can be theoretically proven[9,10] and most finite element software gives a printout of stresses at these points only. Where stresses at the nodes are required then these values are extrapolated from the Gauss points and averaged as discussed in Chapter 11. This is done in the finite element demonstration program DEMON listed in Appendix A.

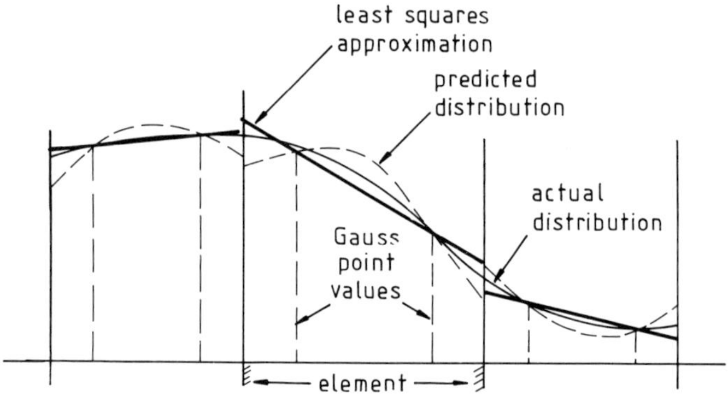

Figure 4.12 Approximation to the actual solution by finite elements and a least squares approximation of the predicted distribution

4.6 EXAMPLES–1

Here we show by means of some simple problems, which are analysed with programs DEMON (see Appendix A) and BEFE, certain characteristics of the finite element method. Particular attention is paid to the question of what accuracy may be expected from an analysis. As will be outlined in Chapter 10 there are several sources of error but the most important one, and the one on which we will focus our attention here, is the error introduced by the discretisation. Discretisation errors are due to the assumptions of geometric and functional variation over each finite element.

4.6.1 Cantilever beam

Consider a cantilever plate of thickness t having a span of 5 m with E=10000 MPa and $\nu = 0$ and subjected to a load of 10 kN/m along its free side. The plate is assumed to be in a state of plane strain. We make several numerical experiments to determine the effect of element aspect ratio, element distortion and the order of the shape function. The computed displacements are compared with the classical beam bending theory which does not take account of displacement due to shear strain, which can be significant for deep beams. The classical beam theory gives a maximum deflection of $u = 0.5/t^3$ mm and a maximum stress of $\sigma_x = 0.3/t^2$ MPa. For deep beams computed values of displacements are expected to be greater. For slender beams the influence of shear displacement should disappear. In the examples we will use the eight-node isoparametric elements of DEMON except when we look at the effect of shape function order where linear elements of program BEFE are used.

Aspect ratio

Here we look at the effect of the element aspect ratio (ratio element length to height) on the results of the analysis. We reduce the thickness of the cantilever from 1m to 0.2m and 0.05m (see Figure 4.13). The results of the analyses are compared in Table 4.1.

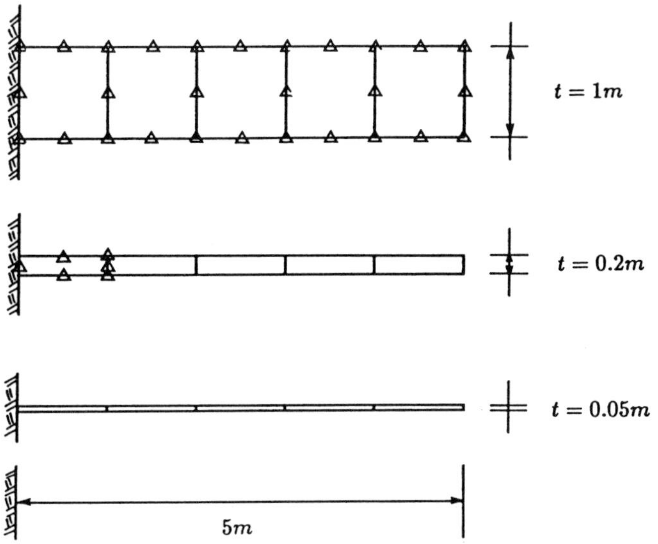

Figure 4.13 Cantilever beam and mesh of parabolic elements used for different aspect ratios

Table 4.1 Influence of aspect ratio

t	Maximum deflection (mm)		Maximum stress (MPa)	
	Computed	Beam theory	Computed	Beam theory
1.0	0.5100	0.5000	0.300	0.300
0.2	62.55	62.50	7.50	7.50
0.05	4000.0	4000.0	120.0	120.0

It can be seen that even for extreme values of aspect ratio excellent results are obtained for this case. As we will see in Chapter 7 for boundary elements, problems are experienced with extreme element aspect ratios there.

Element distortion

To see what effect geometric distortion of an element has on the results a mesh is used where the elements are not rectangular (Figure 4.14). The analysis results from DEMON for $t = 1.0$ are 0.5159 mm for the maximum deflection and 0.303 MPa for the maximum stress. This indicates a negligible influence of element distortion on the results. Indeed meshes with distorted elements are used to test newly developed and programmed elements in practice ('patch test'). With early elements it has been found that some elements work well where they are rectuangular but give errors when they are distorted. The 'patch test' which originates from Bruce Irons is a simple test which should be used to test new software.

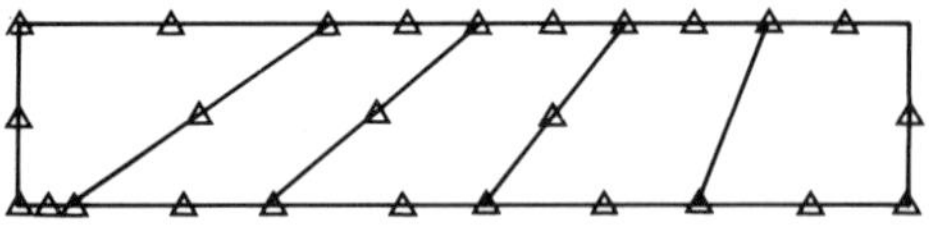

Figure 4.14 Distorted mesh for cantilever beam

Shape function order

Here we investigate the effect of the order of the shape function used for describing the element geometry and the variation of the unknown. The different meshes used for the analysis are shown in Figure 4.15. A value of $t = 1.0$ is assumed for which the theoretical solution for the maximum deflection is 0.500 and for the maximum stress is 0.300. The findings are summarised in Table 4.2.

Table 4.2 Computed values of deflection for different function order and mesh size for a beam with t=1 m

Number of elements	Function order 1 = linear 2 = parabolic	Degrees of freedom	Maximum deflection (mm)	Maximum Stress (MPa)
5	2	56	0.510	0.300
5	1	24	0.339	0.178
20	1	66	0.453	0.253
80	1	200	0.496	0.287

It can be seen that for the five-element mesh linear elements give poor results but this is hardly surprising as the assumption of a linear displacement field implies that, for example, the horizontal stresses are constant along a horizontal line through an element. Doubling the mesh in both directions still gives an error in the displacements of about 10% and in the stresses of 15%.

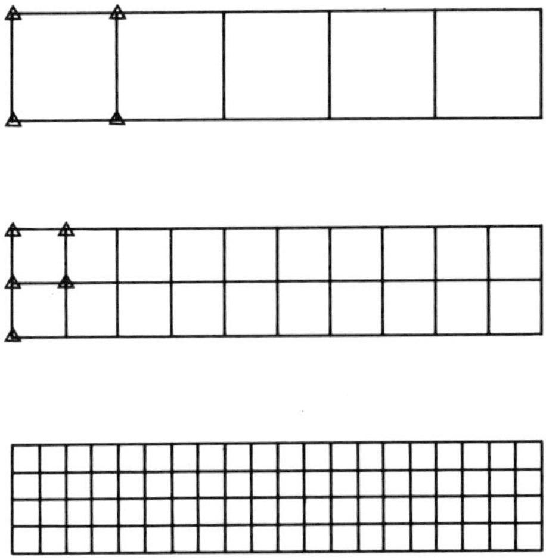

Figure 4.15 Three different meshes of linear elements for cantilever beam

This mesh has ten degrees of freedom more than the five-element mesh with parabolic elements. Doubling the mesh again in both directions brings the error in displacement down to less than 1% and the error in stress to about 4%.

To explain these results consider a small element in pure bending (Figure 4.16). For Poisson's ratio equal to zero the exact displacements are given by

$$u_x = Cxy \tag{4.118}$$

$$u_y = \frac{1}{2}C\left(a^2 - x^2\right) \tag{4.119}$$

where C is a constant and a is half the length of the element. For a linear finite element the displaced shape is approximated by (Figure 4.16c)

$$u_x = Cxy; \quad u_y = 0 \tag{4.120}$$

Thus the error in the solution for the y-displacement is

$$e(u_y) = \frac{1}{2}C_2\left(a^2 - x^2\right) \tag{4.121}$$

and for the shear stresses

$$e(\tau_{xy}) = C_3\left(\frac{\partial u_y}{\partial x}\right) = C_4 x \tag{4.122}$$

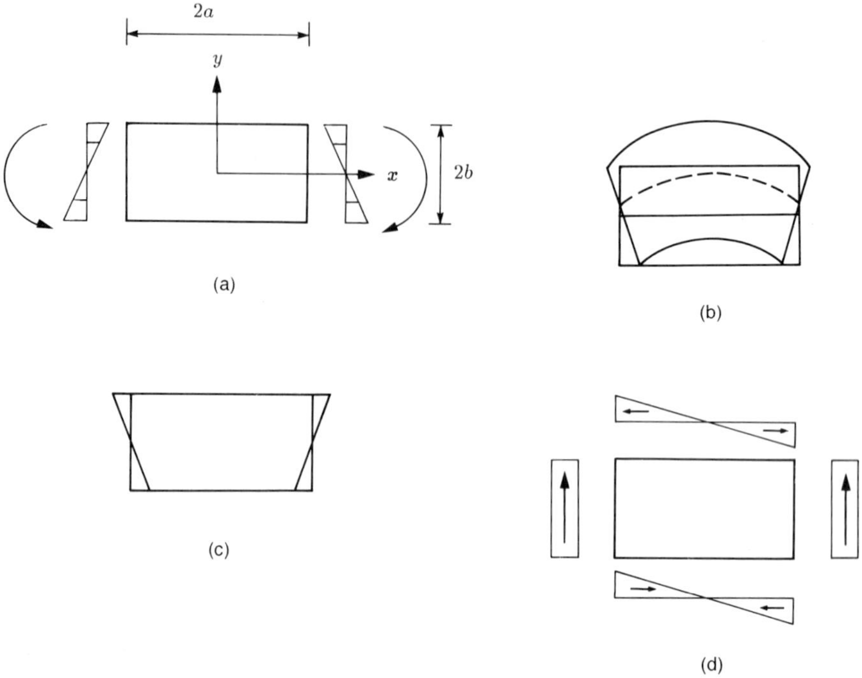

Figure 4.16 An element in pure bending and errors introduced by linear shape functions. (a) Bending stree applied to element. (b) Exact displacements. (c) Displacements—linear element. (d) Error in sheer stress distribution

The distribution of this error is shown in Figure 4.16d. In the case where Poisson's ratio is zero this erroneous distribution of shear stress is the main source of error and this error increases for slender beams. A simple modification to the element was introduced by Wilson[11] in 1971. He found that by adding terms to the displacement approximation the results could be vastly improved for the cantilever problem. The modified shape functions are

$$\begin{aligned} u_x &= \sum N_i u_{xi} + N_5 a_{1x} + N_6 a_{2x} \\ u_y &= \sum N_i u_{yi} + N_5 a_{1y} + N_6 a_{2y} \end{aligned} \tag{4.123}$$

where the additional terms are incompatible modes and the additional shape functions are given by (see Figure 4.17)

$$\begin{aligned} N_5 &= (1 - \xi^2) \\ N_6 &= (1 - \eta^2) \end{aligned} \tag{4.124}$$

They are called incompatible because the unknowns associated with these modes are unique to the element and therefore displacement incompatibilities between elements at interelement boundaries can exist. Because the

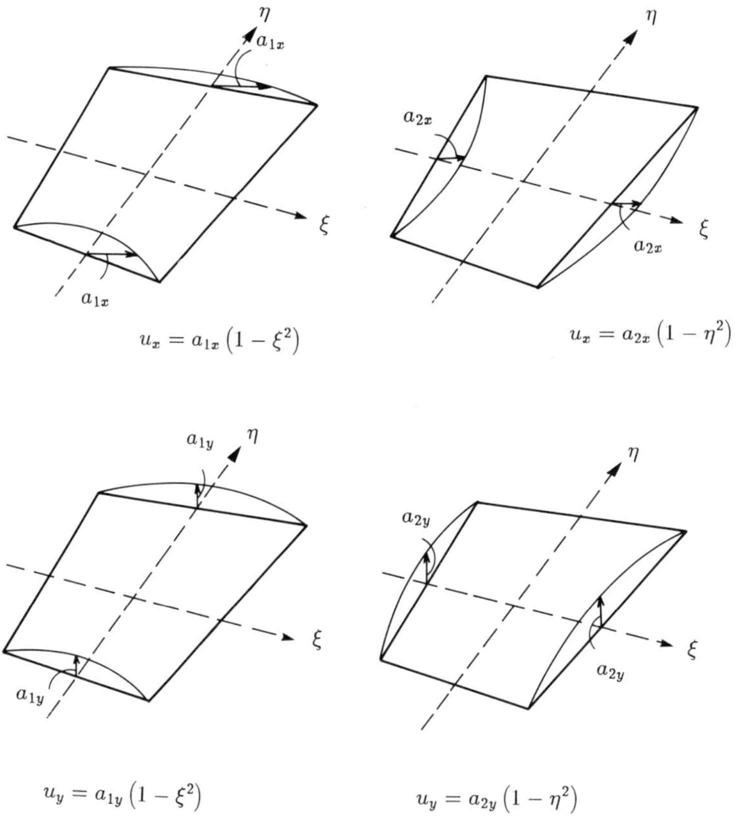

Figure 4.17 Incompatible displacement modes for linear quadrilateral element

incompatible modes only occur in an element the associated degrees of freedom may be eliminated and a stiffness matrix obtained which is of the same size as the original four-noded element.

Thus with a relatively small additional computational effort and without adding any degrees of freedom it is possible to obtain similar accuracy for linear elements as for parabolic ones.

4.6.2 Plate with square hole

Consider a plate with a square hole, subjected to a tensile load at one edge. The exact solution for this problem is not known. Theoretically the stresses are infinite at a corner of the square cutout.

The three different finite element meshes used for analysing the problem are shown in Figure 4.18. Only one quarter of the plate is analysed and symmetry conditions (roller supports) are applied on three sides as shown.

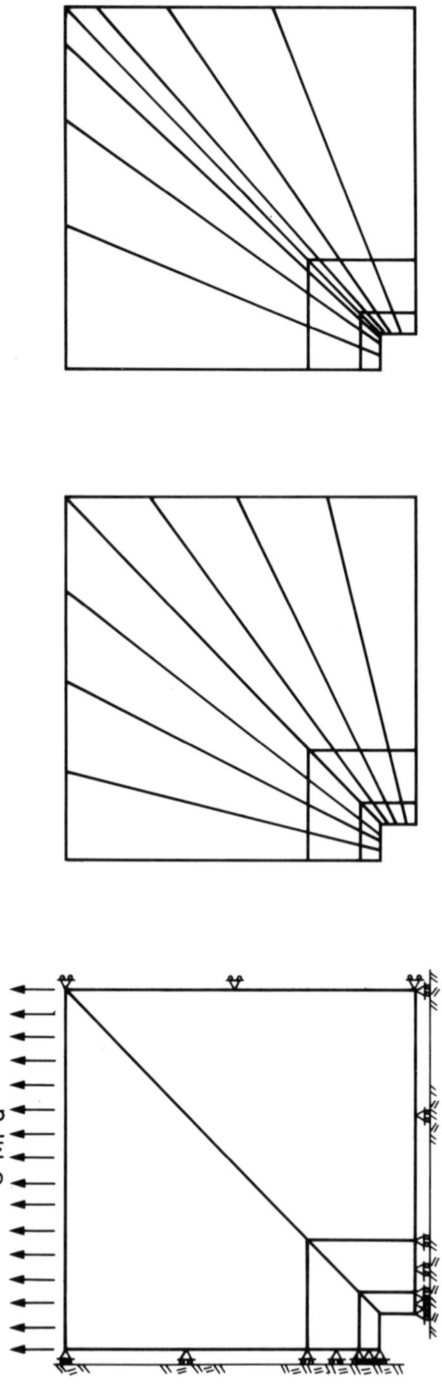

Figure 4.18 Plate with square hole—3 different meshes used

The results are plotted in Figure 4.19 as the distribution of tangential stresses on the boundary C-A. One can see that the results for all meshes are similar. The greatest differences between the results is, as expected, at the corner. Even with the graded fine mesh the finite element analysis is not able to predict the stress concentration at the corner. In fact the stresses drop off as the corner is approached instead of increasing. The same problem is solved with boundary elements in Chapter 7.

4.7 EXAMPLES–2

The examples shown here represent a fairly wide range of problems in mining, civil and mechanical engineering. All the examples are three dimensional and have been analysed with the coupled boundary element/finite element program BEFE and its predecessor MISES3.

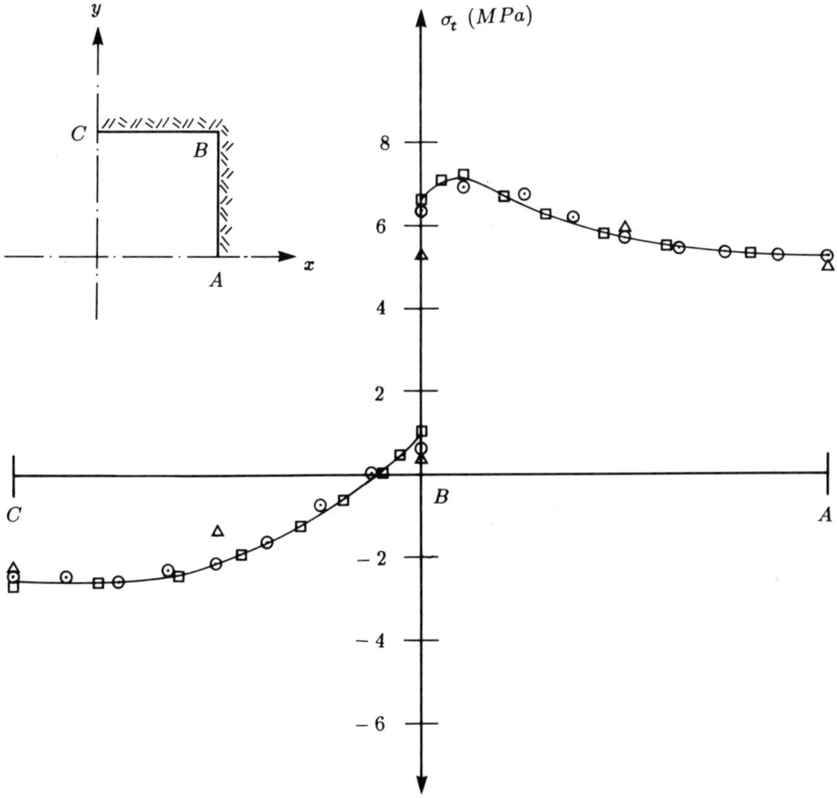

Figure 4.19 Plate with square hole—stress distribution on boundary. (△ two elements, ○ eight elements, □ eight elements (graded mesh))

4.7.1 Analysis of sequential mining

Here the three-dimensional finite element method is used for the analysis of sequential mining excavation. In the particular application discussed here the mineralisation occurs in a 'pipe' which is about 150 m in diameter at 700 m depth and 60 m diameter at 850 m depth.

Figure 4.20a shows a schematic cross section of the mine with the material properties. A perspective view of the mesh used for the analysis is shown

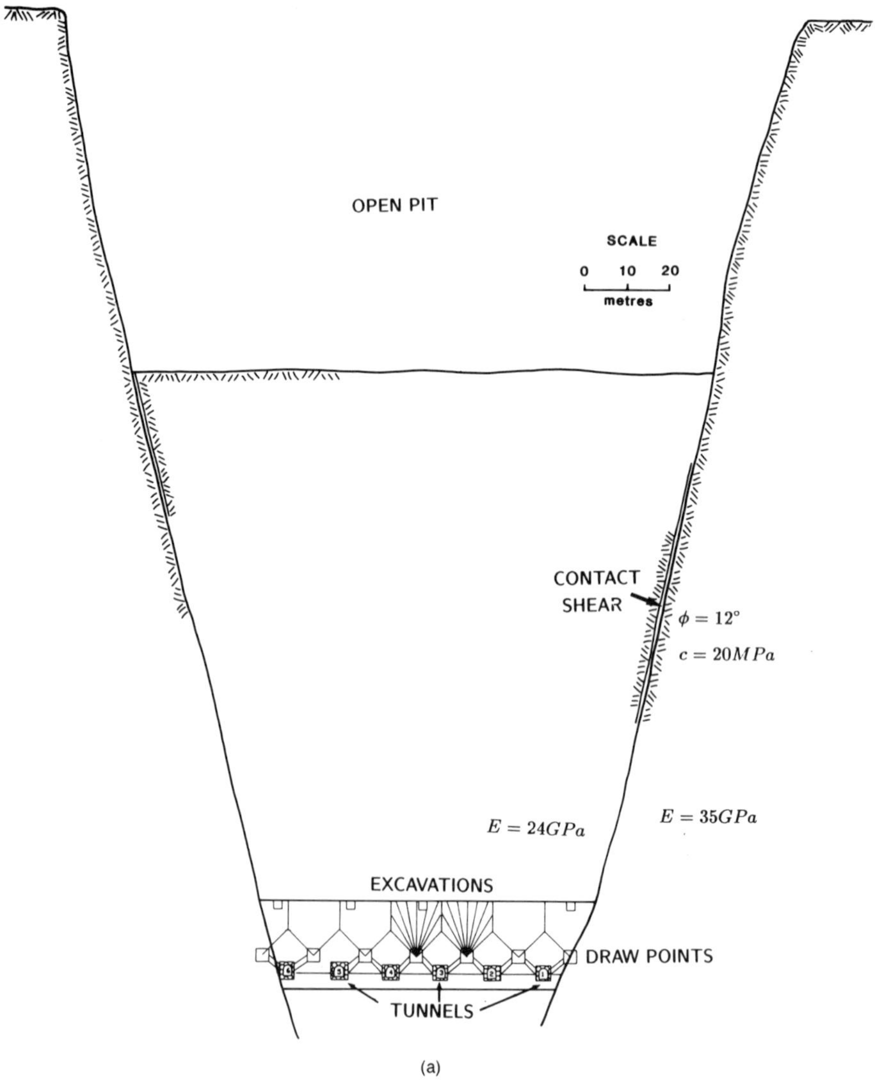

Figure 4.20 (a) Schematic cross section of mine

in Figure 4.20b (see colour section). For the analysis it was assumed that the rock mass which surrounds the pipe is rigid. The fault zone was modelled by joint elements as discussed in the next chapter. The mesh consists of 973 eight-node linear finite elements and has 1368 nodes. The solution for one excavation stage took about 60 seconds using a fully vectorised version of BEFE on a CONVEX C120 mini supercomputer. For the first stage the excavation of drives or tunnels was simulated by removing elements as shown in Figure 4.20b. Subsequently six mining steps were analysed. Figure 4.21 (see colour section) shows contours of minimum principal stress (i.e. maximum compressive stress) on a vertical section through the mesh.

4.7.2 Concrete arch dam

In the design of concrete arch dams it is important to design the shape of the arch in such a way that significant tensile stresses at the base of the arch—where it is founded on rock—are avoided as such stresses may cause the opening of cracks and problems with seepage. The finite element method was used to model the interaction of an arch dam and rock foundation. Figure 4.22 shows the mesh where the concrete arch dam and immediate rock mass is divided into approximately 200 20-node isoparametric 'brick' elements. All five sides of the mesh representing the rock mass were fixed in all directions. The applied loading consisted of gravity and water pressure. Figure 4.23 shows one result of the analysis namely the contours of maximum principal stress at the base of the dam. The area where tension occurs is shaded. The predicted tensile zone is in agreement with observations.

4.7.3 Nuclear reactor cooling pump

In nuclear reactors pumps are used to cool the reactor down in the event of an emergency. In the case of start up the inside of the pump casing is heated to in excess of 275°C whereas the outside is still at room temperature. This temperature differential causes distortions in the casing which in turn give rise to stresses. A finite element analysis was used to determine firstly the temperature distribution at various times after starting and secondly determine the distribution of stresses in the casing.

Figure 4.24 shows one half of the mesh used for the heat flow and stress analyses. Symmetry boundary conditions were applied. Artificial boundary conditions were applied where the horizontal and vertical pipes connect. The mesh consists of about 700 parabolic 20-node elements. For the transient heat flow analysis a temperature of 275°C was prescribed at the inside surface of the casing and a temperature of 20°C at the outside. For the thermal stress analysis the temperature distribution as obtained by the transient heat flow analysis was used to determine the nodal forces in equation (4.105).

Figure 4.25 shows the temperature distribution in the casing after time step 18.

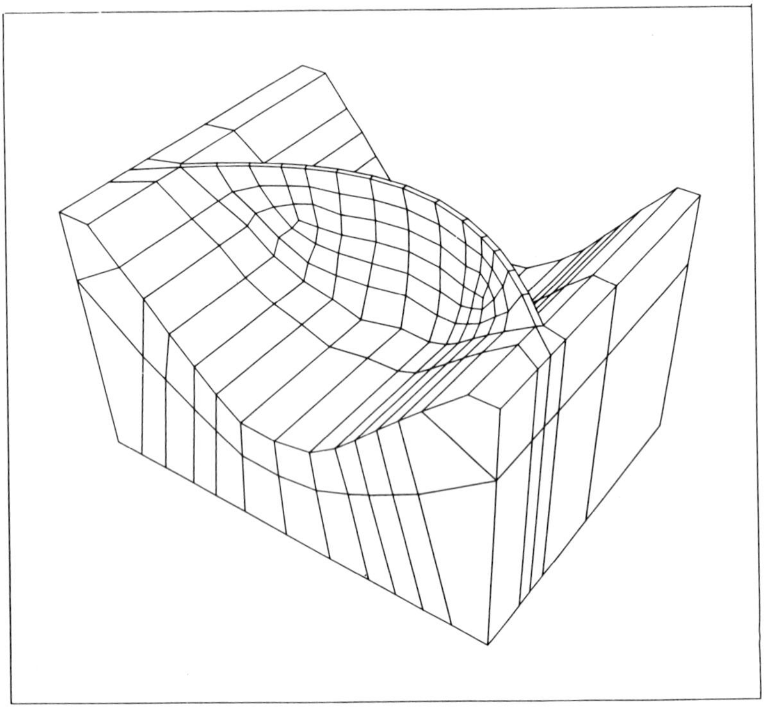

Figure 4.22 Concrete arch dam: axonometric view of finite element mesh for the dam and the surrounding rock. View from the dry side

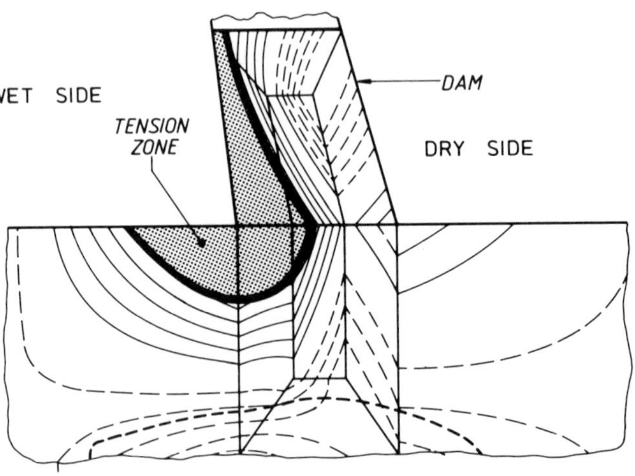

Figure 4.23 Contours of maximum principal stress at base of arch dam (tension zone is shaded)

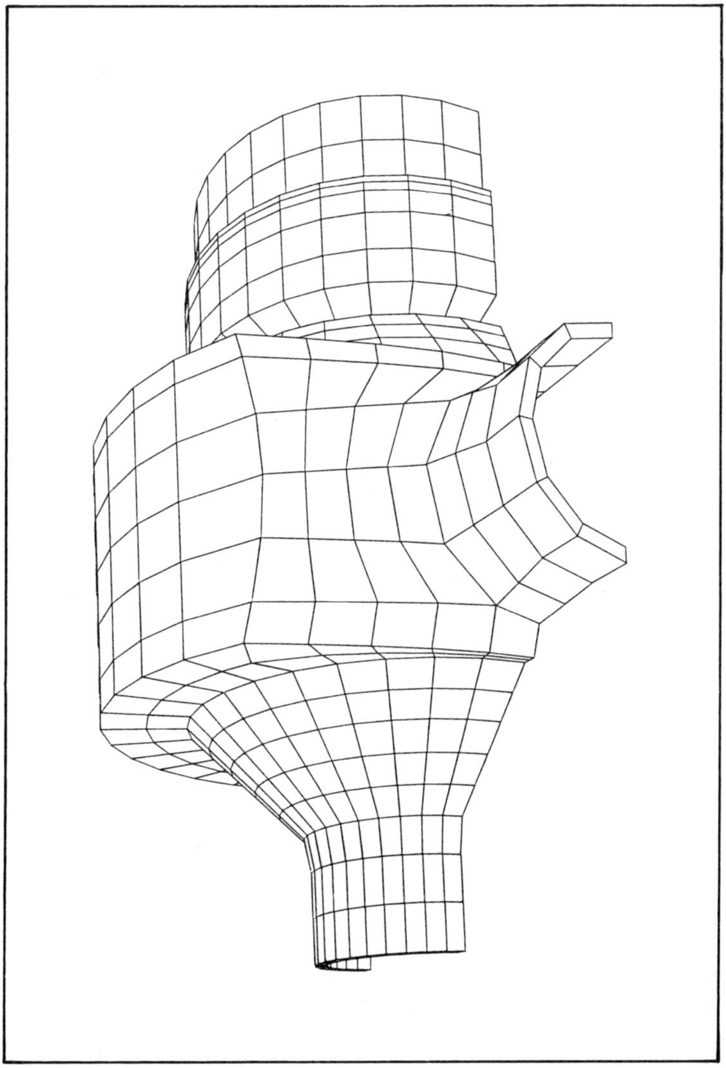

Figure 4.24 Nuclear reactor cooling pump—finite element mesh

The distortion of the casing due to the temperature gradient is shown on a section through the casing in Figure 4.26. The results were used to check the design of the pump housing.

4.8 CONCLUDING REMARKS

In this chapter we have discussed the theoretical background of the finite element method and its numerical implementation.

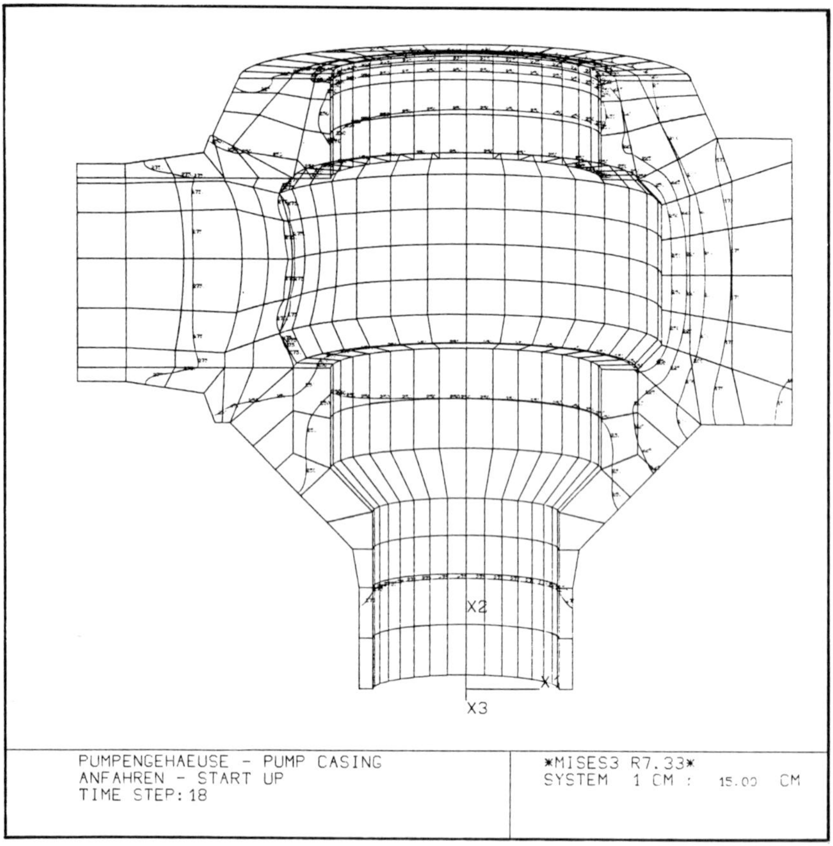

Figure 4.25 Temperature distribution after starting

There are five distinct tasks in a finite element analysis, which correspond to stages in the computer program. These are the following.

(1) Computation of element response matrices.
(2) Computation of element load vectors.
(3) Assembly of element matrices and load vectors.
(4) Solution of system of equations subject to certain boundary conditions and loading.
(5) Computation of results inside elements.

In this chapter an approximate element response matrix has been obtained by assuming a distribution of displacement inside an element. Satisfaction of the differential equations is in general only enforced in an integral sense, i.e. by making the total potential energy a minimum or letting the integral of the

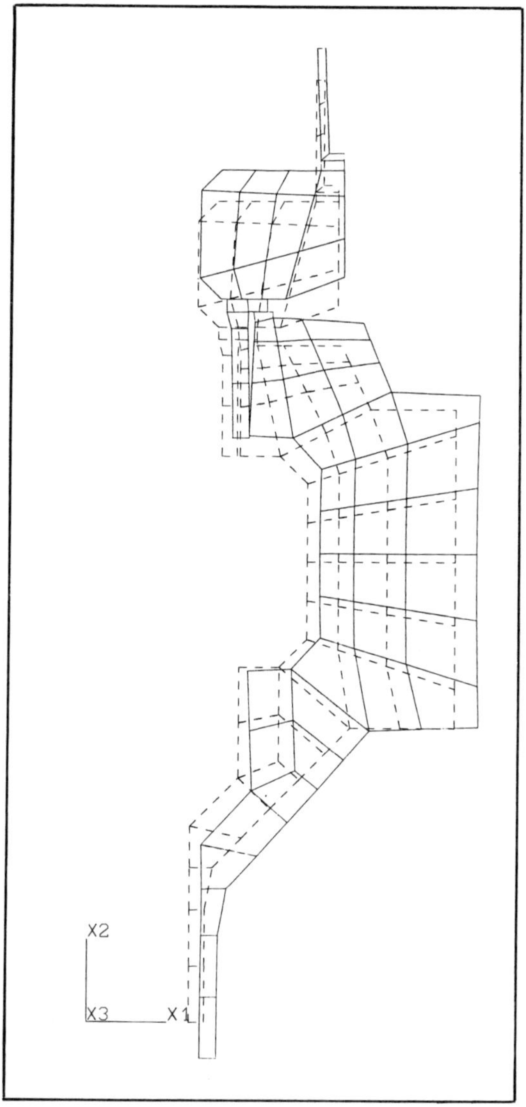

Figure 4.26 Nuclear reactor cooling pump: displaced shape due to temperature increase (dotted line is original shape)

error in satisfaction of the differential equation vanish. Thus the differential equation may not be satisfied inside the element or on element boundaries, i.e. for elasticity problems for example there may be local violation of equilibrium. The basis of the finite element method is that such errors will become rapidly smaller as the number of elements increases, i.e. the solution will *converge* to the right one for increased mesh density.

The central feature of the finite element method is the assumption of locally (element) based shape functions. In this chapter we have chosen to apply these to the displacement field.

There is no reason why the shape functions can not be applied to stress fields. Indeed such *stress based* elements have been developed and have been found to perform well, sometimes better than the displacement based elements. Finally it is possible to apply the shape functions to both the stress and displacement field. In a *hybrid* element we may for example approximate the stress inside the element and the displacement on the boundaries.

The reader is referred to the books cited at the end of this chapter for further reading.

4.9 EXERCISES

4.1 For a triangular element with three nodes the interpolation functions for the displacements are given in terms of global coordinates. For example the shape function for node i is

$$N_i = \frac{(a_i + b_i x + c_i y)}{2\Delta}$$

where a_i, b_i and c_i are constants defined by

$$\begin{aligned} a_i &= x_j y_m - x_m y_j \\ b_i &= y_j - y_m \\ c_i &= x_m - x_j \end{aligned}$$

and Δ is the area of the triangle while x_j, y_j; x_m, y_m are the coordinates of points j and m. Other shape functions are obtained by cyclic permutation.

It can be seen that N_i represent planar surfaces defined by three points as shown in Figure 4.27 and that

$$\frac{\partial N_i}{\partial x} = \text{constant} \qquad \frac{\partial N_i}{\partial y} = \text{constant}$$

This means that the strains inside the triangle are constant thus the name 'constant strain' triangle.

Following the approach in 4.2 derive the element stiffness matrix.

Note: In this particular case we do not need to calculate a jacobian matrix since the shape functions are already given in global coordinates. Also the integration is extremely simple since strains and stresses are constant.

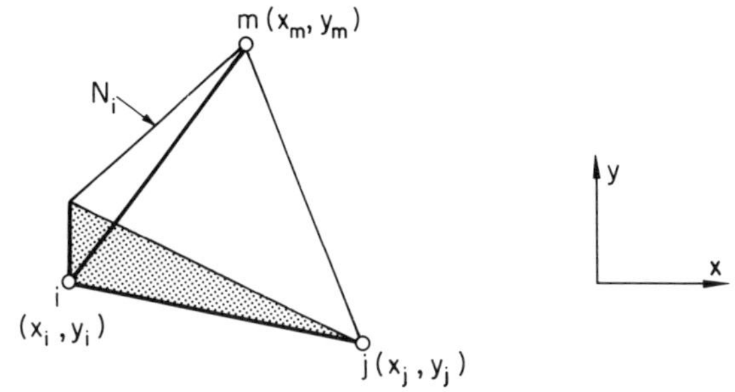

Figure 4.27 Triangular element with shape function N_i

4.2 For three-dimensional problems the strain vector is given by equation (3.8). Derive according to Section 4.2.3 the **B** matrix for the three-dimensional case.

4.3 For axisymmetric problems the strain vector is given by equation (3.9). Derive the **B** matrix for this case.

4.4 Using similar reasoning as in section 4.3 show for an assembly of one-dimensional elements that the assembly process is also valid for potential problems such as heat and fluid flow.

4.5 By performing the differentiation of equation (4.21) using the product formula prove that equation (4.24) is obtained.

4.6 Assume a quadratic curve between $\xi = \pm 1$. Show that a straight line drawn between the function values at the Gauss points for a two-point integration is a least squares approximation to this curve.

4.7 Complete the example of assembly using the frontal technique in Figure 4.11.

4.8 Do the example in Figure 4.11 by assembling the elements in the order 1, 3, 2.

4.9 Using the approximation of the displacement field as shown in equation (4.123) derive the stiffness matrix of the incompatible four-node element.

4.10 REFERENCES

1. Washizu, K. (1975) *Variational Methods in Elasticity and Plasticity* 2nd Edn, Pergamon Press, Oxford.
2. Schied, F. (1968) *Numerical Analysis (Schaum Series)*, McGraw-Hill, New York.
3. Irons, B.M. (1971) Quadrature rules for brick based finite elements. *Int. J. Num. Meth. Eng.*, **3**.
4. Stroud, A.H. and Secrest, D. (1966) *Gaussian Quadrature Formulas*, Prentice-Hall, Englewood Cliffs, New Jersey.

5. Zienkiewicz, O.C. (1977) *The Finite Element Method,* (3rd Edn), McGraw-Hill, New York (Chapter 12).
6. Irons, B.M. (1970) A frontal solution program for finite element analysis. *Int. J. Num. Meth. Eng.*, **2**, 5–32.
7. Beer, G. and Haas, W. (1982) A partitioned frontal solver for finite element analysis. *Int. J. Num. Meth. Eng.*, **18**, 1623–1654.
8. Beer, G. (1981) *A second generation frontal solution program.* Research report No. CE12, Department of Civil Engineering University of Queensland, Australia.
9. Moan, T. (1973) On the local distribution of errors by the finite element approximation. *Theory and Practice in Finite Element Structural Analysis* (eds Y. Yamada and R.H. Gallagher), University of Tokyo Press, Tokyo.
10. Barlow, J. (1976) Optimal stress location in finite element models. *Int. J. Num. Meth. Eng.* **10**, 243–251.
11. Wilson, E.L. (1971) *A static analysis program for three dimensional solid structures.* Report to Denver Mining Research Center, U.S. Department of the Interior, Bureau of Mines.

4.11 RECOMMENDED FURTHER READING

Irons, B.M. and Ahmad, S. (1980) *Techniques of Finite Elements,* Halsted Press, New York.

Hinton, H. and Owen, D.R.J. (1980) *A Simple Guide to Finite Elements,* Pineridge Press, Swansea, U.K.

Zienkiewicz, O.C. and Morgan, K. (1983) *Finite Elements and Approximations,* Wiley, London.

Robinson, J. (1973) *Integrated Theory of Finite Element Methods,* John Wiley, London.

5

Special Finite Elements: Plate/Shell Element, Joint Element

5.1 INTRODUCTION

In this chapter we will discuss special elements which have been developed from the solid elements for specific applications. For the analysis of plates and shells for example, the strain and stress perpendicular to the shell surfaces can be assumed to be zero. An element formulated on this basis has fewer degrees of freedom than the equivalent solid element. Whereas the application of such an element to the analysis of moderately thin plates and shells is economical, the plate/shell formulations become essential if the thickness to element size ratio is very small. This is because the large stiffness coefficients for the compression mode across the thickness can lead to numerical round-off problems in the solution of the system of equations.

The literature on shell elements is extensive and a great number of elements has been proposed ranging from extremely simple elements[1] to sophisticated elements where only one element is needed to give accurate results for, say the problem of a clamped square plate under uniform load[2]. Only the simplest element form will be discussed here with the intention of demonstrating the basic principles involved. For further study the reader is referred to the literature on the subject cited at the end of this chapter.

Another special element is required for situations where discontinuous behaviour occurs between finite elements. These joint elements have zero thickness and allow slip and separation to occur between finite elements.

5.2 PLATE ELEMENTS

To introduce the basic concepts of plate analysis we discuss the 'simplest effective plate bending element yet proposed' (Hughes *et al.*[1]). The linear plate element shown in Figure 5.1 is a degenerate of the linear solid element. The nodes are placed at midsurface of the plate and three degrees of freedom are assumed at each node (two rotations θ_1, θ_2 and one displacement u_z). The element has a total of 12 degrees of freedom, that is half the number for the linear solid element. In the plate formulation we are neglecting membrane action and the stress across the thickness. The displacements as well as the rotations are interpolated from the nodal values in the usual way:

$$\begin{aligned} u_z &= \sum N_i u_{zi} \\ \theta_1 &= \sum N_i \theta_{1i} \\ \theta_2 &= \sum N_i \theta_{2i} \end{aligned} \tag{5.1}$$

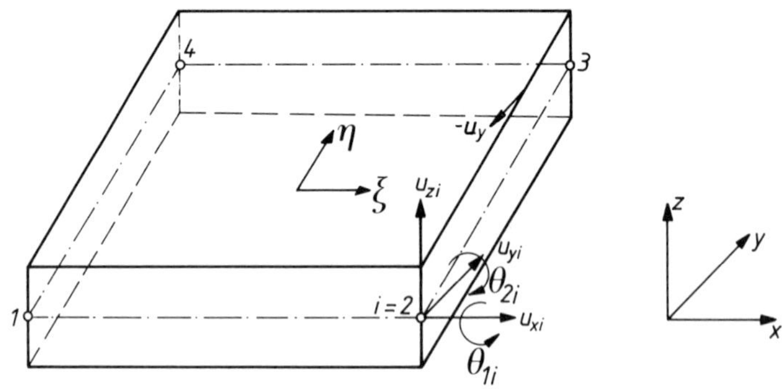

Figure 5.1 Linear plate element

Using the sign conventions in Figure 5.1 and assuming $z = 0$ for the midsurface, the displacements u_x and u_y at any point in the plate can be expressed as:

$$\begin{aligned} u_x &= \theta_2 z &&= \left(\sum N_i \theta_{2i}\right) z \\ u_y &= -\theta_1 z &&= -\left(\sum N_i \theta_{1i}\right) z \end{aligned} \tag{5.2}$$

There are only five strain components since the normal strain in z direction is neglected.

$$\boldsymbol{\epsilon} = \begin{Bmatrix} \epsilon_x \\ \epsilon_y \\ \gamma_{xy} \\ \gamma_{yz} \\ \gamma_{xz} \end{Bmatrix} = \begin{Bmatrix} \partial u_x/\partial x \\ \partial u_y/\partial y \\ \partial u_y/\partial x + \partial u_x/\partial y \\ \partial u_y/\partial z + \partial u_z/\partial y \\ \partial u_x/\partial z + \partial u_z/\partial x \end{Bmatrix} \qquad 5.3$$

It is convenient here to divide the strain vector into 'bending' and 'transverse shear' terms.

$$\boldsymbol{\epsilon}_{\mathrm{B}} = \begin{Bmatrix} \epsilon_x \\ \epsilon_y \\ \gamma_{xy} \end{Bmatrix} \quad \boldsymbol{\epsilon}_{\mathrm{S}} = \begin{Bmatrix} \gamma_{yz} \\ \gamma_{xz} \end{Bmatrix} \qquad 5.4$$

The relationship between strains and displacements can be written as:

$$\begin{aligned} \boldsymbol{\epsilon}_{\mathrm{B}} &= \mathbf{B}_{\mathrm{B}}\boldsymbol{\delta}^e \\ \boldsymbol{\epsilon}_{\mathrm{S}} &= \mathbf{B}_{\mathrm{S}}\boldsymbol{\delta}^e \end{aligned} \qquad (5.5)$$

where

$$\mathbf{B}_{\mathrm{B}} = z \begin{bmatrix} \mathbf{0} & \mathbf{0} & \partial\mathbf{N}/\partial x \\ \mathbf{0} & -\partial\mathbf{N}/\partial y & 0 \\ \mathbf{0} & -\partial\mathbf{N}/\partial x & \partial\mathbf{N}/\partial y \end{bmatrix} \qquad (5.6)$$

and

$$\mathbf{B}_{\mathrm{S}} = \begin{bmatrix} \partial\mathbf{N}/\partial y & -\mathbf{N} & \mathbf{0} \\ \partial\mathbf{N}/\partial x & \mathbf{0} & \mathbf{N} \end{bmatrix} \qquad (5.7)$$

In the above equations:

$$\begin{aligned} \mathbf{N} &= [N_1, N_2, N_3, N_4] \\ \partial\mathbf{N}/\partial x &= [\partial N_1/\partial x, \partial N_2/\partial x, \partial N_3/\partial x, \partial N_4/\partial x] \\ \partial\mathbf{N}/\partial y &= [\partial N_1/\partial y, \partial N_2/\partial y, \partial N_3/\partial y, \partial N_4/\partial y] \end{aligned} \qquad (5.8)$$

and $\mathbf{0}$ is a 4×1 null matrix. The generalised displacement vector is defined by

$$\boldsymbol{\delta}^e = [u_{z1} \ldots u_{z4}, \theta_{11} \ldots \theta_{14}, \theta_{21} \ldots \theta_{24}]^{\mathrm{T}} \qquad (5.9)$$

The stresses in the plate are related to the strains by

$$\begin{aligned} \boldsymbol{\sigma}_{\mathrm{B}} &= \mathbf{D}_{\mathrm{B}}\, \boldsymbol{\epsilon}_{\mathrm{B}} \\ \boldsymbol{\sigma}_{\mathrm{S}} &= \mathbf{D}_{\mathrm{S}}\, \boldsymbol{\epsilon}_{\mathrm{S}} \end{aligned} \qquad (5.10)$$

Here $\mathbf{D}_\mathrm{B}$ is the plane stress elasticity matrix and

$$\mathbf{D}_\mathrm{S} = \begin{bmatrix} G & 0 \\ 0 & G \end{bmatrix} C \tag{5.11}$$

where G is the shear modulus and C is a factor whose purpose is to improve the shear displacement approximation. It can be seen that in matrix $\mathbf{B}_\mathrm{S}$ the transverse shear terms are independent of z, i.e., the transverse shear is assumed to be constant across the thickness whereas in reality the distribution is approximately parabolic. A value of $C = 5/6$ is usually assumed and this is the ratio of the respective strain energies. The computation of the stiffness matrix follows usual steps as discussed for solid finite element, i.e.

$$\mathbf{K} = \iiint \mathbf{B}_\mathrm{B}^\mathrm{T} \mathbf{D} \mathbf{B}_\mathrm{B} \; \mathrm{d}x \; \mathrm{d}y \; \mathrm{d}z \; + \iiint \mathbf{B}_\mathrm{S}^\mathrm{T} \mathbf{D}_\mathrm{S} \mathbf{B}_\mathrm{S} \; \mathrm{d}x \; \mathrm{d}y \; \mathrm{d}z \tag{5.12}$$

The integration in z-direction can be performed analytically. The integrands which appear are

$$\int_{-\frac{t}{2}}^{+\frac{t}{2}} z^2 \; \mathrm{d}z = \frac{t^3}{12} \quad \text{and} \quad \int_{-\frac{t}{2}}^{+\frac{t}{2}} \mathrm{d}z = t \tag{5.13}$$

Integration in the other two directions is performed in the local ξ, η directions using Gauss quadrature. A 2×2 point integration is sufficient for a linear shape function. However, it was found[1] that if both the bending and transverse shear energy terms are integrated in this way the element is much too stiff especially in thin plate situations. It has been found that if the 'transverse shear' term is 'under-integrated' using a 1×1 point rule the element gives excellent results.

A more sophisticated element can be derived using higher order shape functions N_i. If a parabolic shape function is used the element will have 8 nodes and 24 degrees of freedom compared with the equivalent solid element which has 60 degrees of freedom.

5.3 SHELL ELEMENTS

Shells differ from plates by the fact that membrane action is significant and in many cases (i.e. thin shell applications) the dominant mode of carrying the load. In addition shells can no longer be assumed to be planar.

Figure 5.2 shows a distorted linear shell element where a third local axis, ζ, is assumed in the direction normal to the shell surface. The shell midsurface has $\zeta = 0$. Because the membrane action has to be taken into account the element has two additional degrees of freedom u_{xi} and u_{yi} at each node. Also

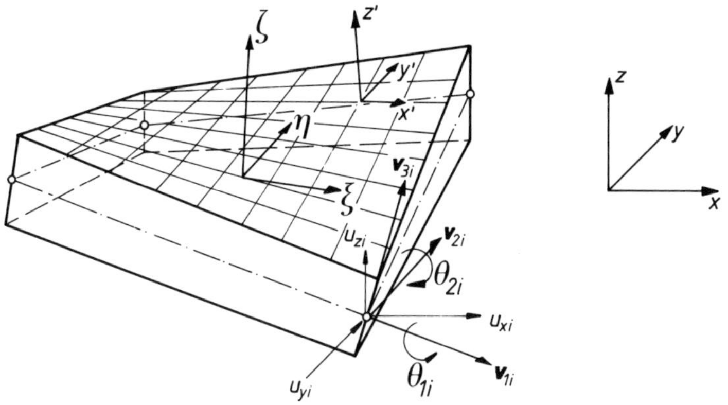

Figure 5.2 Linear shell element

since the shell may now be oriented in a general direction care has to be taken in the definition of the rotations θ_1 and θ_2. A local orthogonal vector system is established at each element node as shown in Figure 5.2.

Vectors $\mathbf{v}_1$ and $\mathbf{v}_2$ are determined in the following way: let $\mathbf{v}_3$ be the unit vector along the thickness of the shell as shown in the figure, then

$$\begin{aligned} \mathbf{v}_1 &= \mathbf{v}_3 \times \mathbf{v}_y \\ \mathbf{v}_2 &= \mathbf{v}_1 \times \mathbf{v}_3 \end{aligned} \tag{5.14}$$

where $\mathbf{v}_y$ is the unit vector in the y-direction. If $\mathbf{v}_3$ is exactly in $\mathbf{v}_y$ direction the cross product (5.14) gives a null vector and in this case $\mathbf{v}_1$ is computed from

$$\mathbf{v}_1 = \mathbf{v}_3 \times \mathbf{v}_x \tag{5.15}$$

where $\mathbf{v}_x$ is a unit vector in the x-direction.

The displacements in the shell can be written in terms of nodal generalised displacements as

$$\begin{Bmatrix} u_x \\ u_y \\ u_z \end{Bmatrix} = \sum N_i \begin{Bmatrix} u_{xi} \\ u_{yi} \\ u_{zi} \end{Bmatrix} - \sum N_i \frac{t}{2} \zeta \mathbf{v}_{2i}\, \theta_{1i} + \sum N_i \frac{t}{2} \zeta \mathbf{v}_{1i}\, \theta_{2i} \tag{5.16}$$

When defining the strains the basic shell assumption that the strains and stresses in the direction of the shell thickness are neglected has to be satisfied. This means that $\boldsymbol{\epsilon}$ has to be defined in a local system of axes x', y', z' where the z'-axis is parallel to the vector normal to the shell surface, $\mathbf{v}_3$, as defined in Section 2.8.

The orthogonal system of axes at point $\xi,\ \eta,\ \zeta$ is established in exactly the same way as the system of local axes at the nodal points for the definition of the rotations. The strains in the local x', y', z' directions, $\boldsymbol{\epsilon}'$, are obtained from the global strains by

$$\boldsymbol{\epsilon}' = \mathbf{T}_{\epsilon}^{-1}\boldsymbol{\epsilon} \tag{5.17}$$

The global strains are defined as

$$\boldsymbol{\epsilon} = \begin{Bmatrix} \epsilon_x \\ \epsilon_y \\ \epsilon_z \\ \gamma_{xy} \\ \gamma_{yz} \\ \gamma_{xz} \end{Bmatrix} \tag{5.18}$$

The strain in the local (shell) coordinate system is defined as:

$$\boldsymbol{\epsilon}' = \begin{Bmatrix} \epsilon_{x'} \\ \epsilon_{y'} \\ 0 \\ \gamma_{x'y'} \\ \gamma_{y'z'} \\ \gamma_{x'z'} \end{Bmatrix} \tag{5.19}$$

Note that $\epsilon_{z'}$ the strain normal to the shell surface is zero here. The strain transformation matrix $\mathbf{T}_{\epsilon}$ has been defined in Chapter 3.

The global strain components can be computed by differentiating the global displacements (5.16)

$$\boldsymbol{\epsilon} = \mathbf{B}\boldsymbol{\delta}^e \tag{5.20}$$

where $\boldsymbol{\delta}^e$ is now given by

$$\boldsymbol{\delta}^e = [u_{x1}\ldots u_{x4},\ u_{y1}\ldots u_{y4},\ u_{z1}\ldots u_{z4},\ \theta_{11}\ldots\theta_{14},\ \theta_{21}\ldots\theta_{24}]^{\mathrm{T}} \tag{5.21}$$

The local strain vector is then defined as

$$\boldsymbol{\epsilon}' = \mathbf{B}'\boldsymbol{\delta}^e \tag{5.22}$$

where

$$\mathbf{B}' = \mathbf{T}_{\epsilon}^{-1}\mathbf{B} \tag{5.23}$$

The strain can again be split up into bending and transverse shear terms

$$\epsilon_{\mathrm{B}} = \mathbf{B}'_{\mathrm{B}}\delta^e \tag{5.24}$$
$$\epsilon_{\mathrm{S}} = \mathbf{B}'_{\mathrm{S}}\delta^e \tag{5.25}$$

where $\mathbf{B}'_{\mathrm{B}}$ and $\mathbf{B}'_{\mathrm{S}}$ are obtained by partitioning the $\mathbf{B}'$ matrix as for the plate element.

The derivation of the element stiffness matrix proceeds in exactly the same manner as for the plate element, i.e. we obtain

$$\mathbf{K} = \iiint \mathbf{B}'^{\mathrm{T}}_{\mathrm{B}}\mathbf{D}\mathbf{B}'_{\mathrm{B}}|J|\ \mathrm{d}\xi\ \mathrm{d}\eta\ \mathrm{d}\xi + \iiint \mathbf{B}'^{\mathrm{T}}_{\mathrm{S}}\mathbf{D}_{\mathrm{S}}\mathbf{B}'_{\mathrm{S}}|J|\ \mathrm{d}\xi\ \mathrm{d}\eta\ \mathrm{d}\xi \tag{5.26}$$

For general shell applications, especially for thick shells the integration across the thickness can no longer be performed analytically and numerical Gauss quadrature is used.

More complicated geometries may be modelled if parabolic shape functions are used[6]. Such an element has 8 nodes and 40 degrees of freedom compared with 20 nodes and 60 degrees of freedom for the generic solid element. The derivation of such higher order elements is identical to the one just presented except that parabolic shape functions N_i are used.

5.4 EXAMPLES AND APPLICATIONS OF PLATE AND SHELL ELEMENTS

Here we will present a few examples of application of the plate and shell element. First we deal with a few simple examples where analytical solutions are available and which allow us to ascertain the accuracy obtainable with a given mesh of elements.

5.4.1 Clamped plate subjected to concentrated load

Figure 5.3 shows the results for a clamped plate subjected to a concentrated load at its centre. The results are given for 4, 16 and 64 elements per quarter. It can be seen that the convergence is quite rapid and that the accuracy of the results even with only a few elements is quite good.

5.4.2 Cylindrical roof

Here the membrane action has been included in the analysis. The different discretisations into eight-node (parabolic) shell elements are shown in

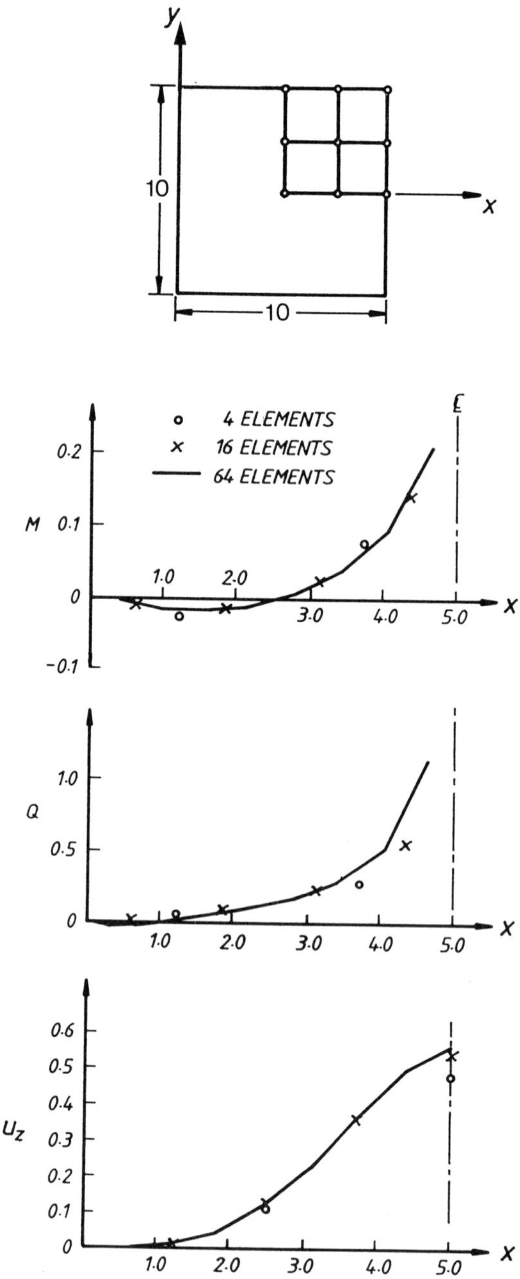

Figure 5.3 Moment (M), shear resultant (Q) and displacement (u_z) plotted against coordinate x for a clamped square plate subjected to a concentrated load

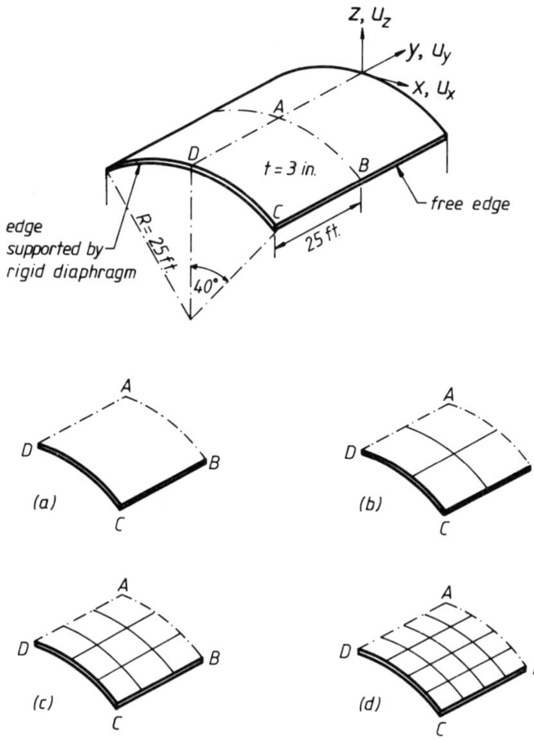

Mesh type	Number of degrees-of-freedom
(a)	23
(b)	76
(c)	159
(d)	272

Figure 5.4 Cylindrical roof. Assumptions and meshes used

Figure 5.4. Only 1/4 of the roof has been considered and boundary conditions applied as shown. Figure 5.5 shows the displacements obtained for the load case self weight and these are compared with an analytical solution. Again, even a coarse mesh gives a fairly good agreement.

5.4.3 Pinched cylinder

In this example a cylinder is subjected to concentrated (pinching) forces. Figure 5.6 shows the problem geometry with the symmetry conditions and the discretisation used for a quarter of the cylinder. Figure 5.7 shows the results of the analysis.

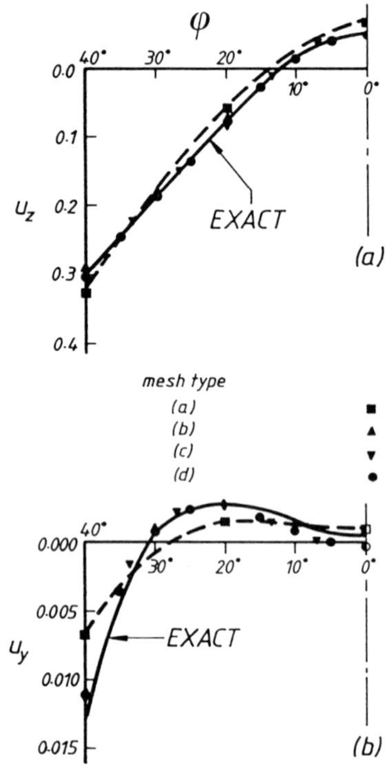

Figure 5.5 (a) Vertical displacement at midsection and (b) axial displacement at support for roof under its own weight

5.4.4 Studio with ferro–concrete roof

This is a practical application to a structural engineering problem and related to the design of a studio with a ferro-concrete roof. Figure 5.8 shows the geometry and the mesh used for the analysis. The mesh consists of eight-node shell elements and truss elements. A result of the analysis namely the deflected shape due to wind loading is shown in Figure 5.9.

5.5 JOINT ELEMENTS

There are some cases where the displacements are discontinuous between two elements, i.e., the displacement values are different at the same node for each of the elements connecting to it. Examples occur in geomechanics where the rock mass is intersected by joints or faults, in mechanical engineering where two machine components are joined together and in civil engineering where there is an interaction between a soil mass and a structure. These

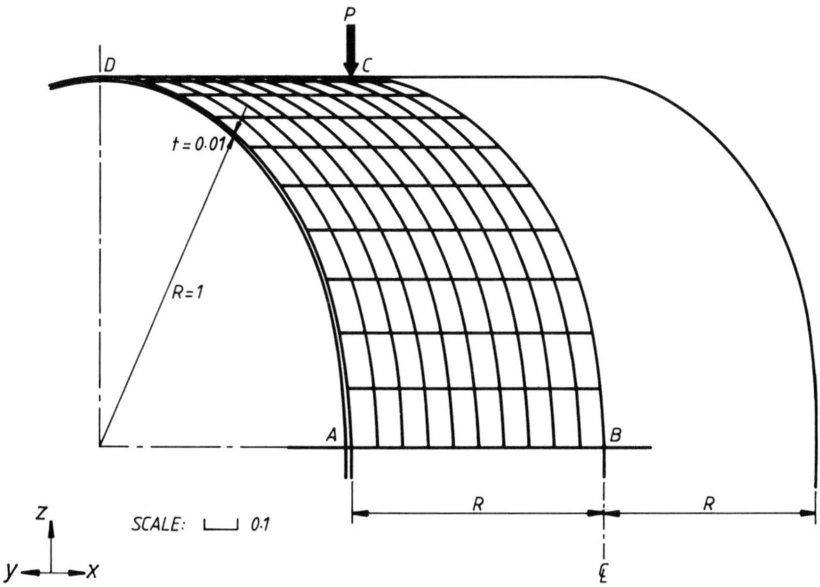

Figure 5.6 Shell test example: pinched cylinder 10 × 10 mesh

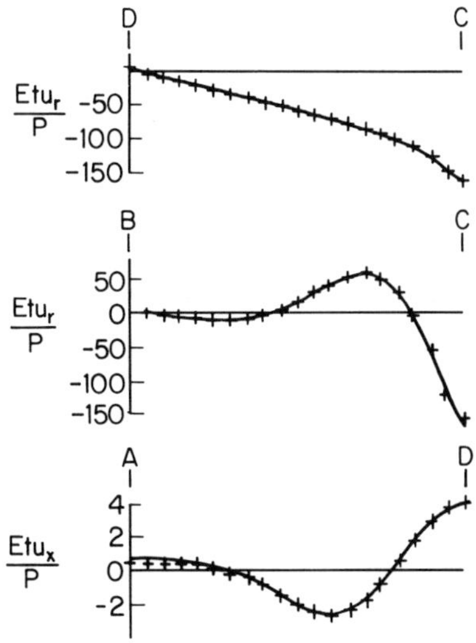

Figure 5.7 Results of analysis for pinched cylinder. (u_r = radial displacement. u_x = axial displacement). (— theory, + 10 × 10 parabolic shell elements)

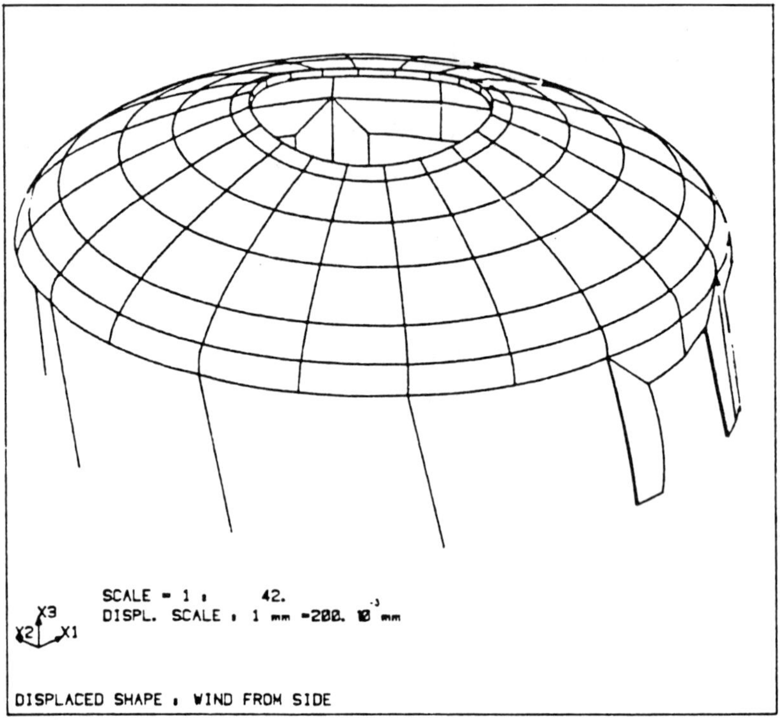

Figure 5.8 Ferro-concrete roof: mesh

problems are dealt with by connecting two elements via special joint elements. Early formulations of joint elements were due to Goodman[9], Ghaboussi[10] and Zienkiewicz[11]. The joint element developed here is an extension of the basic theory to isoparametric two and three-dimensional elements.

5.5.1 Computation of stiffness matrix

A two-dimensional linear joint element is shown in Figure 5.10. The element has four nodes, two of which are connected to the 'top' element and two to the 'bottom' element. There are two possible displacement modes. These are:

Δ_s slip

Δ_n opening/closing

They are defined as relative movements of the 'top' and 'bottom' elements in the local directions x', y':

$$\begin{aligned} \Delta_s &= u_{x'}^{\text{top}} - u_{x'}^{\text{bot}} \\ \Delta_n &= u_{y'}^{\text{top}} - u_{y'}^{\text{bot}} \end{aligned} \tag{5.28}$$

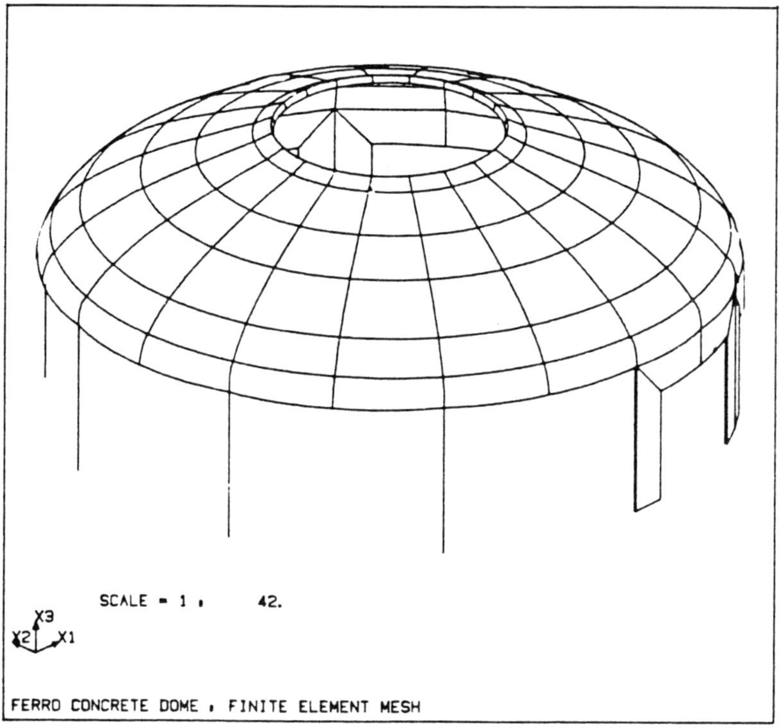

Figure 5.9 Ferro-concrete roof: displaced shape due to wind

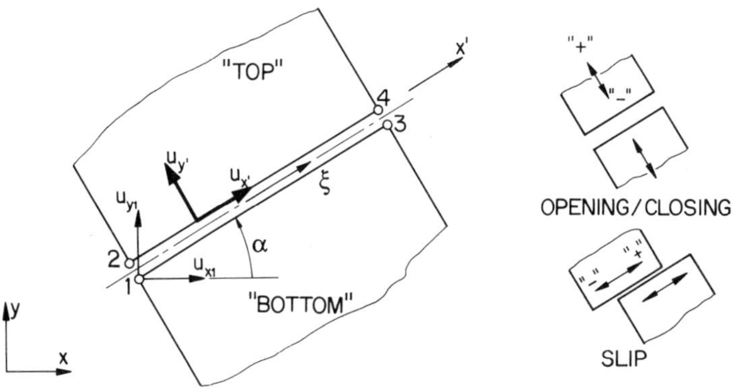

Figure 5.10 Two-dimensional linear joint element

or in matrix notation

$$\boldsymbol{\Delta} = \begin{Bmatrix} \Delta_s \\ \Delta_n \end{Bmatrix} = \begin{Bmatrix} u_{x'} \\ u_{y'} \end{Bmatrix}_{\text{top}} - \begin{Bmatrix} u_{x'} \\ u_{y'} \end{Bmatrix}_{\text{bot}} \tag{5.29}$$

The displacements in the local directions are related to the global displacements by

$$\begin{Bmatrix} u_{x'} \\ u_{y'} \end{Bmatrix} = \mathbf{T}_g^{\mathrm{T}} \begin{Bmatrix} u_x \\ u_y \end{Bmatrix} \tag{5.30}$$

where $\mathbf{T}_g^{\mathrm{T}}$ is the geometrical transformation matrix.

The relative joint displacements can now be expressed in terms of global node displacements as

$$\begin{Bmatrix} \Delta_s \\ \Delta_n \end{Bmatrix} = \mathbf{T}_g^{\mathrm{T}} \left(\begin{Bmatrix} u_x \\ u_y \end{Bmatrix}_{\text{top}} - \begin{Bmatrix} u_x \\ u_y \end{Bmatrix}_{\text{bot}} \right) \tag{5.31}$$

The joint stress pseudo-vector $\mathbf{t}$ consists of the normal stress t_n and the shear stress t_s.

$$\mathbf{t} = \begin{Bmatrix} t_s \\ t_n \end{Bmatrix} \tag{5.32}$$

The stresses are related to the relative displacements by:

$$\mathbf{t} = \mathbf{k}\boldsymbol{\Delta} \tag{5.33}$$

where the constitutive matrix is defined as

$$\mathbf{k} = \begin{bmatrix} k_s & 0 \\ 0 & k_n \end{bmatrix} \tag{5.34}$$

Here k_{n} is the normal and k_{s} the shear stiffness of the joint.

The displacements at a particular point inside the element with the coordinate ξ can be expressed in terms of nodal displacements using the linear interpolation functions N_1, N_2:

$$\begin{Bmatrix} u_x \\ u_y \end{Bmatrix}_{\text{top}} = N_1 \begin{Bmatrix} u_{x2} \\ u_{y2} \end{Bmatrix} + N_2 \begin{Bmatrix} u_{x4} \\ u_{y4} \end{Bmatrix} \tag{5.35}$$

$$\begin{Bmatrix} u_x \\ u_y \end{Bmatrix}_{\text{bot}} = N_1 \begin{Bmatrix} u_{x1} \\ u_{y1} \end{Bmatrix} + N_2 \begin{Bmatrix} u_{x3} \\ u_{y3} \end{Bmatrix} \tag{5.36}$$

where the linear shape functions from Chapter 2 are given by

$$
\begin{aligned}
N_1 &= \frac{1}{2}(1-\xi) \\
N_2 &= \frac{1}{2}(1+\xi)
\end{aligned}
\tag{5.37}
$$

The joint relative displacements can be expressed in terms of the nodal displacements by

$$\boldsymbol{\Delta} = \mathbf{B}\ \mathbf{u}^e \tag{5.38}$$

where

$$\mathbf{u}^e = [\mathbf{u}_1,\ \mathbf{u}_2,\ \mathbf{u}_3,\ \mathbf{u}_4]^{\mathrm{T}} \tag{5.39}$$

and

$$\mathbf{B} = \mathbf{T}_{\mathrm{g}}^{\mathrm{T}}\mathbf{N} \tag{5.40}$$

$$\mathbf{N} = \begin{bmatrix} -N_1 & 0 & N_1 & 0 & -N_2 & 0 & N_2 & 0 \\ 0 & -N_1 & 0 & N_1 & 0 & -N_2 & 0 & N_2 \end{bmatrix}$$

The stiffness matrix for the joint element is obtained in the same manner as for the solid finite elements by minimisation of the total potential energy.

The strain energy is obtained by multiplying the joint stresses with the joint displacements.

$$U = \frac{1}{2}\int_S (t_n\Delta_n + t_s\Delta_s)\ \mathrm{d}S \tag{5.41}$$

Equation 5.41 can be written in matrix notation as

$$U = \frac{1}{2}\int_S \mathbf{t}^{\mathrm{T}}\boldsymbol{\Delta}\ \mathrm{d}S \tag{5.42}$$

Substitution of (5.38) and (5.33) gives for the strain energy

$$U = \frac{1}{2}(\mathbf{u}^e)^{\mathrm{T}}\left(\int \mathbf{B}^{\mathrm{T}}\mathbf{k}\mathbf{B}\ \mathrm{d}S\right)\ \mathbf{u}^e \tag{5.43}$$

The reader may verify that the stiffness matrix of the joint element is obtained as

$$\mathbf{K}^e = \int_{-1}^{+1} \mathbf{B}^{\mathrm{T}}\mathbf{k}\mathbf{B}J \, \mathrm{d}\xi = \int \mathbf{N}^{\mathrm{T}}\mathbf{T}_{\mathrm{g}}\mathbf{k}\mathbf{T}_{\mathrm{g}}^{\mathrm{T}}\mathbf{N}J \, \mathrm{d}\xi \tag{5.44}$$

or

$$\mathbf{K}^e = \int \mathbf{N}^{\mathrm{T}}\mathbf{k}'\mathbf{N}J \, \mathrm{d}\xi \tag{5.45}$$

where

$$\mathbf{k}' = \mathbf{T}_{\mathrm{g}}\mathbf{k}\mathbf{T}_{\mathrm{g}}^{\mathrm{T}} \tag{5.46}$$

and J is the Jacobian (see Chapter 2).

A parabolic joint element which can model curved interfaces can be derived (Figure 5.11) by substituting parabolic shape functions N_1, N_2, N_3 instead of the linear funtions N_1, N_2 in equation (5.37).

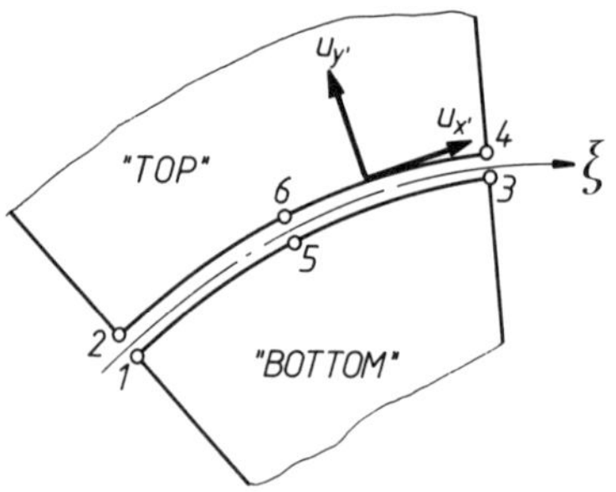

Figure 5.11 Two-dimensional parabolic joint element

The joint element theory can be expanded to three dimensions as shown in Figure 5.12 where two solid elements are connected by a joint element. There are now three possible displacement modes:

Δ_n opening/closing
Δ_{s_1} slip in direction 1
Δ_{s_2} slip in direction 2

The displacement vector $\mathbf{\Delta}$ is now given by

$$\mathbf{\Delta} = \begin{Bmatrix} \Delta_{\mathrm{s}_1} \\ \Delta_{\mathrm{s}_2} \\ \Delta_{\mathrm{n}} \end{Bmatrix} = \begin{Bmatrix} u_{x'} \\ u_{y'} \\ u_{z'} \end{Bmatrix}_{\mathrm{top}} - \begin{Bmatrix} u_{x'} \\ u_{y'} \\ u_{z'} \end{Bmatrix}_{\mathrm{bot}} \tag{5.47}$$

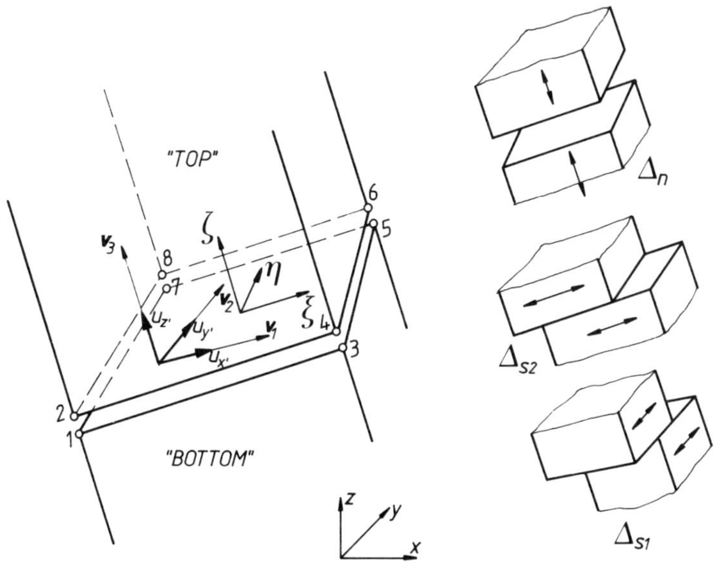

Figure 5.12 Three-dimensional linear joint element

The primed displacements are obtained from the global displacements by the usual geometry transformation

$$\begin{Bmatrix} u_{x'} \\ u_{y'} \\ u_{z'} \end{Bmatrix} = \mathbf{T}_g^{\mathrm{T}} \begin{Bmatrix} u_x \\ u_y \\ u_z \end{Bmatrix} \tag{5.48}$$

There are now three joint stress components one normal and two shear stresses in the directions of $\mathbf{v}_1$ and $\mathbf{v}_2$

$$\mathbf{t} = \begin{Bmatrix} t_{s_1} \\ t_{s_2} \\ t_n \end{Bmatrix} \tag{5.49}$$

The constitutive matrix $\mathbf{k}$ is now defined as

$$\mathbf{k} = \begin{bmatrix} k_{s_1} & 0 & 0 \\ 0 & k_{s_2} & 0 \\ 0 & 0 & k_n \end{bmatrix} \tag{5.50}$$

The displacements at a point with coordinates ξ and η on the top and bottom surface of the joint element can be expressed in terms of nodal displacements as:

$$\begin{aligned}\mathbf{u}_{\text{top}} &= N_1\mathbf{u}_2 + N_2\mathbf{u}_4 + N_3\mathbf{u}_6 + N_4\mathbf{u}_8\\ \mathbf{u}_{\text{bot}} &= N_1\mathbf{u}_1 + N_2\mathbf{u}_3 + N_3\mathbf{u}_5 + N_4\mathbf{u}_7\end{aligned} \tag{5.51}$$

The linear shape functions are taken from Chapter 2 to be

$$\begin{aligned}N_1 &= \frac{1}{4}(1-\xi)(1-\eta)\\ N_2 &= \frac{1}{4}(1+\xi)(1-\eta)\\ N_3 &= \frac{1}{4}(1+\xi)(1+\eta)\\ N_4 &= \frac{1}{4}(1-\xi)(1+\eta)\end{aligned} \tag{5.52}$$

The $\mathbf{B}$ matrix can now be calculated in the same way as before

$$\mathbf{B} = \mathbf{T}_g^{\mathrm{T}}\mathbf{N} \tag{5.53}$$

where

$$\mathbf{N} = [-N_1\mathbf{I},\ N_1\mathbf{I},\ -N_2\mathbf{I},\ N_2\mathbf{I},\ -N_3\mathbf{I},\ N_3\mathbf{I},\ -N_4\mathbf{I},\ N_4\mathbf{I}] \tag{5.54}$$

where $\mathbf{I}$ is a 3×3 unit matrix.

The derivation of the stiffness matrix follows exactly the same steps as for the two-dimensional element.

The formulation can be extended to include parabolic shape functions for modelling of curved interfaces (Figure 5.13). Again, except for different shape functions used and a larger number of nodes identical expressions for the stiffness matrix are obtained.

As shown in reference 7 it is possible to derive a generic joint element which can model two- and three-dimensional interfaces and connect to linear and parabolic elements. With this formulation it is also possible to compute the stiffness matrix of joint elements which connect solid elements to shell elements and shell/plate elements to each other.

5.5.2 Non-linear joint behaviour

In most applications of joint elements the relationship between joint displacement and joint stress is not linear. For example the normal stiffness of the joint can change from an infinite value when it is completely closed to

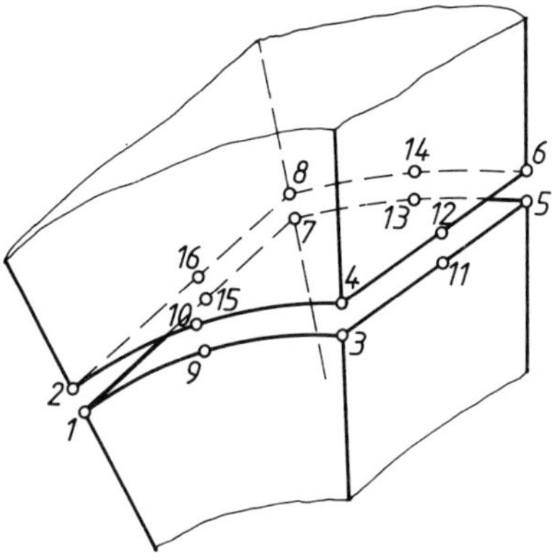

Figure 5.13 Three-dimensional parabolic joint element

a zero value when the joint is opening. In describing the relationship between joint normal stress and displacement one may assume a hyperbolic curve as shown in Figure 5.14a which is asymptotic to a line parallel to the t_n axis with an offset of Δ_{nc}. We may think of this as the 'maximum joint closure', i.e. the maximum amount the joint can close. This may be related to the thickness of any joint infill or to the roughness of the interface. In rock mechanics for example, we may model faults which might be up to 1/2 m thick and may have fill material or rough joint surfaces where asperities have to be crushed before complete closure of the joint. For a joint in compression, i.e. $\Delta_n < 0$ the following relation can be written

$$t_n = \frac{-1}{(\Delta_{nc} + \Delta_n)} + \frac{1}{\Delta_{nc}} \tag{5.55}$$

As can be seen this gives $t_n = \infty$ for $\Delta_n = -\Delta_{nc}$ (Δ_{nc} the maximum joint closure has no sign) and $t_n = 0$ for $\Delta_n = 0$.

The normal stiffness of the joint can be computed by

$$k_n = \frac{\mathrm{d}t_n}{\mathrm{d}\Delta_n} = \frac{1}{(\Delta_{nc} + \Delta_n)^2} \tag{5.56}$$

This means that the stiffness is infinite if $\Delta_n = -\Delta_{nc}$. In a numerical implementation it is unacceptable to have an infinite value for the stiffness. Instead we must limit the value to a peak value k_n^p. It is also desirable to be able to have control over the shape of the curve, i.e. we may want a rapid

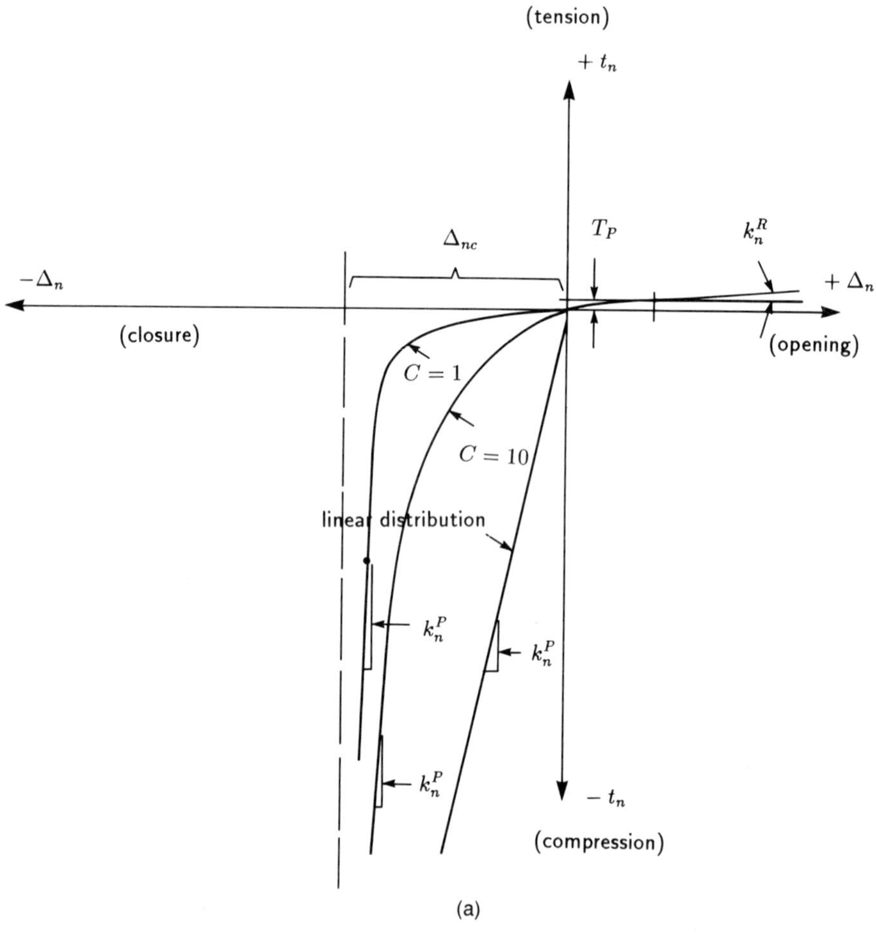

Figure 5.14 (a) Non-linear relationship between joint tractions and displacements

change of stiffness for joints with no infill and a slower change for joints with fill. An alternative relationship can be written as

$$t_n = -\left(\frac{1}{(\Delta_{nc} + \Delta_n)} - \frac{1}{\Delta_{nc}}\right)\frac{C}{k_n^p} \tag{5.57}$$

where C is a curve constant and k_n^p is the peak normal stiffness. Figure 5.14 shows curves for various values of C.

For joint opening the value of stiffness drops rapidly to zero but for a numerical implementation a small value of stiffness must be assumed. For joint opening, i.e. $\Delta_n > 0$ we have

$$t_n = k_n^{\mathrm{R}} \Delta_n \tag{5.58}$$

where k_n^{R} is a residual stiffness.

Summarising the above we can write

$$\begin{aligned} k_{\mathrm{n}} &= \frac{1}{(\Delta_{nc} + \Delta_n)^2} \qquad &\text{for } \Delta_n < 0 \\ k_n &= k_n^{\mathrm{R}} \qquad &\text{for } \Delta_n > 0 \end{aligned} \tag{5.59}$$

The joint behaviour in shear is also non-linear but the change in stiffness will be more rapid. A joint under compression will have very little movement until a critical shear stress is reached when joint slip commences. Using a Mohr–Coulomb yield condition to determine the onset of slip, this occurs when

$$|t_s| = -t_n \tan\phi + c \tag{5.60}$$

where ϕ is the angle of friction, c the cohesion and t_s is the resultant shear stress (i.e. $t_s = \sqrt{t_{s1}^2 + t_{s2}^2}$ for three-dimensional joints). Before this critical stress is reached the slip occurring in the joint plane is elastic and recoverable and the shear stiffness is high. Theoretically for a perfectly smooth joint with zero thickness it is infinite but in practice we assume a finite value to account for example for asperities, infill etc. When the critical shear stress has been reached the shear stiffness theoretically drops to zero until the stress falls below the critical level either by an increase in normal stress or a drop in shear stress. In numerical implementation we have to assume a finite value for the joint stiffness in this case.

Summarising we have

$$\begin{aligned} k_s &= k_s^{\mathrm{P}} \qquad &|t_s| < -t_n \tan\phi + c \\ k_s &= k_s^{\mathrm{R}} \qquad &|t_s| = -t_n \tan\phi + c \end{aligned} \tag{5.61}$$

In classical plasticity theory the shear stress is not allowed to be greater than $t_s^{\mathrm{c}} = -t_n \tan\phi + c$ so a stress correction has to be applied in an incremental solution as explained below.

5.5.3 Solution algorithms for non-linear joints

Non-linear joints are modelled by an incremental method as will be discussed in detail in Chapter 9. Initially all joints are assumed to be elastic and have peak normal and shear stiffness. The loading is applied in increments and at each load increment the stress at points inside each element is checked.

These points may conveniently be taken as the Gauss integration points used for the computation of the stiffness matrix. If the normal stress is reducing then a new value of normal stiffness should have been applied during the load increment. We can calculate the stiffness at the end of the load increment using equations (5.59) and change the value of k_n in equation (5.34) or (5.50) for the recomputation of the joint stiffness matrix for the next load increment. However, because we have kept the stiffness constant throughout the increment we have moved in a tangential direction off the non-linear curve.

Assume that the normal stress at the beginning of the load increment was t_n^1 and that after the load increment has been applied a relative normal displacement of $d\Delta_n^1$ occurs, resulting in a computed normal stress of t_n^2.

This stress is off the t_n, Δ_n curve as shown in Figure 5.14b. A correction of Δt_n^{P} is needed to ensure that the point is on the curve. The value of Δt_n^{P} is computed by subtracting the value t_{n}^2 as obtained by the analysis from the value of $t_n^{2\mathrm{p}}$ obtained by substituting Δ_n^1 into equations (5.55) or (5.57).

A similar procedure is applied to the shear stress. If the shear stress at the end of the increment is larger than the critical value i.e.

$$t_s^2 > t_s^{\mathrm{c}} \tag{5.62}$$

then a correction has to be applied because the stress can at most be equal to but not larger than t_s^{c}. The correction is simply

$$\Delta t_s^{\mathrm{P}} = t_s - t_s^{\mathrm{c}} \tag{5.63}$$

In addition the shear stiffness k_s has to be changed to k_s^{R}.

For three-dimensional joints the resultant plastic shear stress increment Δt_s^{P} has to be split up into x' and y' components $\Delta t_{s1}^{\mathrm{P}}$ and $\Delta t_{s2}^{\mathrm{P}}$. One may write

$$\begin{aligned} \Delta t_{s1}^{\mathrm{P}} &= \Delta t_s^{\mathrm{P}} \frac{t_{s1}}{t_s} \\ \Delta t_{s2}^{\mathrm{P}} &= \Delta t_s^{\mathrm{P}} \frac{t_{s2}}{t_s} \end{aligned} \tag{5.64}$$

5.5.4 Numerical implementation

In a numerical implementation we check each Gauss integration point if a point has reached the elastic limit either in the shear or normal direction. If this is the case we compute the contribution to the equivalent nodal forces for the joint element.

The equivalent nodal forces due to an increment of stress $\Delta \mathbf{t}^{\mathrm{P}}$ inside a joint element are determined by minimising the potential energy.

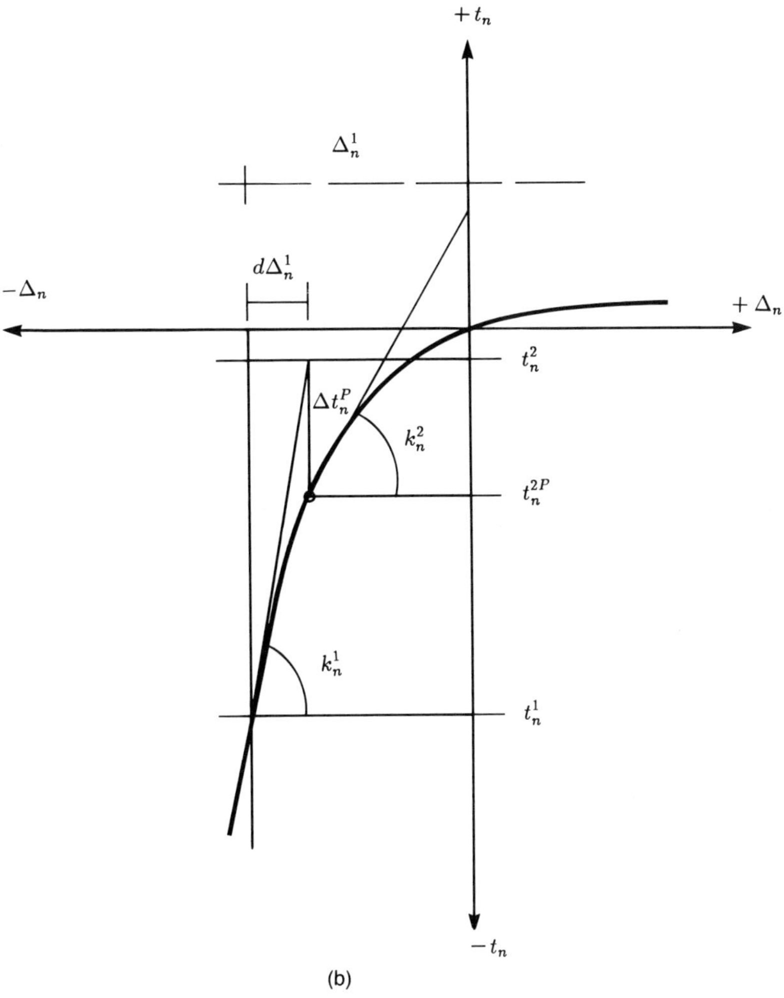

Figure 5.14 (b) This diagram follows the normal stress loading of one point in the joint plane

The work done by the forces acting on the nodes of the joint element is given by

$$W_e = (\mathbf{u}^e)^{\mathrm{T}} \ \Delta\mathbf{F}^e \tag{5.65}$$

where $\mathbf{u}^e$ is as defined previously and $\mathbf{F}^e$ is a vector of nodal forces for element e:

$$\Delta\mathbf{F}^e = [\mathbf{F}_1, \mathbf{F}_2, \mathbf{F}_3, \ldots]^{e\mathrm{T}} \tag{5.66}$$

The internal work done by $\mathbf{\Delta t}^{\mathrm{P}}$ is

$$U = \int_S \mathbf{\Delta}^{\mathrm{T}} \left(\Delta \mathbf{t}^{\mathrm{P}}\right) \, \mathrm{d}S$$
$$= \int_S \left(\mathbf{B}\mathbf{u}^e\right)^{\mathrm{T}} \left(\Delta \mathbf{t}^{\mathrm{P}}\right) \, \mathrm{d}S \tag{5.67}$$

The equivalent nodal point forces for the joint element are given by

$$\Delta \mathbf{F}^e = \int_S \mathbf{B}^{\mathrm{T}} \left(\Delta \mathbf{t}^{\mathrm{P}}\right) \, \mathrm{d}S \tag{5.68}$$

The formula for numerical integration by Gauss quadrature is for a one-dimensional element and L Gauss points

$$\Delta \mathbf{F}^e = \sum_{l=1}^{L} W_l . \mathbf{B}^{\mathrm{T}} (\xi_l) \left(\Delta \mathbf{t}^{\mathrm{P}} (\xi_l)\right) J (\xi_l) \tag{5.69}$$

In the numerical implementation an 'excess stress' $\Delta \mathbf{t}^{\mathrm{P}}$ is computed at each Gauss point (if the point is elastic and not opening or closing this may be zero). For each Gauss point which has changed state a new stiffness coefficient is computed.

Nodal point forces are then computed according to equation (5.69) and a new stiffness matrix according to equation (5.45). The residual nodal point forces are applied at the next iteration or time step.

It is time consuming to recompute the joint stiffness matrix at each iteration step or increment because the system of equations has to be solved each time the global stiffness matrix changes, that is instead of a solution with a new right hand side a complete Gauss elimination has to be performed.

To avoid this we may recompute the element stiffness matrix every second or third time step and then only if large changes have taken place. The convergence of the iterations will be slower but this is compensated by less time spent in equation solving.

Some problems, in particular those involving joint opening after closure, where the joint normal stiffness changes dramatically are very sensitive to the choice of the time increment and small time steps must be chosen.

5.6 EXAMPLES AND APPLICATIONS OF JOINT ELEMENTS

A few applications of the joint element presented here are discussed. The first two applications deal with problems in geomechanics and the modelling of jointed rock. The last example deals with a soil–structure interaction problem.

5.6.1 Modelling of shear behaviour

This example shows how shear displacement of a rock joint can be modelled. Two excavations in rock are separated by a pillar of rock which is transected by a joint with a 30° inclination.

For this analysis a combination of finite and boundary elements is used as will be explained later. The pillar is divided into linear finite elements with linear joint elements connecting the elements above and below the joint.

The material properties assumed were

Rock mass	E = 80000 MPa
	$\nu = 0.3$
Joint	$k_n = 10^7$ MPa/m
	$k_s = 10^7$ MPa/m
	$\phi = 10°$
	c = 0

The pillar was loaded by applying excavation forces to the boundary element mesh. This excavation force was applied in several time steps and the displaced shape of the pillar after the last time step and after equilibrium has been reached is shown in Figure 5.15b. The discontinuous displacement pattern caused by the joint can be clearly seen.

5.6.2 Modelling of joint slip and opening

This is a more complicated problem as both slip and joint opening have to be modelled. Here one of the walls of an excavation which is inclined by 70° to the horizontal (hangingwall) contains joint planes which are parallel to the excavation surface. In the past overbreak has occurred and it was observed that this overbreak was preceded by the opening of the joints. The aim of the analysis was to predict the mechanism of joint opening so that measures could be devised on how to improve the excavation design.

The following material properties were assumed.

Rock mass	E = 80000 MPa	
	$\nu = 0.3$	
Joint planes	$k_n^{\mathrm{P}} = 10^7$ MPa/m	peak
	$k_s^{\mathrm{P}} = 10^7$ MPa/m	values
	$k_n^{\mathrm{R}} = 10$ MPa/m	residual
	$k_s^{\mathrm{R}} = 10$ MPa/m	values
Maximum joint closure	$\Delta_{nc} = 0.01$ mm	
Curve parameter	$C = 1$	
Angle of friction	$\phi = 10°$	
Cohesion	$c = 0$	

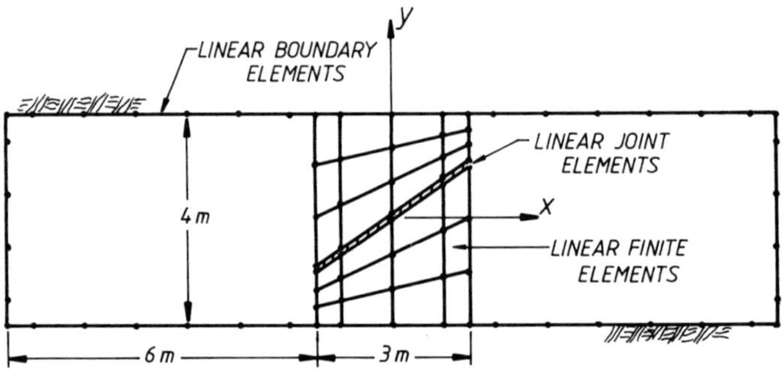

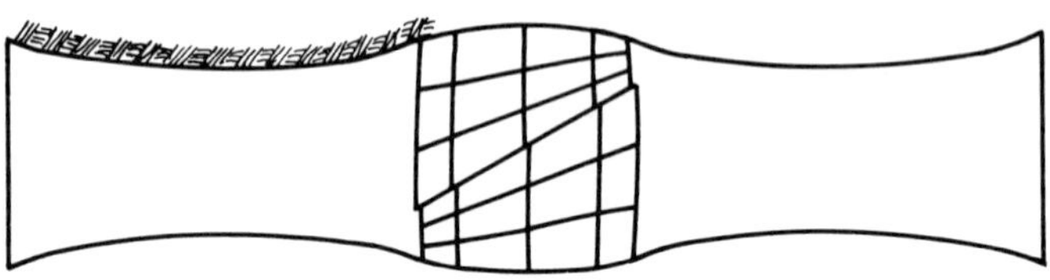

Figure 5.15 A mine excavation with a pillar transected by a joint at 30° (plane strain analysis). (a) Finite element mesh. (b) Displaced shape showing slip on interface

The excavation was simulated by removing the finite elements and generating out of balance forces as explained in Chapter 4. All joint elements were prestressed by the initial stress field previously in the rock mass. At the bottom of the excavation the initial stresses were assumed to be:

$$\begin{aligned}\sigma_x &= -18 \text{ MPa}\\ \sigma_y &= -25 \text{ MPa}\\ \sigma_z &= -36 \text{ MPa}\\ \tau_{xy} &= \tau_{yz} = \tau_{xz} = 0\end{aligned}$$

The joints were assumed to be totally closed initially and to have a peak normal and shear stiffness.

After excavation significant shear stress occurred at the abutment and this was followed by slip. Consequently joints opened near the centre of the excavation.

Figure 5.16b shows the contours of joint opening on the plane located about 1 metre from the excavation surface (i.e. 1 m into the hangingwall.)

(a)

(b)

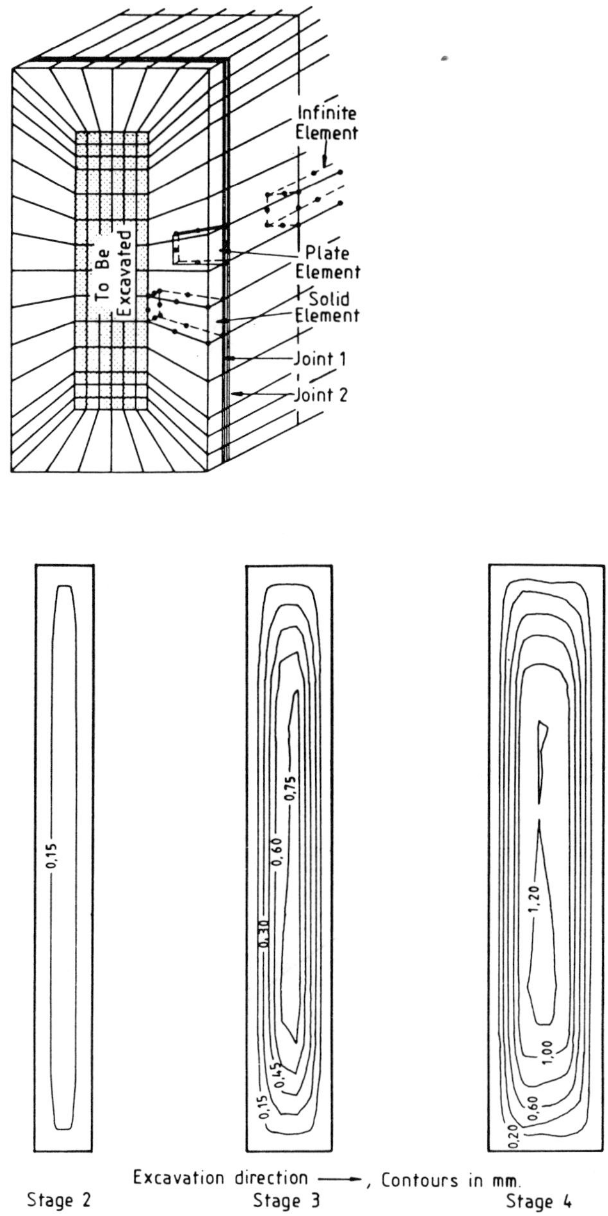

Figure 5.16 Hangingwall of a mine excavation with parallel jointing. (a) Mesh with solid and plate finite elements connected by joint elements. (b) Contours of joint opening at different excavation stages

5.6.3 Modelling of soil structure interaction

This is the classical problem of a plate on an elastic foundation and involves both slip and joint opening.

A plan and section of the mesh used for the analysis is shown in Figure 5.17. The mesh consists of plate elements as discussed previously and solid finite elements. The ratio of elastic modulus of the plate to the one for the elastic foundation was assumed to be 40. The Poisson's ratio is assumed to be zero. A concentrated load P is applied at the middle of the plate. Three cases are considered: (1) the interface between plate and halfspace does not allow slip or separation to develop (rigid interface); (2) separation is allowed but no slip (no slip interface) and (3) the interface allows slip and separation to develop unhindered (smooth surface). The predicted contact pressures for all three cases are shown in Figure 5.17b.

5.7 CLOSURE

In this chapter we have discussed the derivation of finite elements which have been developed for special applications. The plate/shell elements, for example, not only lead to a reduction in the number of degrees of freedom compared with solid elements due to the simplified assumptions of the distribution of strain across the thickness but also to simplifications in data input since a plate/shell element can be described by its geometrical form and thickness only. In addition, these examples allow the analysis of very thin shells where the conventional three-dimensional solid element could lead to numerical difficulties.

Another special element—the joint element—allows discontinuous behaviour to take place between finite elements. This is needed, for example, in situations where no rigid connection exists between elements or group of elements where the material is transected by pre-existing fractures or joints. At an interface the nodal point displacements are no longer unique and different values are obtained for the elements of 'top' and 'bottom' of the joint. This is because differential slip and separation is allowed to take place. With these special elements the range of application of the finite element method is greatly enhanced.

5.8 EXERCISES

5.1 Name examples where shell elements cannot be used, i.e., where the shell assumptions lead to errors in the results.

5.2 For the linear shell element determine the $\mathbf{B}$ matrix in equation (5.20).

5.3 For the linear shell element determine $\mathbf{B}_{\mathrm{B}}$ and $\mathbf{B}_{\mathrm{S}}$ by partitioning $\mathbf{B}$.

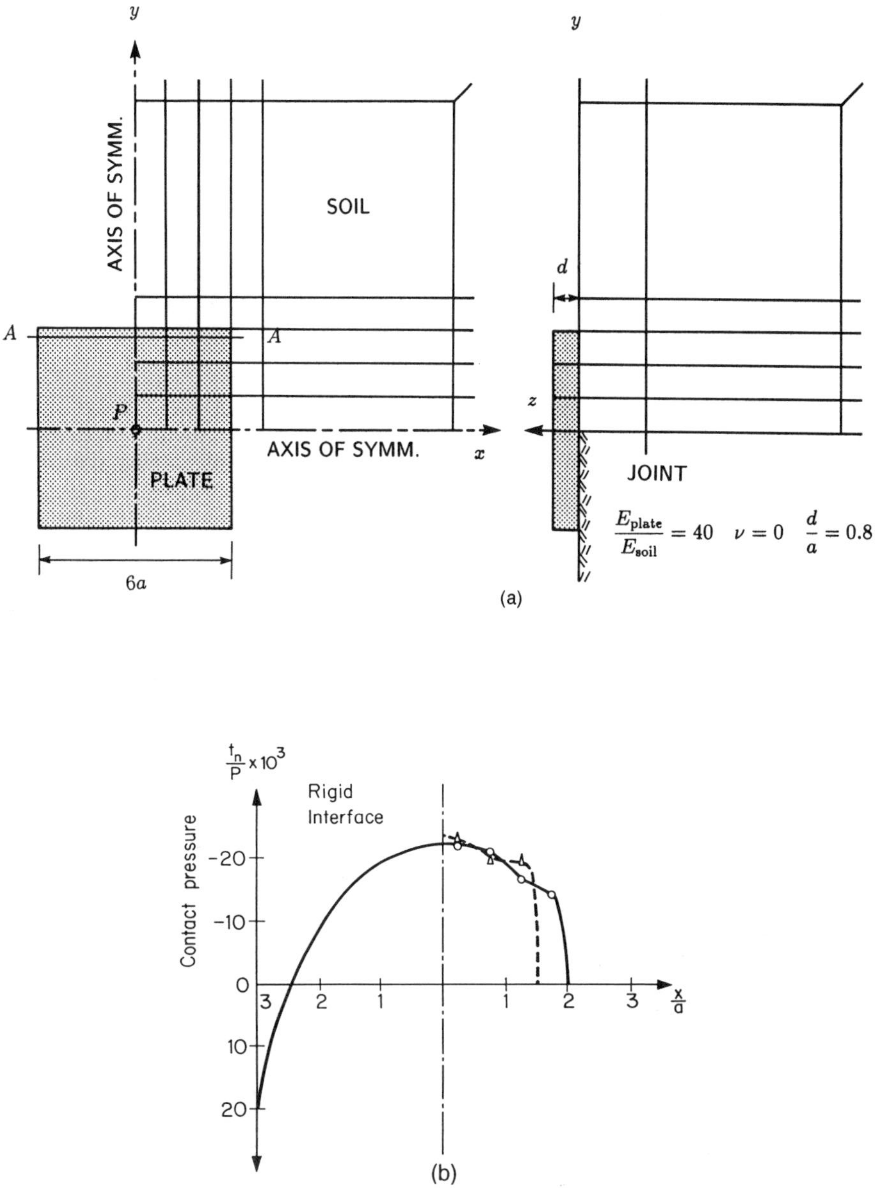

Figure 5.17 Plate on elastic foundation. (a) Finite element mesh with plate and solid elements and (b) contact pressure t_n at the interface due to concentrated load P. (◯ no slip interface △ smooth interface)

5.4 Show how the value $C = 5/6$ in equation (5.11) is obtained.

5.5 List the minimum input data required to specify a shell element:
(a) if it has uniform thickness,
(b) if its thickness varies.

5.6 Derive the matrix $\mathbf{N}$ in equation (5.40) for a parabolic joint element.

5.7 Derive the matrix $\mathbf{N}$ in equation (5.54) for a three-dimensional parabolic joint element.

5.9 REFERENCES

1. Hughes, R.J.R., Taylor, R.L. and Kanoknukulchai, W. (1977). A simple and efficient element for plate bending, *Int. J. Num. Meth. Eng.*, **11**, 1529–1543.
2. Jirousek, T. and Nicos, L. (1977). A powerful finite element for plate bending, *Comp. Meths. Appl. Mech. Eng.*, **12**, 77–96.
3. Gallagher, R. H. (1975). Shell elements, Proc. *First World Congress on Finite Element Methods in Structural Mechanics*, Bournemouth, U.K.
4. Argyris, J.H., *et al.* (1977). A simple triangular facet shell element with applications to linear and nonlinear equilibrium and elastic stability problems, *Comput. Meths. Appl. Mech. Eng.*, **10**, 371–403.
5. Hinton, E., *et al.* (1984). *Finite Element Software for Plates and Shells*, Swansea: Pineridge Press.
6. Ahmad, S., Irons, B.M. and Zienkiewicz, O.C. (1970). Analysis of thick and thin shell structures by curved finite elements, *Int. J. Num. Meth. Eng.*, **2**, 419–451.
7. Beer, G. (1985). A joint/interface element for finite element analysis, *Int. J. Num. Meth. Eng.*, **21**, 585–600.
8. Beer, G., Cowling, R. and Bywater, S. (1985). A novel three-dimensional joint element with application to the modelling of mining excavations. *Proceedings of International Symposium on Fundamentals of Rock Joints, Centek*, Sweden, pp. 357–365.
9. Goodman, R.E., *et al.* (1968). A model for the mechanics of jointed rocks, *J. ASCE*, Vol. 94, 637–659.
10. Ghaboussi, T., *et al.* (1968). Finite elements for rock joints and interfaces, *J. ASCE*, **99**, No. SM3.
11. Zienkiewicz, *et al.* (1970). *Analysis of nonlinear problems in rock mechanics with particular reference to jointed rock systems.* 2nd ISRM Conf., Belgrade, 8–14.
12. Beer, G., Cowling, R. and Bywater, S. (1985). *Numerical modelling of shale hanging walls at Mount Isa Mine, Australia.* Proc. 5th International Conference on Numerical Methods in Geomechanics, Balkema.

5.10. RECOMMENDED FURTHER READING

Hinton, E., *et al.* (1984) *Finite Element Software for Plates and Shells* Pineridge Press, Swansea.

Schellekens J.C.J. (1990) *Interface Elements in Finite Element Analysis* TU Delft, Faculty of Engineering Rpt. no 25-2-90-5-17.

6

Boundary Element Methods—Basic Principles

Willst du ins Unendliche schreiten
Geh nur im Endlichen nach allen Seiten.

J.W. Goethe

6.1 INTRODUCTION

In boundary element methods, approximations are made only on the boundary (or surface in three-dimensional analysis) of the domain to be analysed. At every interior point of the domain the governing partial differential equations, which for elasticity for example are the equations of equilibrium and compatibility, are satisfied exactly. This contrasts with the finite element method, in which there is not in general pointwise satisfaction of the governing differential equations, resulting for example in the computation of a displacement field to which corresponds a residual body force distribution, not present in the original problem to be solved. In consequence of approximations being made only on the boundary, no interior mesh of elements need be specified, and in a numerical implementation there are no unknowns associated with interior points of the domain.

Boundary element methods can be applied only if a fundamental solution of the governing partial differential equations is known. A fundamental solution is a function that satisfies the differential equation with zero right side (body force in the case of elasticity) at every point of an infinite domain except at one point known as the source or load point, at which the right side is infinite. For Laplace's equation, the fundamental solution is the temperature distribution in a material of unit conductivity, due to a steady concentrated source of heat; for elasticity, it is the displacement field due to a concentrated load acting on an infinite elastic domain. Generally, fundamental solutions are known for homogeneous materials, whether isotropic or not, and so boundary element methods are applicable to the analysis of homogeneous and piecewise homogeneous two- and three-dimensional domains. The bending of plates can also be analysed by boundary element methods, but it is uncertain whether shells can economically be so analysed because of the difficulty of computing fundamental solutions. Furthermore, it is not yet known whether non-linear behaviour such as elastoplasticity can economically be analysed by boundary element methods.

There exist two basic types of boundary element method, the indirect and direct methods, and other methods in which fundamental solutions are used. The precursor of boundary element methods was the Trefftz method[1], proposed in 1926 to solve problems of St Venant torsion in which the governing partial differential equation is Laplace's equation. In the Trefftz method, an approximate solution of the boundary value problem is obtained by the superposition of fundamental solutions the source points of which are located outside the domain, and it only remains to adjust the intensities of the sources or loads to obtain the best possible agreement between the boundary conditions satisfied by the approximate solution and the actual boundary conditions. The greater the number of source points, the better the agreement that can be achieved. The principle is elegant but its implementation can be problematical, as there is in general no guarantee that, as the number of source points outside the domain is increased, the solutions obtained will converge to the exact solution. Much depends upon the choice of source points, which for practical geometries is beyond the competence of the average user of the method, and difficult to automate within a computer program. In any case, it can be demonstrated (see next section) that there exist problems for which convergence to the exact solution can never be attained.

In indirect boundary element methods, the solution is again sought by superposition but instead of the sources being located at a finite number of points outside the domain they are distributed continuously over its boundary. The intensity of the distribution, which generally varies from point to point of the boundary, is usually known as the density function; the partial differential equation is automatically satisfied at every interior point of the domain, and all that is required is to satisfy the boundary conditions by suitable choice of that function. It is found that for the boundary conditions to be satisfied, the density function must be the solution of an integral equation over the boundary, the form of which depends upon the type of boundary condition. It can be shown that, subject to certain surface and boundary condition smoothness conditions, a solution of the integral equation does exist. Indeed, solutions probably do exist for problems in which these conditions are not met but rigorous proofs that this is the case have not yet been developed.

The boundary integral equation cannot in general be solved exactly; instead, an approximate solution is obtained by proposing some variation of the density function over the boundary in terms of a finite number of parameters. Once the integral equation has been solved, physically meaningful results at boundary and interior points of the domain are computed by integration over the boundary. This integration replaces the summation of the effects of the finite number of sources in the Trefftz method.

In direct boundary element methods, an integral equation is obtained from the divergence theorem, which states that the integral over a domain of the divergence of a vector field equals the integral over the boundary of the domain of the outward normal component of that field. Taking steady state heat conduction as an analogy, the divergence theorem states that the rate at which heat is created within a domain (by a chemical reaction for example) is equal to the rate at which heat leaves the domain by crossing its boundary. By making appropriate substitutions into the divergence theorem, formulae such as Green's symmetric identity and the Maxwell–Betti reciprocal theorem are obtained. Structural engineers may recall that the latter theorem may be used in the computation of influence lines for elastic structures. By taking one of the two arbitrary functions in such a formula to be the function to be computed, and the other to be a fundamental solution with the source point located on the boundary, an integral equation over the boundary is obtained. This is known as the boundary integral equation of the direct method because the functions appearing in the equation are physically meaningful functions (such as displacement and traction for elasticity) rather than fictitious density functions. Solution of the integral equation therefore immediately yields the desired results on the boundary. If results at interior points of the domain are required, they can then be computed from formulae obtained in the same way as the integral equation, but with the source point of the fundamental solution located at interior points.

In this chapter the basic principles only of boundary elements are described. Numerical implementations, practical in the sense that they are reasonably accurate and computationally efficient, are discussed in Chapter 7 and a demonstration program is presented in Appendix B. An attempt is made to keep the mathematics simple but inevitably the reader must be prepared to understand concepts which may be unfamiliar, such as integral equations.

6.2 FUNDAMENTAL SOLUTIONS

By various mathematical techniques, fundamental solutions of a wide range of partial differential equations have been derived (see Appendix C for some of these solutions). In Chapters 6 and 7 we shall restrict attention to those for Laplace's equation and for isotropic elasticity, as these serve to illustrate most aspects of the mathematical analysis and numerical implementation of boundary element methods. The fundamental solution of Laplace's equation in three dimensions is

$$U(P,Q) = \frac{1}{4\pi r} \tag{6.1}$$

where (see Figure 6.1) r is distance from P to Q. Note that the fundamental solution has two arguments: P, the source point, and Q, the observation point. A physical interpretation of equation (6.1) is that if there is a unit steady state source of heat at P in an infinite medium with unit conductivity and zero temperature at infinity, then the temperature at the point Q is $U(P,Q)$. The function $U(P,Q)$ satisfies Laplace's equation in three dimensions at every point except P. Differentiating equation (6.1) with respect to x, the first cartesian coordinate of the observation point, we obtain

$$\frac{\partial U}{\partial x} = \frac{x_0 - x}{4\pi r^3} \tag{6.2}$$

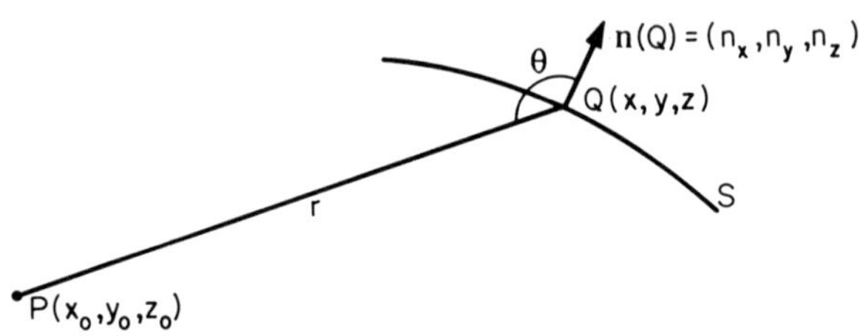

Figure 6.1 Notation for fundamental solution of Laplace's equation

Similarly, we can obtain derivatives with respect to y and z at the observation point Q. In the Neumann boundary value problem we are given the normal derivative of the function which is to satisfy Laplace's equation. The normal derivative of the fundamental solution is (see Figure 6.1)

$$T(P,Q) = \frac{\partial U}{\partial \mathbf{n}} = n_x \frac{\partial U}{\partial x} + n_y \frac{\partial U}{\partial y} + n_z \frac{\partial U}{\partial z} = \frac{\cos\theta}{4\pi r^2} \tag{6.3}$$

where θ is the angle between the line QP and the outward normal. Note that $T(P,Q)$ is a function of the coordinates of P and Q, and the normal to the boundary S at Q.

The fundamental solution of Laplace's equation in the xy plane (i.e. in two dimensions) may be obtained by integrating that defined by equation (6.1) between the limits $z_0 = -\infty$ and $z_0 = \infty$. This gives

$$U(P,Q) = \frac{1}{2\pi} \ln \frac{1}{r} \tag{6.4}$$

A physical interpretation of this result is that if unit heat per second is generated per unit length of an infinitely long line passing through P and perpendicular to the xy plane, in an infinite (three-dimensional) medium with unit conductivity and zero temperature at unit distance from that line, then the temperature at the point Q is $U(P,Q)$. Note that it is now impossible to specify zero temperature at infinity, and that the physical interpretation depends upon the unit of distance measurement. Differentiating equation (6.4) with respect to x, we obtain

$$\frac{\partial U}{\partial x} = \frac{x_0 - x}{2\pi r^2} \tag{6.5}$$

Similarly, we can differentiate with respect to y at the observation point Q. The normal derivative of the fundamental solution is now (see Figure 6.1)

$$T(P,Q) = \frac{\partial U}{\partial \mathbf{n}} = n_x \frac{\partial U}{\partial x} + n_y \frac{\partial U}{\partial y} = \frac{\cos\theta}{2\pi r} \tag{6.6}$$

For three-dimensional elasticity, the fundamental solution is the displacement at a point Q of the infinite elastic space due to a unit concentrated load at P. Since there are displacements in the x, y and z directions at Q and the load at P could act in any of those three cartesian directions, the fundamental solution has nine components:

$$\begin{bmatrix} u_x(Q) \\ u_y(Q) \\ u_z(Q) \end{bmatrix} = \begin{bmatrix} U_{xx}(P,Q) & U_{yx}(P,Q) & U_{zx}(P,Q) \\ U_{xy}(P,Q) & U_{yy}(P,Q) & U_{zy}(P,Q) \\ U_{xz}(P,Q) & U_{yz}(P,Q) & U_{zz}(P,Q) \end{bmatrix} \begin{bmatrix} f_x(P) \\ f_y(P) \\ f_z(P) \end{bmatrix} \tag{6.7}$$

where f_x, f_y and f_z are cartesian components of the load at P. This may be written more compactly as

$$\mathbf{u}(Q) = \mathbf{U}^{\mathrm{T}}(P,Q)\,\mathbf{f}(P) \tag{6.8}$$

$\mathbf{U}$ is symmetric with respect to both its arguments and its subscripts. The first leading diagonal component is given by[2]

$$U_{xx}(P,Q) = \frac{(1+\nu)}{8\pi E(1-\nu)r}\left[(3-4\nu) + r_x^2\right] \tag{6.9}$$

whereas typical off-diagonal components are given by

$$U_{xy}(P,Q) = U_{yx}(P,Q) = \frac{(1+\nu)}{8\pi E(1-\nu)r}\, r_x r_y \tag{6.10}$$

In equations (6.9) and (6.10), E and ν are Young's modulus and Poisson's ratio, and r_x and r_y are direction cosines of the line from Q to P:

$$\begin{aligned} r_x &= \frac{x_0 - x}{r} \\ r_y &= \frac{y_0 - y}{r} \end{aligned} \tag{6.11}$$

Formulae for U_{yy} and U_{zz} are obtained by replacing the subscript x in equation (6.9) by y and z, and formulae for U_{yz} and U_{zx} are obtained by replacing the subscripts x and y in equation (6.10) by y and z, and z and x. Strains corresponding to the fundamental solution may be computed by differentiating components of $\mathbf{U}$ with respect to x, y and z and substituting into equations (3.7). By then substituting the expressions obtained into Hooke's Law (equation (3.21)), stress at the point Q may be computed:

$$\boldsymbol{\sigma}(Q) = \mathbf{D}\,\mathbf{B}\,\mathbf{U}^{\mathrm{T}}(P,Q)\,\mathbf{f}(P) = \mathbf{S}^{\mathrm{T}}(P,Q)\,\mathbf{f}(P) \tag{6.12}$$

Note that the differentiations implied by $\mathbf{B}$ (see Section 1.3) are with respect to the coordinates of the point Q. Coefficients of $\mathbf{S}$ are given in Appendix C. The most commonly occurring boundary condition of elasticity is that in which the traction vector $\mathbf{t}(Q)$ is given, where

$$\mathbf{t}(Q) = \mathbf{N}(Q)\boldsymbol{\sigma}(Q) \tag{6.13}$$

in which

$$\mathbf{N}(Q) = \begin{bmatrix} n_x & 0 & 0 & n_y & 0 & n_z \\ 0 & n_y & 0 & n_x & n_z & 0 \\ 0 & 0 & n_z & 0 & n_y & n_x \end{bmatrix} \tag{6.14}$$

By substituting equation (6.12) into equation (6.13), it is found that the traction acting on the surface S at Q (see Figure 6.1) due to a concentrated load at P is given by

$$\mathbf{t}(Q) = \mathbf{T}^{\mathrm{T}}(P,Q)\,\mathbf{f}(P) \tag{6.15}$$

where $\mathbf{T}$ is the sum of two parts, one symmetric and the other antisymmetric about the leading diagonal:

$$\mathbf{T} = \mathbf{T}^{\mathrm{S}} + \mathbf{T}^{\mathrm{A}} \tag{6.16}$$

The first leading diagonal component of $\mathbf{T}^{\mathrm{S}}$ is given by[2]

$$T^{\mathrm{S}}_{xx}(P,Q) = \frac{1}{8\pi(1-\nu)r^2}\left[(1-2\nu) + 3r_x^2\right]\cos\theta \tag{6.17}$$

whereas typical off-diagonal terms are given by

$$T^{\mathrm{S}}_{xy}(P,Q) = T^{\mathrm{S}}_{yx}(P,Q) = \frac{3}{8\pi(1-\nu)r^2} r_x r_y \cos\theta \tag{6.18}$$

where θ is the angle shown in Figure 6.1. Leading diagonal components of the antisymmetric part are zero by definition. Typical off-diagonal components are given by

$$T^{\mathrm{A}}_{xy}(P,Q) = -T^{\mathrm{A}}_{yx}(P,Q) = \frac{(1-2\nu)}{8\pi(1-\nu)r^2}(n_x r_y - n_y r_x) \tag{6.19}$$

Other components of $\mathbf{T}^{\mathrm{S}}$ and $\mathbf{T}^{\mathrm{A}}$ are computed by appropriate substitution of subscripts as before.

The fundamental solution for plane strain in the xy plane may be obtained by integrating that defined by equation (6.7) between the limits $z_0 = -\infty$ and $z_0 = \infty$. For plane strain $f_z(P) = 0$ and $u_z(Q) = 0$, so this gives

$$\begin{bmatrix} u_x(Q) \\ u_y(Q) \end{bmatrix} = \begin{bmatrix} U_{xx}(P,Q) & U_{yx}(P,Q) \\ U_{xy}(P,Q) & U_{yy}(P,Q) \end{bmatrix} \begin{bmatrix} f_x(P) \\ f_y(P) \end{bmatrix} \tag{6.20}$$

For the isotropic case, $\mathbf{U}$ is again symmetric with respect to both its arguments and its subscripts. The first leading diagonal component is given by[2]

$$U_{xx}(P,Q) = \frac{(1+\nu)}{4\pi E(1-\nu)} \left((3-4\nu)\ln\frac{1}{r} + r_x^2 \right) \tag{6.21}$$

whereas the off-diagonal components are given by

$$U_{xy}(P,Q) = U_{yx}(P,Q) = \frac{(1+\nu)}{4\pi E(1-\nu)} r_x r_y \tag{6.22}$$

The component U_{yy} is computed by replacing the subscript x by y in equation (6.21). Strains and stresses corresponding to the fundamental solution may be computed in the same way as for the three-dimensional case, and the traction acting on the boundary S at Q due to a concentrated load at P is given by equation (6.15), in which now the vectors $\mathbf{t}(Q)$ and $\mathbf{f}(P)$ each contain only two coefficients, and $\mathbf{T}$ contains four coefficients instead of nine. It is found that[2]

$$\begin{aligned} T^{\mathrm{S}}_{xx}(P,Q) &= \frac{1}{4\pi(1-\nu)r}\left[(1-2\nu) + 2r_x^2\right]\cos\theta \\ T^{\mathrm{S}}_{xy}(P,Q) = T^{\mathrm{S}}_{yx}(P,Q) &= \frac{1}{2\pi(1-\nu)r} r_x r_y \cos\theta \\ T^{\mathrm{A}}_{xy}(P,Q) = -T^{\mathrm{A}}_{yx}(P,Q) &= \frac{(1-2\nu)}{4\pi(1-\nu)r}(n_x r_y - n_y r_x) \end{aligned} \tag{6.23}$$

Fundamental solutions for heat flow through an anisotropic domain, and for the most general case of an anisotropic elastic material with 21 independent elastic constants, can also be obtained. For an anisotropic elastic material, **U** and **T** are given only as integrals which must be evaluated numerically[3]. Solutions are also available for axisymmetric heat flow and elasticity[4], and for thin plates subject to bending[5]. In addition, fundamental solutions which satisfy governing partial differential equations in a halfspace and certain boundary conditions on the surface of the halfspace have been developed. Typical of such solutions is the Mindlin[6] solution, which is the displacement of an isotropic elastic halfspace with zero surface traction, due to an embedded concentrated load. Some fundamental solutions are tabulated in Appendix C.

6.3 TREFFTZ METHODS

As an easy introduction to boundary element and related techniques, let us consider the solution of a boundary value problem for Laplace's equation in two dimensions by the Trefftz method. Let us consider the steady state temperature distribution in a plate of infinite extent and unit conductivity, in which there is a hole of unit radius. Let us suppose that at infinity the temperature gradient is zero in the x direction and 1 in the y direction, and that conduction in the plate is the sole means of heat transfer so that no heat enters or leaves the hole at any point of its circumference. In an analysis of this problem by the finite element method it would be necessary either to truncate the mesh at some distance from the hole, or to use infinite finite elements[7]. In the Trefftz method, we aim to synthesise the solution by placing sources of the type described in the preceding section at a number of points outside the domain to be analysed, i.e. inside the hole.

Whereas the fundamental solution given by equation (6.4) tends to infinity only as the logarithm of distance from the hole, the temperature distribution to be synthesised tends linearly to infinity in the y direction. Therefore, it is not possible to find the solution directly. We seek instead a solution as the sum of two parts, these being the temperature distribution that would exist if the hole were not there, and a perturbation of that solution chosen so that the sum of the two parts satisfies the boundary conditions on the edge of the hole:

$$u(Q) = u^{(1)}(Q) + u^{(2)}(Q) \tag{6.24}$$

By analogy with the theory of ordinary differential equations, we call $u^{(1)}(Q)$ the particular integral and $u^{(2)}(Q)$ the complementary function. We take

$$u^{(1)}(Q) = y \tag{6.25}$$

and compute $u^{(2)}(Q)$ by the Trefftz method. The boundary condition on the circumference of the hole is the Neumann boundary condition

$$\frac{\partial u}{\partial \mathbf{n}} = 0 \tag{6.26}$$

Since on that boundary $\partial u^{(1)}/\partial \mathbf{n} = -\sin\phi$ (see Figure 6.2), we must have

$$\frac{\partial u^{(2)}}{\partial \mathbf{n}} = \sin\phi \tag{6.27}$$

for equation (6.26) to be satisfied.

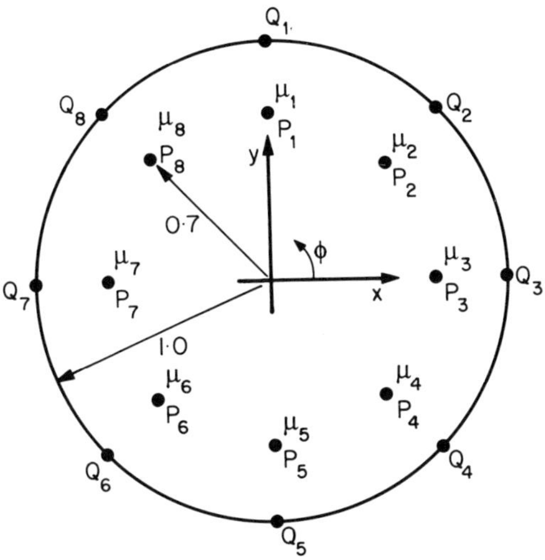

Figure 6.2 Sources and collocation points for analysis of plate with hole

Let us propose eight sources μ_1 to μ_8 located at 45° intervals of ϕ and at a radius of 0.7 from the centre of the hole, and aim to satisfy the boundary condition at the collocation points Q_1 to Q_8 (see Figure 6.2). The number and disposition of source points is of course arbitrarily chosen. In addition, satisfaction of the boundary conditions at a discrete set of points is not necessarily the optimal strategy: it could be advantageous to specify instead that the error in satisfaction of boundary conditions be orthogonal to a set of trial functions. It is in principle necessary to construct and solve a system of eight simultaneous equations in the eight unknowns μ_1–μ_8. Fortunately, the problem possesses symmetries, by virtue of which $\mu_3 = \mu_7 = 0$, $\mu_8 = \mu_2$, $\mu_4 = \mu_6 = -\mu_2$ and $\mu_5 = -\mu_1$. There are therefore only two unknowns, μ_1 and μ_2. By the same symmetry, if the boundary conditions are satisfied at points Q_1 and Q_2, then so are they at the other collocation points. Therefore, we write equations stating that equation (6.27) is satisfied at Q_1 and Q_2, in which the unknowns are μ_1 and μ_2:

$$\begin{aligned}\sum_{i=1}^{8} T(P_i, Q_1)\,\mu_i &= \frac{\partial u^{(2)}}{\partial \mathbf{n}}(Q_1)\\ \sum_{i=1}^{8} T(P_i, Q_2)\,\mu_i &= \frac{\partial u^{(2)}}{\partial \mathbf{n}}(Q_2)\end{aligned} \tag{6.28}$$

where $T(P,Q)$ is computed according to equation (6.6). Values of $T(P_i, Q_1)$ and $T(P_i, Q_2)$ are given in Table 6.1.

Table 6.1 Normal derivatives of fundamental solutions at collocation points

Point	$T(P_i, Q_1)$	$T(P_i, Q_2)$	Source
P_1	0.530 52	0.160 73	μ_1
P_2	0.160 73	0.530 52	μ_2
P_4	0.095 94	0.106 81	$-\mu_2$
P_5	0.093 62	0.095 94	$-\mu_1$
P_6	0.095 94	0.093 62	$-\mu_2$
P_8	0.160 73	0.106 81	μ_2

Taking account of symmetry, we obtain the following simultaneous equations:

$$\begin{aligned}0.436\,90\,\mu_1 + 0.129\,58\,\mu_2 &= 1.0000\\ 0.064\,79\,\mu_1 + 0.436\,90\,\mu_2 &= 0.7071\end{aligned} \tag{6.29}$$

The solution of these equations is $\mu_1 = 1.892, \mu_2 = 1.338$. The value of $u^{(2)}(Q)$ can now be computed at any point of the domain, by writing

$$u^{(2)}(Q) = \sum_{i=1}^{8} U(P_i, Q)\,\mu_i \tag{6.30}$$

where $U(P,Q)$ is computed according to equation (6.4). This yields for example $u^{(2)}(Q_1) = 0.863$ and $u^{(2)}(Q_2) = 0.611$, and therefore by equations (6.24) and (6.25) $u(Q_1) = 1.863$ and $u(Q_2) = 1.318$ compared with exact values[8] of 2.000 and 1.414, respectively. The computed values of temperature at Q_1 and Q_2 are therefore in error by about 7%. Since half the result in both cases is the particular integral, a better indication of accuracy might be that the computed values of $u^{(2)}(Q)$ are in error by about 14%.

There is unfortunately no guarantee that, as the number of source points is increased and so larger systems of simultaneous equations are solved, the solution computed by the Trefftz method converges to the exact solution of a problem. Suppose for example we wish to solve Laplace's equation in a circular domain, subject to the Dirichlet boundary condition

$$u(Q) = \ln \frac{1}{r_0} \tag{6.31}$$

where r_0 is the distance from Q on the boundary S to a point x_0 shown in Figure 6.3, and let us suppose that sources are distributed around the contour S'. The exact solution of the problem is $u(Q) = \ln \frac{1}{r_0}$ throughout the domain. A function satisfying Laplace's equation can be regarded as (say) the real part of an analytic function of the complex variable $z = x + \mathrm{i}y$, and this function can be analytically continued outside the domain shown in Figure 6.3. In the present example the analytic continuation has the real part $\ln \frac{1}{r_0}$ and so tends to infinity as the point x_0 is approached. It is impossible to obtain by any linear combination of sources on S' a function that is infinite at x_0, however large the number of sources. Therefore in this case the solution computed according to the Trefftz method cannot converge to the exact solution as the number of sources tends to infinity. Convergence could be achieved if S' were taken closer to S than x_0, or if a source were placed at x_0, but for practical problems the locations of poles of the analytic continuation of the solution are not known in advance. The difficulty illustrated by this example is most likely to occur at internal radii of the boundary, which unfortunately are often areas of interest as it is there that in elasticity for example there are severe stress concentrations.

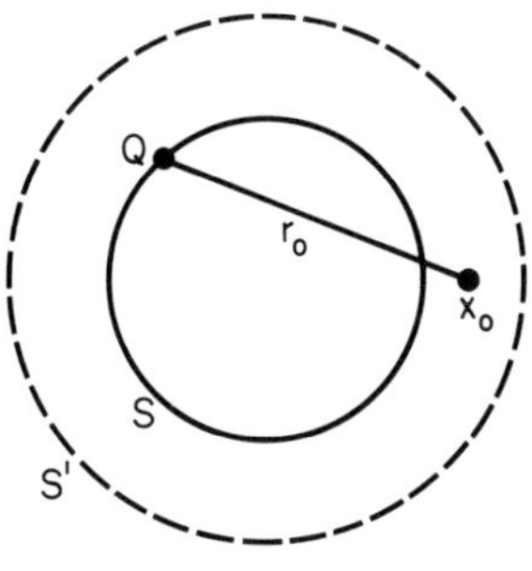

Figure 6.3 Boundary value problem for circular domain

6.4 THE INDIRECT BOUNDARY ELEMENT METHOD FOR LAPLACE'S EQUATION

In the Trefftz method for Laplace's equation, we seek a solution in the form

$$u(Q) = \sum_{i=1}^{n} U(P_i, Q)\, \mu_i \tag{6.32}$$

where $\mu_1, \mu_2, \ldots, \mu_n$ are sources concentrated at the points $P_1, P_2, \ldots, P_n$ outside the domain. In the indirect boundary element method, we seek instead a solution in the form

$$u(Q) = \int_S U(P,Q)\,\mu(P)\,\mathrm{d}S_P \tag{6.33}$$

where the density function $\mu(P)$ is the intensity of sources continuously† distributed over the boundary S of the domain. In three-dimensional problems, $\mathrm{d}S$ means 'with respect to surface area' and symbolises a double integral; in two-dimensional problems it means 'with respect to arc length'. The subscript P means that in the integration the point P moves over the boundary whilst the point Q stays still. Taking the analogy of heat flow, whereas the units of μ_i in equation (6.32) might be calories per second, those of $\mu(P)$ in equation (6.33) would be calories per second per square metre. Since $U(P,Q)$ satisfies the governing partial differential equation everywhere except at P, $u(Q)$ as defined by equation (6.33) satisfies the differential equation at all interior points of the domain, but not on its surface S. Equation (6.33) is referred to as the integral representation.

As in the Trefftz method, it now only remains to satisfy the boundary conditions. When considering how to do this, it should be borne in mind that the solution must satisfy the governing partial differential equation at all points inside the domain and also on the surface S. The integral representation does not do this, and it follows that for the Dirichlet and Neumann boundary conditions it is necessary to equate the limiting values as Q on S is approached from inside the domain (if such values exist) of $u(Q)$ and $\partial u/\partial \mathbf{n}(Q)$ as defined by equation (6.33) to the given boundary data.

For the Dirichlet condition, we must write (see Figure 6.4)

$$\lim_{Q' \to Q} \int_S U(P,Q')\,\mu(P)\,\mathrm{d}S_P = \bar{u}(Q) \tag{6.34}$$

where $\bar{u}(Q)$ is the given value of u at the point Q on S. It can be shown[9] that

$$\lim_{Q' \to Q} \int_S U(P,Q')\,\mu(P)\,\mathrm{d}S_P = \int_S U(P,Q)\,\mu(P)\,\mathrm{d}S_P \tag{6.35}$$

and so the condition that must be satisfied is

$$\int_S U(P,Q)\,\mu(P)\,\mathrm{d}S_P = \bar{u}(Q) \tag{6.36}$$

† 'Continuously' in the loose sense of covering the entire surface area, rather than in the strict mathematical sense according to which $\mu(P)$ would be a continuous function. It is not intended to imply anything about the continuity of $\mu(P)$ at this stage.

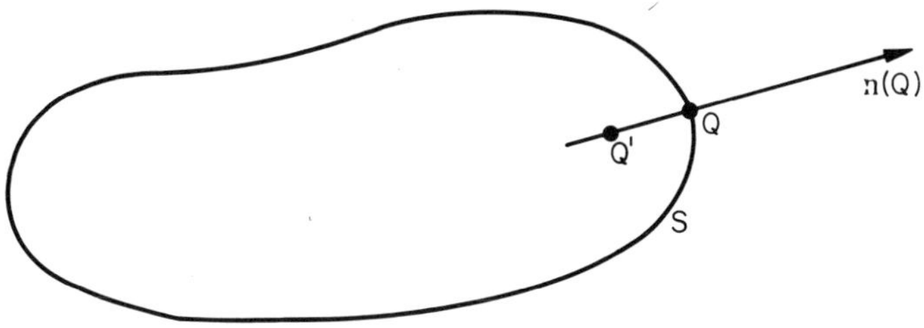

Figure 6.4 Satisfaction of boundary condition in the indirect method

for all points Q on S. In this case, therefore, the result is the same as that which would have been obtained had the fact that the integral representation does not satisfy the governing partial differential equation on S been ignored. Equation (6.36) is known as an integral equation of the first kind. The function $U(P,Q)$ is called the kernel because it characterises the equation. From equations (6.1) or (6.4) it is evident that $U(P,Q)$ tends to infinity as P approaches Q but, since it is only of order $1/r$ on the surface of a three-dimensional domain or of order $\ln\frac{1}{r}$ on the boundary of a two-dimensional domain, it is Riemann integrable. There is no general theory of integral equations of the first kind even for bounded kernels. However, in practice solutions are found to exist provided the unit of distance is chosen such that the distance r from P to Q is always less than 1, and indeed some theory on the existence of a solution of equation (6.36) has recently been developed[10].

For the Neumann condition, we must set the limiting value as Q' approaches Q (see Figure 6.4) of the derivative in the direction $\mathbf{n}(Q)$ of $u(Q')$ as defined by equation (6.33) to its given value $\bar{t}(Q)$. The function $U(P,Q')$ is bounded and so it is permissible to differentiate under the integral sign:

$$\frac{\partial u}{\partial \mathbf{n}}(Q') = \int_S T(P,Q')\,\mu(P)\,\mathrm{d}S_P \tag{6.37}$$

where $T(P,Q')$ is given by equation (6.3) or (6.6). We now write

$$\lim_{Q'\to Q} \frac{\partial u}{\partial \mathbf{n}}(Q') = \bar{t}(Q) \tag{6.38}$$

To illustrate (though not to determine for the general case) the behaviour of the function $\partial u/\partial\mathbf{n}(Q')$ as Q' approaches and passes through S at Q, let us take Q to be on a straight part of the boundary of a two-dimensional domain, and let us suppose that over this straight part of the boundary $\mu(P)$ is constant and therefore equal to $\mu(Q)$. We denote by $S(Q,\eta)$ the straight

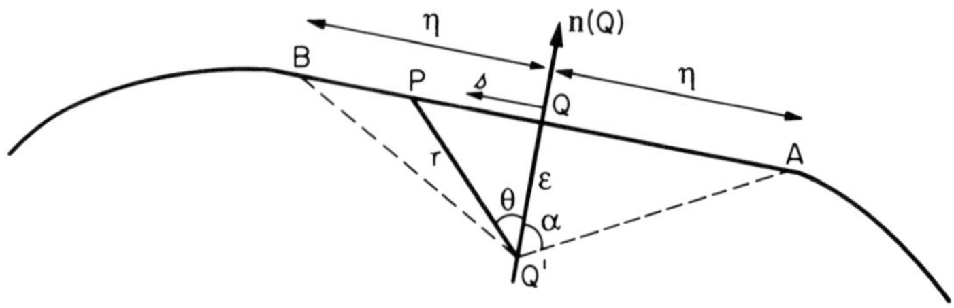

Figure 6.5 Limit of integral over AB as Q' approaches Q

part AB of the boundary shown in Figure 6.5, and by ϵ the distance QQ'. Then

$$\frac{\partial u}{\partial \mathbf{n}}(Q') = I_1(Q') + I_2(Q') \tag{6.39}$$

where

$$I_1(Q') = \int_{S(Q,\eta)} T(P,Q')\,\mu(P)\,\mathrm{d}S_P \tag{6.40}$$

and

$$I_2(Q') = \int_{S-S(Q,\eta)} T(P,Q')\,\mu(P)\,\mathrm{d}S_P. \tag{6.41}$$

It is easy to prove that the function $I_2(Q')$ varies continuously as Q' approaches and passes through S. When Q' is inside the domain as shown in Figure 6.5,

$$I_1(Q') = \int_{-\eta}^{\eta} \frac{\cos\theta}{2\pi r}\,\mu(P)\,\mathrm{d}s = \mu(Q)\int_{-\alpha}^{\alpha} \frac{1}{2\pi}\,\mathrm{d}\theta \tag{6.42}$$

where $\alpha = \tan^{-1}\frac{\eta}{\epsilon}$†. By integrating and substituting limits we find that

$$I_1(Q') = \mu(Q)\cdot\frac{\alpha}{\pi} \tag{6.43}$$

and so, since α tends to $\pi/2$ as Q' approaches (but never reaches) Q,

$$\lim_{Q'\to Q} I_1(Q') = \frac{1}{2}\,\mu(Q) \tag{6.44}$$

† Change of variable in equation (6.42): $s = \epsilon\tan\theta$, $\mathrm{d}s/\mathrm{d}\theta = \epsilon\sec^2\theta$ and $r = \epsilon\sec\theta$.

When Q' is at Q, $T(P, Q') = 0$ for all P in $S(Q, \eta)$, and so $I_1(Q) = 0$. Therefore,

$$\lim_{Q' \to Q} \frac{\partial u}{\partial \mathbf{n}}(Q') = \int_S T(P, Q)\, \mu(P)\, \mathrm{d}S_P + \frac{1}{2}\, \mu(Q) \tag{6.45}$$

Substituting this result into equation (6.38) gives

$$\frac{1}{2}\mu(Q) + \int_S T(P, Q)\, \mu(P)\, \mathrm{d}S_P = \bar{t}(Q) \tag{6.46}$$

which is the integral equation for the Neumann boundary condition. Equation (6.46) can be shown to be valid under a wider range of conditions than those considered in the preceding discussion; it is valid if the boundary S and given function $\bar{t}(Q)$ are regular (sufficiently smooth) according to the definitions of Kellogg[11], or if S is a Lyapunov surface and $\bar{t}(Q)$ is Hölder continuous[9]. For the purpose of numerical analysis the significance of these restrictions is that the point Q cannot be located at an edge or corner, or at any point at which $\bar{t}(Q)$ is discontinuous. Edges and discontinuous $\bar{t}(Q)$ occur frequently in engineering analysis. Equation (6.46) is known as an integral equation of the second kind, because the unknown function appears both under the integral sign and in a free term (the term $\frac{1}{2}\mu(Q)$). As the point P on S approaches Q, for three-dimensional problems the kernel $T(P, Q)$ tends to infinity as $1/r$ and for two-dimensional problems it remains bounded, so it is always Riemann integrable. Subject to the surface and function smoothness conditions referred to above, the Fredholm theory of integral equations of the second kind applies and it can be shown that solutions $\mu(Q)$ exist[9]. It should be borne in mind however that in the analysis leading to this conclusion it is supposed that $\mu(Q)$ is bounded and indeed continuous, whereas analogy with the distribution of electrical charge over a conducting surface suggests that there may exist solutions of equation (6.46) in which $\mu(Q)$ tends to infinity at edges, corners and discontinuities of $\bar{t}(Q)$. The generality of the method may therefore be greater than that suggested by existing mathematical analysis.

In the simplest implementation of the indirect boundary element method of solution of Laplace's equation, the boundary S is represented by flat triangular elements in three dimensions, or straight line elements in two, and it is supposed that over each of these elements $\mu(Q)$ is constant. Simultaneous equations for the values of $\mu(Q)$ are obtained by taking the point Q in equations (6.36) or (6.46) to be located at the centroid of each of the elements in turn. Let there be p elements $S_1,\ S_2,\ \ldots\ ,\ S_p$. Then for a Dirichlet problem

$$\sum_{b=1}^{p} \Delta U_{ab}\, \mu(Q_b) = \bar{u}(Q_a), \qquad a = 1, 2, \ldots, p \tag{6.47}$$

where Q_b is the centroid of element S_b, and

$$\Delta U_{ab} = \int_{S_b} U(P, Q_a)\, \mathrm{d}S_P \tag{6.48}$$

For a Neumann problem, the simultaneous equations are

$$\frac{1}{2}\mu(Q_a) + \sum_{b=1}^{p} \Delta T_{ab}\, \mu(Q_b) = \bar{t}(Q_a), \qquad a = 1, 2, \dots, p \tag{6.49}$$

where

$$\Delta T_{ab} = \int_{S_b} T(P, Q_a)\, \mathrm{d}S_P \tag{6.50}$$

Since in equation (6.49) the point Q_a is at the centroid of an element, the surface smoothness conditions for validity of that equation are always satisfied.

The solution of a Neumann boundary value problem contains an additive arbitrary constant (i.e. if $u(Q)$ is a solution then so is $u(Q) + k$). For a finite domain, the matrix of equation coefficients is singular in the limit as the number of elements tends to infinity and it is necessary for example to take $u(Q)$ to be zero at one of the elemental centroids; equation (6.47) is then written at that centroid and equation (6.49) at all the others. For an infinite domain, the integral representation can model at infinity $u(Q) = 0$ in three dimensions, or $u(Q) = \ln \frac{1}{r_0} \int_S \mu(P)\, \mathrm{d}S_P$ where r_0 is distance from an arbitrarily chosen reference point in two dimensions, but not a non-zero constant value. Therefore, the solution found lacks the additive arbitrary constant of the general solution of the Neumann boundary value problem, and the matrix of equation coefficients is not singular in the limit as the number of elements tends to infinity.

It is possible to evaluate analytically the integrals ΔU_{ab} and ΔT_{ab}: note that since the elements are straight or flat, $T(P, Q_a) = 0$ everywhere on the element S_a and so in equation (6.49), $\Delta T_{aa} = 0$ and leading diagonal coefficients all equal 1/2. Often the boundary condition is of the Dirichlet type over part of S and Neumann over the rest, so we write equation (6.47) at some element centroids and equation (6.49) at others. The matrix coefficients of equation (6.49) are dimensionless in the sense that they have the same numerical values regardless of the choice of the unit of distance. This is not true of equation (6.47), and for a problem with Dirichlet boundary conditions it is important that the unit of distance for the purpose of constructing the system of equations be chosen such that the possibility of singularity of the matrix is avoided. To illustrate this possibility, let us consider the

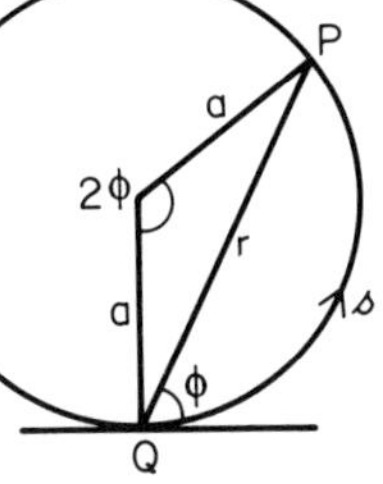

Figure 6.6 Domain for which Dirichlet problem is to be solved

solution of a Dirichlet problem for the circular domain shown in Figure 6.6. Let us suppose that the solution of equation (6.36) contains an arbitrary constant k. This implies that

$$\int_S U(P,Q).k \, \mathrm{d}S_P = 0 \tag{6.51}$$

for all Q on S and therefore (see equation (6.4) and Figure 6.6)

$$\int_S \ln r \, \mathrm{d}s = 0 \tag{6.52}$$

where s is arc length. From Figure 6.6 we see that $s = 2a\phi$ and $r = 2a\sin\phi$, so equation (6.52) can be rewritten as follows:

$$\int_0^\pi \ln(2a\sin\phi).2a \, \mathrm{d}\phi = 0 \tag{6.53}$$

or, dividing by $2a$,

$$\int_0^\pi (\ln 2a + \ln\sin\phi) \, \mathrm{d}\phi = 0 \tag{6.54}$$

Noting that[12] the integral of $\ln\sin\phi$ from 0 to π is $-\pi\log 2$, we find that equation (6.54) is satisfied if $a = 1$. Therefore, for a circular domain of unit radius, the solution of equation (6.36) is not unique, and in the limit as the number of boundary elements tends to infinity the solution of equation (6.47) is not unique either. The practical solution of this problem is to use, for the purpose of constructing the system of equations, a unit of distance equal to the length of the diagonal of a rectangle which encloses the entire boundary S, or if S is of infinite extent than that part of S which is modelled by finite boundary elements (see Chapter 7). For three-dimensional problems it is not clear whether the same difficulty can arise, but in practice the unit of distance is taken to be the length of the diagonal of a cuboid enclosing S because this also ensures that, for a mixed boundary value problem (part Dirichlet, part Neumann), all the matrix coefficients are roughly in the same order of magnitude. Subject to this precaution having been taken, the system of simultaneous equations may reliably and economically be solved by Gaussian elimination (in matrix algebra terms, bifactorisation), using the first equation to eliminate the first unknown, the second equation to eliminate the second unknown, and so on. Note that the non-sequential use of equations in forward reduction results in an unacceptable input–output overhead if the matrix is held on a disc file, so the ability to bifactorise the matrix sequentially is essential. Due to the lack of a free term which contributes to the leading diagonal and the somewhat greater magnitude of off-diagonal coefficients of the matrix the accumulation of roundoff error during simultaneous equation solution is higher for equation (6.47) than for equation (6.49), but has never been known to be excessive if arithmetic is done to CDC single precision (14 significant figures) or IBM double precision (16 significant figures).

Once the elemental values of the density function have been computed, the solution at any interior or boundary point may be obtained from the integral representation:

$$u(Q) = \sum_{b=1}^{p} \Delta U_b \, \mu(Q_b) \tag{6.55}$$

where

$$\Delta U_b = \int_{S_b} U(P,Q) \, \mathrm{d}S_P \tag{6.56}$$

6.5 THE INDIRECT METHOD FOR ELASTICITY

The treatment of elasticity is very similar to that of Laplace's equation, and the reader is encouraged to make comparisons with section 6.4. The scalar unknown is replaced by the displacement vector $\mathbf{u}$. The most commonly encountered boundary conditions are the displacement boundary condition in which the displacement is given (analogous to Dirichlet), the traction boundary condition in which the traction vector $\mathbf{t}$ is given (analogous to Neumann), and mixed conditions in which the given data consists partly of displacement components and partly of traction components (no equivalent for Laplace's equation). True displacement boundary conditions do not occur in nature because they would imply the existence of an infinitely stiff foundation. In engineering analysis, however, mixed conditions in particular are quite common because they can be taken to exist on planes of symmetry of structure and loading, thereby enabling the domain on only one side of a plane to be analysed.

We adopt the integral representation

$$\mathbf{u}(Q) = \int_S \mathbf{U}^{\mathrm{T}}(P,Q)\boldsymbol{\mu}(P)\,\mathrm{d}S_P \tag{6.57}$$

where the density function $\boldsymbol{\mu}(P)$ is now the intensity of a fictitious layer of force which may be imagined to act on the boundary S embedded in the infinite elastic space. For three-dimensional problems, for example,

$$\boldsymbol{\mu}(P) = \begin{bmatrix} \mu_x(P) \\ \mu_y(P) \\ \mu_z(P) \end{bmatrix} \tag{6.58}$$

The displacement field defined by equation (6.57) satisfies the governing partial differential equations of elasticity at every interior point of the domain. It can be shown[13,14] that for the displacement boundary condition the integral equation that must be satisfied is

$$\int_S \mathbf{U}^{\mathrm{T}}(P,Q)\boldsymbol{\mu}(P)\,\mathrm{d}S_P = \bar{\mathbf{u}}(Q) \tag{6.59}$$

for all points Q on S, where $\bar{\mathbf{u}}(Q)$ is the given value of displacement. By a procedure similar to that described in section 6.4, it can be shown that for the traction boundary condition the integral equation is[13,14]

$$\frac{1}{2}\boldsymbol{\mu}(Q) + \int_S \mathbf{T}^{\mathrm{T}}(P,Q)\boldsymbol{\mu}(P)\,\mathrm{d}S_P = \bar{\mathbf{t}}(Q) \tag{6.60}$$

for all points Q on S, where $\bar{\mathbf{t}}(Q)$ is the given traction vector. The proofs of equations (6.59) and (6.60) are similar to those of equations (6.36) and (6.46) of the preceding section, but there are important differences in the surface and functional smoothness requirements, and in the interpretation of the integral on the left side of equation (6.60). S must now be a Lyapunov surface and $\bar{\mathbf{t}}(Q)$ Hölder continuous[9], and the Cauchy principal value of the integral of $\mathbf{T}^{\mathrm{T}}(P,Q)\boldsymbol{\mu}(P)$ must be taken because in the antisymmetric component of $\mathbf{T}$ (equation (6.19) or (6.23)) there is no factor $\cos\theta$ (see Figure 6.1), and so that component tends to infinity as $1/r^2$ for three-dimensional problems or $1/r$ for plane strain as P on S approaches Q. The Cauchy principal value is the arbitrarily chosen definition

$$\int_S \mathbf{T}^{\mathrm{T}}(P,Q)\boldsymbol{\mu}(P)\,\mathrm{d}S_P \equiv \lim_{\epsilon\to 0}\int_{S-S(Q,\epsilon)} \mathbf{T}^{\mathrm{T}}(P,Q)\boldsymbol{\mu}(P)\,\mathrm{d}S_P \tag{6.61}$$

where the region of exclusion $S(Q,\epsilon)$ is in three dimensions that part of S which lies inside a sphere of radius ϵ centred at Q, and in two dimensions is

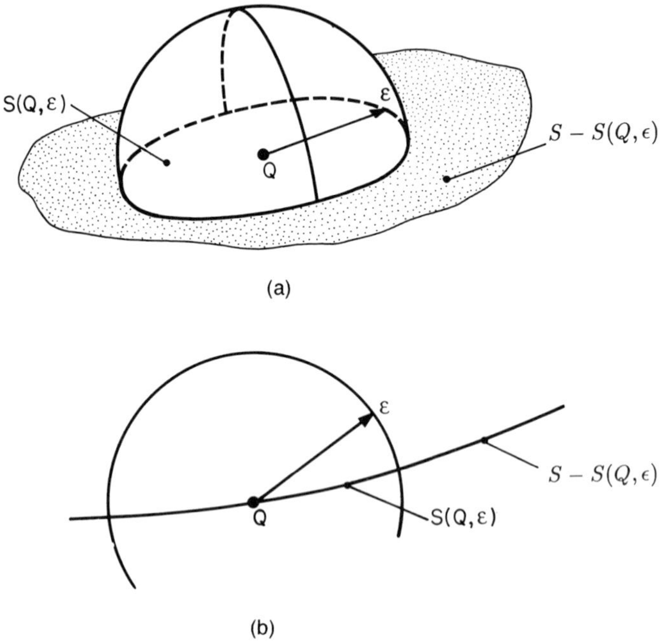

Figure 6.7 Definition of Cauchy principal value. (a) Three dimensions, (b) two dimensions

that part of S which lies inside a circle of radius ϵ centred at Q (see Figure 6.7). It can be shown that, subject to Hölder smoothness of $\boldsymbol{\mu}(P)$, the limiting value defined by equation (6.61) exists for a Lyapunov surface S. It does not in general exist if the point Q is on an edge or corner, or at any point at which $\boldsymbol{\mu}(Q)$ is discontinuous. Since the Cauchy principal value must be computed, at least a principal part of the integral must be evaluated analytically rather than numerically. Subject to the surface and functional smoothness conditions referred to above, the Fredholm theory of integral equations of the second kind can be extended to apply to equation (6.60) (which is known as a singular integral equation), and it can be shown that solutions $\boldsymbol{\mu}(Q)$ exist[14].

In the simplest numerical implementation, the boundary S is represented by flat triangular elements in three dimensions[13], or straight line elements in plane strain[15], and it is supposed that over each of these elements the fictitious force density $\boldsymbol{\mu}(Q)$ is constant. Simultaneous equations for the values of $\boldsymbol{\mu}(Q)$ are obtained by taking the point Q in equation (6.59) or (6.60) to be located at the centroid of each of the elements in turn. For a displacement boundary value problem we write

$$\sum_{b=1}^{p} \Delta\mathbf{U}_{ab}\boldsymbol{\mu}(Q_b) = \bar{\mathbf{u}}(Q_a), \qquad a = 1, 2, \ldots, p \tag{6.62}$$

where

$$\Delta\mathbf{U}_{ab} = \int_{S_b} \mathbf{U}^{\mathrm{T}}(P, Q_a)\, \mathrm{d}S_P \tag{6.63}$$

For a traction boundary value problem the simultaneous equations are

$$\frac{1}{2}\boldsymbol{\mu}(Q_a) + \sum_{b=1}^{p} \Delta\mathbf{T}_{ab}\boldsymbol{\mu}(Q_b) = \bar{\mathbf{t}}(Q_a), \qquad a = 1, 2, \ldots, p \tag{6.64}$$

where

$$\Delta\mathbf{T}_{ab} = \int_{S_b} \mathbf{T}^{\mathrm{T}}(P, Q_a)\, \mathrm{d}S_P \tag{6.65}$$

Since in equation (6.64) the point Q_a is at the centroid of an element, the surface smoothness conditions for validity of that equation are always satisfied. The solution of a traction boundary value problem contains six arbitrary constants in three dimensions, and three in plane strain, each of the constants corresponding to a rigid body displacement. For a finite domain, it is necessary for example to specify either six or three components of element centroidal displacements, these being chosen in such a way as to render unique the solution of the boundary value problem. This necessitates the writing of equation (6.62) at certain centroids. For an infinite domain, the solution defined by the integral representation lacks the arbitrary constants referred to above, and there is no need to specify the values of any components of element centroidal displacement in advance of simultaneous equation solution; equation (6.62) is therefore not required.

The integrals $\Delta\mathbf{U}_{ab}$ and $\Delta\mathbf{T}_{ab}$ may be evaluated analytically. The kernel $\mathbf{T}^{\mathrm{T}}(P, Q_a)$ is non-zero on the element S_a, and the Cauchy principal value of its integral must be computed. This is only necessary in the three-dimensional case; for a line element in two dimensions the integral of $\mathbf{T}^{\mathrm{T}}(P, Q_a)$ is zero because Q_a is halfway between the ends. The matrix coefficients of equation (6.64) are dimensionless, but those of equation (6.62) depend upon both the unit of distance and the unit of force. As in the case of Laplace's equation with the Dirichlet boundary condition, there exists at least for two-dimensional problems the possibility that the matrix of equation coefficients is singular. To avoid this danger, and to ensure that for a problem with both displacement and traction boundary conditions all matrix coefficients are roughly in the same order of magnitude, a unit of distance equal to the length of the diagonal of a rectangle or cuboid enclosing S is adopted, and also the unit of force is chosen such that Young's modulus equals 1. Subject to these precautions having been taken, the system of simultaneous equations may be solved in the same way as for Laplace's equation (Section 6.4).

Once the elemental values of the density function have been computed, the displacement at any interior or boundary point may be obtained from the integral representation:

$$\mathbf{u}(Q) = \sum_{b=1}^{p} \Delta\mathbf{U}_b \boldsymbol{\mu}(Q_b) \tag{6.66}$$

where

$$\Delta\mathbf{U}_b = \int_{S_b} \mathbf{U}^{\mathrm{T}}(P, Q)\, \mathrm{d}S_P \tag{6.67}$$

Stresses at any interior point may likewise be computed by integration over the boundary:

$$\boldsymbol{\sigma}(Q) = \sum_{b=1}^{p} \Delta\mathbf{S}_b \boldsymbol{\mu}(Q_b) \tag{6.68}$$

where

$$\Delta\mathbf{S}_b = \int_{S_b} \mathbf{S}^{\mathrm{T}}(P, Q)\, \mathrm{d}S_P \tag{6.69}$$

In equation (6.69), $\mathbf{S}(P, Q)$ is the function defined by equation (6.12).

6.6 THE DIRECT BOUNDARY ELEMENT METHOD FOR LAPLACE'S EQUATION

According to the divergence theorem[8],

$$\int_V \boldsymbol{\nabla}\mathbf{a}\, \mathrm{d}V = \int_S \mathbf{a} \cdot \mathbf{n}\, \mathrm{d}S \tag{6.70}$$

where $\mathbf{a}$ is any continuously differentiable vector field, and $\mathbf{n}$ is the unit outward normal to the surface S of the domain V. If the cartesian components of $\mathbf{a}$ are (a_x, a_y, a_z) then

$$\boldsymbol{\nabla}\mathbf{a} = \frac{\partial a_x}{\partial x} + \frac{\partial a_y}{\partial y} + \frac{\partial a_z}{\partial z} \tag{6.71}$$

and

$$\mathbf{a} \cdot \mathbf{n} = a_x n_x + a_y n_y + a_z n_z \tag{6.72}$$

In the analogy of steady state heat flow $\mathbf{a}$ is the heat flux vector, the component a_x for example being the flow in calories per second per square metre in the x direction. The divergence $\nabla\mathbf{a}$ is the rate of production of heat inside V in calories per second per cubic metre, $\mathbf{a}\cdot\mathbf{n}$ is the rate at which heat crosses S in calories per second per square metre, and equation (6.70) states that the heat produced per second in the domain V equals the heat that leaves V per second by crossing S.

The formula leading to the boundary integral equation for Laplace's equation is derived by making certain substitutions in the divergence theorem. Let us replace $\mathbf{a}$ by the product $\phi\boldsymbol{\nabla}\psi$ where ϕ and ψ are arbitrary scalar fields, and according to the usual vector notation

$$\boldsymbol{\nabla}\psi = \left(\frac{\partial\psi}{\partial x}, \frac{\partial\psi}{\partial y}, \frac{\partial\psi}{\partial z}\right) \tag{6.73}$$

Then (see Appendix D) we obtain the result

$$\int_V \left(\phi\nabla^2\psi + \boldsymbol{\nabla}\phi\cdot\boldsymbol{\nabla}\psi\right)\,\mathrm{d}V = \int_S \phi\frac{\partial\psi}{\partial\mathbf{n}}\,\mathrm{d}S \tag{6.74}$$

This equation is known as Green's first identity. Let us now rewrite equation (6.74) with ϕ and ψ interchanged:

$$\int_V \left(\psi\nabla^2\phi + \boldsymbol{\nabla}\psi\cdot\boldsymbol{\nabla}\phi\right)\,\mathrm{d}V = \int_S \psi\frac{\partial\phi}{\partial\mathbf{n}}\,\mathrm{d}S \tag{6.75}$$

By subtracting equation (6.75) from equation (6.74), we obtain the result known as Green's symmetric identity:

$$\int_V \left(\phi\nabla^2\psi - \psi\nabla^2\phi\right)\,\mathrm{d}V = \int_S (\phi\frac{\partial\psi}{\partial\mathbf{n}} - \psi\frac{\partial\phi}{\partial\mathbf{n}})\,\mathrm{d}S \tag{6.76}$$

Let us take ϕ to be the solution u of the boundary value problem for Laplace's equation. Then, since $\nabla^2 u = 0$ in V,

$$\int_V u(Q)\nabla^2\psi(Q)\,\mathrm{d}V_Q = \int_S \left(u(Q)\frac{\partial\psi}{\partial\mathbf{n}}(Q) - \psi(Q)\frac{\partial u}{\partial\mathbf{n}}(Q)\right)\,\mathrm{d}S_Q \tag{6.77}$$

By taking $\psi(Q)$ to be the fundamental solution $U(P,Q)$ where P lies inside V, and modifying the domain for which equation (6.77) is written to ensure satisfaction of functional smoothness requirements (see Appendix D), we obtain an expression for the solution u at an interior point in terms of u and $\partial u/\partial\mathbf{n}$ on the boundary. In the notation of equation (6.3),

$$u(P) = \int_S \left(U(P,Q)\frac{\partial u}{\partial\mathbf{n}}(Q) - T(P,Q)u(Q)\right)\,\mathrm{d}S_Q \tag{6.78}$$

This result is known as the Poisson formula.

If we now take $\psi(Q)$ to be the fundamental solution $U(P,Q)$ where P lies on S and modify the domain as before (see Appendix D), we obtain the result

$$C(P)u(P) + \int_S T(P,Q)u(Q)\,\mathrm{d}S_Q = \int_S U(P,Q)\frac{\partial u}{\partial \mathbf{n}}(Q)\,\mathrm{d}S_Q \qquad (6.79)$$

where $C(P)$ is given by equation (D.24), or for the two dimensional case by equation (D.26).

Equation (6.79) is the boundary integral equation of the direct method for Laplace's equation. The computational procedure is to solve this equation for the unknown boundary data, then if any results at interior points are required to apply the Poisson formula, equation (6.78). The method is called the direct method because physically meaningful results (for the surface) are obtained immediately upon solution of the integral equation. This has been claimed as an advantage over the indirect method, but that advantage is offset by the need to evaluate integrals involving given boundary data during the construction of simultaneous equations approximating to equation (6.79). It could indeed be that the indirect method is potentially more efficient than the direct method if results are to be computed at a large number of interior points of the domain, because the form of the integral representation, equation (6.33), is simpler than that of the Poisson formula, equation (6.78). Such factors, however, are of only minor significance in determining which of the two methods should be used; other considerations, detailed in Chapter 7, are more important. We note at this stage only that in the direct method the existence of a solution of the integral equation is guaranteed, provided the problem to be solved is actually soluble (i.e. for a finite domain subject to the Neumann boundary condition over its entire surface, the given data integrates to zero); and that by contrast with the integral equation of the indirect method for the Neumann problem, equation (6.79) can be written at points P on edges and corners of the surface.

In the simplest numerical implementation[16], the boundary S is represented by flat triangular elements in three dimensions, or straight line elements in two, and it is supposed that over each of these elements $u(P)$ and $\frac{\partial u}{\partial \mathbf{n}}(P)$ are constant. Simultaneous equations for the unknown values of $u(P)$ and $\frac{\partial u}{\partial \mathbf{n}}(P)$ are obtained by taking the point P in equation (6.79) to be located at the centroid of each of the elements in turn. Let there be p elements $S_1,\ S_2, \dots,\ S_p$. Then, since $C(P) = 1/2$ on a smooth part of S,

$$\frac{1}{2}u(P_a) + \sum_{b=1}^{p} \Delta T_{ab} u(P_b) = \sum_{b=1}^{p} \Delta U_{ab} \frac{\partial u}{\partial \mathbf{n}}(P_b), \qquad a = 1, 2, \dots, p \qquad (6.80)$$

where P_b is the centroid of element S_b, and

$$\Delta T_{ab} = \int_{S_b} T(P_a, Q)\, \mathrm{d}S_Q \tag{6.81}$$

$$\Delta U_{ab} = \int_{S_b} U(P_a, Q)\, \mathrm{d}S_Q \tag{6.82}$$

The integrals ΔT_{ab} and ΔU_{ab} can be evaluated analytically: note that since $T(P_a, Q) = 0$ everywhere on the element S_a, $\Delta T_{aa} = 0$. Because the divergence theorem applies to a domain V enclosed by a surface S, the foregoing development of the direct method applies to finite domains only; generalisation to infinite domains is described in Section 6.11. For a finite domain and the Neumann boundary condition everywhere on S, the matrix of equation coefficients is singular if the integrals ΔT_{ab} are evaluated exactly and it is necessary for example to take $u(P)$ to be zero over one of the elements. For a Dirichlet boundary value problem, equation (6.79) is an integral equation of the first kind with kernel $U(P, Q)$, in which integration is with respect to the second argument Q. Since $U(P, Q) = U(Q, P)$, this equation can be rewritten in the form

$$\int_S U(Q, P) \frac{\partial u}{\partial \mathbf{n}}(Q)\, \mathrm{d}S_Q = f(P) \tag{6.83}$$

where $f(P)$ is a known function. This equation is of the same form as equation (6.36) and so, as is demonstrated in section 6.4, its solution may not be unique. To avoid the possibility of singularity of the matrix of equation coefficients in equation (6.80), therefore, the unit of distance is chosen in the same way as for the indirect method: for two-dimensional problems, it is the length of the diagonal of a rectangle which encloses the boundary S; for three-dimensional problems, it is the length of the diagonal of a cuboid enclosing S. All the matrix coefficients are then in roughly the same order of magnitude even for a mixed boundary value problem, and the system of simultaneous equations may reliably and economically be solved by Gaussian elimination, using the first equation to eliminate the first unknown, the second equation to eliminate the second unknown, and so on.

Once equation (6.80) has been solved for unknown values of $u(P_a)$ and $\partial u/\partial \mathbf{n}(P_a)$, the solution at any interior point may be obtained by the Poisson formula:

$$u(P) = \sum_{b=1}^{p} \Delta U_b \frac{\partial u}{\partial \mathbf{n}}(P_b) - \sum_{b=1}^{p} \Delta T_b u(P_b) \tag{6.84}$$

where

$$\Delta U_b = \int_{S_b} U(P,Q)\,\mathrm{d}S_Q \tag{6.85}$$

and

$$\Delta T_b = \int_{S_b} T(P,Q)\,\mathrm{d}S_Q \tag{6.86}$$

The tangential derivative of u at the centroid of a boundary element may be computed by differencing the values of u at centroids of adjacent elements. It would be possible to compute the gradient of u at interior points of V by differencing values of u given by equation (6.84), but we usually compute it directly:

$$\boldsymbol{\nabla} u(P) = \sum_{b=1}^{p} \Delta \mathbf{S}_b \frac{\partial u}{\partial \mathbf{n}}(P_b) - \sum_{b=1}^{p} \Delta \mathbf{R}_b u\,(P_b) \tag{6.87}$$

where

$$\Delta \mathbf{S}_b = \int_{S_b} \boldsymbol{\nabla} U(P,Q)\,\mathrm{d}S_Q \tag{6.88}$$

and

$$\Delta \mathbf{R}_b = \int_{S_b} \boldsymbol{\nabla} T(P,Q)\,\mathrm{d}S_Q \tag{6.89}$$

In equations (6.88) and (6.89), the differentiations implied by the operator $\boldsymbol{\nabla}$ are with respect to the coordinates of the point P. The coefficients ΔS_b and ΔR_b, shown as matrices in the notation of this text, are of dimensions 1×3 in three dimensions or 1×2 in two and therefore effectively are vectors.

6.7 THE DIRECT METHOD FOR ELASTICITY

Let $\mathbf{b}^{(1)}$ and $\mathbf{t}^{(1)}$ be the body force vector and traction vector corresponding to the arbitrary displacement field $\mathbf{u}^{(1)}$, and let $\mathbf{b}^{(2)}$ and $\mathbf{t}^{(2)}$ be the body force vector and traction vector corresponding to the arbitrary displacement field $\mathbf{u}^{(2)}$. Then according to Betti's theorem (see Appendix E)

$$\int_V \left(-\mathbf{b}^{(1)} \cdot \mathbf{u}^{(2)} + \mathbf{b}^{(2)} \mathbf{u}^{(1)}\right) \mathrm{d}V = \int_S \left(\mathbf{t}^{(1)} \cdot \mathbf{u}^{(2)} - \mathbf{t}^{(2)} \cdot \mathbf{u}^{(1)}\right) \mathrm{d}S \tag{6.90}$$

Equation (6.90) is also known as the Maxwell–Betti reciprocal theorem, and is well known to structural engineers as the statement that, for an elastic structure, the sum of displacements due to a first set of loads times a second set of loads equals the sum of displacements due to the second set of loads times the first set of loads. For Betti's theorem to be true, both $\mathbf{u}^{(1)}$ and $\mathbf{u}^{(2)}$ must be twice continuously differentiable. In the usual applications of Betti's theorem in structural analysis the smoothness requirement is of no importance, but in the derivation of boundary integral equations it is essential that it be respected.

We take $\mathbf{u}^{(1)}$ to be the solution $\mathbf{u}$ of the boundary value problem of elasticity under consideration, this being a problem in which there is no body force. Since $\mathbf{b}^{(1)} = \mathbf{0}$, this yields the result

$$\int_V \mathbf{b}^{(2)}(Q)\mathbf{u}(Q)\,\mathrm{d}V_Q = \int_S \left(\mathbf{u}^{(2)}(Q)\mathbf{t}(Q) - \mathbf{t}^{(2)}(Q)\mathbf{u}(Q)\right)\,\mathrm{d}S_Q \tag{6.91}$$

where $\mathbf{t}$ is traction due to the displacement field $\mathbf{u}$ and $\mathbf{t}^{(2)}$ (Q).

By analogy with the procedure for Laplace's equation, we take $\mathbf{u}^{(2)}(Q)$ to be the kernels $\mathbf{U}(P,Q)$ and $\mathbf{T}$ (P, Q) where P is an interior point of V, and modify the domain for which equation (6.91) is written in order to ensure satisfaction of functional smoothness requirements (see Appendix E). In this way we obtain a formula for the displacement vector at an arbitrary interior point of V:

$$\mathbf{u}(P) = \int_S \left(\mathbf{U}(P,Q)\mathbf{t}(Q) - \mathbf{T}(P,Q)\mathbf{u}(Q)\right)\,\mathrm{d}S_Q \tag{6.92}$$

This result is known as Somigliana's identity, and is analogous to the Poisson formula, equation (6.78), for Laplace's equation.

If we now take $\mathbf{u}^{(2)}(Q)$ to be the fundamental solution $\mathbf{U}(P,Q)$ where P lies on S and modify the domain as before (see Appendix E), we obtain the result

$$\mathbf{C}(P)\mathbf{u}(P) + \int_S \mathbf{T}(P,Q)\mathbf{u}(Q)\,\mathrm{d}S_Q = \int_S \mathbf{U}(P,Q)\mathbf{t}(Q)\,\mathrm{d}S_Q \tag{6.93}$$

where $\mathbf{C}(P)$ is given by equation (E.18). Once equation (6.93), the boundary integral equation of the direct method for elasticity, has been solved for the unknown boundary data, results at interior points may be computed by Somigliana's identity, equation (6.92).

In the simplest numerical implementation[17], the boundary S is represented by flat triangular elements in three dimensions, or straight line elements in two, and it is supposed that over each of these elements $\mathbf{u}(P)$ and $\mathbf{t}(P)$ are constant. The point P in equation (6.93) is taken to be located at the centroid of each of the elements $S_1, S_2, \ldots, S_p$ in turn. Since $\mathbf{C}(P) = \mathbf{I}/2$ on a smooth part of S, this yields the simultaneous equations

$$\frac{1}{2}\mathbf{u}(P_a) + \sum_{b=1}^{p} \Delta\mathbf{T}_{ab}\mathbf{u}(P_b) = \sum_{b=1}^{p} \Delta\mathbf{U}_{ab}\mathbf{t}(P_b), \qquad a = 1, 2, \ldots, p \tag{6.94}$$

where P_b is the centroid of element S_b, and

$$\Delta\mathbf{T}_{ab} = \int_{S_b} \mathbf{T}(P_a, Q)\, \mathrm{d}S_Q \tag{6.95}$$

$$\Delta\mathbf{U}_{ab} = \int_{S_b} \mathbf{U}(P_a, Q)\, \mathrm{d}S_Q \tag{6.96}$$

The integrals $\Delta\mathbf{T}_{ab}$ and $\Delta\mathbf{U}_{ab}$ can be evaluated analytically, it being necessary to take the Cauchy principal value of $\Delta\mathbf{T}_{aa}$. The results of this section apply to finite domains only; infinite domains are considered in section 6.11. For a finite domain with traction boundary conditions everywhere on S, the matrix of equation coefficients is singular if the integrals $\Delta\mathbf{T}_{ab}$ are evaluated exactly. This is a consequence of arbitrariness in the solution of the boundary value problem and may be avoided by specifying sufficient constraints on displacement to prevent rigid body translation and rotation. For a problem in which displacements are specified everywhere on S, a singular matrix of equation coefficients can be obtained even though there is no arbitrariness in the solution of the boundary value problem itself. As in the case of the Dirichlet problem for Laplace's equation, this possibility may be avoided by appropriate choice of the unit of distance for the purpose of construction of the system of equations. To ensure that for a boundary value problem the solution of which is unique the solution of the integral equation is also unique, and that all matrix coefficients are roughly in the same order of magnitude even for a mixed boundary value problem, the unit of distance is taken to be the length of the diagonal of a rectangle or cuboid enclosing S, and the unit of force is chosen such that Young's modulus equals 1. The system of equations may then be solved by Gaussian elimination as described previously.

Once equation (6.94) has been solved for the unknown boundary data, the solution at any interior point may be obtained by Somigliana's identity:

$$\mathbf{u}(P) = \sum_{b=1}^{p} \Delta\mathbf{U}_b\mathbf{t}(P_b) - \sum_{b=1}^{p} \Delta\mathbf{T}_b\mathbf{u}(P_b) \tag{6.97}$$

where

$$\Delta\mathbf{U}_b = \int_{S_b} \mathbf{U}(P, Q)\, \mathrm{d}S_Q \tag{6.98}$$

and

$$\Delta \mathbf{T}_b = \int_{S_b} \mathbf{T}(P,Q)\,\mathrm{d}S_Q \tag{6.99}$$

Usually the most important results of an elastic analysis are the stresses rather than the displacements. The tangential strain at the centroid of a boundary element may be computed by differencing the displacements at centroids of adjacent elements. Stresses can then be computed by Hooke's law from tangential strain and the traction vector. It would be possible to compute stresses at interior points by differencing displacements for strains, but it is the usual practice to compute them directly as follows:

$$\boldsymbol{\sigma}(P) = \sum_{b=1}^{p} \Delta \mathbf{S}_b \mathbf{t}(P_b) - \sum_{b=1}^{p} \Delta \mathbf{R}_b \mathbf{u}(P_b) \tag{6.100}$$

where

$$\Delta \mathbf{S}_b = \int_{S_b} \mathbf{D}\,\mathbf{B}\,\mathbf{U}(P,Q)\,\mathrm{d}S_Q \tag{6.101}$$

and

$$\Delta \mathbf{R}_b = \int_{S_b} \mathbf{D}\,\mathbf{B}\,\mathbf{T}(P,Q)\,\mathrm{d}S_Q \tag{6.102}$$

Note that in equations (6.101) and (6.102), the differentiations implied by the operator $\mathbf{B}$ are with respect to the coordinates of the point P.

6.8 THE DISPLACEMENT DISCONTINUITY METHOD FOR LAPLACE'S EQUATION

It is often desired to solve boundary value problems for domains with cracks. A crack has two surfaces S^+ and S^- on which boundary conditions are to be satisfied, and these surfaces are coincident in space. In Figure 6.8, the surfaces S^+ and S^- are shown separated for the purpose of illustration only. The boundary conditions on the surfaces of a crack usually are Neumann for Laplace's equation, or traction given for elasticity.

Let us consider Laplace's equation, to be solved by the indirect method. If a solution of the form

$$u(Q) = \int_{S_0+S^++S^-} U(P,Q)\,\mu(P)\,\mathrm{d}S_P \tag{6.103}$$

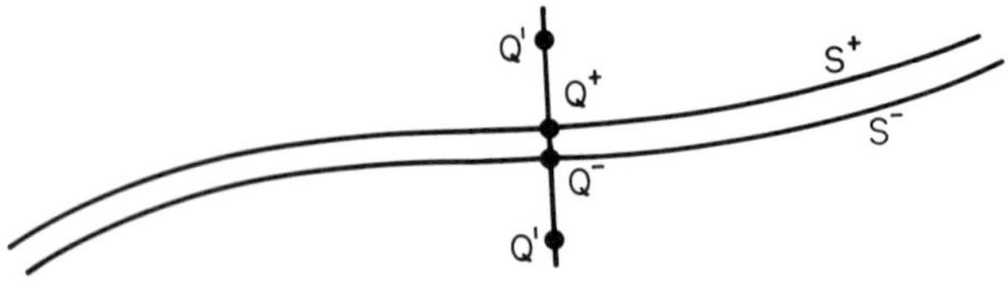

Figure 6.8 Surfaces of a crack, and the interior point Q'

where S_0 is the rest of the boundary S of the domain is proposed then, because the limiting value of $u(Q')$ as Q' approaches Q^+ (see Figure 6.8) equals $u(Q^+)$ and that as Q' approaches Q^- equals $u(Q^-)$, and the values $u(Q^+)$ and $u(Q^-)$ are equal, it is impossible to model a discontinuity of the solution across the crack. Therefore, the indirect method as described in Section 6.4 cannot be used to solve problems for domains with cracks.

Let us now consider the solution of Laplace's equation by the direct method. In the derivation of the boundary integral equation at the point P^+ on S^+, we take the neighbourhood $v(P^+, \epsilon)$ to be a complete sphere of radius ϵ centred at P^+ as shown in Figure 6.9, rather than part of a sphere as in Figure D.4. Upon making the same substitutions in Green's symmetric identity as in section 6.6, and then letting ϵ tend to zero, we obtain the integral equation

$$\frac{1}{2}\left(u(P^+) + u(P^-)\right) + \int_{S_0+S^++S^-} T(P^+, Q)\, u(Q)\, \mathrm{d}S_Q = \int_{S_0+S^++S^-} U(P^+, Q)\, \frac{\partial u}{\partial \mathbf{n}}(Q)\, \mathrm{d}S_Q \tag{6.104}$$

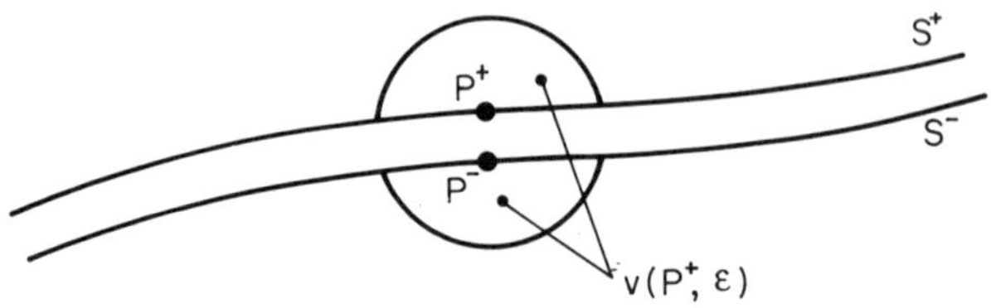

Figure 6.9 The neighbourhood $v(P^+, \epsilon)$

if S^+ and S^- are smooth at P^+ and P^-. Repeating this procedure for the point P^-, we obtain the equation

$$\frac{1}{2}\left(u(P^+) + u(P^-)\right) + \int_{S_0+S^++S^-} T(P^-, Q)\, u(Q)\, \mathrm{d}S_Q = \int_{S_0+S^++S^-} U(P^-, Q)\, \frac{\partial u}{\partial \mathbf{n}}(Q)\, \mathrm{d}S_Q \tag{6.105}$$

But $T(P^+, Q) = T(P^-, Q)$ and $U(P^+, Q) = U(P^-, Q)$. Therefore, equations (6.104) and (6.105) are identical and if in a numerical implementation provision is made for $u(P^+)$ and $u(P^-)$ to be unequal, then there will be fewer linearly independent equations than there are unknowns. The direct method as described in Section 6.6 cannot therefore be used to solve problems for domains with cracks.

In the indirect formulation, solutions which are discontinuous across cracks can be modelled if along the cracks we distribute not sources of intensity $\mu(P)$ but source dipoles of intensity $\delta(P)$. A unit source dipole is the limiting case as h tends to zero of a pair of sources of intensities $1/h$ and $-1/h$, separated by a distance h as shown in Figure 6.10; it gives rise to the singular solution of Laplace's equation

$$\lim_{h \to 0} \left(\frac{U(P_1, Q) - U(P_2, Q)}{h} \right) = \frac{\partial U(P, Q)}{\partial s(P)} \tag{6.106}$$

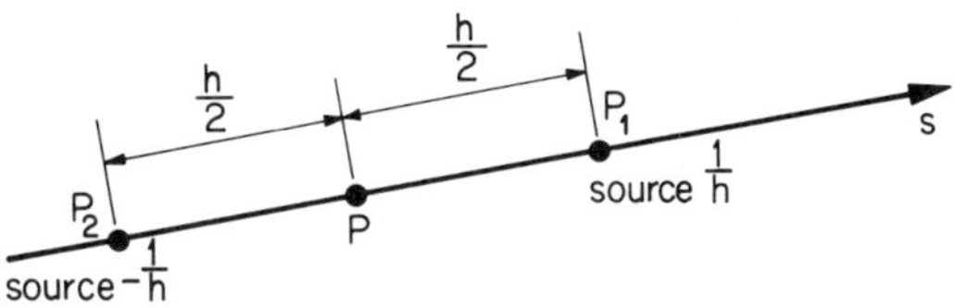

Figure 6.10 Unit source dipole

where s is the distance measured along the axis of the dipole. In the indirect formulation for cracks the axial direction of dipoles is taken to be the outward normal direction $\mathbf{n}(P)$ to S^- as shown in Figure 6.11. Since $U(P, Q) = U(Q, P)$,

$$\frac{\partial U(P, Q)}{\partial \mathbf{n}(P)} = \frac{\partial U(Q, P)}{\partial \mathbf{n}(P)} = T(Q, P) \tag{6.107}$$

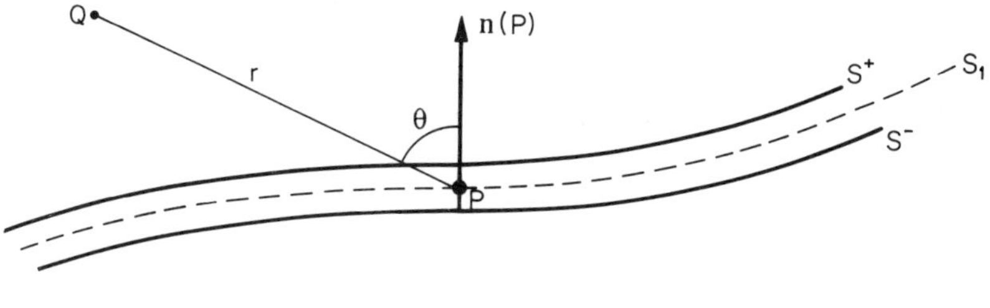

Figure 6.11 Fundamental solution for crack

where $T(Q, P)$ is given either by equation (6.3) or by equation (6.6). We propose a solution of the form

$$u(Q) = \int_{S_0} U(P, Q)\, \mu(P)\, \mathrm{d}S_P + \int_{S_1} T(Q, P)\, \delta(P)\, \mathrm{d}S_P \qquad (6.108)$$

where the density function $\delta(P)$ is the intensity of source dipoles distributed over a surface S_1 which coincides with the cracks[9]. To see how $u(Q)$ varies near S_1, we refer to results obtained earlier for integrals with respect to the second argument of the kernel T. In equations (6.78) and (6.79), the Poisson formula and the boundary integral equation of the direct method, the limiting value of the integral of $U(P', Q)\frac{\partial u}{\partial \mathbf{n}}(Q)$ as P' inside V approaches the point P on S is the same as the integral of $U(P, Q)\frac{\partial u}{\partial \mathbf{n}}(Q)$. The limiting value of $u(P')$ also equals its value $u(P)$ on S. By comparing equations (6.78) and (6.79) it may be seen that, if P is on a smooth part of S and so $C(P) = 1/2$,

$$\lim_{P' \to P} \int_S T(P', Q)\, u(Q)\, \mathrm{d}S_Q = \int_S T(P, Q)\, u(Q)\, \mathrm{d}S_Q - \frac{1}{2} u(P) \qquad (6.109)$$

In equation (6.108), the integral over S_0 is continuous as Q passes through S_1. By replacing u by δ, P' by Q', P by Q and Q by P in equation (6.109), we obtain the result (see Figure 6.8)

$$\lim_{Q' \to Q^-} \int_{S_1} T(Q', P)\, \delta(P)\, \mathrm{d}S_P = \int_{S_1} T(Q, P)\, \delta(P)\, \mathrm{d}S_P - \frac{1}{2}\delta(Q) \qquad (6.110)$$

In this equation, Q is a point which coincides with Q^+ and Q^-, and the point Q' is taken to approach Q^- because $\mathbf{n}(P)$ is defined to be the outward normal to S^-. It can similarly be proved that (Figure 6.8)

$$\lim_{Q' \to Q^+} \int_{S_1} T(Q', P)\, \delta(P)\, \mathrm{d}S_P = \int_{S_1} T(Q, P)\, \delta(P)\, \mathrm{d}S_P + \frac{1}{2}\delta(Q) \qquad (6.111)$$

Therefore, for $u(Q)$ as defined by equation (6.108),

$$u(Q^+) - u(Q^-) = \delta(Q) \qquad (6.112)$$

so it is possible to model solutions which are discontinuous across cracks.

For the Neumann boundary condition on S^+ and S^-, it is necessary to set the limiting values as Q' approaches Q^+ and Q^- of the normal derivatives of $u(Q')$ to their given values $\bar{t}(Q^+)$ and $\bar{t}(Q^-)$. Let us consider the limiting value as Q' approaches Q^- of $\frac{\partial u(Q')}{\partial \mathbf{n}(Q)}$, which must equal $\bar{t}(Q^-)$ since $\mathbf{n}(Q)$ is defined to be the outward normal to S^-. By differentiating equation (6.108) in the direction $\mathbf{n}(Q)$ at the point Q', we find that

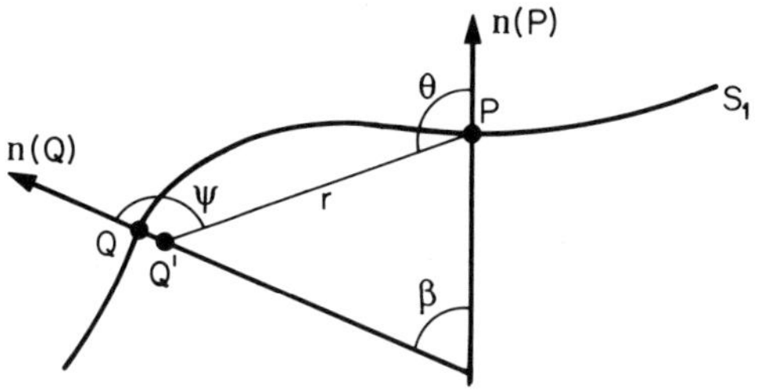

Figure 6.12 Notation for definition of normal derivative of kernel T

$$\frac{\partial u}{\partial \mathbf{n}(Q)}(Q') = \int_{S_0} T(P, Q')\,\mu(P)\,\mathrm{d}S_P + \int_{S_1} \frac{\partial T(Q', P)}{\partial \mathbf{n}(Q)}\,\delta(P)\,\mathrm{d}S_P \tag{6.113}$$

where (see Figure 6.12) for a two-dimensional problem

$$\frac{\partial T(Q', P)}{\partial \mathbf{n}(Q)} = \frac{1}{2\pi r^2}(\cos\beta + 2\cos\theta\cos\psi) \tag{6.114}$$

For the purpose of illustration let us suppose Q to be on a straight part of S_1, on which $\delta(P)$ is constant and therefore equal to $\delta(Q)$, as shown in Figure 6.13. We denote by $S_1(Q, \eta)$ the straight part AB of S_1, and by ϵ the distance QQ'. We take $\frac{\partial u}{\partial \mathbf{n}(Q)}(Q')$ to be the sum of the two parts $I_1(Q')$ and $I_2(Q')$, where

$$I_1(Q') = \int_{S_1(Q,\eta)} \frac{\partial T(Q', P)}{\partial \mathbf{n}(Q)}\,\delta(P)\,\mathrm{d}S_P \tag{6.115}$$

$I_2(Q')$ varies continuously as Q' approaches and passes through S_1 at Q, and by equation (6.114)

$$I_1(Q') = \delta(Q) \cdot \int_{-\eta}^{\eta} \frac{1}{2\pi r^2}(1 - 2\cos^2\psi)\,\mathrm{d}s \tag{6.116}$$

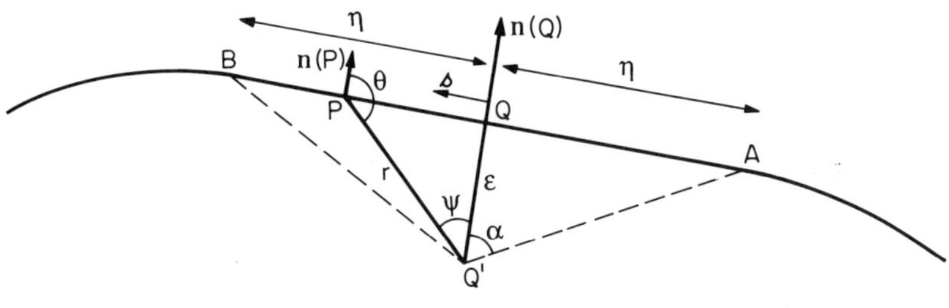

Figure 6.13 The point Q on a straight part of S_1

By integrating and then letting ϵ tend to zero, it is found that

$$\lim_{Q' \to Q^-} I_1(Q') = -\frac{\delta(Q)}{\pi\eta} \tag{6.117}$$

Similarly, by considering the limiting value as Q' approaches Q^+ of $\frac{\partial u}{\partial \mathbf{n}(Q)}(Q')$, it can be shown that

$$\lim_{Q' \to Q^+} I_1(Q') = -\frac{\delta(Q)}{\pi\eta} \tag{6.118}$$

The limiting values of $\frac{\partial u}{\partial \mathbf{n}(Q)}(Q')$ as Q' approaches Q^- and Q^+ therefore are equal. These results apply if Q is on a straight part of the crack, over which $\delta(P)$ is constant. Provided S_1 is sufficiently smooth and $\delta(P)$ varies sufficiently smoothly over S_1, limiting values of $I_1(Q')$ and therefore of $\frac{\partial u}{\partial \mathbf{n}(Q)}(Q')$ as Q' approaches Q^- and Q^+ still exist, and are equal. Since the outward normal direction at Q^+ is opposite to that at Q^-, this implies that the integral representation defined by equation (6.108) is capable of modelling solutions which are discontinuous across the crack, and for which $\bar{t}(Q^+) = -\bar{t}(Q^-)$. The integral equation to be satisfied on S_1 is

$$\lim_{Q' \to Q^-} \left(\int_{S_0} T(P, Q')\, \mu(P)\, \mathrm{d}S_P + \int_{S_1} \frac{\partial T(Q', P)}{\partial \mathbf{n}(Q)}\, \delta(P)\, \mathrm{d}S_P \right) = \bar{t}(Q^-) \tag{6.119}$$

Note that not even a Cauchy principal value of the integral over S_1 of $\frac{\partial T(Q^-, P)}{\partial \mathbf{n}(Q)} \delta(P)$ exists. The first and second integrals of equation (6.108) are known as a single layer potential and a double layer potential respectively. If it is required to solve problems for domains with cracks for which $\bar{t}(Q^+) \neq -\bar{t}(Q^-)$, then the source density $\mu(P)$ is taken to be distributed over S_1 as well as S_0.

In the simplest implementation of the displacement discontinuity method in two dimensions, the surface S_1 is represented by straight line elements, and it is supposed that over each of these elements $\delta(P)$ is constant. Simultaneous equations for the values of $\delta(P)$ are obtained by taking the point Q^- in equation (6.119) to be located at the centroid of each of the elements in turn. Let there be p_0 elements $S_1, S_2, \ldots, S_{p_0}$ on S_0, and $p - p_0$ elements $S_{p_0+1}, S_{p_0+2}, \ldots, S_p$ on S_1. Then the equation obtained by taking Q^- to be at the centroid Q_a of an element of length $2\eta_a$ on S_1 is

$$\sum_{b=1}^{p_0} \Delta T_{ab}\, \mu(Q_b) + \sum_{\substack{b=p_0+1 \\ b \neq a}}^{p} \Delta T_{ab}^{Q}\, \delta(Q_b) - \frac{\delta(Q_a)}{\pi\eta_a} = \bar{t}(Q_a^-) \tag{6.120}$$

where ΔT_{ab} is given by equation (6.50) and

$$\Delta T_{ab}^{Q} = \int_{S_b} \frac{\partial T(Q_a, P)}{\partial \mathbf{n}(Q_a)} \, \mathrm{d}S_P \tag{6.121}$$

In the analogy of heat flow in which $u(Q)$ is temperature for example, S_1 could represent a thin layer of non-conducting material embedded in the domain bounded by S_0. In this case the boundary conditions on S_1 would be $\bar{t}(Q^+) = \bar{t}(Q^-) = 0$. If $u(Q)$ were piezometric potential in a soil mass with unit permeability in which there were steady state groundwater flow, then S_1 could represent a sheet pile, and again the boundary conditions on S_1 would be $\bar{t}(Q^+) = \bar{t}(Q^-) = 0$.

6.9 THE DISPLACEMENT DISCONTINUITY METHOD FOR ELASTICITY

Neither the indirect method as described in Section 6.5 nor the direct method as described in Section 6.7 can be used to solve problems for domains with cracks across which displacements are discontinuous. Such solutions can however be modelled if along the cracks we distribute a double layer of force, analogous to the double layer potential of Section 6.8. We adopt the same notation and definition of $\mathbf{n}(Q)$ on S_1 as in that section, and propose a solution of the form

$$\mathbf{u}(Q) = \int_{S_0} \mathbf{U}(P, Q)^{\mathrm{T}} \boldsymbol{\mu}(P) \, \mathrm{d}S_P + \int_{S_1} \mathbf{T}(Q, P) \boldsymbol{\delta}(P) \, \mathrm{d}S_P \tag{6.122}$$

where $\mathbf{T}(Q, P)$, which is given by either equations (6.16) to (6.19) or equations (6.16) and (6.23), is the displacement at Q of an infinite elastic space due to a combination of force dipoles at P. Each of these force dipoles is the limiting case as h tends to zero of a pair of equal and opposite forces of magnitude $1/h$, separated by a distance h. The combination of force dipoles for three dimensional problems is sometimes referred to as a hexapole, and that for plane strain is known as a quadrapole. By an analysis similar to that of Section 6.8, it can be shown that for $\mathbf{u}(Q)$ as defined by equation (6.122)

$$\mathbf{u}(Q^+) - \mathbf{u}(Q^-) = \boldsymbol{\delta}(Q) \tag{6.123}$$

For the traction boundary condition on S^+ and S^-, it is necessary to set the limiting values, as Q' approaches Q^+ and Q^-, of the tractions due to $\mathbf{u}(Q)$ to their given values $\bar{\mathbf{t}}(Q^+)$ and $\bar{\mathbf{t}}(Q^-)$. By differentiating equation (6.122) with respect to the coordinates of the point Q', and substituting into the strain–displacement and stress–strain relations it can be shown that

$$\mathbf{t}(Q') = \int_{S_0} \mathbf{T}(P, Q')^{\mathrm{T}} \boldsymbol{\mu}(P)\, \mathrm{d}S_P + \int_{S_1} \mathbf{Z}(Q', P) \boldsymbol{\delta}(P)\, \mathrm{d}S_P \qquad (6.124)$$

where

$$\mathbf{Z}(Q', P) = \mathbf{N}(Q)\ \mathbf{D}\ \mathbf{B}\ \mathbf{T}(Q', P) \qquad (6.125)$$

In equation (6.124), $\mathbf{t}(Q')$ is the traction acting on a plane perpendicular to $\mathbf{n}(Q)$ and passing through the point Q' (see Figure 6.12), and $\mathbf{T}(P, Q')$ is computed by reference to the normal direction $\mathbf{n}(Q)$. In equation (6.125), $\mathbf{N}(Q)$ is given by equation (6.14) and the differentiations of $\mathbf{B}$ are with respect to the coordinates of the point Q'. Coefficients of $\mathbf{Z}$ are given in Appendix C. By an analysis similar to that of Section 6.8 it can be shown that, subject to certain smoothness conditions, limiting values of $\mathbf{t}(Q')$ as Q' approaches the point Q^- on S^- or the point Q^+ on S^+ do exist, and that $\mathbf{t}(Q^+) = -\mathbf{t}(Q^-)$. The integral equation to be satisfied on S_1 is then

$$\lim_{Q' \to Q^-} \left(\int_{S_0} \mathbf{T}(P, Q')^{\mathrm{T}} \boldsymbol{\mu}(P)\, \mathrm{d}S_P + \int_{S_1} \mathbf{Z}(Q', P) \boldsymbol{\delta}(P)\, \mathrm{d}S_P \right) = \bar{\mathbf{t}}(Q^-) \qquad (6.126)$$

Note that not even a Cauchy principal value of the integral over S_1 of $\mathbf{Z}(Q, P)\boldsymbol{\delta}(P)$ exists, and it is strictly in the sense of a limiting value as Q' approaches Q^- that the second integral of equation (6.126) is to be understood.

By a lengthy derivation it can be shown that if the point Q lies on a straight part $S_1(Q, \eta)$ of S_1 as shown in Figure 6.13, and $\boldsymbol{\delta}(P)$ is constant over $S_1(Q, \eta)$, then

$$\lim_{Q' \to Q^-} \int_{S_1(Q,\eta)} \mathbf{Z}(Q', P) \boldsymbol{\delta}(P)\, \mathrm{d}S_P = \mathbf{z}(Q) \qquad (6.127)$$

where for plane strain

$$\mathbf{z}(Q) = \frac{-E}{2\pi(1-\nu^2)\eta} \begin{Bmatrix} n_y(Q)\delta_x(Q) + n_x(Q)\delta_y(Q) \\ n_y(Q)\delta_y(Q) - n_x(Q)\delta_x(Q) \end{Bmatrix} \qquad (6.128)$$

Note that in equation (6.127), $\mathbf{Z}(Q', P)$ is computed by reference to the outward normal directions $\mathbf{n}(P)$ and $\mathbf{n}(Q)$ to S^-. In equation (6.128), E and ν are Young's modulus and Poisson's ratio, respectively. In the simplest implementation of the displacement discontinuity method for plane strain, the surface S_1 is represented by straight line elements, and it is supposed that over each of these elements $\boldsymbol{\delta}(P)$ is constant. Simultaneous equations for the values of $\boldsymbol{\delta}(P)$ are obtained by taking the point Q to be located at the centroid of each of the elements in turn and using equation (6.127) to integrate over the element of which Q is the centroid.

The method as described in this section applies only to problems for which $\mathbf{t}(Q^+) = -\mathbf{t}(Q^-)$. Practical problems are almost invariably of this type, but if it were required to consider the general case then it would only be necessary to take the simple layer of force $\boldsymbol{\mu}(P)$ to be distributed over S_1 as well as S_0. The displacement discontinuity method is often used to compute stresses in the neighbourhood of slots created by the mining of tabular (thin) orebodies. If the aperture of the slot is small compared with its other dimensions then accurate results are obtained everywhere except near the edges or abutment of the slot where a more detailed analysis is required.

6.10 BODY FORCES

An elastic body is generally subject to load distributed through its volume, known as body force. The two most commonly encountered body forces are that due to gravity, and a body force distribution known as centripetal force which represents the effect of inertial forces due to rotation. Body forces are taken into account by taking the displacement $\mathbf{u}$ to be the sum of a particular integral $\mathbf{u}^{(1)}$ and a complementary function $\mathbf{u}^{(2)}$:

$$\mathbf{u} = \mathbf{u}^{(1)} + \mathbf{u}^{(2)} \tag{6.129}$$

where $\mathbf{u}^{(1)}$ is a function that satisfies the governing partial differential equation but not necessarily the boundary conditions. Over a part of the boundary on which displacement is specified, the boundary condition for the complementary function is therefore

$$\mathbf{u}^{(2)} = \bar{\mathbf{u}} - \mathbf{u}^{(1)} \tag{6.130}$$

and over a part of the boundary on which traction is specified, the boundary condition for the complementary function is

$$\mathbf{t}^{(2)} = \bar{\mathbf{t}} - \mathbf{t}^{(1)} \tag{6.131}$$

where $\mathbf{t}^{(1)}$ is traction due to $\mathbf{u}^{(1)}$, and $\mathbf{t}^{(2)}$ is traction due to $\mathbf{u}^{(2)}$. The complementary function satisfies the partial differential equations of equilibrium with zero body force term, and therefore can be computed by solving a boundary integral equation as described in Sections 6.5, 6.7 or 6.9. Particular integrals are easy to find because they do not have to be determined with regard to boundary conditions. For an arbitrary body force distribution $\mathbf{b}(P)$ acting on the domain V we may propose

$$\mathbf{u}^{(1)}(Q) = \int_V \mathbf{U}(P,Q)^{\mathrm{T}}\mathbf{b}(P)\,\mathrm{d}V_P \tag{6.132}$$

where $\mathbf{U}(P, Q)$ is given by equations (6.9) and (6.10) or by equations (6.21) and (6.22). Evaluation of the particular integral in this form requires integration over an interior mesh of elements, and therefore somewhat detracts from the advantage of boundary element methods in not requiring an interior mesh. Very often, however, the use of a particular integral of completely general form can be avoided. For gravitational load in the direction $-z$, the particular integral may be taken as

$$\begin{aligned} u_x^{(1)} &= 0 \\ u_y^{(1)} &= 0 \\ u_z^{(1)} &= \frac{(1+\nu)(1-2\nu)\rho g z^2}{2E(1-\nu)} \end{aligned} \tag{6.133}$$

where ρ is density and g is acceleration due to gravity. For centripetal force corresponding to rotation about the third axis z' of a local cartesian system (x', y', z'), the particular integral may be taken to be

$$\begin{aligned} u_{r'}^{(1)} &= -\frac{(1+\nu)(1-2\nu)\omega^2 \rho r'^3}{8E(1-\nu)} \\ u_{\theta'}^{(1)} &= 0 \\ u_{\mathbf{z}'}^{(1)} &= 0 \end{aligned} \tag{6.134}$$

where $r' = \sqrt{x'^2 + y'^2}$, $\theta' = \tan^{-1}(y'/x')$ and ω is angular velocity in radians per second. Global components may be computed by the transformations described in Chapter 3.

6.11 INFINITE REGIONS

In some fields of application, it is convenient to take the domain to be analysed to be of infinite extent. For example, when computing stresses in the neighbourhood of a cavity deep underground, we suppose that the rock mass extends an infinite distance in all directions rather than attempt to model the entire Earth's crust. If the dimensions of the cavity are significant compared with its depth below the surface, then account has to be taken of the existence of the zero traction boundary condition at the surface and so a halfspace has to be analysed. In Section 6.7, the direct method is derived for a finite domain V bounded by a surface S, this being the case to which the divergence theorem applies. In this section, the direct method for Laplace's equation and elastostatics is generalised to infinite domains with finite or infinite surface area.

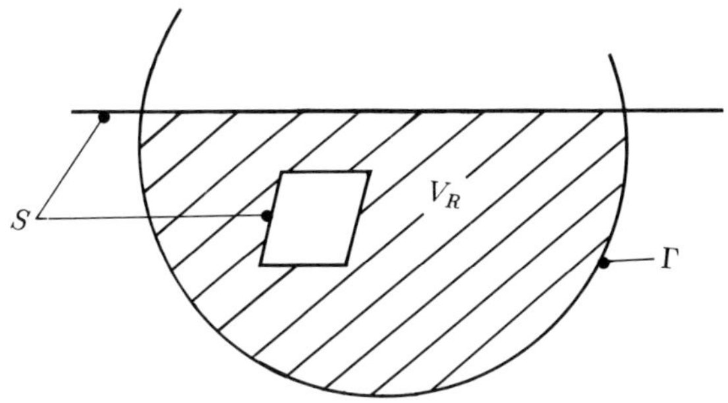

Figure 6.14 The finite domain V_R

In the generalisation, the infinite domain V in three dimensions is regarded as the limiting case of the finite domain V_R bounded by the actual surface S and the surface Γ of a sphere of radius R centred at the origin of coordinates, as R tends to infinity. In Figure 6.14 is shown a typical infinite domain the surface S of which consists of two parts one of which is of infinite extent, and the surface Γ. We consider first Laplace's equation. Writing equation (6.79) for the domain V_R, we obtain

$$C(P)u(P) + \int_{S_R+\Gamma_R} T(P,Q)u(Q)\ \mathrm{d}S_Q = \int_{S_R+\Gamma_R} U(P,Q)\frac{\partial u}{\partial \mathbf{n}}(Q)\ \mathrm{d}S_Q \tag{6.135}$$

where S_R is that part of S which lies inside the sphere, and Γ_R is that part of Γ which lies inside the domain V. Let us now suppose that the point P lies inside a sphere of radius R_0 centred at the origin of coordinates, and that $R \geq 2R_0$, so that the distance r from P to a point Q on Γ_R is equal to at least $R/2$. Let us also suppose that, on the surface Γ_R,

$$\begin{aligned} |u(Q)| &\leq \frac{K}{R} \\ \left|\frac{\partial u}{\partial \mathbf{n}}(Q)\right| &\leq \frac{K}{R^2} \end{aligned} \tag{6.136}$$

where K is a constant. If Q is on Γ_R then by equation (6.1), equation (6.3) and the inequality $r \geq R/2$, we have

$$\begin{aligned} |T(P,Q)| &\leq \frac{1}{\pi R^2} \\ |U(P,Q)| &\leq \frac{1}{2\pi R} \end{aligned} \tag{6.137}$$

The area of Γ_R is not greater than $4\pi R^2$, so by inequalities (6.136) and (6.137)

$$\left|\int_{\Gamma_R} T(P,Q)u(Q)\,\mathrm{d}S_Q\right| \le \frac{1}{\pi R^2}\cdot\frac{K}{R}.4\pi R^2 = \frac{4K}{R}$$
$$\left|\int_{\Gamma_R} U(P,Q)\frac{\partial u}{\partial \mathbf{n}}(Q)\,\mathrm{d}S_Q\right| \le \frac{1}{2\pi R}\cdot\frac{K}{R^2}.4\pi R^2 = \frac{2K}{R} \qquad (6.138)$$

From these results it is evident that in the limit as R tends to infinity and S_R becomes the surface S,

$$C(P)u(P) + \int_S T(P,Q)u(Q)\,\mathrm{d}S_Q = \int_S U(P,Q)\frac{\partial u}{\partial \mathbf{n}}(Q)\,\mathrm{d}S_Q \qquad (6.139)$$

provided the point P lies within a sphere of radius R_0 centred at the origin of coordinates. But the choice of R_0 is arbitrary, so equation (6.139) is valid for P located at any finite distance from the origin. It would be possible to prove equation (6.139) subject to the looser conditions $|u(Q)| \le KR^{-\alpha}$ and $\left|\frac{\partial u}{\partial \mathbf{n}}\right| \le KR^{-(1+\alpha)}$ where $\alpha > 0$, but there is no reason to do this because solutions of Laplace's equation which satisfy these conditions but not inequalities (6.136) are not encountered in practice. If inequalities (6.136) are not satisfied, then the solution must be taken as the sum of a particular integral $u^{(1)}$ and complementary function $u^{(2)}$ where $u^{(2)}$ does satisfy inequality (6.136), as is done in the Trefftz solution of Section 6.3. For a two-dimensional domain, Γ is taken to be a circle centred at the origin, and it can similarly be shown that equation (6.139) is valid if inequalities (6.136) are satisfied. Particular integrals which ensure the satisfaction of inequalities (6.136) by the complementary function are easy to find, and indeed in most cases can be expressed as polynomials as illustrated in Section 6.10.

For two- and three-dimensional elasticity, it can likewise be shown that equation (6.93), the boundary integral equation of the direct formulation, is valid provided displacement and traction satisfy the inequalities

$$|\mathbf{u}(Q)| \le \frac{K}{R}$$
$$|\mathbf{t}(Q)| \le \frac{K}{R^2} \qquad (6.140)$$

where K is a constant. Again the solution is in general taken as the sum of a particular integral and a complementary function, the particular integral being chosen such that the complementary function satisfies inequalities (6.140). Particular integrals are easy to find: for example if stress tends to some limiting value at infinity then the particular integral is taken to be the displacement field corresponding to a uniform state of stress, equal to the given limiting value. In the general case in which body forces are present, the particular integral is the sum of a component due to body force and a component chosen to ensure that the complementary function satisfies inequalities (6.140).

6.12 SUMMARY AND CONCLUSIONS

In principle, boundary element methods are a very attractive means of solving boundary value problems encountered in engineering practice. The need to define a mesh of elements through the interior of the domain is eliminated, and the system of simultaneous equations to be solved is smaller than that of the finite element method. Practical boundary element methods can only be developed, however, if a fundamental solution of the governing partial differential equation is known, and can be computed economically. There remain therefore many types of problems which can be solved by the finite element method but not by the boundary element method. The finite element method will therefore never be totally supplanted by boundary element methods. In addition, for those problems to which boundary element methods are applicable, although the matrix of equation coefficents is smaller than that generated by the finite element method, it is neither banded nor symmetric about the leading diagonal and also the cost of matrix construction per coefficient is higher. The advantages of boundary element methods are therefore not so great as might at first be supposed.

The indirect method may be seen as a refinement of the Trefftz method, or alternatively as a generalisation to the multidimensional case of the solution of an ordinary differential equation by the construction of a complementary function. The arbitrary constants in a complementary function are analogous to the density function, and the simultaneous equations for those constants are analogous to the boundary integral equation. Proofs of the existence of solutions of the proposed form (see for example equation (6.33)) have been developed only for domains with smooth surfaces, whereas in engineering practice domains with edges and corners often must be considered. There is little doubt that solutions do exist for domains with only piecewise smooth boundaries, but the density function probably tends to infinity as edges and corners are approached even if the solution of the boundary value problem itself exhibits no singular behaviour. However there appears to be no information in the literature on the form of these singularities. Without such information, it is difficult to propose numerical approximations that will model reasonably well the variation of density functions near edges and corners and thereby minimise the need for refinement of the boundary element mesh in those areas. Mesh refinement is inconvenient and increases the number of unknowns to be computed.

It is evident from the derivation detailed in Section 6.4 that, for the Neumann boundary condition, the integral equation is only valid if the argument Q of the free term does not lie on an edge or corner of the boundary. In the very simple numerical implementation outlined in that section, Q is placed at the centroid of each of the boundary elements in turn and so the boundary condition is satisfied exactly at each centroid, subject to the assumption that the density function $\mu(P)$ is constant over each element. The requirement that Q lies on a smooth part of S is automatically satisfied in this nodal collocation scheme. However if linear or higher order variation of the density function is proposed over each element, and so there are nodes at the extremities or vertices of elements, then in general a nodal collocation scheme cannot be applied because that scheme would require the point Q to be located at nodes which could lie on an edge or corner of the boundary. If elements with higher order variation of the density function are to be used, therefore, an alternative to nodal collocation must be proposed.

In Section 6.5 it is stated that, for the usual traction boundary condition of elasticity, the Cauchy principal value of the integral appearing in the boundary integral equation must be evaluated. This is a simple matter in the numerical implementation outlined in that section. However if curved elements or elements with higher order variation of the density function are to be used and so it becomes necessary to resort to numerical integration, care must be taken to ensure that numerical integration schemes do correctly evaluate Cauchy principal values.

A physical interpretation of the theoretical basis of the direct method is difficult to achieve and probably of little practical use. In one dimension, Green's symmetric identity and Betti's theorem reduce to applications of integration by parts, and integration by parts derives from the formula for differentiation of a product and the fundamental theorem of integral calculus. The fact that in the direct method the results computed by solving the boundary integral equation are of physical significance obviates the need to consider the question of existence of a solution of the integral equation: if the boundary value problem has a solution, then so does the integral equation. Boundary integral equations of the direct method are valid even if the argument P of the free term lies on an edge or corner of the boundary. Therefore it is possible to use a collocation scheme in which the point P is taken to be located at nodes of extremities or vertices of boundary elements, as would be the case if linear or higher order variation of unknown functions over each element were proposed. It is stated in Appendix E that the Cauchy principal value of one of the integrals appearing in the boundary integral equation of elasticity must be evaluated. It is possible, by considering rigid body translations of the domain to be analysed, to avoid computing separately either the Cauchy principal value or the coefficient $\mathbf{C}(P)$ of the free term. This is of little consequence in the numerical procedure described in Section 6.7, but simplifies the implementation of curved elements or elements with linear or higher order variation of unknown functions.

It should now be clear to the reader that the indirect and direct methods are derived in totally different ways, and that it is essential to ensure that a numerical implementation of either method is consistent with the manner in which the method is derived. Some authors have felt the need to develop unified derivations of all boundary element methods or even of all boundary element and finite element methods. This may be mathematically elegant but it invariably obscures theoretical detail pertinent to the implementation of one or more of the methods.

Except for the displacement discontinuity method, which solves a special type of problem efficiently even with elements over each of which the unknown function is taken to be constant ('constant elements'), the simple numerical implementations outlined in this chapter are uncompetitive with finite elements. As a rough guide, the efficiency of a method is improved if the number of unknowns to be computed to solve a given problem to a given accuracy is reduced. For example, if the cost of computation of each equation coefficient is doubled but the number of unknowns is halved, the cost of matrix construction is halved and that of bifactorisation by Gaussian elimination is divided by eight. Clearly a considerable increase in cost of computation of each matrix coefficient is acceptable in return for a reduction in the number of unknowns, especially for large problems in which the cost of simultaneous equation solution is the dominant cost. Let us consider the boundary element meshes shown in Figure 6.15, in which the unknown function is taken to be constant, to vary linearly and to vary quadratically over each element. Dashes represent ends of elements, and triangles represent nodes with which are associated unknowns. The results obtained using linear elements should be more accurate than those obtained using constant elements because the linear interpolate over each element can model more accurately the actual variation of the unknown function. The number of unknowns is the same for both meshes, and so the computing cost would be about the same. Linear elements are therefore expected to be more efficient than constant elements. For the mesh of quadratic elements shown in Figure 6.15c, in which there are half as many elements as in the other two meshes, the number of unknowns is the same as for the linear elements but whereas over two linear elements the unknown function is represented by a linear interpolate over each element, in the single quadratic element which replaces them it is represented by an interpolate with no discontinuity of derivative. To what extent this may be an advantage is difficult to predict. However, isoparametric quadratic elements model the geometry of curved boundaries very well, so are expected to possess a considerable advantage over linear elements if the domain to be analysed has curved boundaries.

Once constant elements are abandoned, the numerical implementation of boundary element methods becomes considerably more difficult than that of the finite element method. For the reasons stated earlier in this section, curved elements and higher order functional variation are easier to develop for the direct method than for the indirect method; therefore, more progress to date

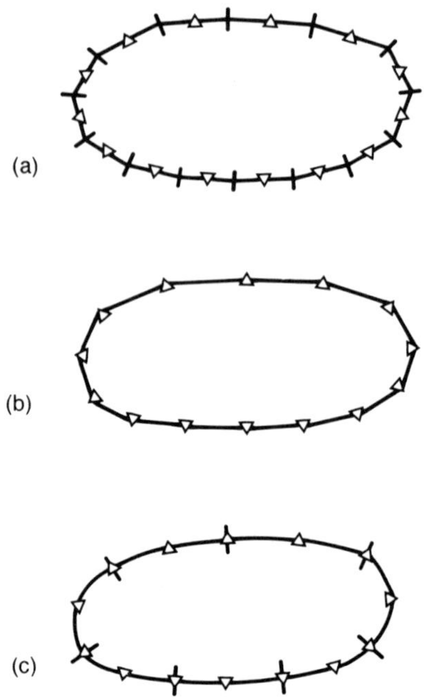

Figure 6.15 Boundary element meshes. (a) Constant elements. (b) Linear elements. (c) Quadratic elements

has been made in the efficient implementation of the direct method than of the indirect method. In the next chapter, the principles of implementation of higher order elements for two- and three-dimensional analysis are explained in some detail and examples of application are shown.

6.13 REFERENCES

1. Trefftz, E. (1926) Ein Gegenstück zum Ritzschen Verfahren *Proc. 2nd Int. Congress in Applied Mechanics*, Zürich, p. 131.
2. Kelvin, Lord and Tait (1879) *Natural Philosophy*, 2nd edn.
3. Synge, J.L. (1957) *The Hypercircle in Mathematical Physics*, Cambridge University Press, Cambridge, p. 411.
4. Rizzo, F.J. and Shippy, D.J. (1979) A boundary integral approach to potential and elasticity problems for axisymmetric bodies with arbitrary boundary conditions. *Mech. Res. Commun.*, **6**, 99–103.
5. Stern, M. (1979) A general boundary integral formulation for the numerical solution of plate bending problems. *Int. J. Solids Struct.*, **15**, 769–82.
6. Mindlin, R.D. (1936) Force at a point in the interior of a semi-infinite solid. *Physics*, **7**, 195–202.
7. Bettess, P. (1977) Infinite elements. *Int. J. Num. Meth. Eng.*, **11**, 53–64.
8. Kreyszig, E. (1988) *Advanced Engineering Mathematics*, 6th edn, Wiley, New York.

9. Smirnoff, V.I. (1964) *A Course of Higher Mathematics*, Vol 4, Pergamon Press, London.
10. Stephan, E.P. and Wendland, W.L. (1984) *An augmented Galerkin procedure for the boundary integral method applied to two-dimensional screen and crack problems.* Technische Hochschule Darmstadt. Preprint No. 802, February 1984.
11. Kellogg, O.D. (1929) *Foundations of Potential Theory.* Springer, Berlin.
12. Abramowitz, M. and Stegun, I.A. (1965) *Handbook of Mathematical Functions.* Dover, New York.
13. Watson, J.O. (1972) *The analysis of thick shells with holes by integral representation of displacement.* PhD. thesis, University of Southampton.
14. Kupradze, V.D. (1965) *Potential Methods in the Theory of Elasticity.* (Transl. from Russian by Israel Program for Scientific Transl. Jeursalem.)
15. Hoek, E. and Brown, E.T. (1980) *Underground Excavations in Rock.* I.M.M., London.
16. Jaswon, M.A. (1963) Integral equation methods in potential theory–1 *Proc. R. Soc.*, **273(A)**, 23–32.
17. Rizzo, F.J. (1967) An integral equation approach to boundary value problems of classical elastostatics. *Q. Appl. Math.*, **25**, 83–95.

7

Boundary Element Methods—Practical Implementation

Everything takes half as long again,
costs 30 percent more, and can
be improved by 10 percent.

Anon

7.1 INTRODUCTION

In Chapter 6 the need for boundary elements with higher order functional variation is explained, and it is stated that the development of such elements is more straightforward for the direct method than for indirect methods. Accordingly, the practical implementations described in this chapter are of linear and quadratic variation elements for the direct method.

The first aspect of practical implementation discussed is the representation of the boundary, and of functions over the boundary, by means of elements. This requires isoparametric elements, and infinite elements to model domains with surfaces of infinite extent. Once such a representation has been proposed, the value of an unknown function at any point on the boundary can be expressed in terms of its values at a set of nodes on the boundary. The next step is to obtain a system of simultaneous equations, referred to here as the discrete system, in terms of unknown nodal values. The usual way in which this is done is by nodal collocation, in which the argument of the free term of the integral equation is taken to be located at each of the nodes in turn. However, there are alternative techniques in which an attempt is made to satisfy the integral equation not only at the nodes but also elsewhere on the boundary; these techniques, known as methods of moments, are described and compared with nodal collocation in Section 7.2.

Perhaps the greatest problem posed by the introduction of higher order functional variation and isoparametric elements is that of evaluation of the integrals of kernel–shape function products of which equation coefficients of the discrete system are composed. Analytical integration is generally no longer possible, and it is necessary to use quadrature formulae to integrate functions which tend to infinity in the neighbourhood of the first argument of the kernels, to integrate over infinite elements, and to integrate reliably and economically over other parts of the boundary functions which vary rapidly over certain elements and slowly over others. In the integration schemes presented in Sections 7.4 and 7.5, Gaussian quadrature formulae are adapted to evaluate accurately all weakly singular integrals over elements or parts of elements adjacent to the first argument of the kernels, and integrals over infinite elements. Strongly singular integrals, that is integrals of which only a Cauchy principal value exists, are never explicitly computed and nor is the coefficient of the free term. Instead, the sum of these contributions to the discrete system is computed by consideration of rigid body translations in elastostatics or of a uniform increase in potential for Laplace's equation. Note that for Laplace's equation there are no strongly singular integrals to evaluate, but for uniformity with the treatment of elastostatics and to avoid the need to compute the coefficient of the free term from the angle subtended by the domain at a node, the implicit integration procedure is retained. Requirements of economy and consistent accuracy are met by choosing the orders of Gaussian quadrature formulae according to the estimated rapidity of variation of the function to be integrated. If the orders of Gaussian formulae were not varied in this way, then it would be necessary to impose restrictions on the length to breadth ratios of elements representing surfaces in three dimensions, and to limit by means of mesh refinement the angle subtended by an element at any node or other point at which the first argument of the kernels is to be located. Especially where there are close parallel boundaries, or in three dimensions where there are parts of the surface over which the solution varies very much more rapidly in one direction than in another, this could result in an unacceptable increase in the number of unknowns. Finally, it is only by variation of the orders of Gaussian quadrature formulae that accurate results can be obtained by the Poisson formula or Somigliana's identity at interior points close to the boundary in relation to the dimensions of boundary elements.

If the domain to be analysed consists of more than one material, then because fundamental solutions that can be economically computed are valid only for a homogeneous domain it is necessary to divide the domain into subregions each of which is homogeneous and to construct a discrete system for each subregion. For this purpose the interfaces between subregions are represented by elements, as well as the surface of the domain. There are twice as many unknowns at a node on an interface as there are at a node on the surface, and the unknowns at a node on an interface are common to the discrete systems of the adjacent subregions. The resulting system of simultaneous equations may be solved by regarding the subregions as substructures, for which first the unknowns local to each subregion are eliminated and then a system in terms of unknowns at interface nodes only is solved. Whether or not there are subregions, the matrix of equation coefficients is non-symmetric. For problems of practical size an out of core solver is required because it is usually not possible to hold the matrix in byte or word addressable memory. For large three-dimensional problems of elasticity, simultaneous equation solution is the most costly part of the analysis in terms of computer time. The cost of such an analysis can be considerably reduced by an iterative solution technique in which only approximate lower and upper triangular factors of the matrix of equation coefficients are computed.

The reader will appreciate that the practical implementation of boundary element methods has necessitated the development of several specialised numerical techniques, especially for numerical integration. These techniques are continually being generalised and improved, and this chapter serves only as an introduction to the emerging discipline of numerical implementation of boundary element methods.

7.2 ISOPARAMETRIC ELEMENTS AND DISCRETE SYSTEMS

Let us take the boundary S to be represented by p isoparametric linear or quadratic elements $S_1, S_2, \ldots, S_p$ as described in Chapter 2. A system of simultaneous equations in terms of nodal values of given and unknown data on the boundary may be obtained by taking the argument P of the free term in equation (6.79) for Laplace's equation or equation (6.93) for elasticity to be located at each of the q nodes $P_1, P_2, \ldots, P_q$ in turn, and substituting the parametric interpolates defined by equation (2.38). By writing equation (6.79) at the node P_a, we obtain the equation of nodal collocation

$$C(P_a)u(P_a) + \sum_{b=1}^{p} \int_{S_b} T(P_a, Q)u(Q)\ \mathrm{d}S_Q = \sum_{b=1}^{p} \int_{S_b} U(P_a, Q)\frac{\partial u}{\partial \mathbf{n}}(Q)\ \mathrm{d}S_Q \tag{7.1}$$

into which we may substitute the parametric interpolates for $u(Q)$ and $\frac{\partial u}{\partial \mathbf{n}}(Q)$ to give an equation in terms of nodal values of u and $\frac{\partial u}{\partial \mathbf{n}}$:

$$C(P_a)u(P_a) + \sum_{b=1}^{p}\sum_{c=1}^{n(b)} \Delta T_{ac}^{b} u_c^b = \sum_{b=1}^{p}\sum_{c=1}^{n(b)} \Delta U_{ac}^{b} t_c^b \tag{7.2}$$

In this equation,

$$\Delta T_{ac}^{b} = \int_{S_b} T(P_a, Q) N_c(Q)\, \mathrm{d}S_Q \tag{7.3}$$

$$\Delta U_{ac}^{b} = \int_{S_b} U(P_a, Q) N_c(Q)\, \mathrm{d}S_Q \tag{7.4}$$

and u_c^b and t_c^b are the values of u and $\frac{\partial u}{\partial \mathbf{n}}$ at node c of element b. The coefficient $C(P_a)$ may be computed according to equation (D.24) or (D.26), and in principle a system of q simultaneous equations in terms of q unknown nodal values of u or $\frac{\partial u}{\partial \mathbf{n}}$ may be obtained.

The procedure outlined above cannot be applied without modification to equation (6.93) because the integral on the left side of that equation exists only as a Cauchy principal value, and so the coefficient

$$\Delta \mathbf{T}_{ac}^{b} = \int_{S_b} \mathbf{T}(P_a, Q) N_c(Q)\, \mathrm{d}S_Q$$

does not exist if node c of element b is the node P_a and that node is shared by two or more elements as shown in Figure 7.1. Noting that all the coefficients which cannot be computed individually are coefficients of $\mathbf{u}(P_a)$, we write equation (6.93) at node P_a with the parametric interpolates substituted for $\mathbf{u}(Q)$ and $\mathbf{t}(Q)$ as follows:

$$\mathbf{A}(P_a)\mathbf{u}(P_a) + \sum_{b=1}^{p} \sum_{\substack{c=1 \\ d(b,c)\neq a}}^{n(b)} \Delta \mathbf{T}_{ac}^{b} \mathbf{u}_c^b = \sum_{b=1}^{p} \sum_{c=1}^{n(b)} \Delta \mathbf{U}_{ac}^{b} \mathbf{t}_c^b \tag{7.5}$$

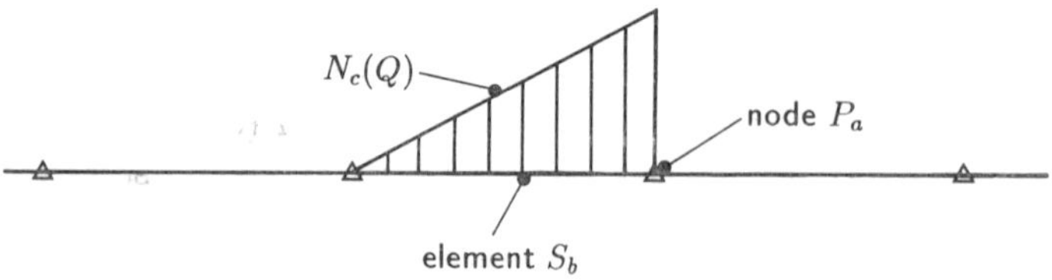

Figure 7.1 Case for which $\Delta \mathbf{T}_{ac}^{b}$ does not exist (linear elements)

In this equation,

$$\Delta \mathbf{T}_{ac}^{b} = \int_{S_b} \mathbf{T}(P_a, Q) N_c(Q)\, \mathrm{d}S_Q \tag{7.6}$$

$$\Delta \mathbf{U}_{ac}^{b} = \int_{S_b} \mathbf{U}(P_a, Q) N_c(Q)\, \mathrm{d}S_Q \tag{7.7}$$

and $\mathbf{u}_c^b$ and $\mathbf{t}_c^b$ are the values of $\mathbf{u}$ and $\mathbf{t}$ at the node c of element b; $d(b,c)$ is the global node number of node c of element b. The coefficient $\mathbf{A}(P_a)$ is given by

$$\mathbf{A}(P_a) = \mathbf{C}(P_a) + \sum_{\substack{b=1 \\ d(b,c)=a}}^{p} \sum_{c=1}^{n(b)} \Delta \mathbf{T}^b_{ac} \tag{7.8}$$

It would be possible to compute $\mathbf{A}(P_a)$ directly according to equation (7.8) and equation (E.18) for $\mathbf{C}(P_a)$, but in practice it is much easier to evaluate it implicitly by consideration of rigid body translations. For a finite domain, we substitute into equation (7.5) the solutions $\mathbf{u} = (1, 0)$ and $(0, 1)$ in two dimensions or $(1, 0, 0)$, $(0, 1, 0)$ and $(0, 0, 1)$ in three. Since for these displacement fields $\mathbf{t} = \mathbf{0}$, we obtain the result[1]

$$\mathbf{A}(P_a) = -\sum_{\substack{b=1 \\ d(b,c)\neq a}}^{p} \sum_{c=1}^{n(b)} \Delta \mathbf{T}^b_{ac} \tag{7.9}$$

Isoparametric elements may also be used to represent the boundary of an infinite domain with boundary of finite extent. The coefficient $\mathbf{A}(P_a)$ may be computed in this case by consideration of the displacement fields $(1, 0)$ and $(0, 1)$ in two dimensions or $(1, 0, 0)$, $(0, 1, 0)$ and $(0, 0, 1)$ in three, but since these displacements do not satisfy inequalities (6.140) it is necessary to determine from first principles the expression for $\mathbf{A}(P_a)$. Let us consider the finite domain bounded by S and the surface Γ of a sphere of radius R in three dimensions, or a circle Γ of radius R in two, centred at the origin of coordinates as shown in Figure 7.2. By considering rigid body displacements of this domain, we obtain the result

$$\mathbf{A}(P_a) = -\sum_{\substack{b=1 \\ d(b,c)\neq a}}^{p} \sum_{c=1}^{n(b)} \Delta \mathbf{T}^b_{ac} - \int_{\Gamma} \mathbf{T}(P_a, Q)\, \mathrm{d}S_Q \tag{7.10}$$

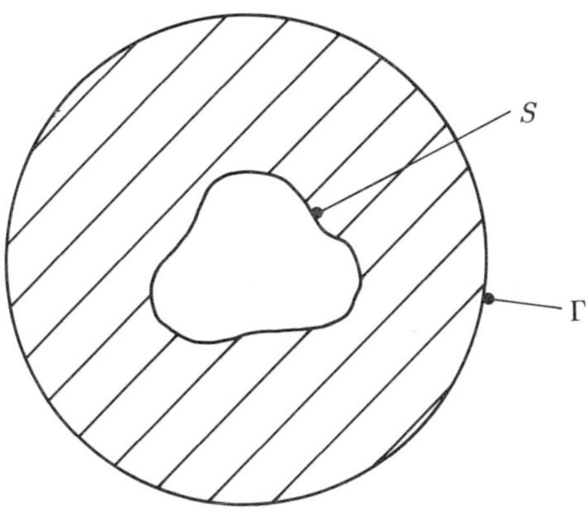

Figure 7.2 Finite domain for implicit computation of $\mathbf{A}(P_a)$

It is easily shown that

$$\lim_{R\to\infty} \int_{\Gamma} \mathbf{T}(P_a, Q)\, \mathrm{d}S_Q = -\mathbf{I} \tag{7.11}$$

where $\mathbf{I}$ is the unit matrix. Therefore, for an infinite domain with boundary of finite extent[2],

$$\mathbf{A}(P_a) = \mathbf{I} - \sum_{b=1}^{p} \sum_{\substack{c=1 \\ d(b,c)\neq a}}^{n(b)} \Delta\mathbf{T}_{ac}^{b} \tag{7.12}$$

As an alternative to the computation of the coefficient $\mathbf{A}(P_a)$ according to equation (7.9) or (7.12), it is possible to consider instead rigid body translations of a subdomain bounded by elements of which P_a is a node, and an arbitrarily chosen auxiliary boundary Ω_a as illustrated in Figure 7.3. Equations (7.9) and (7.12) may then be replaced by the single formula[3]

$$\mathbf{A}(P_a) = -\sum_{\substack{b=1 \\ P_a \in S_b}}^{p} \sum_{\substack{c=1 \\ d(b,c)\neq a}}^{n(b)} \Delta\mathbf{T}_{ac}^{b} - \int_{\Omega_a} \mathbf{T}(P_a, Q)\, \mathrm{d}S_Q \tag{7.13}$$

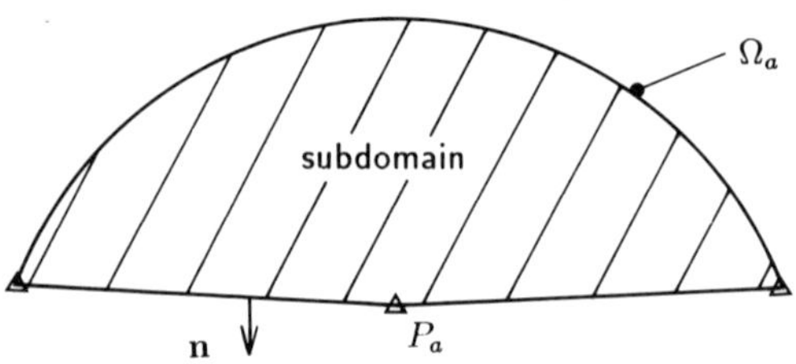

Figure 7.3 The auxiliary boundary Ω_a

This increases the cost of computation because it is now necessary to construct and integrate over the boundary Ω_a, but the extra cost is only a linear function of the number of boundary elements and so is of little significance for large problems. There is no advantage in this approach for the problems described in this section but in the analysis of infinite domains with boundaries of infinite extent as discussed in Section 7.3, the use of equation (7.13) results in simplification of analysis and implementation. The primary motivation however for the construction of an auxiliary boundary is to facilitate the implementation of Hermitian cubic and singular elements and general crack elements, which are outside the scope of this introductory text.

Nodal collocation is only one way in which a discrete system can be obtained. It may be argued that, rather than aiming to satisfy the boundary integral equation at each of the nodes with which are associated unknowns, it would be better to satisfy the equation as well as possible not only at the nodes but also at all other points of the boundary. In practice, with a solution defined in terms of only a finite number of unknowns, this means choosing the solution such that the integral over S of the product of each of a set of linearly independent trial functions and the error in satisfaction of the equation is zero[4]. It is then said that the error is orthogonal to each of the trial functions over S. A method of this type is known as a method of moments and clearly there is an infinite number of such methods because there is an infinite number of ways in which to choose the trial functions. Nodal collocation can be seen as the method obtained if the trial function $f_a(P)$ is taken to be infinite at node P_a and zero elsewhere, and to integrate to 1 over S (a Dirac delta function). Perhaps the best method in principle is the Galerkin method, in which the trial function $f_a(P)$ is zero over all elements except those of which P_a is a node. Over such an element, $f_a(P)$ is taken to be equal to the shape function which equals 1 at P_a. Typical trial functions of the Galerkin method are shown in Figure 7.4. The discrete system is now

$$\sum_{\substack{e=1\\P_a\in S_e}}^{p}\int_{S_e} f_a(P)\mathbf{C}(P)\mathbf{u}(P)\,\mathrm{d}S_P+\sum_{\substack{e=1\\P_a\in S_e}}^{p}\int_{S_e} f_a(P)\sum_{b=1}^{p}\int_{S_b}\mathbf{T}(P,Q)\mathbf{u}(Q)\,\mathrm{d}S_Q\,\mathrm{d}S_P$$
$$=\sum_{\substack{e=1\\P_a\in S_e}}^{p}\int_{S_e} f_a(P)\sum_{b=1}^{p}\int_{S_b}\mathbf{U}(P,Q)\mathbf{t}(Q)\,\mathrm{d}S_Q\,\mathrm{d}S_P,\quad a=1,2,\ldots,q \qquad (7.14)$$

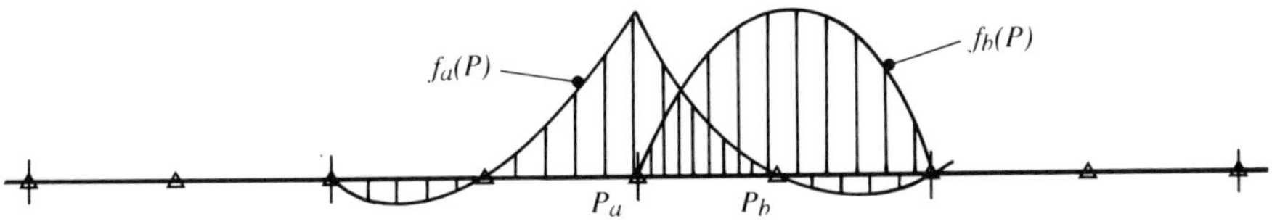

Figure 7.4 Typical trial functions of the Galerkin method (quadratic elements)

in which most of the terms of the outer summation are omitted by virtue of the condition $P_a \in S_e$. It is in general impossible to perform the integration with respect to P analytically, so a quadrature formula must be proposed according to which the integral of a function $\phi(P)$ is approximated as follows:

$$\int_{S_e}\phi(P)\,\mathrm{d}S_P\approx\sum_{j=1}^{m}W_j^e\phi(P_j^e) \qquad (7.15)$$

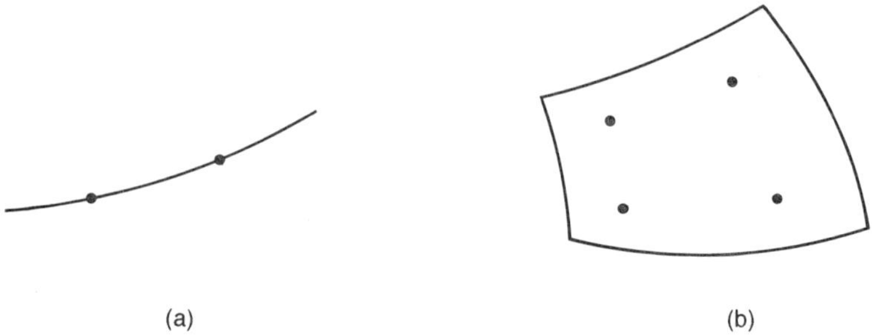

Figure 7.5 Typical sampling points for the Galerkin method

where W_j^e are weights and P_j^e are sampling points. It is usual to propose a Gaussian quadrature formula,[5] and for such a formula all the sampling points lie inside the element, away from the ends in two dimensions or the sides in three dimensions (see Figure 7.5). For a one-dimensional element, the weight W_j^e is the product of the Gauss weight and the Jacobian $J(P_j^e)$:

$$W_j^e = A_j J(P_j^e) \tag{7.16}$$

For a two-dimensional element, A_j in equation (7.16) is the product of two Gauss weights. Effectively, rows of the discrete system are linear combinations of the equations of collocation $\mathbf{E}_j^e = \mathbf{0}$ at the sampling points:

$$\sum_{\substack{e=1 \\ d(e,f)=a}}^{p} \sum_{f=1}^{n(e)} \sum_{j=1}^{m} W_j^e N_f(P_j^e)\mathbf{E}_j^e = \mathbf{0}, \quad a = 1, 2, \ldots, q \tag{7.17}$$

where

$$\mathbf{E}_j^e = \mathbf{A}(P_j^e)\mathbf{u}(P_j^e) + \sum_{c=1}^{n(e)} \delta\mathbf{T}_{ejc}\mathbf{u}_c^e + \sum_{\substack{b=1 \\ b\neq e}}^{p} \sum_{c=1}^{n(b)} \Delta\mathbf{T}_{ejc}^b\mathbf{u}_c^b - \sum_{b=1}^{p} \sum_{c=1}^{n(b)} \Delta\mathbf{U}_{ejc}^b\mathbf{t}_c^b \tag{7.18}$$

In this equation,

$$\Delta\mathbf{T}_{ejc}^b = \int_{S_b} \mathbf{T}(P_j^e, Q) N_c(Q)\, \mathrm{d}S_Q \tag{7.19}$$

$$\Delta\mathbf{U}_{ejc}^b = \int_{S_b} \mathbf{U}(P_j^e, Q) N_c(Q)\, \mathrm{d}S_Q \tag{7.20}$$

$$\delta\mathbf{T}_{ejc} = \int_{S_e} \mathbf{T}(P_j^e, Q) \left(N_c(Q) - N_c(P_j^e)\right)\, \mathrm{d}S_Q \tag{7.21}$$

Since P_j^e cannot be on an edge or corner of the boundary, $\mathbf{C}(P_j^e) = \frac{1}{2}\mathbf{I}$. Also, the shape functions sum to 1 at the point P_j^e and so the coefficient $\mathbf{A}(P_j^e)$ is given by

$$\mathbf{A}(P_j^e) = \frac{1}{2}\mathbf{I} + \int_{S_e} \mathbf{T}(P_j^e, Q)\, \mathrm{d}S_Q \tag{7.22}$$

The coefficient $\mathbf{A}(P_j^e)$ may more conveniently be evaluated by consideration of rigid body translations. For a finite domain, substitution into equation (7.18) of the solutions $\mathbf{u} = (1,0)$ and $(0,1)$ in two dimensions or $(1,0,0)$, $(0,1,0)$ and $(0,0,1)$ in three yields the result

$$\mathbf{A}(P_j^e) = -\sum_{c=1}^{n(e)} \delta\mathbf{T}_{ejc} - \sum_{\substack{b=1\\ b\neq e}}^{p} \sum_{c=1}^{n(b)} \Delta\mathbf{T}_{ejc}^{b} \tag{7.23}$$

For infinite domains the same considerations apply as for nodal collocation, and it is also possible to evaluate $\mathbf{A}(P_j^e)$ by considering rigid body displacements of a subdomain bounded by an auxiliary contour and just the element S_e. The displacement vector $\mathbf{u}(P_j^e)$ can be expressed in terms of nodal displacements of the element S_e:

$$\mathbf{u}(P_j^e) = \sum_{c=1}^{n(e)} N_c(P_j^e)\mathbf{u}_c^e \tag{7.24}$$

In the Galerkin method the computing cost depends upon the number of sampling points, and to avoid a singular matrix of equation coefficients it is necessary to have at least as many sampling points as there are nodes:

$$p.j \geq q \tag{7.25}$$

This however is only a necessary condition: in practice it is possible, especially in three-dimensional analysis, to have satisfaction of inequality (7.25) and still obtain a singular matrix because although globally there are enough sampling points, there is a shortage of them locally such that there is a subset of equations which is constructed from the equations of collocation at an insufficient number of sampling points. It is not easy to ensure the matrix will be non-singular, especially for three-dimensional analysis in which there is irregular mesh topology and where there are constraints on displacement (see Section 7.6), without taking $p.j$ to be considerably greater than q. For example, for a typical mesh of eight-node quadrilateral elements, the 2×2 Gaussian quadrature formula the sampling points of which are shown in Figure 7.5b yields approximately $p.j = 1.3q$ but certainly does not guarantee a non-singular matrix. To do that under all possible conditions would require a 3×3 formula (a 4×2 formula would be sufficient but is undesirable unless the length of the element is considerably greater than its width), yielding approximately $p.j = 3q$ which is very uneconomical. Maybe it is possible by judicious variation of the order of Gaussian formula to achieve $p.j$ in the range 1.5 to 2.0, but this is not a practical proposition unless it can be economically automated within a computer program. The conclusion is therefore that for three-dimensional problems the cost of construction of the discrete system by the Galerkin method is between 1.5 and 2 times that of nodal collocation. To be worthwhile, therefore, the method must yield significantly better results for a given degree of mesh refinement than nodal collocation.

Further insight into the relative merits of the nodal collocation and Galerkin methods may be obtained by considering what happens if in the Galerkin method the sampling points are taken to be at the nodes of the element. If this is done for a linear element, this is equivalent to replacing Gaussian quadrature by trapezoidal rule in the integration with respect to P. For quadratic elements, it amounts to replacing Gaussian quadrature by Simpson's rule. But if the sampling points are the node points, rows of the discrete system defined by equation (7.17) are the equations of collocation at the nodes multiplied by scalar factors, and so the solution obtained by the Galerkin method is the same as that obtained by nodal collocation. Nodal collocation as described earlier in this section is therefore equivalent to the Galerkin method in which integration with respect to P is carried out by trapezoidal rule for linear elements, and by Simpson's rule for quadratic elements. Since both the two point Gaussian formula and Simpson's rule are fourth-order accurate, that is they both integrate exactly a polynomial of the form $a_0 + a_1\xi + a_2\xi^2 + a_3\xi^3$, it is doubtful whether for quadratic elements results obtained by the Galerkin method with two-point Gaussian formula converge with mesh refinement to the exact solution any more rapidly than those obtained by nodal collocation.

7.3 INFINITE ELEMENTS

Infinite elements are required to represent the boundary of regions of the type shown in Figure 6.14. The boundary of such a region is represented by isoparametric elements near the region of interest, generally with mesh gradation so that element size increases with distance from the centre of the region. The continuation of the boundary to infinity is represented by infinite elements adjacent to the outermost isoparametric elements of the mesh as illustrated in Figure 7.6.

For the solution of Laplace's equation in two dimensions, it is necessary to model over an infinite element functions $u(Q)$ and $\frac{\partial u}{\partial \mathbf{n}}(Q)$ satisfying inequalities (6.136). For those inequalities to be satisfied, the particular integral must be chosen such that for the complementary function u

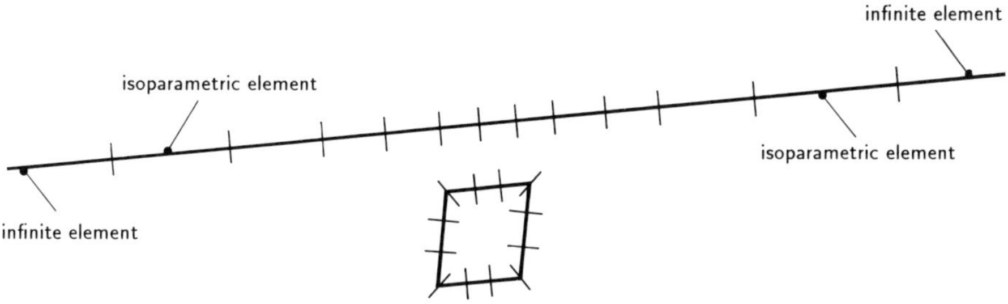

Figure 7.6 Mesh of isoparametric and infinite elements

$$\int_S \frac{\partial u}{\partial \mathbf{n}}(Q)\mathrm{d}S_Q = 0 \tag{7.26}$$

which in the analogy of heat flow specifies zero net inflow of heat to the region. Under this condition, the values of $u(Q)$ and $\frac{\partial u}{\partial \mathbf{n}}(Q)$ tend towards those due to a source dipole located at the origin of coordinates (see equation (6.106)), as distance R from the origin tends to infinity. From equations (6.107) and (6.114) it may be seen that $u(Q)$ and $\frac{\partial u}{\partial \mathbf{n}}(Q)$ therefore tend to zero as $1/R$ and $1/R^2$ respectively, and so we should take on an infinite element

$$\begin{aligned} u(Q) &= \frac{A}{R} \\ \frac{\partial u}{\partial \mathbf{n}}(Q) &= \frac{B}{R^2} \end{aligned} \tag{7.27}$$

where A and B are constants. In practice, this is improved upon by proposing that $u(Q)$ and $\frac{\partial u}{\partial \mathbf{n}}(Q)$ be taken proportional to the inverse and the inverse square of distance $r_0(Q)$ from a reference point P_0 located near the centre of the region modelled by isoparametric elements:

$$\begin{aligned} u(Q) &= u(Z)\frac{r_0(Z)}{r_0(Q)} \\ \frac{\partial u}{\partial \mathbf{n}}(Q) &= \frac{\partial u}{\partial \mathbf{n}}(Z)\left(\frac{r_0(Z)}{r_0(Q)}\right)^2 \end{aligned} \tag{7.28}$$

where Z is the end of the infinite element (see Figure 7.7).

Figure 7.7 $u(Q)$ and $\frac{\partial u}{\partial \mathbf{n}}(Q)$ over an infinite element in two dimensions

In the analysis of plane strain, displacement $\mathbf{u}(Q)$ and traction $\mathbf{t}(Q)$ must satisfy inequalities (6.140) and therefore the particular integral must be chosen such that for the complementary function

$$\int_S \mathbf{t}(Q)\,\mathrm{d}S_Q = \mathbf{0} \tag{7.29}$$

The values of $\mathbf{u}(Q)$ and $\mathbf{t}(Q)$ then tend towards those due to a force dipole located at the origin of coordinates as distance R from the origin tends to infinity, and therefore they tend to zero as $1/R$ and $1/R^2$, respectively. Displacement and traction are therefore taken to be proportional to the inverse and the inverse square of distance $r_0(Q)$ from the reference point P_0:[6]

$$\begin{aligned} \mathbf{u}(Q) &= \mathbf{u}(Z)\frac{r_0(Z)}{r_0(Q)} \\ \mathbf{t}(Q) &= \mathbf{t}(Z)\left(\frac{r_0(Z)}{r_0(Q)}\right)^2 \end{aligned} \tag{7.30}$$

In three-dimensional analysis two types of infinite element are used:[6] an element over which as in the two-dimensional case $u(Q)$ or $\mathbf{u}(Q)$ tends to zero as R tends to infinity, and an element over which $u(Q)$ or $\mathbf{u}(Q)$ is independent of the intrinsic coordinate ξ shown in Figure 7.8. These are called the asymptotic element and the plane strain element, respectively. An infinite element in three dimensions adjoins an isoparametric element along its baseline $\xi = -1$, and other infinite elements along its infinitely long sides

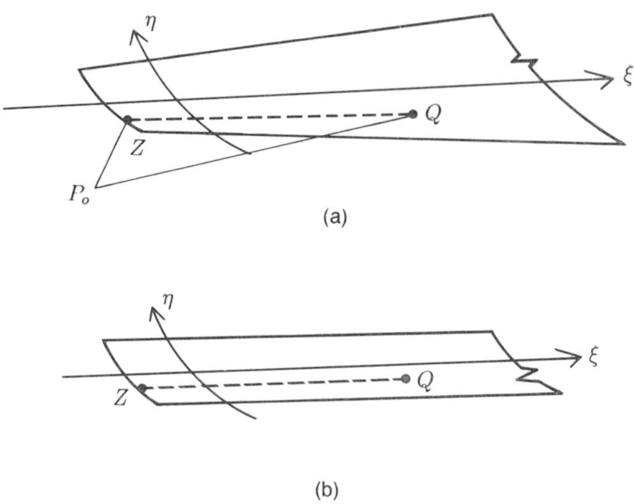

Figure 7.8 The asymptotic and plane strain infinite elements in three dimensions

$\eta = \pm 1$. Lines η = constant are straight, but lines ξ = constant may be curved if the isoparametric elements have at least quadratic functional variation. Whereas the sides $\eta = 1$ and $\eta = -1$ of an asymptotic element may diverge, those of a plane strain element must be parallel.

For Laplace's equation in three dimensions, equation (7.26) need not be satisfied. The values of $u(Q)$ and $\frac{\partial u}{\partial \mathbf{n}}(Q)$ tend towards those due to a source located at the origin of coordinates, as distance R from the origin tends to infinity. From equations (6.1) and (6.3) it may be seen that $u(Q)$ and $\frac{\partial u}{\partial \mathbf{n}}(Q)$ therefore tend to zero as $1/R$ and $1/R^2$, respectively, so we take for an asymptotic element the variation of $u(Q)$ and $\frac{\partial u}{\partial \mathbf{n}}(Q)$ specified by equation (7.28), where Z is now the point on the baseline with the same η coordinate as Q. Therefore, for an asymptotic element S_b,

$$\begin{aligned} u(Q) &= \frac{r_0(Z)}{r_0(Q)} \sum_{c=1}^{n(b)} N_c(\eta) u_c^b \\ \frac{\partial u}{\partial \mathbf{n}}(Q) &= \left(\frac{r_0(Z)}{r_0(Q)}\right)^2 \sum_{c=1}^{n(b)} N_c(\eta) t_c^b \end{aligned} \tag{7.31}$$

where $N_c(\eta)$ are one dimensional shape functions and n is the number of nodes on the baseline (2 for linear variation, 3 for quadratic). Over a plane strain element,

$$\begin{aligned} u(Q) &= \sum_{c=1}^{n(b)} N_c(\eta) u_c^b \\ \frac{\partial u}{\partial \mathbf{n}}(Q) &= \sum_{c=1}^{n(b)} N_c(\eta) t_c^b \end{aligned} \tag{7.32}$$

Asymptotic elements are used to model surfaces such as that of a halfspace, as shown in Figure 7.9a. Plane strain elements model surfaces of infinite extent over which the variation of the solution in the direction in which the surface extends to infinity tends to zero as distance R from the origin tends to infinity. The most commonly occurring example of such a surface is that of a long tunnel, as shown in Figure 7.9b. Near the end of the tunnel there is three-dimensional variation of the solution, but far from the end there is negligible variation in the axial direction.

For three-dimensional elasticity, equation (7.29) need not be satisfied by the complementary function and the values of $\mathbf{u}(Q)$ and $\mathbf{t}(Q)$ tend towards those due to a concentrated load at the origin of coordinates, as distance R from the origin tends to infinity. From equations (6.9) to (6.11) and (6.17) to (6.19) it may be seen that $\mathbf{u}(Q)$ and $\mathbf{t}(Q)$ therefore tend to zero as $1/R$ and $1/R^2$, respectively, so we take for an asymptotic element S_b

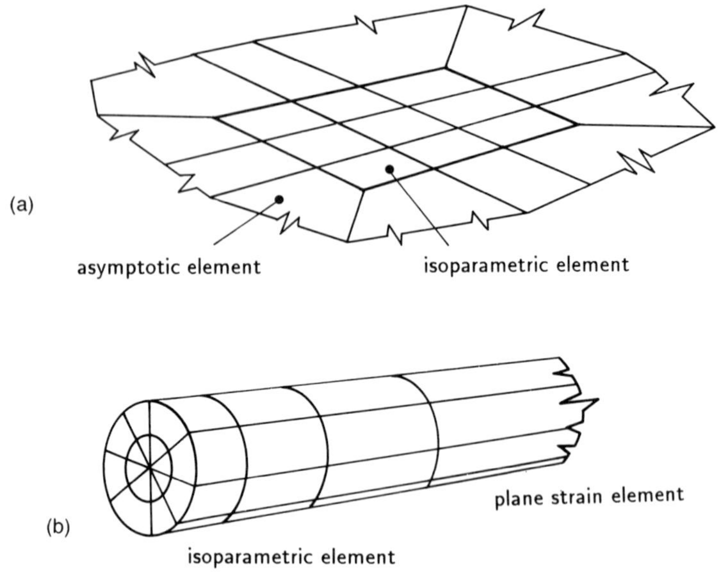

Figure 7.9 Three dimensional meshes with infinite elements: (a) halfspace, (b) tunnel

$$\begin{aligned}\mathbf{u}(Q) &= \frac{r_0(Z)}{r_0(Q)} \sum_{c=1}^{n(b)} N_c(\eta)\mathbf{u}_c^b \\ \mathbf{t}(Q) &= \left(\frac{r_0(Z)}{r_0(Q)}\right)^2 \sum_{c=1}^{n(b)} N_c(\eta)\mathbf{t}_c^b\end{aligned} \tag{7.33}$$

where Z is the point on the baseline with the same η coordinate as Q, $N_c(\eta)$ are one-dimensional shape functions and n is the number of nodes on the baseline. Over a plane strain element,

$$\begin{aligned}\mathbf{u}(Q) &= \sum_{c=1}^{n(b)} N_c(\eta)\mathbf{u}_c^b \\ \mathbf{t}(Q) &= \sum_{c=1}^{n(b)} N_c(\eta)\mathbf{t}_c^b\end{aligned} \tag{7.34}$$

The circumstances in which asymptotic and plane strain elements are used are the same as for Laplace's equation. Note that the so-called plane strain elements are actually capable of modelling a combination of plane strain and antiplane strain (warping), in which there are non-zero displacements in the direction in which the surface extends to infinity. The elements may therefore be used in the analysis of stress around a tunnel the axis of which is not a principal direction of the pre-mining stress field, or in the analysis of stress in a shaft of non-circular cross section.

In Section 7.2, the equation of collocation at node P_a is written in terms of an explicitly computed component and an implicitly computed component. In equation (7.5) for elasticity, the term $\mathbf{A}(P_a)\mathbf{u}(P_a)$ is the implicitly computed component, and the other terms are the explicitly computed component. If S_b is an infinite element in three dimensions, then in equation (7.5) n is the number of nodes on the baseline and instead of equations (7.6) and (7.7) we have

$$\Delta\mathbf{T}^b_{ac} = \int_{S_b} \mathbf{T}(P_a, Q) f_u(Q) N_c(\eta)\ \mathrm{d}S_Q \tag{7.35}$$

$$\Delta\mathbf{U}^b_{ac} = \int_{S_b} \mathbf{U}(P_a, Q) f_t(Q) N_c(\eta)\ \mathrm{d}S_Q \tag{7.36}$$

For an asymptotic element

$$\begin{aligned} f_u(Q) &= \frac{r_0(Z)}{r_0(Q)} \\ f_t(Q) &= \left(\frac{r_0(Z)}{r_0(Q)}\right)^2 \end{aligned} \tag{7.37}$$

and for a plane strain element in three dimensions $f_u(Q) = f_t(Q) = 1$. If S_b is an infinite element in two dimensions, then in equation (7.5) $n = 1$ and

$$\Delta\mathbf{T}^b_{ac} = \int_{S_b} \mathbf{T}(P_a, Q) f_u(Q)\ \mathrm{d}S_Q \tag{7.38}$$

$$\Delta\mathbf{U}^b_{ac} = \int_{S_b} \mathbf{U}(P_a, Q) f_t(Q)\ \mathrm{d}S_Q \tag{7.39}$$

where $f_u(Q)$ and $f_t(Q)$ are given by equation (7.37).

The coefficient $\mathbf{A}(P_a)$ may be computed by considering rigid body translation of either the domain V_R shown in Figure 6.14 in the limit as R tends to infinity, or a finite subdomain as illustrated in Figure 7.3.[6] The consideration of rigid body translation of V_R is quite complicated as account has to be taken of the fact that inequalities (6.140) are not satisfied on the boundary of that region, part of which is modelled by infinite elements. The simpler approach is to consider translation of a finite subdomain. For a node P_a which is not a node of an infinite element, equations (7.5) and (7.13) apply as before. If P_a is a node of one or more infinite elements then each of those elements is divided into a finite part S_b^F adjacent to P_a and an infinite part S_b^I as shown in Figure 7.10. In place of equation (7.5) we write

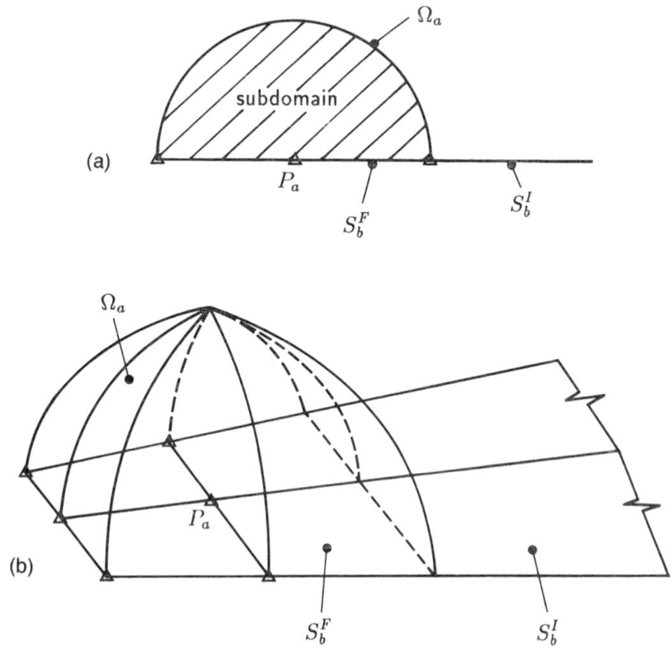

Figure 7.10 The auxiliary boundary Ω_a for a node of an infinite element: (a) in two dimensions, (b) in three dimensions

$$\mathbf{A}(P_a)\mathbf{u}(P_a) + \sum_{\substack{b=1 \\ d(b,c)\neq a}}^{p'} \sum_{c=1}^{n(b)} \Delta \mathbf{T}_{ac}^b \mathbf{u}_c^b + \sum_{\substack{b=p'+1 \\ P_a \notin S_b}}^{p} \sum_{c=1}^{n(b)} \Delta \mathbf{T}_{ac}^b \mathbf{u}_c^b + \sum_{\substack{b=p'+1 \\ P_a \in S_b \\ d(b,c)\neq a}}^{p} \sum_{c=1}^{n(b)} \Delta_F \mathbf{T}_{ac}^b \mathbf{u}_c^b$$

$$+ \sum_{\substack{b=p'+1 \\ P_a \in S_b}}^{p} \sum_{c=1}^{n(b)} \Delta_I \mathbf{T}_{ac}^b \mathbf{u}_c^b = \sum_{b=1}^{p} \sum_{c=1}^{n(b)} \Delta \mathbf{U}_{ac}^b \mathbf{t}_c^b \tag{7.40}$$

where $S_1, S_2, \ldots, S_{p'}$ are isoparametric elements and $S_{p'+1}$ to S_p are infinite elements. If S_b is an infinite element in three dimensions, then in equation (7.40)

$$\Delta_F \mathbf{T}_{ac}^b = \int_{S_b^F} \mathbf{T}(P_a, Q) f_u(Q) N_c(\eta) \; \mathrm{d}S_Q \tag{7.41}$$

$$\Delta_I \mathbf{T}_{ac}^b = \int_{S_b^I} \mathbf{T}(P_a, Q) f_u(Q) N_c(\eta) \; \mathrm{d}S_Q \tag{7.42}$$

If S_b is an infinite element in two dimensions, then

$$\Delta_{\mathrm{F}}\mathbf{T}^b_{ac} = \int_{S_b^F} \mathbf{T}(P_a, Q) f_u(Q)\, \mathrm{d}S_Q \tag{7.43}$$

$$\Delta_{\mathrm{I}}\mathbf{T}^b_{ac} = \int_{S_b^I} \mathbf{T}(P_a, Q) f_u(Q)\, \mathrm{d}S_Q \tag{7.44}$$

where $f_u(Q)$ is given by equation (7.37). By considering rigid body translations of the finite subdomain shown in Figure 7.10b, we obtain for the three-dimensional case the result[6]

$$\mathbf{A}(P_a) = -\sum_{\substack{b=1\\P_a\in S_b}}^{p'} \sum_{\substack{c=1\\d(b,c)\neq a}}^{n(b)} \Delta\mathbf{T}^b_{ac} - \sum_{\substack{b=p'+1\\P_a\in S_b}}^{p} \int_{S_b^F} \left[1 - f_u(Q) \sum_{\substack{c=1\\d(b,c)=a}}^{n(b)} N_c(\eta)\right] \mathbf{T}(P_a, Q)\, \mathrm{d}S_Q$$
$$- \int_{\Omega_a} \mathbf{T}(P_a, Q)\, \mathrm{d}S_Q \tag{7.45}$$

For the two-dimensional case shown in Figure 7.10a,

$$\mathbf{A}(P_a) = -\sum_{\substack{b=1\\P_a\in S_b}}^{p'} \sum_{\substack{c=1\\d(b,c)\neq a}}^{n(b)} \Delta\mathbf{T}^b_{ac} - \sum_{\substack{b=p'+1\\P_a\in S_b}}^{p} \int_{S_b^F} [1 - f_u(Q)]\, \mathbf{T}(P_a, Q)\, \mathrm{d}S_Q$$
$$- \int_{\Omega_a} \mathbf{T}(P_a, Q)\, \mathrm{d}S_Q \tag{7.46}$$

Equation (7.40) is the equation of collocation at a node of an infinite element, for two- or three-dimensional elasticity. Results similar to equations (7.40), (7.45) and (7.46) may be obtained for Laplace's equation, and a generalisation to methods of moments is also possible.

7.4 INTEGRATION SCHEMES FOR TWO-DIMENSIONAL ANALYSIS

In this section an integration scheme for the equations of nodal collocation only is described, but the generalisation to methods of moments is straightforward. To avoid repetition, only plane strain is considered: any scheme for plane strain is also valid for Laplace's equation. The integrals to be evaluated numerically are the coefficients $\Delta\mathbf{T}^b_{ac}$, $\Delta\mathbf{U}^b_{ac}$, $\Delta_F\mathbf{T}^b_{ac}$ and $\Delta_I\mathbf{T}^b_{ac}$ appearing in equations (7.5) and (7.40), and other terms on the right sides of equations (7.13) and (7.46). Gaussian quadrature formulae are used; they are the most efficient means of numerical integration available provided the functions to be integrated are of the form for which the formulae are designed, but are very inaccurate if improperly used. The g point Gauss–Legendre formula

$$\int_{-1}^{1} \phi(\beta)\, \mathrm{d}\beta \approx \sum_{s=1}^{g} B_s^g \phi(\beta_s^g) \tag{7.47}$$

integrates exactly any function $\phi(\beta) = a_0 + a_1\beta + \ldots + a_{2g-1}\phi^{2g-1}$, so yields accurate results if the function to be integrated may reasonably be approximated by a polynomial in the independent variable β. Since there are functions with logarithmic singularities to be integrated and the function $\ln\beta$ cannot accurately be approximated by a polynomial in β near $\beta = 0$, we also require the Gaussian formula

$$\int_{0}^{1} \phi(\gamma) \ln\frac{1}{\gamma}\, \mathrm{d}\gamma \approx \sum_{s=1}^{g} C_s^g \phi(\gamma_s^g) \tag{7.48}$$

which integrates exactly any function $\phi(\gamma)\ln\frac{1}{\gamma}$ where $\phi(\gamma) = a_0 + a_1\gamma + \ldots + a_{2g-1}\gamma^{2g-1}$. The weights B_s^g and C_s^g and offsets β_s^g and γ_s^g are tabulated by Stroud and Secrest[5] for a wide range of values of g. The requirements to be satisfied by the integration scheme are as follows.

1. Integrations must be between the limits -1 and 1 or 0 and 1, depending on the formula used.
2. The integrand must vary approximately as a polynomial in β, or as a polynomial in γ multiplied by $\ln\frac{1}{\gamma}$.
3. The value of g should be such as to ensure adequate precision. Unnecessarily high values of g should be avoided though, as they increase the cost of computation.

For the purpose of development of the integration scheme, we consider separately integration over elements of which P_a is a node, integration over the auxiliary boundary Ω_a, and off-diagonal integration by which is meant integration over elements of which P_a is not a node. The cost of the first two of these three categories of integrations increases only linearly with problem size as defined by the number of elements of the mesh, so for large problems most of the computer time required for construction of the equations of nodal collocation is spent on off-diagonal integration. Care should therefore be taken to make the off-diagonal integration as economical as possible, but it is not worthwhile to go to great lengths to optimise the efficiency of integration over elements of which P_a is a node, or over Ω_a.

We consider first off-diagonal integration over isoparametric elements, using the formula (7.47). The functions to be integrated are products of a kernel, a shape function and a Jacobian. Generally the Jacobian does not vary much over an element. Shape functions vary equally rapidly with respect to the intrinsic coordinates over all elements, but the kernels vary more rapidly over elements which are close to the node P_a compared with their length than over elements which are far from P_a. Clearly a greater number of Gauss points is required for some elements than for others: the problem is to develop a criterion that will ensure roughly the same precision of integration everywhere. Stroud and Secrest[5] give upper bounds for error in terms of an upper bound for the absolute value of the $2g$th derivative of the integrand. Suppose that

$$\left|\frac{\mathrm{d}^{2g}\phi}{\mathrm{d}\beta^{2g}}\right| \le H_g \tag{7.49}$$

Then, very nearly, the upper bound for error is

$$\left|\int_{-1}^{1} \phi(\beta)\,\mathrm{d}\beta - \sum_{k=1}^{g} B_k^g \phi(\beta_k^g)\right| \le \frac{4H_g}{2^{2g}(2g)!} \tag{7.50}$$

In a computer it is only possible to specify, by means of DATA statements for example, the Gauss weights and offsets for a finite range of values of g. If the required precision of integration over an element cannot be obtained by taking the highest available value of g, then it is necessary to subdivide that element into two or more intervals, over each of which one of the available formulae is then applied. The integration scheme must therefore provide for the subdivision of an element into more than one interval. In the simple scheme described here, each of the intervals into which an element is subdivided corresponds to an equal range of the intrinsic coordinate ξ. Let there be G intervals. Then we write

$$\int_{-1}^{1} \phi(\xi)\,\mathrm{d}\xi \approx \frac{1}{G}\sum_{F=1}^{G}\sum_{k=1}^{g(F)} B_k^{g(F)} \phi\left(\xi_k^{g(F)}\right) \tag{7.51}$$

in which the factor $1/G$ is the Jacobian $\mathrm{d}\xi/\,\mathrm{d}\beta$ and (see Figure 7.11)

$$\xi_k^{g(F)} = \frac{1}{G}\left[G - 2F + 1 + \beta_k^{g(F)}\right] \tag{7.52}$$

The strategy in principle is to use inequality (7.50) to determine appropriate values of G and $g(1), g(2), \ldots, g(G)$ in equation (7.51). This requires the evaluation of upper bounds H_g as defined by inequality (7.49), which is of course impractical for the integrals in question. Therefore, we consider instead a representative simplified problem.[2] Since the kernel $\mathbf{T}$ varies more rapidly than $\mathbf{U}$ at least over elements near to P_a, the integration scheme is determined by the need to evaluate accurately the coefficients $\Delta\mathbf{T}_{ac}^b$. We

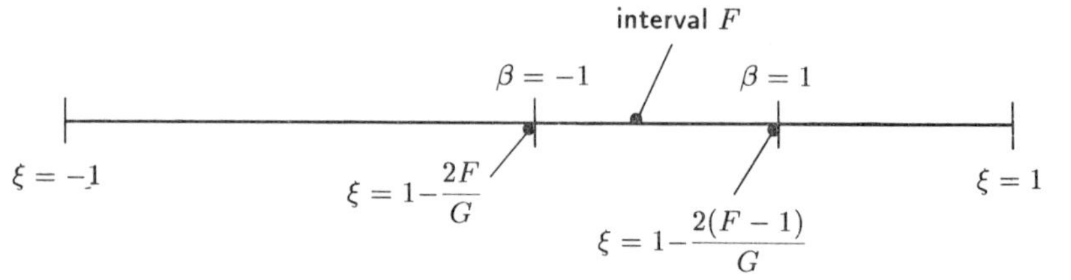

Figure 7.11 Mapping of one interval onto an isoparametric element

ignore the variation of the Jacobian and of the shape functions, and take the variation of the function $1/r$ to be representative of that of the kernel $\mathbf{T}$. The problem is then further simplified by considering not the actual interval but an interval of the same length L as the actual interval, lying on a straight line passing through the node P_a as shown in Figure 7.12. The minimum distance R from P_a to the interval is taken to be the same in the simplified problem as for the actual interval. By differentiating $1/r$ $2g$ times with respect to β and then substituting $r = R$ we obtain the approximate result

$$H_g = \left(\frac{L}{2}\right)^{2g} \frac{(2g)!}{R^{2g+1}} \tag{7.53}$$

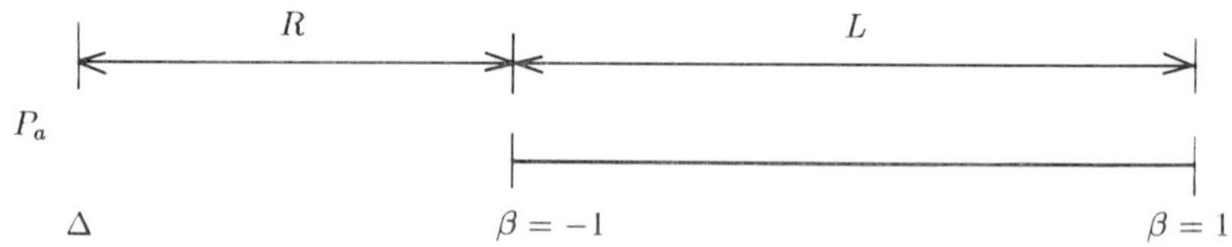

Figure 7.12 Representative interval of integration

Substituting into inequality (7.50), we obtain the approximate upper bound

$$\left| \int_{-1}^{1} \frac{1}{r}(\beta)\, \mathrm{d}\beta - \sum_{k=1}^{g} B_k^g \frac{1}{r}(\beta_k^g) \right| \leq \left(\frac{L}{4}\right)^{2g} \cdot \frac{4}{R^{2g+1}} \tag{7.54}$$

We take as a measure of precision of integration the ratio between the upper bound (7.54) and an upper bound on the integral of $1/r$ over the interval. We write

$$\left| \int_{-1}^{1} \frac{1}{r}(\beta)\, \mathrm{d}\beta \right| \leq \frac{2}{R} \tag{7.55}$$

Therefore we require

$$\left(\frac{L}{4}\right)^{2g} \cdot \frac{4}{R^{2g+1}} \leq \frac{2}{R}.\epsilon \tag{7.56}$$

where ϵ is a constant to be decided in the light of experience. This result may be rearranged for computational convenience as follows:

$$\frac{L}{R} \leq 4\left(\frac{\epsilon}{2}\right)^{\frac{1}{2g}} \tag{7.57}$$

Let L_b be the length of element S_b, R_b be the minimum distance from a node P_a to S_b, and $g_{\max}$ be the highest available value of g. Since for the purpose of choosing the integration scheme the Jacobian is taken to be constant over the element, we may write for an interval

$$L = \frac{L_b}{G} \tag{7.58}$$

The value of G must be chosen such that satisfactory precision of integration can be attained for the interval in which the point of S_b nearest to P_a lies, using a Gaussian formula with no more than $g_{\max}$ points. By substituting equation (7.58) and the values $R = R_b$ and $g = g_{\max}$ into equation (7.57), we obtain after some rearrangement the result

$$G \geq \frac{L_b}{4R_b}\left(\frac{2}{\epsilon}\right)^{\frac{1}{2g_{\max}}} \tag{7.59}$$

The value of G is therefore taken to be the smallest integer satisfying equation (7.59). Once G has been computed, the values of $g(1), g(2), \ldots, g(G)$ are determined according to equation (7.57), and $\Delta\mathbf{T}^b_{ac}$ and $\Delta\mathbf{U}^b_{ac}$ are computed according to the formulae

$$\Delta\mathbf{T}^b_{ac} \approx \frac{1}{G}\sum_{F=1}^{G}\sum_{k=1}^{g(F)} B_k^{g(F)}\mathbf{T}\left(P_a, Q(\xi_k^{g(F)})\right) N_c(\xi_k^{g(F)})J(\xi_k^{g(F)}) \tag{7.60}$$

$$\Delta\mathbf{U}^b_{ac} \approx \frac{1}{G}\sum_{F=1}^{G}\sum_{k=1}^{g(F)} B_k^{g(F)}\mathbf{U}\left(P_a, Q(\xi_k^{g(F)})\right) N_c(\xi_k^{g(F)})J(\xi_k^{g(F)}) \tag{7.61}$$

where J is the Jacobian $\mathrm{d}s/\mathrm{d}\xi$.

For off-diagonal integration over an infinite element, the intrinsic coordinate ξ varies from -1 to infinity rather than from -1 to 1 as required by the integration scheme for an isoparametric element. In program BOUND (see Appendix B), integration is carried out with respect to the coordinate $\bar{\xi}$ defined by the transformation[6]

$$\xi = \frac{3\bar{\xi}+1}{1-\bar{\xi}} \tag{7.62}$$

If $\bar{\xi} = -1$ then $\xi = -1$ and as $\bar{\xi}$ tends to 1, ξ tends to infinity; indeed, the range of values $-1 \leq \xi < \infty$ maps onto the range $-1 \leq \bar{\xi} < 1$ as shown in Figure 7.13. By equations (7.38) and (7.39),

$$\Delta \mathbf{T}_{ac}^{b} = \int_{-1}^{1} \mathbf{T}\left(P_a, Q\left(\xi\right)\right) f_u\left(Q\left(\xi\right)\right) J\left(\xi\right) \frac{\mathrm{d}\xi}{\mathrm{d}\bar{\xi}} \cdot \mathrm{d}\bar{\xi} \tag{7.63}$$

$$\Delta \mathbf{U}_{ac}^{b} = \int_{-1}^{1} \mathbf{U}\left(P_a, Q\left(\xi\right)\right) f_t\left(Q\left(\xi\right)\right) J\left(\xi\right) \frac{\mathrm{d}\xi}{\mathrm{d}\bar{\xi}} \cdot \mathrm{d}\bar{\xi} \tag{7.64}$$

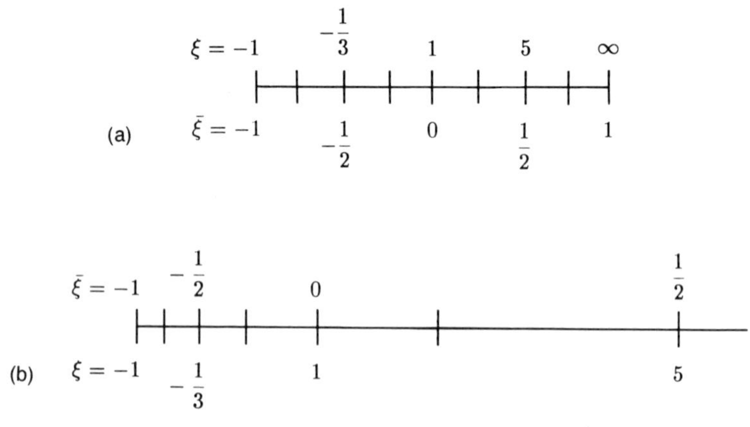

Figure 7.13 Correspondence between (a) in $\bar{\xi}$ space and (b) in ξ space

where $J(\xi)$ is the Jacobian $\mathrm{d}s/\mathrm{d}\xi$, and $\mathrm{d}\xi/\mathrm{d}\bar{\xi}$ is obtained by differentiating equation (7.62):

$$\frac{\mathrm{d}\xi}{\mathrm{d}\bar{\xi}} = \frac{4}{(1-\bar{\xi})^2} \tag{7.65}$$

Let us now consider the behaviour of the integrands as distance r from P_a to Q tends to infinity. Over an infinite element the Jacobian $J(\xi)$ is constant and so the intrinsic coordinate ξ tends to infinity as r. By equation (7.62), $\xi(1-\overline{\xi}) = 3\overline{\xi}+1$ and therefore $(1-\overline{\xi})$ tends to zero as $1/r$. Therefore, the Jacobian $\mathrm{d}\xi/\mathrm{d}\overline{\xi}$ tends to infinity as r^2. In equation (7.63), both the kernel $\mathbf{T}$ and the asymptotic shape function f_u tend to zero as $1/r$. Therefore the product of $\mathbf{T}$, f_u, J and $\mathrm{d}\xi/\mathrm{d}\overline{\xi}$ remains bounded as r tends to infinity. Not only does it remain bounded but also its variation with respect to $\overline{\xi}$ is such that it may reasonably be approximated by a polynomial in $\overline{\xi}$, so Gauss–Legendre formulae may be used. $\Delta\mathbf{T}^b_{ac}$ may accurately be computed according to the formula

$$\Delta\mathbf{T}^b_{ac} \approx \frac{1}{G}\sum_{F=1}^{G}\sum_{k=1}^{g(F)} B_k^{g(F)}\mathbf{T}\left(P_a, Q(\overline{\xi}_k^{g(F)})\right) f_u\left(Q(\overline{\xi}_k^{g(F)})\right) J(\overline{\xi}_k^{g(F)})\frac{\mathrm{d}\xi}{\mathrm{d}\overline{\xi}}\left(\overline{\xi}_k^{g(F)}\right) \tag{7.66}$$

where

$$\overline{\xi}_k^{g(F)} = \frac{1}{G}[G-2F+1+\beta_k^{g(F)}] \tag{7.67}$$

Reasonable values of G and $g(1), g(2), \ldots, g(G)$ may be determined by considering integration over an isoparametric element which coincides with the part $-1 \leq \xi \leq 1$ of the infinite element, rather than integration over the entire element. In equation (7.64) the kernel $\mathbf{U}$ tends to infinity as $\ln r$, whereas the asymptotic shape function f_t tends to zero as $1/r^2$. Therefore the product of $\mathbf{U}$, f_t, J and $\mathrm{d}\xi/\mathrm{d}\overline{\xi}$ tends to infinity as $\ln r$, and it is necessary to apply the Gaussian formula with logarithmic weight function, equation (7.48), over the first of the G intervals of integration. Note that $\ln\frac{1}{\gamma}$ and therefore the product $\phi(\gamma)\ln\frac{1}{\gamma}$ equals zero at $\gamma=1$, so equation (7.48) integrates accurately only functions that are logarithmically singular at one end of the interval and zero at the other. Therefore, $\Delta\mathbf{U}^b_{ac}$ is computed according to the formula

$$\begin{aligned}
\Delta\mathbf{U}^b_{ac} \approx \frac{1}{G}\Bigg\{ & 2\sum_{l=1}^{g(1)} C_l^{g(1)}\mathbf{U}_L\left(P_a, Q\left(\overline{\xi}_l^{g(1)}\right)\right) f_t\left(Q\left(\overline{\xi}_l^{g(1)}\right)\right) J\left(\overline{\xi}_l^{g(1)}\right)\frac{\mathrm{d}\xi}{\mathrm{d}\overline{\xi}}\left(\overline{\xi}_l^{g(1)}\right) \\
& + \sum_{k=1}^{g(1)} B_k^{g(1)}\left[\mathbf{U}\left(P_a, Q\left(\overline{\xi}_k^{g(1)}\right)\right) - \mathbf{U}_L\left(P_a, Q\left(\overline{\xi}_k^{g(1)}\right)\right)\ln\frac{1}{\gamma_k^{g(1)}}\right] \\
& \cdot f_t\left(Q\left(\overline{\xi}_k^{g(1)}\right)\right) J\left(\overline{\xi}_k^{g(1)}\right)\frac{\mathrm{d}\xi}{\mathrm{d}\overline{\xi}}\left(\overline{\xi}_k^{g(1)}\right) + \sum_{F=2}^{G}\sum_{k=1}^{g(F)} B_k^{g(F)}\mathbf{U}\left(P_a, Q\left(\overline{\xi}_k^{g(F)}\right)\right) \\
& \cdot f_t\left(Q\left(\overline{\xi}_k^{g(F)}\right)\right) J\left(\overline{\xi}_k^{g(F)}\right)\frac{\mathrm{d}\xi}{\mathrm{d}\overline{\xi}}\left(\overline{\xi}_k^{g(F)}\right)\Bigg\}
\end{aligned} \tag{7.68}$$

in which if $G = 1$ then the last term is omitted, and according to equations (6.21) and (6.22)

$$\mathbf{U}_L(P,Q) = \frac{-(1+\nu)(3-4\nu)}{4\pi E(1-\nu)}\mathbf{I} \tag{7.69}$$

Note that $\mathbf{U}_L$ is negative because the leading diagonal components of the kernel $\mathbf{U}$ tend to minus infinity as r tends to infinity and $\bar{\xi}$ tends to 1. In the first term of equation (7.68),

$$\bar{\xi}_l^{g(1)} = 1 - \frac{2\gamma_l^{g(1)}}{G} \tag{7.70}$$

and the factor 2 arises because the absolute value of the Jacobian $\mathrm{d}\bar{\xi}/\mathrm{d}\gamma$ is $2/G$. In the second term of equation (7.68), which is the integral over the first interval of the bounded component according to the Gauss–Legendre formula,

$$\gamma_k^{g(1)} = \frac{1}{2}\left(1 - \beta_k^{g(1)}\right) \tag{7.71}$$

For integration over an isoparametric element of which P_a is a node, the element is divided into either one or two subelements, as shown in Figure 7.14 for the case of quadratic variation. For each subelement we define a coordinate v which varies from -1 to 1 and denote by v_0 the value of v at P_a. For quadratic variation, v and v_0 are as given in Table 7.1. The coefficients $\Delta\mathbf{T}_{ac}^{b}$ are evaluated for an element of which P_a is a node only if $d(b,c) \neq a$, and so in equation (7.6) $|N_c(Q)| \leq Kr$ where K is a constant. The kernel $\mathbf{T}$ tends to infinity as $1/r$ as the point Q approaches P_a, so the product $\mathbf{T}(P_a,Q)N_c(Q)$ remains bounded and may accurately be integrated over a subelement by means of Gauss–Legendre formulae:

$$\int_{S_b^e} \mathbf{T}(P_a,Q)N_c(Q)\,\mathrm{d}S_Q$$

$$\approx \frac{J_e}{G}\sum_{F=1}^{G}\sum_{k=1}^{g(F)} B_k^{g(F)}\mathbf{T}\left(P_a, Q\left(v_k^{g(F)}\right)\right)N_c\left(v_k^{g(F)}\right)J\left(v_k^{g(F)}\right) \tag{7.72}$$

in which $J_e = \frac{\mathrm{d}\xi}{\mathrm{d}v}$, $J = \frac{\mathrm{d}s}{\mathrm{d}\xi}$ and

$$v_k^{g(F)} = \frac{1}{G}\left[G - 2F + 1 + \beta_k^{g(F)}\right] \tag{7.73}$$

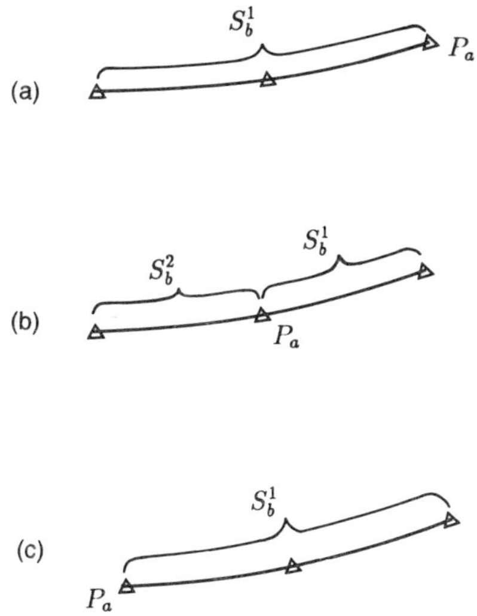

Figure 7.14 Subelements of a quadratic element in two dimensions. (a) Node P_a at ξ=1 (b) Node P_a at ξ=0. (c) Node P_a at ξ=−1

Table 7.1 Values of v and v_0 for quadratic element

P_a at $\xi = 1$ S_b^1	P_a at $\xi = 0$ S_b^1	 S_b^2	P_a at $\xi = -1$ S_b^1
$v = \xi$ $v_0 = 1$	$v = 2\xi + 1$ $v_0 = -1$	$v = 2\xi - 1$ $v_0 = 1$	$v = \xi$ $v_0 = -1$

A reasonable value of G may be determined by considering integration over an isoparametric element of length equal to that of the subelement, and of minimum distance from node P_a equal to its length, so that equation (7.59) becomes

$$G \geq \frac{1}{4}\left(\frac{2}{\epsilon}\right)^{\frac{1}{2g_{\max}}} \tag{7.74}$$

G is taken to be the smallest integer satisfying this equation. We then propose $g(1) = g(2) = \ldots = g(G) = g'$ where g' is the smallest integer for which

$$G \geq \frac{1}{4}\left(\frac{2}{\epsilon}\right)^{\frac{1}{2g'}} \tag{7.75}$$

In equation (7.7) for an element of which P_a ia a node, the shape function $N_c(Q)$ may equal 1 at P_a, and the kernel $\mathbf{U}$ tends to infinity as $\ln \frac{1}{r}$ as the point

Q approaches P_a. Therefore it is necessary to apply the Gaussian quadrature formula with logarithmic shape function over the interval adjacent to P_a. Let us denote by F_0 the number of this interval: $F_0 = 1$ if $v_0 = 1$, or G if $v_0 = -1$. Then in the same way as for off-diagonal integration over an infinite element we write

$$\int_{S_b^e} \mathbf{U}(P_a, Q) N_c(Q)\, \mathrm{d}S_Q$$

$$\approx \frac{J_e}{G} \left\{ 2 \sum_{l=1}^{g(F_0)} C_l^{g(F_0)} \mathbf{U}_L \left(P_a, Q(v_l^{g(F_0)}) \right) N_c \left(v_l^{g(F_0)} \right) J(v_l^{g(F_0)}) \right.$$

$$+ \sum_{k=1}^{g(F_0)} B_k^{g(F_0)} \left[\mathbf{U} \left(P_a, Q \left(v_k^{g(F_0)} \right) \right) - \mathbf{U}_L \left(P_a, Q(v_k^{g(F_0)}) \right) \ln \frac{1}{\gamma_k^{g(F_0)}} \right]$$

$$.N_c \left(Q(v_k^{g(F_0)}) \right) J(v_k^{g(F_0)})$$

$$\left. + \sum_{\substack{F=1 \\ F \neq F_0}}^{G} \sum_{k=1}^{g(F)} B_k^{g(F)} \mathbf{U} \left(P_a, Q(v_k^{g(F)}) \right) N_c \left(Q(v_k^{g(F)}) \right) J(v_k^{g(F)}) \right\} \tag{7.76}$$

where

$$\mathbf{U}_L(P, Q) = \frac{(1+\nu)(3-4\nu)}{4\pi E(1-\nu)} \mathbf{I} \tag{7.77}$$

In the first term of equation (7.76),

$$v_l^{g(F_0)} = v_0 \left(1 - \frac{2\gamma_l^{g(F_0)}}{G} \right) \tag{7.78}$$

and the factor 2 arises because the absolute value of the Jacobian $\mathrm{d}v/\mathrm{d}\gamma$ is $2/G$. In the second term of equation (7.76),

$$\gamma_k^{g(F_0)} = \frac{1}{2} \left(1 - \beta_k^{g(F_0)} \right) \tag{7.79}$$

The values of $G, g(1), g(2), \ldots, g(G)$ are taken to be the same as for equation (7.72).

For two-dimensional analysis, the finite part S_b^F of an infinite element may conveniently be taken to be the part $-1 \leq \xi \leq 1$. The coefficients $\Delta_F \mathbf{T}_{ac}^b$ in equation (7.40) are never evaluated because an infinite element in two dimensions has only one node, which must be P_a. It therefore only remains to evaluate for an infinite element of which P_a is a node the coefficients $\Delta_I \mathbf{T}_{ac}^b$ and $\Delta \mathbf{U}_{ac}^b$ in equation (7.40), and the integral over S_b^F in equation (7.46). We consider $\Delta \mathbf{U}_{ac}^b$ as the sum of $\Delta_F \mathbf{U}_{ac}^b$ and $\Delta_I \mathbf{U}_{ac}^b$ where

$$\Delta_F \mathbf{U}^b_{ac} = \int_{S_b^F} \mathbf{U}(P_a, Q) f_t(Q) \, \mathrm{d}S_Q \tag{7.80}$$

$$\Delta_I \mathbf{U}^b_{ac} = \int_{S_b^I} \mathbf{U}(P_a, Q) f_t(Q) \, \mathrm{d}S_Q \tag{7.81}$$

Integrals over S_b^F are evaluated by taking S_b^F to be a subelement with node P_a at $\xi = -1$ as shown in Figure 7.14c. Since $|1 - f_u(Q)| \leq Kr$ where K is a constant, the integral over S_b^F in equation (7.46) may accurately be evaluated by the application of equation (7.72) with $N_c(Q)$ replaced by $(1 - f_u(Q))$. Similarly the coefficients $\Delta_F \mathbf{U}^b_{ac}$ may be evaluated by the application of equation (7.76) with $N_c(Q)$ replaced by $f_t(Q)$. G and $g(1), g(2), \ldots, g(G)$ are chosen in the same way as for equation (7.72). Integrals over S_b^I are evaluated according to the procedure for off-diagonal integration over an infinite element, with the number of intervals G for the entire element S_b taken to be twice that for integration over S_b^I. From Figure 7.13a it may be seen that S_b^I, which is the interval $1 \leq \xi < \infty$, maps onto the interval $0 \leq \bar{\xi} < 1$ and so the intervals $1, 2, \ldots, G/2$ lie in S_b^I and the intervals $G/2+1, \ldots, G$ lie in S_b^F. $\Delta_I \mathbf{T}^b_{ac}$ and $\Delta_I \mathbf{U}^b_{ac}$ are therefore integrated according to equations (7.66) and (7.68), with $F = 1, 2, \ldots, G/2$ and the last term of equation (7.68) omitted if $G/2 = 1$.

Finally, in equations (7.13) and (7.46) it is necessary to evaluate the integral of $\mathbf{T}(P_a, Q)$ over the auxiliary boundary Ω_a. Since the value of this integral is independent of the geometric form of Ω_a and computational efficiency is of secondary importance, the auxiliary boundary may be constructed from isoparametric elements. In program BOUND (see Appendix B) these elements are taken to lie approximately on a circular or spiral boundary as shown in Figure 7.15, the number of elements being such that no element subtends an angle greater than one radian at P_a. The integration scheme for each element is determined in the same way as for off-diagonal integration over an isoparametric element, and the integral of $\mathbf{T}(P_a, Q)$ computed according to equation (7.60).

7.5 INTEGRATION SCHEMES FOR THREE DIMENSIONAL ANALYSIS

Only the equations of nodal collocation for elasticity are considered in this section. The generalisation to methods of moments is fairly straightforward, and any integration scheme that is valid for elasticity is also valid for Laplace's equation. The integrals to be evaluated numerically are those appearing in equations (7.5) and (7.40), and the terms on the right sides of equations (7.13) and (7.45). Gauss–Legendre formulae are used throughout. The $g \times h$ product formula

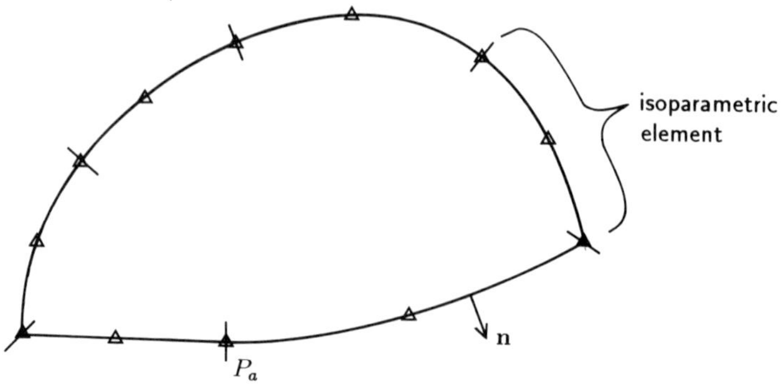

Figure 7.15 Representation of auxiliary boundary by isoparametric elements

$$\int_{-1}^{1}\int_{-1}^{1}\phi(\beta,\gamma)\ \mathrm{d}\beta\ \mathrm{d}\gamma \approx \sum_{s=1}^{g}\sum_{t=1}^{h} B_s^g B_t^h \phi(\beta_s^g,\beta_t^h) \tag{7.82}$$

integrates exactly any function $\phi(\beta,\gamma) = [a_0+a_1\beta+\ldots+a_{2g-1}\beta^{2g-1}][b_0+b_1\gamma+\ldots+b_{2h-1}\gamma^{2h-1}]$. The requirements that must be satisfied by the integration scheme are the same as for two-dimensional analysis, and again we consider separately integration over elements of which P_a is a node, integration over the auxiliary boundary Ω_a, and off-diagonal integration. For larger problems most of the computer time required for construction of the equations of nodal collocation is spent on off-diagonal integration. Isoparametric elements may be quadrilateral or triangular, and infinite elements may have asymptotic or plane strain functional variation.

We consider first off-diagonal integration over quadrilateral elements. To ensure that satisfactory precision of integration can be maintained it is necessary to enable elements to be subdivided into more than one interval of integration in either or both of the intrinsic coordinate directions. We therefore propose the scheme[1]

$$\Delta \mathbf{T}_{ac}^b \approx \sum_{E=1}^{G}\sum_{F=1}^{H} I_{EF}\left(\mathbf{T}(P_a,Q)N_c(Q)\right) \tag{7.83}$$

$$\Delta \mathbf{U}_{ac}^b \approx \sum_{E=1}^{G}\sum_{F=1}^{H} I_{EF}\left(\mathbf{U}(P_a,Q)N_c(Q)\right) \tag{7.84}$$

where

$$I_{EF}(\phi) = \frac{1}{GH}\sum_{s=1}^{g(E,F)}\sum_{t=1}^{h(E,F)} B_s^{g(E,F)} B_t^{h(E,F)} \phi\left(\xi_s^{g(E,F)},\eta_t^{h(E,F)}\right) .J\left(\xi_s^{g(E,F)},\eta_t^{h(E,F)}\right) \tag{7.85}$$

In equation (7.85), J is the Jacobian $\mathrm{d}A/\mathrm{d}(\xi, \eta)$ and

$$\xi_s^{g(E,F)} = \frac{1}{G}\left[G - 2E + 1 + \beta_s^{g(E,F)}\right] \tag{7.86}$$

$$\eta_t^{h(E,F)} = \frac{1}{H}\left[H - 2F + 1 + \beta_t^{h(E,F)}\right] \tag{7.87}$$

The intersection of interval E in the ξ direction and interval F in the η direction forms the area interval EF as shown in Figure 7.16. G and H are taken to be the smallest integers satisfying the equations analogous to equation (7.59)

$$G \geq \frac{L_b^{(\xi)}}{4R_b}\left(\frac{2(2g_{\max}+1)}{\epsilon}\right)^{\frac{1}{2g_{\max}}} \tag{7.88}$$

$$H \geq \frac{L_b^{(\eta)}}{4R_b}\left(\frac{2(2g_{\max}+1)}{\epsilon}\right)^{\frac{1}{2g_{\max}}} \tag{7.89}$$

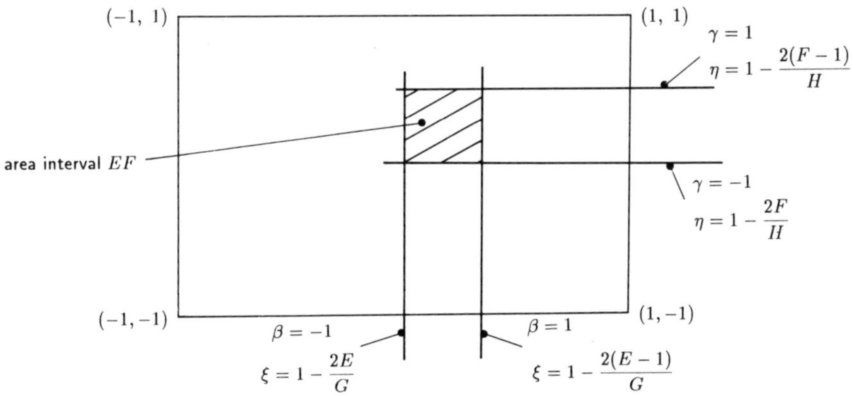

Figure 7.16 Area interval EF of isoparametric quadrilateral element

where $L_b^{(\xi)}$ and $L_b^{(\eta)}$ are the lengths of element S_b in the ξ and η directions respectively. The values of $g(E, F)$ and $h(E, F)$ are then determined for each area interval according to the equations analogous to equation (7.57)

$$\frac{L^{(\xi)}}{R} \leq 4\left(\frac{\epsilon}{2(2g+1)}\right)^{\frac{1}{2g(E,F)}} \tag{7.90}$$

$$\frac{L^{(\eta)}}{R} \leq 4\left(\frac{\epsilon}{2(2h+1)}\right)^{\frac{1}{2h(E,F)}} \tag{7.91}$$

where $L^{(\xi)}$ and $L^{(\eta)}$ are the lengths of the area interval EF in the ξ and η directions, and R is the minimum distance from P_a to the interval. Determination of the integration scheme according to inequalities (7.88) to (7.91) results not only in a greater number of Gauss points for elements nearer to P_a compared with their size, but also in a greater number of Gauss points in the direction of the longer dimension of an element than in that of its shorter dimension. This enables adequate precision of integration to be maintained economically, even if the length of an element is several times its width.

The natural choice of intrinsic coordinate system for a triangular element is the area coordinate system shown in Figure 7.17a in which each of the coordinates A_1, A_2 and A_3 varies from 1 at a vertex of the element to zero on the opposite side. Ideally for off-diagonal integration the quadrature formula used should integrate exactly certain polynomials in the area coordinates. Such formulae, known as Hammer formulae, do exist but those so far published integrate exactly only polynomials of the same degree in each of the area coordinates, and therefore cannot easily be incorporated into a scheme to evaluate economically integrals over elements which are significantly longer in one direction than in another. In addition, triangular elements of a mesh are usually greatly outnumbered by quadrilateral elements, so the improvement in performance resulting from the use of a special integration scheme for triangular elements is not worth the extra programming effort. We therefore adopt an aesthetically less appealing scheme in which product formulae are used, integration being carried out with respect to linear functions of the coordinates ξ and η shown in Figure 7.17b. The relation between (ξ, η) and the area coordinates is

$$\xi = 2A_1 - 1 \tag{7.92}$$

$$\eta = \frac{A_2 - A_3}{A_2 + A_3} \tag{7.93}$$

The Jacobian $\mathrm{d}A/\ \mathrm{d}(\xi, \eta)$ tends to zero as ξ tends to 1 and so the optimal product formula for the triangular area intervals adjacent to vertex 1 would be the product of the Gauss–Legendre formula in the η direction and the Gauss–Radeau (with weight function $1-\beta$) in the ξ direction, but again the potential improvement of efficiency is not worth the extra programming effort. We therefore carry out off-diagonal integration over a triangular element according to equations (7.83) and (7.84).

For off-diagonal integration over an infinite element, integration is carried out with respect to the coordinates $(\overline{\xi}, \eta)$ where $\overline{\xi}$ is given by equation (7.62). By equations (7.35) and (7.36)[6],

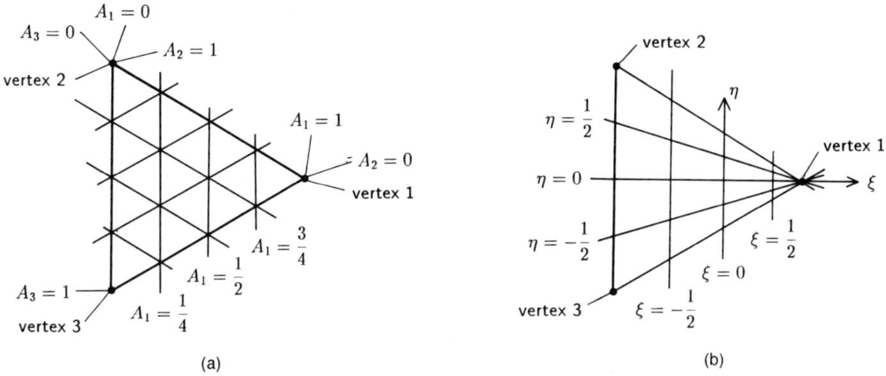

Figure 7.17 Coordinate systems for isoparametric triangular element. (a) Area coordinates. (b) Coordinates for product formula integration

$$\Delta \mathbf{T}^b_{ac} = \int_{-1}^{1}\int_{-1}^{1} \mathbf{T}\left(P_a, Q\left(\xi,\eta\right)\right) f_u\left(Q\left(\xi,\eta\right)\right) N_c\left(\eta\right) J\left(\xi,\eta\right) \frac{\mathrm{d}\xi}{\mathrm{d}\overline{\xi}}\, \mathrm{d}\overline{\xi}\, \mathrm{d}\eta \tag{7.94}$$

$$\Delta \mathbf{U}^b_{ac} = \int_{-1}^{1}\int_{-1}^{1} \mathbf{U}\left(P_a, Q\left(\xi,\eta\right)\right) f_t\left(Q\left(\xi,\eta\right)\right) N_c\left(\eta\right) J\left(\xi,\eta\right) \frac{\mathrm{d}\xi}{\mathrm{d}\overline{\xi}}\, \mathrm{d}\overline{\xi}\, \mathrm{d}\eta \tag{7.95}$$

In equation (7.94) the kernel $\mathbf{T}$ tends to zero as $1/r^2$, $N_c(\eta)$ is bounded, and the Jacobian $\frac{\mathrm{d}\xi}{\mathrm{d}\overline{\xi}}$ tends to infinity as r^2. For an element with asymptotic functional variation, $f_u(Q)$ tends to zero as $1/r$ and because the sides $\eta = \pm 1$ need not be parallel, the Jacobian $J(\xi,\eta)$ can tend to infinity as r. Therefore, the product of $\mathbf{T}, f_u, N, J$ and $\frac{\mathrm{d}\xi}{\mathrm{d}\overline{\xi}}$ remains bounded as r tends to infinity. For an element with plane strain functional variation $f_u(Q)$ does not tend to zero as r tends to infinity, but since the sides $\eta = \pm 1$ are parallel the Jacobian $J(\xi,\eta)$ remains bounded; therefore, the product of $\mathbf{T}, f_u, N, J$ and $\mathrm{d}\xi/\mathrm{d}\overline{\xi}$ still remains bounded as r tends to infinity. For both types of infinite elements, then, $\Delta \mathbf{T}^b_{ac}$ is computed according to the formula

$$\Delta \mathbf{T}^b_{ac} \approx \sum_{E=1}^{G}\sum_{F=1}^{H} \overline{I}_{EF}\left(\mathbf{T}\left(P_a, Q\right) f_u(Q) N_c\left(\eta\right)\right) \tag{7.96}$$

where

$$\overline{I}_{EF}(\phi) = \frac{1}{GH} \sum_{s=1}^{g(E,F)} \sum_{t=1}^{h(E,F)} B_s^{g(E,F)} B_t^{h(E,F)} \phi\left(\overline{\xi}_s^{g(E,F)}, \eta_t^{h(E,F)}\right) . J\left(\overline{\xi}_s^{g(E,F)}, \eta_t^{h(E,F)}\right) \frac{\mathrm{d}\xi}{\mathrm{d}\overline{\xi}}\left(\overline{\xi}_s^{g(E,F)}\right) \tag{7.97}$$

in which

$$\overline{\xi}_s^{g(E,F)} = \frac{1}{G}\left[G - 2E + 1 + \beta_s^{g(E,F)}\right] \tag{7.98}$$

and $\eta_t^{h(E,F)}$ is given by equation (7.87). Reasonable values of $G, H, g(1,1), \ldots$ and $h(1,1), \ldots$ may be determined by considering integration over an isoparametric quadrilateral element which coincides with the part $-1 \leq \xi \leq 1$ of the infinite element.

In equation (7.95) the kernel $\mathbf{U}$ tends to zero as $1/r$, $N_c(\eta)$ is bounded, and the Jacobian $\mathrm{d}\xi/\mathrm{d}\overline{\xi}$ tends to infinity as r^2. For an element with asymptotic functional variation, $f_t(Q)$ tends to zero as $1/r^2$ and the Jacobian $J(\xi,\eta)$ can tend to infinity as r. Therefore, the product of $\mathbf{U}$, f_t, N, J and $\mathrm{d}\xi/\mathrm{d}\overline{\xi}$ remains bounded as r tends to infinity and $\Delta\mathbf{U}_{ac}^b$ is computed according to the formula

$$\Delta\mathbf{U}_{ac}^b \approx \sum_{E=1}^{G}\sum_{F=1}^{H} \overline{I}_{EF}\left(\mathbf{U}(P_a,Q)f_t(Q)N_c(\eta)\right) \tag{7.99}$$

For an infinite element with plane strain functional variation, $f_t(Q)$ does not tend to zero as r tends to infinity, and the Jacobian $J(\xi,\eta)$ remains bounded. Therefore, the product of $\mathbf{U}, f_t, N, J$ and $\mathrm{d}\xi/\mathrm{d}\overline{\xi}$ tends to infinity as r and $\Delta\mathbf{U}_{ac}^b$ cannot be evaluated according to equation (7.99). However, according to equation (7.29), in the analysis of plane strain the particular integral must be chosen such that the surface traction due to the complementary function integrates to zero. Plane strain elements of a three dimensional boundary element mesh model the surfaces either of openings of uniform cross section or of prismatic parts of the domain, extending to infinity. We arbitrarily define for each of the surfaces of this type an axis as shown in Figure 7.18 and denote by Q' the foot of the perpendicular from the point Q to the axis. Then since by equation (7.29) the integral of traction with respect to arc length around the contour C shown in Figure 7.18 is zero,

$$\begin{aligned}\Delta\mathbf{U}_{ac}^b = \int_{-1}^{1}\int_{-1}^{1}&\Big(\mathbf{U}(P_a,Q(\xi,\eta)) - \mathbf{U}(P_a,Q')\Big)\\ &\cdot f_t\left(Q(\xi,\eta)\right)N_c(\eta)J(\xi,\eta)\frac{\mathrm{d}\xi}{\mathrm{d}\overline{\xi}}\,\mathrm{d}\overline{\xi}\,\mathrm{d}\eta\end{aligned} \tag{7.100}$$

By the mean value theorem, $(\mathbf{U}\left(P_a,Q(\xi,\eta)\right) - \mathbf{U}(P_a,Q'))$ is equal to the distance from Q to Q' multiplied by the derivative of the kernel $\mathbf{U}$ in the direction QQ' at some point on the line QQ'. Since that derivative tends to zero as $1/r^2$ where r is distance from P_a to Q, $(\mathbf{U}\left(P_a,Q\left(\xi,\eta\right)\right) - \mathbf{U}\left(P_a,Q'\right))$ also tends to zero as $1/r^2$. Therefore, the product of $(\mathbf{U}\left(P_a,Q\left(\xi,\eta\right)\right) - \mathbf{U}(P_a,Q'))$, f_t, N, J and $\mathrm{d}\xi/\mathrm{d}\overline{\xi}$ remains bounded as r tends to infinity and $\Delta\mathbf{U}_{ac}^b$ is computed according to the formula

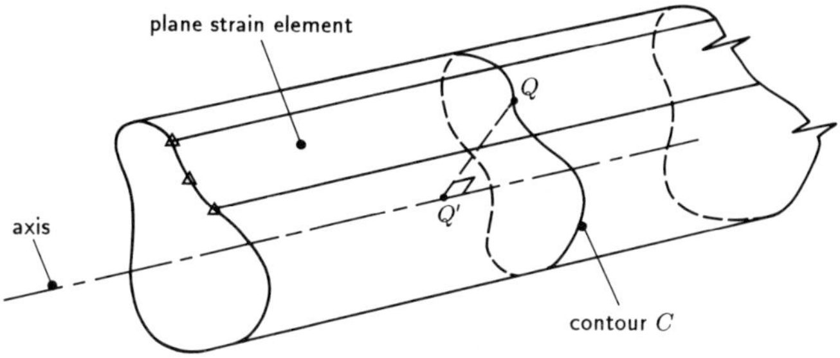

Figure 7.18 Surface modelled by infinite elements with plane strain functional variation

$$\Delta \mathbf{U}^b_{ac} \approx \sum_{E=1}^{G} \sum_{F=1}^{H} \overline{I}_{EF} \left(\left\{ \mathbf{U}(P_a, Q) - \mathbf{U}(P_a, Q') \right\} f_t(Q) N_c(\eta) \right) \tag{7.101}$$

Note that if in equation (7.101) $\{\mathbf{U}(P_a, Q) - \mathbf{U}(P_a, Q')\}$ were replaced by $\mathbf{U}(P_a, Q)$, the computed values of $\Delta \mathbf{U}^b_{ac}$ would be significantly in error.

For integration over a quadrilateral element of which P_a is a node, the element is divided into either two or three subelements, as shown in Figure 7.19 for the case of quadratic variation[1]. For each subelement we define coordinates v and w according to the transformation

$$\xi = \sum_{c=1}^{4} L_c(v, w) \xi_c \tag{7.102}$$

$$\eta = \sum_{c=1}^{4} L_c(v, w) \eta_c \tag{7.103}$$

where L_c are the bilinear shape functions defined by equation (2.17), (ξ_3, η_3) and (ξ_4, η_4) are the intrinsic coordinates of the ends of the side of the subelement opposite P_a, and (ξ_1, η_1) and (ξ_2, η_2) are equal to the intrinsic coordinates of P_a. Typical coordinate systems (v, w) are shown in Figure 7.19. For the subelement the coordinates (v, w) of which are shown in Figure 7.19a, $(\xi_3, \eta_3) = (1, -1)$, $(\xi_4, \eta_4) = (1, 1)$ and $(\xi_1, \eta_1) = (\xi_2, \eta_2) = (-1, -1)$. For that shown in Figure 7.19b, $(\xi_3, \eta_3) = (1, 1)$, $(\xi_4, \eta_4) = (-1, 1)$ and $(\xi_1, \eta_1) = (\xi_2, \eta_2) = (0, -1)$. Both v and w range from -1 to 1 over a subelement. At P_a $w = -1$ and v is indeterminate, and as P_a is approached the Jacobian of the transformation from (v, w) to (ξ, η) tends to zero:

$$J_e(Q) = \left| \frac{\partial(\xi, \eta)}{\partial(v, w)} \right| \leq Kr \tag{7.104}$$

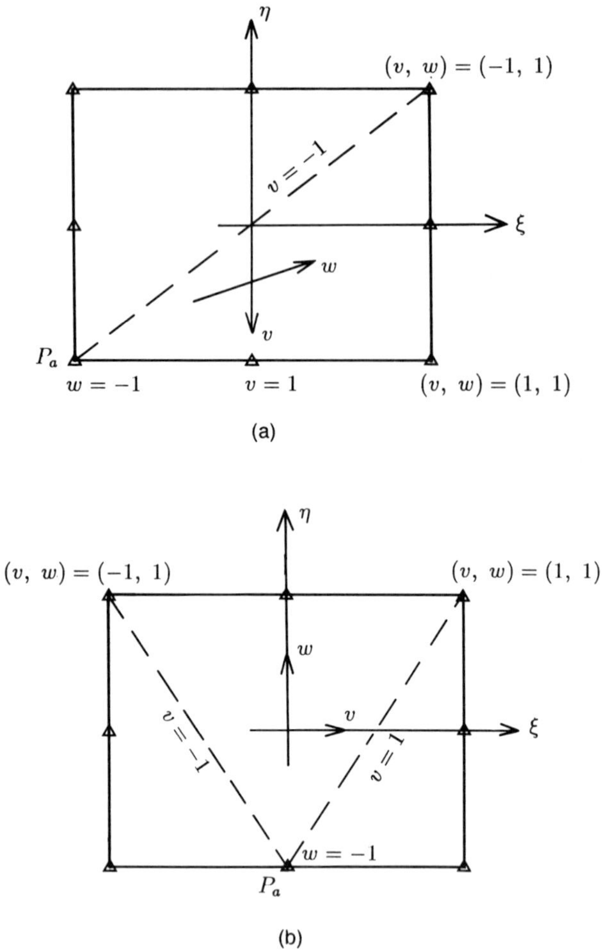

Figure 7.19 Subelements of a quadratic quadrilateral element in three dimensions. (a) P_a at corner. (b) P_a on side

where K is a constant and r is the distance from P_a.

The coefficients $\Delta \mathbf{T}^b_{ac}$ are evaluated for an element of which P_a is a node only if $d(b,c) \neq a$ and so in equation (7.6), $|N_c(Q)| \leq Kr$. The kernel $\mathbf{T}$ tends to infinity as $1/r^2$ as the point Q approaches P_a, so the product $\mathbf{T}\,(P_a, Q)\,N_c(Q) J_e(Q) J(Q)$ is bounded and may accurately be integrated over a subelement by means of Gauss–Legendre formulae:

$$\int_{S^e_b} \mathbf{T}(P_a, Q) N_c(Q)\, \mathrm{d}S_Q \approx \sum_{E=1}^{G} \sum_{F=1}^{H} I^*_{EF}\left(\mathbf{T}\,(P_a, Q)\, N_c(Q)\right) \tag{7.105}$$

where

$$I^*_{EF}(\phi) = \frac{1}{GH} \sum_{s=1}^{g(E,F)} \sum_{t=1}^{h(E,F)} B_s^{g(E,F)} B_t^{h(E,F)} \phi\left(v_s^{g(E,F)}, w_t^{h(E,F)}\right)$$
$$.J\left(v_s^{g(E,F)}, w_t^{h(E,F)}\right) J_e\left(v_s^{g(E,F)}, w_t^{h(E,F)}\right) \qquad (7.106)$$

in which

$$v_s^{g(E,F)} = \frac{1}{G}\left[G - 2E + 1 + \beta_s^{g(E,F)}\right] \qquad (7.107)$$
$$w_s^{h(E,F)} = \frac{1}{H}\left[H - 2F + 1 + \beta_t^{h(E,F)}\right] \qquad (7.108)$$

Reasonable values of G, H, $g(E,F)$ and $h(E,F)$ may be determined in a manner similar to the determination of G and $g(F)$ in equation (7.72). In equation (7.7) for an element of which P_a is a node, the shape function $N_c(Q)$ may equal 1 at P_a, and the kernel $\mathbf{U}$ tends to infinity as $1/r$ as the point Q approaches P_a. Therefore, the product $\mathbf{U}(P_a, Q)\, N_c(Q) J_e(Q) J(Q)$ is bounded and may accurately be integrated over a subelement by means of Gauss–Legendre formulae:

$$\int_{S_b^e} \mathbf{U}(P_a, Q) N_c(Q)\, dS_Q \approx \sum_{E=1}^{G} \sum_{F=1}^{H} I^*_{EF}\left(\mathbf{U}(P_a, Q)\, N_c(Q)\right) \qquad (7.109)$$

For integration over a triangular element of which P_a is a node, the element may be divided into either one or two subelements, as shown in Figure 7.20 for the case of quadratic variation. The coordinates (v, w) of a subelement are defined in terms of its area coordinates (A_1', A_2', A_3') by the transformation

$$v = \frac{A_2' - A_3'}{A_2' + A_3'} \qquad (7.110)$$

$$w = 1 - 2A_1' \qquad (7.111)$$

The subelement area coordinates are in turn defined in terms of the area coordinates of the element; in Figure 7.20a for example $(A_1', A_2', A_3') = (A_1, A_2, A_3)$, and for the subelement the coordinates (v, w) of which are shown in Figure 7.20b $(A_1', A_2', A_3') = (2A_1, A_2 - A_1, A_3)$. Then at P_a $w = -1$ and v is indeterminate, and

$$\left|\frac{\partial A}{\partial(v, w)}\right| \leq Kr \qquad (7.112)$$

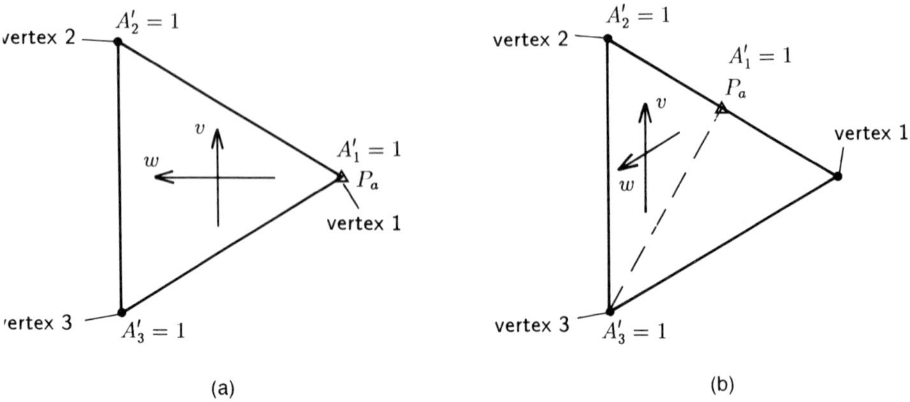

Figure 7.20 Subelements of a quadratic triangular element in three dimensions. (a) P_a at corner. (b) P_a on side

Coefficients $\Delta\mathbf{T}_{ac}^b$ and $\Delta\mathbf{U}_{ac}^b$ are evaluated according to equations (7.105) and (7.109), in which

$$I_{EF}^*(\phi) = \frac{1}{GH} \sum_{s=1}^{g(E,F)} \sum_{t=1}^{h(E,F)} B_s^{g(E,F)} B_t^{h(E,F)} \phi\left(v_s^{g(E,F)}, w_t^{h(E,F)}\right) \cdot \frac{\partial A}{\partial(v,w)}\left(v_s^{g(E,F)}, w_t^{h(E,F)}\right) \tag{7.113}$$

For both quadrilateral and triangular elements of which P_a is a node, therefore, integrals of functions that tend to infinity as $1/r$ are evaluated by the introduction of a coordinate system for which the Jacobian tends to zero as P_a is approached.

The finite part S_b^F of an infinite element of which P_a is a node is taken to be the part $-1 \leq \xi \leq 1$. In equation (7.40), the coefficients $\Delta_F\mathbf{T}_{ac}^b$ and $\Delta_F\mathbf{U}_{ac}^b$ as defined by equations (7.41) and (7.80), respectively, are computed in the same way as $\Delta\mathbf{T}_{ac}^b$ and $\Delta\mathbf{U}_{ac}^b$ for a quadrilateral element of which P_a is a node. In the same way as for an element in two dimensions, integrals over S_b^I are evaluated according to the procedure for off-diagonal integration over an infinite element, with the number of intervals G in the ξ direction for the entire element S_b taken to be twice that for integration over S_b^I. $\Delta_I\mathbf{T}_{ac}^b$ and $\Delta_I\mathbf{U}_{ac}^b$ are therefore computed according to equations (7.96) and either (7.99) or (7.101), with $E = 1, 2, \ldots, G/2$. The second term on the right side of equation (7.45) is evaluated in the same way as $\Delta\mathbf{T}_{ac}^b$ for a quadrilateral element of which P_a is a node.

The auxiliary boundary Ω_a may be constructed from quadrilateral and triangular elements. For the evaluation of integrals appearing in equations (7.13) and (7.45), the integration scheme for each of these elements is determined in the same way as for off-diagonal integration, and the integral of $\mathbf{T}(P_a, Q)$ computed according to equation (7.83).

7.6 ASSEMBLY OF DISCRETE SYSTEMS

The construction of a system of simultaneous equations in terms of unknown parameters at the q nodes of a mesh is now considered. For Laplace's equation, a system of q equations is obtained by substituting given boundary data into the equations analogous to equations (7.5) for elasticity

$$A(P_a)u(P_a) + \sum_{\substack{b=1 \\ d(b,c)\neq a}}^{p} \sum_{c=1}^{n(b)} \Delta T_{ac}^b u_c^b = \sum_{b=1}^{p} \sum_{c=1}^{n(b)} \Delta U_{ac}^b t_c^b , \quad a = 1, 2, \ldots, q \tag{7.114}$$

Since u is uniquely defined at each node of the mesh, this equation may be rewritten as follows:

$$\sum_{b=1}^{p} \sum_{c=1}^{n(b)} \Delta_1 T_{ac}^b u_c^b = \sum_{b=1}^{p} \sum_{c=1}^{n(b)} \Delta U_{ac}^b t_c^b , \quad a = 1, 2, \ldots, q \tag{7.115}$$

where $\Delta_1 T_{ac}^b = \Delta T_{ac}^b$ for $d(b,c) \neq a$ and for $d(b,c) = a$ the coefficients $\Delta_1 T_{ac}^b$ are chosen in any way that satisfies the relations

$$\sum_{\substack{b=1 \\ d(b,c)=a}}^{p} \sum_{c=1}^{n(b)} \Delta_1 T_{ac}^b = A(P_a) , \quad a = 1, 2, \ldots, q \tag{7.116}$$

For the general case of a mixed boundary value problem, the mesh of elements is arranged so that the boundary condition over any one element of the mesh is either Dirichlet over the entire element or Neumann over the entire element. In addition, it is arranged that any discontinuities in specified values of $\partial u/\partial \mathbf{n}$ coincide with sides or ends of boundary elements. Discontinuities in specified values of u are not allowed. Under these conditions the function u is continuous not only within each element but also from an element to an adjacent element. The function $\partial u/\partial \mathbf{n}$ may be discontinuous from one element to an adjacent element, for any of the following reasons: the normal direction is not continuous, specified values are not continuous, or u is given on one element and $\partial u/\partial \mathbf{n}$ on the other. Let $D(b) = 0$ if $\partial u/\partial \mathbf{n}$ is given on element b (Neumann condition) and $D(b) = 1$ if u is given (Dirichlet), and let $E(a) = 0$ if u at node a is unknown and $E(a) = 1$ if it is known. Then equation (7.115) may be rewritten as follows with all unknown integrals on the left side, and all known integrals on the right side:

$$\sum_{b=1}^{p}\sum_{c=1}^{n(b)} \Delta_1 T_{ac}^b E^* \left(d(b,c)\right) u_c^b - \sum_{b=1}^{p}\sum_{c=1}^{n(b)} \Delta U_{ac}^b D(b) t_c^b$$
$$= \sum_{b=1}^{p}\sum_{c=1}^{n(b)} \Delta U_{ac}^b D^*(b) t_c^b - \sum_{b=1}^{p}\sum_{c=1}^{n(b)} \Delta_1 T_{ac}^b E\left(d(b,c)\right) u_c^b\,, \quad a = 1, 2, \dots, q \tag{7.117}$$

where $D^* = 1 - D$ and $E^* = 1 - E$.

Let us now consider the case in which the boundary of the domain has no edges or corners. The function u is continuous so the first term on the left side of equation (7.117) may be written in terms of unique nodal values. Within contiguous groups of elements which model parts of the boundary subject to the Dirichlet condition, $\partial u/\partial \mathbf{n}$ is continuous from one element to the next and the second term on the left side may also be written in terms of unique nodal values. Therefore,

$$\sum_{v=1}^{q} \Delta T_{av} E^*(v) u(P_v) - \sum_{v=1}^{q} \Delta U_{av} E(v) t(P_v)$$
$$= \sum_{b=1}^{p}\sum_{c=1}^{n(b)} \Delta U_{ac}^b D^*(b) t_c^b - \sum_{b=1}^{p}\sum_{c=1}^{n(b)} \Delta_1 T_{ac}^b E\left(d(b,c)\right) u_c^b\,, \quad a = 1, 2, \dots, q \tag{7.118}$$

where

$$\Delta T_{av} = \sum_{\substack{b=1 \\ d(b,c)=v}}^{p}\sum_{c=1}^{n(b)} \Delta_1 T_{ac}^b \tag{7.119}$$

$$\Delta U_{av} = \sum_{\substack{b=1 \\ d(b,c)=v}}^{p}\sum_{c=1}^{n(b)} \Delta U_{ac}^b D(b) \tag{7.120}$$

Equation (7.118) is a system of q simultaneous equations which may be solved for the q unknowns $u(P_v)$ and $t(P_v)$. For mixed boundary value problems, it is important to assign contributions to the matrix and load vector of the system exactly as prescribed by equations (7.118) to (7.120). For example, for the elements shown in Figure 7.21, $\Delta U_{a4} = \Delta U_{a2}^3 + \Delta U_{a1}^4$ but $\Delta U_{a3} = \Delta U_{a1}^3$, $\Delta_1 T_{a2}^2$ is multiplied by the given value u_2^2 and subtracted from the load vector, and ΔU_{a2}^2 is multiplied by the given value t_2^2 and added to the load vector. Note that t_2^2, which is the limiting value of $\partial u/\partial \mathbf{n}$ as the node P_3 is approached

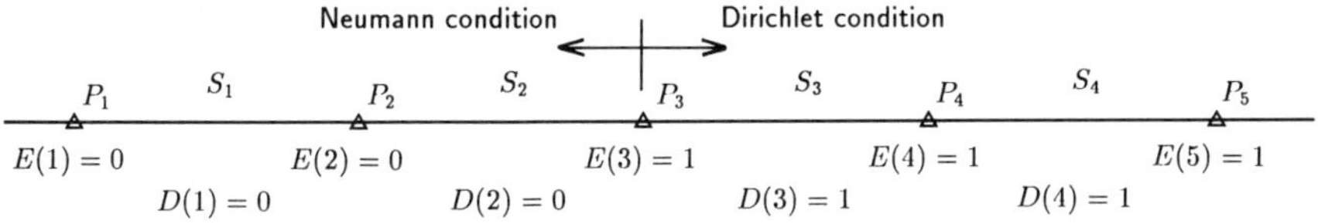

Figure 7.21 Typical mixed boundary conditions, linear elements in two dimensions

from within the element S_2, is in general not the same as the unknown $t(P_3)$ which is the limiting value of $\partial u/\partial \mathbf{n}$ as P_3 is approached from within the element S_3. This procedure yields good results everywhere except near the node P_3, where in the exact solution there is a component of u which varies as the square root of distance from P_3 and a component of $\partial u/\partial \mathbf{n}$ on S_3 which varies as the inverse of the square root of distance from P_3. Since the diameter of the zone in which results are inaccurate is roughly equal to the length of the elements adjacent to P_3, results may be improved by local mesh refinement. Alternatively, accurate results near P_3 may be obtained either by introducing singular shape functions on elements S_2 and S_3, or for quadratic elements by placing the third nodes of elements S_2 and S_3 at a distance from P_3 equal to one quarter of the length of the element.

For domains with edges and corners, equations (7.118) to (7.120) may usually be applied without modification. The only case in which accuracy may be lost is that in which elements on which u is given do not all have the same normal direction at a node P_v, as illustrated in Figure 7.22. The simplest procedure is to take all unknown limiting values of $\partial u/\partial \mathbf{n}$ as P_v is approached to be equal to $t(P_v)$. The results so obtained will usually be inaccurate near the node P_v, but the extent of the zone of inaccuracy can be reduced by local mesh refinement. Alternatively, $t(P_v)$ can be taken to be the derivative of u in the average normal direction $\bar{\mathbf{n}}$ shown in Figure 7.22. We may then write for an element S_b the node c of which is at P_v

$$t_c^b = \sec\theta\left(t(P_v) - \sin\theta\frac{\partial u}{\partial s}\right) \tag{7.121}$$

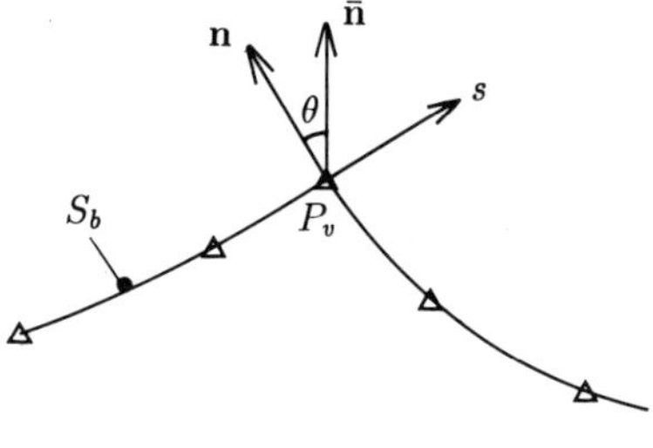

Figure 7.22 Average normal direction at node P_v

where θ is the angle shown in Figure 7.22. In equation (7.121), since u is known on S_b, $\partial u/\partial s$ may be expressed in terms of given nodal values of u.

Very often, typically because account is taken of a symmetry of the problem to be solved, an edge or corner of the boundary coincides with a change of boundary condition as illustrated in Figure 7.23. Usually if there is more than one element on which u is given adjacent to a corner node P_v, then all such elements have the same normal direction at P_v and equations (7.118) to (7.120) may be applied without modification. If the angle γ measured through the domain shown in Figure 7.23 is less than or equal to 90°, the exact solution is not singular at P_v and accurate results are obtained near the edge or corner.

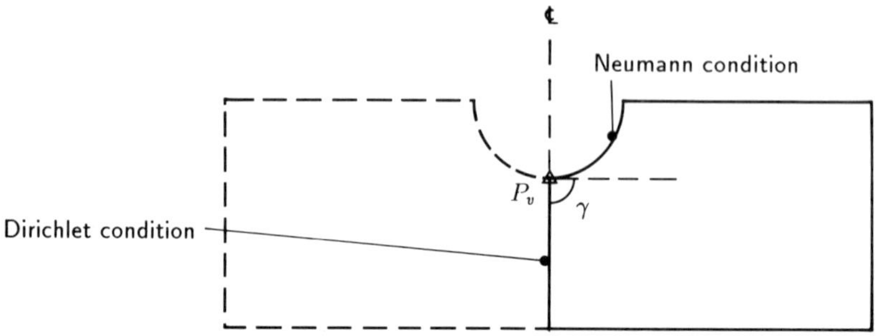

Figure 7.23 Change of boundary condition at corner of domain

If the Neumann boundary condition applies over the entire boundary of a finite domain, there is an additive arbitrary constant in the exact solution. This is reflected in the fact that if the coefficients $A(P_a)$ are computed according to the equation analogous to equation (7.9)

$$A(P_a) = -\sum_{b=1}^{p} \sum_{\substack{c=1 \\ d(b,c)\neq a}}^{n(b)} \Delta T_{ac}^{b}\,, \quad a = 1, 2, \ldots, q \tag{7.122}$$

then the matrix of equation coefficients is exactly singular in the absence of round-off error in floating point arithmetic. For the solution to be unique and therefore possible to compute by Gaussian elimination, the value of u at an arbitrarily chosen node is specified. The usual practice is to set to zero a nodal value $u(P_z)$ by modifying one row of the system of equations as shown in Figure 7.24. If instead the coefficients $A(P_a)$ are computed according to the equation analogous to equation (7.13)

$$\begin{bmatrix} \Delta T_{11} & \Delta T_{12} & \dots & \Delta T_{1z} & \dots & \Delta T_{1q} \\ \Delta T_{21} & \Delta T_{22} & \dots & \Delta T_{2z} & \dots & \Delta T_{2q} \\ \vdots & \vdots & & \vdots & & \vdots \\ 0 & 0 & \dots & 1 & \dots & 0 \\ \vdots & \vdots & & \vdots & & \vdots \\ \Delta T_{q1} & \Delta T_{q2} & \dots & \Delta T_{qz} & \dots & \Delta T_{qq} \end{bmatrix} \begin{bmatrix} u(P_1) \\ u(P_2) \\ \vdots \\ u(P_z) \\ \vdots \\ u(P_q) \end{bmatrix} = \begin{bmatrix} b_1 \\ b_2 \\ \vdots \\ 0 \\ \vdots \\ b_q \end{bmatrix}$$

Figure 7.24 Elimination of arbitrariness of solution of Neumann boundary value problem

$$A(P_a) = -\sum_{\substack{b=1 \\ P_a \in S_b}}^{p} \sum_{\substack{c=1 \\ d(b,c) \neq a}}^{n(b)} \Delta T_{ac}^{b} - \int_{\Omega_a} T(P_a, Q) \, \mathrm{d}S_Q \ , \quad a = 1, 2, \dots, q \tag{7.123}$$

then the matrix of equation coefficients is not quite singular because of error in Gaussian quadrature. The consequence of this is that, especially if the elements adjacent to the node P_z happen to be relatively small, significant spurious local variations of u may be computed near node P_z which effectively acts as a point source or sink. This can spoil the accuracy of results near P_z, so if $A(P_a)$ are computed according to equation (7.123) then the coefficients ΔT_{av} should be adjusted so that

$$\sum_{v=1}^{q} \Delta T_{av} = 0 \ , \quad a = 1, 2, \dots, q \tag{7.124}$$

prior to bifactorisation of the matrix. The adjustments being small, it does not much matter how they are carried out and a satisfactory procedure is to add or subtract the same adjustment to every coefficient of a row of the matrix.

For elasticity a system of either $3q$ or $2q$ simultaneous equations is obtained by substituting given boundary data into equation (7.5). Since displacement is uniquely defined at each node of the mesh, we may write

$$\sum_{b=1}^{p} \sum_{c=1}^{n(b)} \Delta_1 \mathbf{T}_{ac}^{b} \mathbf{u}_c^b = \sum_{b=1}^{p} \sum_{c=1}^{n(b)} \Delta \mathbf{U}_{ac}^{b} \mathbf{t}_c^b \ , \quad a = 1, 2, \dots, q \tag{7.125}$$

where $\Delta_1 \mathbf{T}_{ac}^b = \Delta \mathbf{T}_{ac}^b$ for $d(b,c) \neq a$ and for $d(b,c) = a$ the coefficients $\Delta_1 \mathbf{T}_{ac}^b$ are chosen in any way that satisfies the relations

$$\sum_{\substack{b=1 \\ d(b,c)=a}}^{p} \sum_{c=1}^{n(b)} \Delta_1 \mathbf{T}_{ac}^{b} = \mathbf{A}(P_a)\ , \quad a = 1, 2, \ldots, q \tag{7.126}$$

In three-dimensional analysis, the given data over a part of the boundary may be all three components of traction, two of traction and one of displacement, one of traction and two of displacement, or all three components of displacement. In practice it is rare that more than one component of displacement is specified over a part of the boundary. The mesh of elements is arranged so that within any one element the same components of traction or displacement are given at every point, and so that any discontinuities of given traction coincide with sides or ends of elements. The directions in which traction or displacement are given are not necessarily the global cartesian directions, and in addition if components of displacement of two or more elements adjacent to a node are given then the directions of those components may be neither parallel with nor perpendicular to one another. For example, let us suppose that it is desired to compute stress due to centripetal force in the mechanical component shown in Figure 7.25a. By symmetry it is only necessary to consider one sixth of the structure as shown in Figure 7.25b. On AB the y component of displacement equals zero, whereas on AE the component in a direction at 60° to the y-axis is zero. At the point A therefore, the unknown nodal parameters in a three-dimensional analysis must be the limiting value as A is approached of traction on AB in the direction $(0, 1, 0)$, the limiting value of traction on AE in the direction $(-0.866, 0.5, 0)$, and displacement in the direction $(0, 0, 1)$.

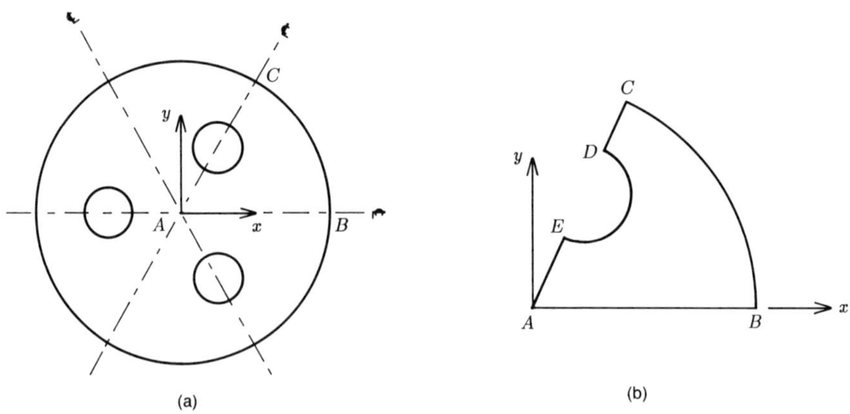

Figure 7.25 Problem with cyclic symmetry. (a) Problem geometry. (b) Domain to be analysed

To take account of the possible range of boundary conditions we define a local coordinate system at each node, such that local components $\mathbf{a}'$ of a vector are given in terms of global components $\mathbf{a}$ by[1]

$$\mathbf{a}' = \mathbf{L}(v)\mathbf{a} \tag{7.127}$$

The local axes at a node are the directions in which displacements or tractions are given on elements adjacent to the node. For example, for a three-dimensional analysis of the problem shown in Figure 7.25, we take for a node at the point A

$$\mathbf{L}(A) = \begin{bmatrix} 0 & 1 & 0 \\ -0.866 & 0.5 & 0 \\ 0 & 0 & 1 \end{bmatrix} \tag{7.128}$$

Note that $\mathbf{L}(A)$ is not a unitary matrix. Often there is arbitrariness in the choice of $\mathbf{L}(v)$: in particular, if all three components of traction are given on every element adjacent to the node P_v, then we may take $\mathbf{L}(v) = \mathbf{I}$. We rewrite equation (7.125) in terms of local components of the vectors $\mathbf{u}_c^b$ and $\mathbf{t}_c^b$:

$$\sum_{b=1}^{p}\sum_{c=1}^{n(b)} \Delta_1'\mathbf{T}_{ac}^b \mathbf{u}_c'^b = \sum_{b=1}^{p}\sum_{c=1}^{n(b)} \Delta'\mathbf{U}_{ac}^b \mathbf{t}_c'^b \,, \quad a = 1, 2, \ldots, q \tag{7.129}$$

where

$$\Delta_1'\mathbf{T}_{ac}^b = \mathbf{L}(a)\,\Delta_1\mathbf{T}_{ac}^b\mathbf{L}^{-1}(d(b,c)) \tag{7.130}$$
$$\Delta'\mathbf{U}_{ac}^b = \mathbf{L}(a)\,\Delta\mathbf{U}_{ac}^b\mathbf{L}^{-1}(d(b,c)) \tag{7.131}$$

Note the premultiplication by $\mathbf{L}$, the purpose of which is to ensure that if $\Delta_1\mathbf{T}_{ac}^b$ and $\Delta\mathbf{U}_{ac}^b$ are diagonally dominant, then so are $\Delta_1'\mathbf{T}_{ac}^b$ and $\Delta'\mathbf{U}_{ac}^b$.

Displacement is continuous from an element to an adjacent element, but each local component of the traction vector may be discontinuous either because the normal direction is not continuous, specified values are not continuous, or the local component of displacement is given on one element and that of traction is given on the other. Let $\mathbf{D}(b)$ be a matrix the off-diagonal coefficients of which are zero, and the diagonal coefficient j of which equals zero if local component j of traction is given on S_b or one if local component j of displacement is given. Similarly, let $\mathbf{E}(a)$ be a matrix the off-diagonal coefficients of which are zero, and the diagonal coefficient j of which equals zero if local component j of displacement at node P_a is unknown or one if local component j of traction is unknown. Then equation (7.129) may be rewritten as follows with all unknown integrals on the left side, and all known integrals on the right side:

$$\sum_{b=1}^{p}\sum_{c=1}^{n(b)} \Delta_1'\mathbf{T}_{ac}^b \mathbf{E}^*\left(d(b,c)\right)\mathbf{u}_c'^b - \sum_{b=1}^{p}\sum_{c=1}^{n(b)} \Delta'\mathbf{U}_{ac}^b \mathbf{D}(b)\mathbf{t}_c'^b$$
$$= \sum_{b=1}^{p}\sum_{c=1}^{n(b)} \Delta'\mathbf{U}_{ac}^b \mathbf{D}^*(b)\mathbf{t}_c'^b - \sum_{b=1}^{p}\sum_{c=1}^{n(b)} \Delta_1'\mathbf{T}_{ac}^b \mathbf{E}\left(d(b,c)\right)\mathbf{u}_c'^b\,, \quad a = 1, 2, \ldots, q \tag{7.132}$$

where $\mathbf{D}^* = \mathbf{I} - \mathbf{D}$ and $\mathbf{E}^* = \mathbf{I} - \mathbf{E}$.

We consider first the case of a domain with no edges or corners. Displacement is continuous, and within contiguous groups of elements which model parts of the boundary over which displacement in a particular direction is given, traction in that direction is continuous from one element to the next. Therefore, we may write both terms on the left side of equation (7.132) in terms of unique nodal values:

$$\sum_{v=1}^{q} \Delta'\mathbf{T}_{av}\mathbf{E}^*(v)\,\mathbf{u}'(P_v) - \sum_{v=1}^{q} \Delta'\mathbf{U}_{av}\mathbf{E}(v)\mathbf{t}'(P_v)$$
$$= \sum_{b=1}^{p}\sum_{c=1}^{n(b)} \Delta'\mathbf{U}_{ac}^b \mathbf{D}^*(b)\mathbf{t}_c'^b - \sum_{b=1}^{p}\sum_{c=1}^{n(b)} \Delta_1'\mathbf{T}_{ac}^b \mathbf{E}\left(d(b,c)\right)\mathbf{u}_c'^b\,, \quad a = 1, 2, \ldots, q \tag{7.133}$$

where

$$\Delta'\mathbf{T}_{av} = \sum_{b=1}^{p}\sum_{\substack{c=1\\ d(b,c)=v}}^{n(b)} \Delta_1'\mathbf{T}_{ac}^b \tag{7.134}$$

$$\Delta'\mathbf{U}_{av} = \sum_{b=1}^{p}\sum_{\substack{c=1\\ d(b,c)=v}}^{n(b)} \Delta'\mathbf{U}_{ac}^b \mathbf{D}(b) \tag{7.135}$$

Equation (7.133) is a system of either $3q$ or $2q$ simultaneous equations which may be solved for the unknown components of displacement and traction $\mathbf{u}'(P_v)$ and $\mathbf{t}'(P_v)$. Just as for Laplace's equation computed results are inaccurate near a node such as P_3 in Figure 7.21, so for elasticity results are inaccurate near a node adjoined by elements with different constraint matrices $\mathbf{D}(b)$. In theory, the solution is singular at such a node: the extent of the zone of inaccuracy may be reduced by local mesh refinement, or alternatively accurate results may be obtained by introducing singular shape functions on the adjacent elements. Midside node shifting can also yield accurate results if the theoretical singularity is of the type encountered at the root of a crack, for which the singular component of displacement varies as the square root of distance from the singular point in all three crack or notch opening modes[7].

For domains with edges and corners, equations (7.133) to (7.135) may usually be applied without modification. Accuracy may be lost if elements on which displacement is given in a particular direction do not all have the same normal direction at a node P_v. Either all limiting values of traction in the direction in which displacement is given can be taken to be equal to a unique nodal value, or a substitution analogous to equation (7.121) may be derived from Hooke's law. In reality such a condition never occurs as it would require the existence of an infinitely stiff foundation. In idealisations of real problems it can arise in the modelling of a very soft material such as rubber bonded to a material such as steel which is taken to be perfectly rigid, or in a detailed analysis of a small part of a structure or mechanical component in which it is assumed that displacements are zero on an arbitrarily chosen boundary passing through the material as shown in Figure 7.26. At the corners B and C of the arbitrarily chosen boundary $ABCD$, all strains are zero according to the idealisation and so all limiting values of traction as the corners are approached are zero; therefore, there is no error in taking all limiting values at a corner to be equal to a unique nodal value. Usually, except for the artificial case illustrated in Figure 7.26, edges and corners are either not subject to any constraint on displacement or they coincide with a change of boundary condition as shown in Figure 7.23 for Laplace's equation. For elasticity we have in place of the Neumann condition all components of traction given, and in place of the Dirichlet condition displacement normal to and traction tangential to the boundary given. Under these conditions, if the angle γ shown in Figure 7.23 does not exceed 90°, the exact solution is not singular at the node P_v on the edge or corner and equations (7.133) to (7.135) yield accurate results near P_v. Accurate results are also obtained for the problem shown in Figure 7.25, in which effectively the angle γ equals 60°.

If all components of the traction vector are given over the entire boundary of a finite domain, there are six arbitrary rigid body displacements in the exact solution of a three-dimensional problem and three in that of a problem of plane strain. To render the solution unique we fix either six or three suitably chosen components of nodal displacement, by modification of rows of the system of

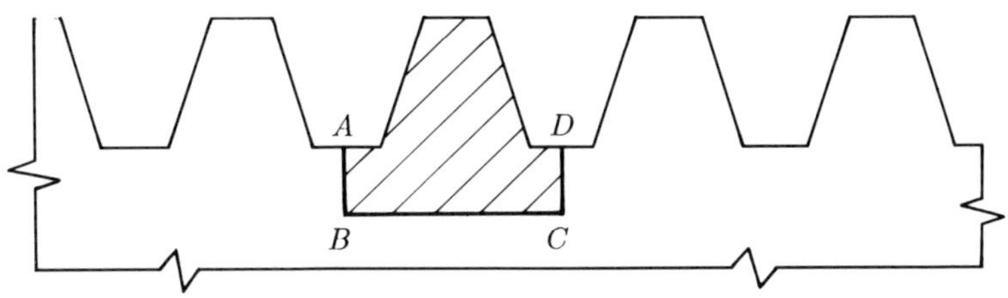

Figure 7.26 Analysis of small part of structure

equations in the same way as indicated in Figure 7.24 for Laplace's equation. If the coefficients $\mathbf{A}(P_a)$ are computed according to equation (7.9), then the system of equations admits rigid body translations but not necessarily rigid body rotations without traction on the boundary. If the coefficients $\mathbf{A}(P_a)$ are computed according to equation (7.13), then due to error in Gaussian quadrature the system will admit neither rigid body translations nor rigid body rotations without traction. In the latter case and also to some extent in the former, this can result in the computation of spurious stresses near nodes at which displacements are specified, as if those nodes were acting as supports. As in the case of Laplace's equation, these erroneous results can be eliminated by carrying out minor adjustments of the matrix prior to bifactorisation.

7.7 SUBREGIONS

It is in practice only feasible to determine fundamental solutions for a homogeneous domain. Very often the domain to be analysed consists of several homogeneous parts or subregions, each with different material properties; that is, it is piecewise homogeneous. For any one of the subregions it is possible to determine a fundamental solution and write an integral equation over the boundary of that subregion. In the direct method for elasticity, the integral equation for a subregion is written in terms of surface displacement and traction and also displacement and traction on the interfaces of the subregion with other subregions[1]. The surface and interfaces are represented by boundary elements, and a discrete system is constructed. Since both displacement and traction are unknown on interfaces, this system has fewer equations than unknowns. We therefore solve simultaneously the discrete systems for all subregions, in conjunction with compatibility and equilibrium conditions on the interfaces. These conditions need not be written explicitly as additional equations to be solved: instead, it is only necessary when constructing the global system of equations to assign the same matrix column for the coefficients of a component of displacement at an interface node in the systems for each of the adjacent subregions, and to assign the same matrix column for the coefficients of a component of interface traction at a node but change the sign of the coefficients in one of the systems. The change of sign of coefficients of traction is required because, whereas displacements of the subregions adjacent to an interface node are equal, the tractions acting upon them are equal and opposite by virtue of Newton's third law of motion.

Let us consider equation (7.132) for a subregion, for the case of a domain with no edges or corners and smooth interfaces between subregions. Displacement is continuous from an element to any adjacent element, and within contiguous groups of elements which model parts of the surface over which displacement in a particular direction is given, traction in that direction is continuous from one element to the next. Traction is also continuous from one element to the next over each of the interfaces between subregions, but not from a surface element to an adjoining interface element or between elements of different interfaces. Let there be j interfaces between the subregion and other subregions, and let us denote by $\mathbf{t}'(P_{vi})$ the traction acting on interface i at a node P_v, and by $\mathbf{t}'(P_v)$ the traction acting on the surface at a node P_v. Then we write equation (7.132) in terms of unique unknown nodal values of displacement, surface traction and interface traction as follows.

$$\sum_{v=1}^{q} \Delta'\mathbf{T}_{av}\mathbf{E}^*(v)\,\mathbf{u}'(P_v) - \sum_{v=1}^{q} \Delta'\mathbf{U}_{av}\mathbf{E}(v)\mathbf{t}'(P_v) - \sum_{i=1}^{j}\sum_{v=1}^{q} \Delta'\mathbf{U}_{avi}\mathbf{t}'(P_{vi})$$
$$= \sum_{b=1}^{p}\sum_{c=1}^{n(b)} \Delta'\mathbf{U}^b_{ac}\mathbf{D}^*(b)\mathbf{t}'^b_c - \sum_{b=1}^{p}\sum_{c=1}^{n(b)} \Delta'_1\mathbf{T}^b_{ac}\mathbf{E}\,(d(b,c))\,\mathbf{u}'^b_c\,, \quad a = 1, 2, \ldots, q \tag{7.136}$$

In this equation, $\mathbf{E}(v) = \mathbf{0}$ and therefore $\mathbf{E}^*(v) = \mathbf{I}$ for an interface node which is not a node of any surface element, and at such a node we arbitrarily take the direction cosine matrix $\mathbf{L} = \mathbf{I}$ (see equation (7.127)). For an interface element, we take $\mathbf{D} = \mathbf{0}$ and also $\mathbf{D}^* = \mathbf{0}$. The coefficients $\Delta'\mathbf{T}_{av}$ and $\Delta'\mathbf{U}_{av}$ are given by equations (7.134) and (7.135), and

$$\Delta'\mathbf{U}_{avi} = \sum_{b=1}^{p}\sum_{c=1}^{n(b)} \Delta'\mathbf{U}^b_{ac}\mathbf{D}_i(b) \tag{7.137}$$

where $\mathbf{D}_i(b) = \mathbf{I}$ if the element S_b lies on interface i, and $\mathbf{0}$ otherwise.

The remarks on domains with edges and corners made in Section 7.6 apply equally to piecewise homogeneous domains with edges and corners on the surface and interfaces. The only further case requiring special consideration is that in which three or more interfaces meet at a point, or two or more interfaces and the surface meet at a point, as illustrated in Figure 7.27. In these circumstances, there are fewer equations than unknowns in the global system of equations. Very often the exact solution of the equations of elasticity is singular at the node shared by three or more subregions, but a solution which is only locally inaccurate may be obtained by reducing by one the order of variation of traction over each of the neighbouring elements in one (or possibly in three dimensions two) of the intrinsic coordinate directions, and substituting for interface traction unknowns at the shared node in terms of traction at other nodes of an element. For example for the quadratic elements shown in Figure 7.27 we write

$$\mathbf{t}^{(2)} = 2\mathbf{t}^{(3)} - \mathbf{t}^{(1)} \tag{7.138}$$

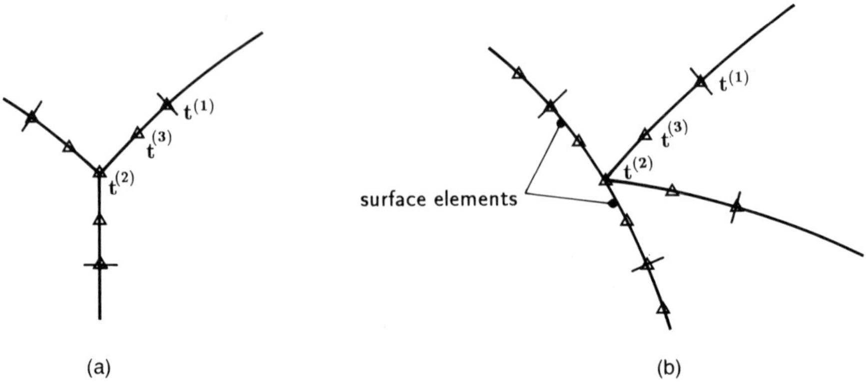

Figure 7.27 Nodes shared by three subregions. (a) Node not on surface. (b) Node on surface

which is equivalent to extrapolating for $\mathbf{t}^{(2)}$ in terms of $\mathbf{t}^{(1)}$ and $\mathbf{t}^{(3)}$. The limiting values of interface traction as the shared node is approached are thereby eliminated from the system of equations. No reduction of order of variation is required over surface elements adjacent to a shared node such as that shown in Figure 7.27b, even if surface tractions are unknown over those elements.

Particular integrals to take account of such factors as body force are defined separately for each subregion. Generally, the particular integrals for adjacent subregions do not satisfy compatibility of displacement or equilibrium of traction on the interface between the subregions, so in order that the sum of the particular integral and complementary function may satisfy such conditions it is necessary to introduce into the complementary function a discontinuity of displacement and a distributed load on the interface. This merely results in additional terms on the right side of equation (7.136).

In the implementation outlined here for the direct method, there are six unknowns per interface node for three-dimensional elasticity, and four for plane strain. Compatibility of displacement and equilibrium of traction are ensured at every point of an interface, subject to the approximation of the variation of displacement and traction over each element implied by the choice of shape functions. By comparison, in a displacement formulation finite element analysis there are three or two unknowns per node on the interface between two different materials, and there is pointwise satisfaction of the compatibility condition but not of equilibrium.

It has been noted earlier that problems in which the material properties vary in a general manner from one point to another cannot usually be solved by boundary element methods. This is of no great consequence because such problems are rarely considered if only because there usually does not exist sufficiently detailed input data. For example most rock masses are markedly inhomogeneous, but the cost of determining local variations of material properties by testing a sufficient number of borehole samples would be prohibitive.

7.8 SIMULTANEOUS EQUATION SOLUTION

The guiding principles according to which simultaneous equations approximating to boundary integral equations are solved are similar for Laplace's equation and elasticity, and for the direct and indirect methods; in this section, particular reference is made to the equations obtained by the direct method for elasticity, these being equations (7.133) for a homogeneous domain and (7.136) for a piecewise homogeneous domain divided into subregions. The techniques described here are based on Gaussian elimination, which in matrix algebra terms is bifactorisation of the matrix of equation coefficients. There are other possibilities such as Jacobi or Gauss–Seidel iteration, but although these have been demonstrated to converge for a variety of test problems it is not known whether they will do so in all circumstances. Cramer's rule and matrix inversion require orders of magnitude more computer time than Gaussian elimination for problems of practical size.

The requirements to be satisfied by a solution scheme are firstly that it be reliable, then that it be capable of execution on commonly available computer systems, and finally that the computer times required for execution be not excessive. It is found that provided appropriate units of distance and stress are chosen for construction of the system of equations (see Chapter 6) and arithmetic is done to at least CDC or Cray single precision (about 14 significant figures), there is negligible accumulation of roundoff error during forward reduction and back substitution. IBM single precision (about 6 significant figures) may indeed be adequate for smaller problems. Since Gaussian elimination does not depend upon the convergence of any iteration scheme, it therefore satisfies the reliability requirement.

We now consider how to ensure that the solution scheme can be implemented on commonly available computer systems. Although the number of equations to be solved is generally far less than in the finite element method, the matrix of equation coefficients for a single subregion is fully populated and not symmetric about the leading diagonal. In the finite element method the number of coefficients to be computed, stored and operated upon during equation solution is around nm (see Figure 7.28a) where n is the number of unknowns and m is the semi-bandwidth if a constant bandwidth sliding triangle solver is used, and somewhat less for a frontal solver depending upon mesh topology. This is usually a larger number of coefficients than can be held in central (byte or word addressable) memory, so the unreduced stiffness matrix (sliding triangle solver only) and triangularised matrix are held on

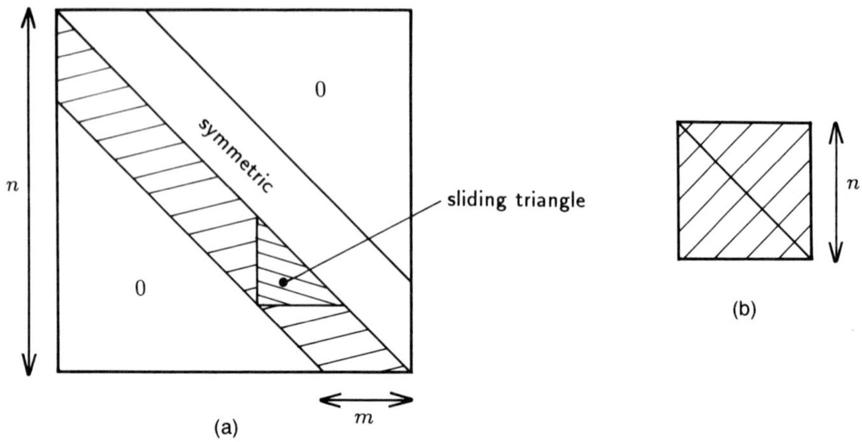

Figure 7.28 Matrices of equation coefficients obtained by the finite and boundary element methods. (a) Finite elements. (b) Boundary elements, single subregion

disc files; however, because the system is banded and symmetric it is possible to devise solvers which only require that about $m^2/2$ coefficients be held in central memory at any one time, without incurring unacceptably high input–output overhead costs. During execution of a sliding triangle solver, the file on which the unreduced matrix is held is read only once and that on which the triangularised matrix is held is written and read once; for a frontal solver only the triangularised matrix is written to a disc file. In boundary element analyses the circumstances are different. It may well be possible to hold the n^2 matrix coefficients in central memory, but then we set a hard upper limit on the size of problem that may be solved on a given computer system. For some systems this limit may be quite high due to virtual memory. However, if virtual memory is used too liberally then, because in Gaussian elimination (in unmodified form) for a fully populated matrix it is inevitable that very large changes of operand address occur in short periods of time, the paging overhead may be excessive. Paging overhead is time spent by the computer moving data between its real central memory and automatically generated files on disc.

For these reasons it is desirable to develop a solver for a fully populated non-symmetric matrix, in which the matrix is held on a disc file. The obvious way to do this is to hold the matrix of equation coefficients on a direct access file, one record per equation. This suits the matrix construction procedure, in which for each value of the subscript a in equation (7.133), integrations are carried out over each element in turn and the kernel–shape function product integrals $\Delta_1'\mathbf{T}_{ac}^b$ and $\Delta'\mathbf{U}_{ac}^b$ (see also equations (7.134) and (7.135)) are scattered horizontally into the matrix rows. The forward reduction then proceeds as follows.

eliminate first unknown

- 1 Read into central memory the first equation.
- 1.2 Read the second equation, subtract from it a proportion of the first equation, rewrite to file.
- 1.3 Read the third equation, subtract from it a proportion of the first equation, rewrite to file.
- .
- .
- 1.n Read the last equation, subtract from it a proportion of the first equation, rewrite to file.

eliminate second unknown

- 2 Read into central memory the second equation.
- 2.3 Read the third equation, subtract from it a proportion of the second equation, rewrite to file.
- 2.4 Read the fourth equation, subtract from it a proportion of the second equation, rewrite to file.
- .
- .
- 2.n Read the last equation, subtract from it a proportion of the second equation, rewrite to file.

eliminate penultimate unknown

- $n-1$ Read into central memory the penultimate equation.
- $(n-1)n$ Read the last equation, subtract from it a proportion of the penultimate equation, rewrite to file.

This works, but is slow because it requires the transfer of a large amount of data to and from the disc files. The problem is that for each read and rewrite of an equation, only a modest amount of arithmetic is done on its coefficients.

The amount of data transferred to and from disc is greatly reduced in a block solver, in which each time an equation is read from and rewritten to the file, several unknowns are eliminated. The principle of a block solver is illustrated in Figure 7.29. Space is reserved in central memory for a block of equations referred to here as the 'slow block' and for a single equation referred to as the 'fast block'. The sequence of operations is as follows[8].

eliminate unknowns 1 to n_1 (Figure 7.29a)	1	Read into the slow block equations 1 to n_1.
	1.A.2	Eliminate unknown 1 from equations 2 to n_1.
	1.A.3	Eliminate unknown 2 from equations 3 to n_1.
	.	
	.	
	1.$A.n_1$	Eliminate unknown $n_1 - 1$ from equation n_1.
	1.B	Rewrite equations 1 to n_1 to file.
	1.$C.(n_1 + 1)$	Read into the fast block equation $n_1 + 1$, eliminate from it unknowns 1 to n_1 and rewrite to file.
	1.$C.(n_1 + 2)$	Read into the fast block equation $n_1 + 2$, eliminate from it unknowns 1 to n_1 and rewrite to file.
	.	
	.	
	1.$C.n$	Read into the fast block equation n, eliminate from it unknowns 1 to n_1 and rewrite to file.

eliminate unknowns $n_1 + 1$ to n_2 (Figure 7.29b)	2	Read into the slow block equations $n_1 + 1$ to n_2.
	2.$A.(n_1 + 2)$	Eliminate unknown $n_1 + 1$ from equations $n_1 + 2$ to n_2.
	2.$A.(n_1 + 3)$	Eliminate unknown $n_1 + 2$ from equations $n_1 + 3$ to n_2.
	.	
	.	
	2.$A.n_2$	Eliminate unknown $n_2 - 1$ from equation n_2.
	2.B	Rewrite equations $n_1 + 1$ to n_2 to file.
	2.$C.(n_2 + 1)$	Read into the fast block equation $n_2 + 1$, eliminate from it unknowns $n_1 + 1$ to n_2 and rewrite to file.
	2.$C.(n_2 + 2)$	Read into the fast block equation $n_2 + 2$, eliminate from it unknowns $n_1 + 1$ to n_2 and rewrite to file.
	.	
	.	
	2.$C.n$	Read into the fast block equation n, eliminate from it unknowns $n_1 + 1$ to n_2 and rewrite to file.

eliminate unknowns $n_{m-1}+1$ to $n-1$ (Figure 7.29c)	m	Read into the slow block equations $n_{m-1}+1$ to n.
	$m.A.(n_{m-1}+2)$	Eliminate unknown $n_{m-1}+1$ from equations $n_{m-1}+2$ to n.
	$m.A.(n_{m-1}+3)$	Eliminate unknown $n_{m-1}+2$ from equations $n_{m-1}+3$ to n.
	.	
	.	
	$m.A.n$	Eliminate unknown $n-1$ from equation n.
	$m.B$	Rewrite equations $n_{m-1}+1$ to n to file.

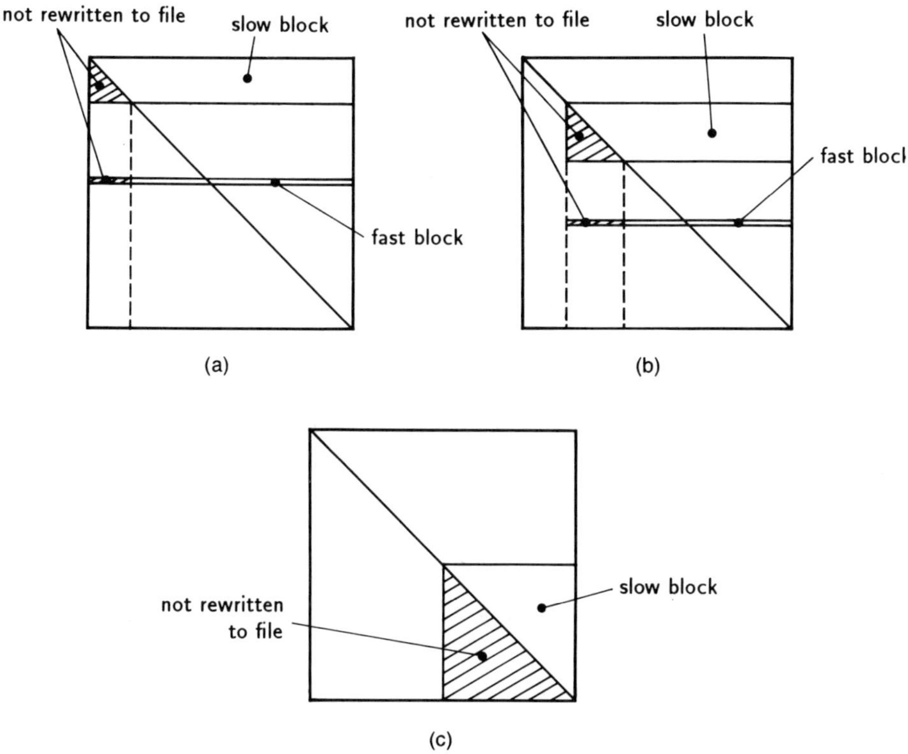

Figure 7.29 Operation of block solver. (a) First slow block. (b) Second slow block. (c) Last slow block

The steps $1.A.2$ to $1.A.n_1$ involve a substantial amount of arithmetic, and during execution of these steps there are no transfers of data between central memory and disc. Each of the steps $1.C.(n_1 + 1)$ to $1.C.n$ also involves a substantial amount of arithmetic. Approximately, the amount of data read from and written to the file during elimination of the first n_1 unknowns is less than in the unblocked solution scheme by a factor of n_1. Similarly, the amount of data transferred during elimination of the next $n_2 - n_1$ unknowns is reduced by a factor of about $n_2 - n_1$, and so on.

Note that equation coefficients in the shaded areas in Figure 7.29 are not rewritten to the file, so the partially reduced equations become shorter and the number of equations that can be held in the slow block increases from one slow block to the next. The blocking factor $n_i - n_{i-1}$ therefore improves as the forward reduction proceeds. Note also that if the number of equations n is small enough for the entire matrix to fit into the slow block, then only steps m to $m.B$ are executed and there are no transfers of data between central memory and disc during forward reduction. As an example, let us suppose that $n = 1000$ (so there are one million equation coefficients) and that the slow block can hold 50 000 coefficients. Then there would be 12 slow blocks, as shown in Table 7.2. With a minimum blocking factor of 50 the input–output overhead would be very reasonable, and the system of equations is solved using array space in central memory amounting to only 5% of matrix size.

On most computer systems it is desirable to minimise not only the amount of data transferred between central memory and disc but also the number of transfers, so it is the usual practice to block the equations several to a direct access file record. The fast block then becomes an array equal in length to one file record, and the length of the slow block is taken to be that of an integral number of file records. Further enhancements are possible: the matrix may be partitioned vertically to improve the blocking factor, and on some computer systems central processing and input–output may be performed simultaneously by means of double buffering. Generally, the input-output overhead is a greater problem on faster (especially vectorising and parallel processing) systems, but on such systems if they are well designed there is more extensive provision for accelerated transfer of data between central

Table 7.2 Block solver performance, 1000 equations and 50 000 coefficient slow block

Slow block	Coefficients in one equation	Equations in slow block	Slow block	Coefficients in one equation	Equations in slow block
1	1000	50	7	652	76
2	950	52	8	576	86
3	898	55	9	490	102
4	843	59	10	388	128
5	784	63	11	260	260
6	721	69	12	68	68

memory and disc. Note that if it is decided to hold the entire matrix in virtual memory, then adoption of the sequence of eliminations of the block solver can greatly reduce the paging overhead as it ensures that a large amount of arithmetic is carried out on each page of matrix coefficients brought into real memory.

In this discussion of the solution of a system of equations of the form $\mathbf{A}\{\mathbf{x}\} = \{\mathbf{b}\}$, no reference has yet been made to arithmetic operations on the load vector $\{\mathbf{b}\}$. If coefficients of the load vector are appended to their respective matrix rows on the file on which the matrix of equation coefficients is held, forward reduction of the load vector may be carried out together with that of the matrix. Back substitution then requires that the file, on which is now held the upper triangular factor $\mathbf{U}$ and the forward reduced load vector, be read in reverse order of record number. If it is desired to carry out forward reduction of the load vector as an operation separate from bifactorisation of the matrix, then it is necessary to write to another disc file coefficients of the lower triangular factor $\mathbf{L}$ where $\mathbf{A} = \mathbf{LU}$, these coefficients being the ratios by which equations are multiplied upon subtraction from other equations during forward reduction (the pivots). The file on which $\mathbf{L}$ is held, which may be a sequential file, is then rewound and read in ascending order of record number during forward reduction of the load vector. Back substitution then proceeds as before, except in that the forward reduced load vector is not read from the file on which the upper triangular factor is held. Note that if during operation of a block solver the pivots are written to the sequential file in the order in which they are computed then they are not conveniently arranged by $\mathbf{L}$ matrix column on that file, so when carrying out forward reduction of the load vector it is necessary to perform the eliminations in the same order as that in which they were performed during forward reduction of the matrix. Note also that since $\mathbf{A}$ is not symmetric, $\mathbf{L}$ is not the transpose of $\mathbf{U}$.

The blocking of eliminations brings the time spent transferring data to and from disc files under control but does nothing to reduce the number of arithmetic operations carried out. The central processing time (that is, time spent on arithmetic) required to bifactorise a matrix is approximately proportional to n^3 where n is the order of the system, and that required for the rest of a boundary element analysis is roughly proportional to n^2. As problem size is increased, a stage is reached beyond which the time required for simultaneous equation solution is greater than that taken by any other part of the analysis. For example, for three-dimensional problems of elasticity analysed using isoparametric quadratic elements, simultaneous equation solution time is greater than the time required to construct the matrix of equation coefficients if the number of boundary elements exceeds around 150, depending upon machine architecture and efficiency of the integration scheme. In engineering practice meshes of over 200 such elements are common, so it is desirable to develop simultaneous equation solution techniques in which fewer arithmetic operations are carried out. The only way to achieve this in principle is by iterative solution, in which successive approximations to the exact solution of the equations are computed. The convergence of Jacobi or Gauss–Seidel iteration is uncertain because in some circumstances the first iterated solution vector is very different from the exact solution. Divergence of the iteration is particularly likely if the surface or boundary includes close parallel parts such as the surfaces of a narrow slot or thin shell, giving rise to off-diagonal matrix coefficients with large absolute values.

Finite difference equations may be solved reliably and economically by a method of successive approximations known as the multigrid method, in which a solution is obtained first for a coarse grid, then for successively finer grids. The solution obtained for the first grid serves as a first estimate of that for the second, and so on. In the multigrid method as applied to boundary elements, analyses are performed using successively finer meshes of elements as illustrated in Figure 7.30. In the first analysis the simultaneous equations are solved by Gaussian elimination, but in subsequent analyses they are solved iteratively starting from the solution of the previous analysis as a first estimate. In the first analysis, the system of equations to be solved is relatively small, and in subsequent analyses the iterative solution converges because the first estimate is reasonably close to the exact solution of the equations. In each successive mesh the dimensions of the elements are halved, so in principle the number of elements is doubled in two dimensions and quadrupled in three. To date the method has not found wide application. Probably the reason is that, due to computer system limitations, the maximum allowable number of elements is usually less than four times the minimum required to yield any meaningful solution at all, for problems in which simultaneous equation solution takes longer than any other part of the analysis.

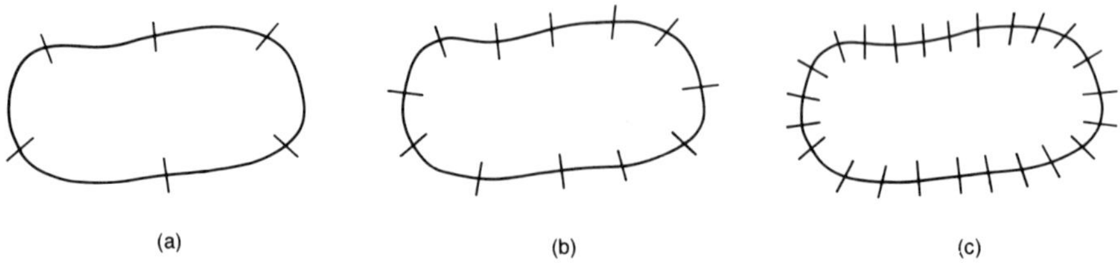

Figure 7.30 Multigrid analysis. (a) First mesh. (b) Second mesh. (c) Third mesh

In other approximate or iterative solution schemes, only one mesh is considered and advantage is taken of the following factors.

1. The kernels $\mathbf{T}(P_a, Q)$ and (for three dimensional analysis) $\mathbf{U}(P_a, Q)$ tend to zero with increasing distance from P_a to Q. In the equations obtained by writing the integral equation at the node P_a, therefore, coefficients of unknowns at nodes far from P_a are relatively small.

2. The differences between values of unknowns at nodes which are close to one another are usually small.

In view of the first factor it might appear that a reasonably accurate first estimate of the solution could be obtained by ignoring all matrix coefficients corresponding to pairs of nodes separated by more than some threshold distance, and bifactorising the resulting sparsely populated matrix. This however is generally not so, because the small coefficients in question are very numerous and collectively have an important effect on the solution. Given though that the differences between the values of the unknowns at neighbouring nodes are usually small, a good first estimate can be obtained by defining small groups of neighbouring nodes (lumps) and, if a group is far from P_a, taking the solution to be constant within the group. If the constant value of the solution is taken to be its value at some node Q_g near the centre of the group, then equations with a reduced number of non-zero coefficients may be constructed as illustrated in Figure 7.31, where (for three-dimensional elastic analysis) the 3×3 submatrix $\mathbf{B}^*$ is given by

$$\mathbf{B}^* = \sum_{i=1}^{l} \mathbf{B}_i \tag{7.139}$$

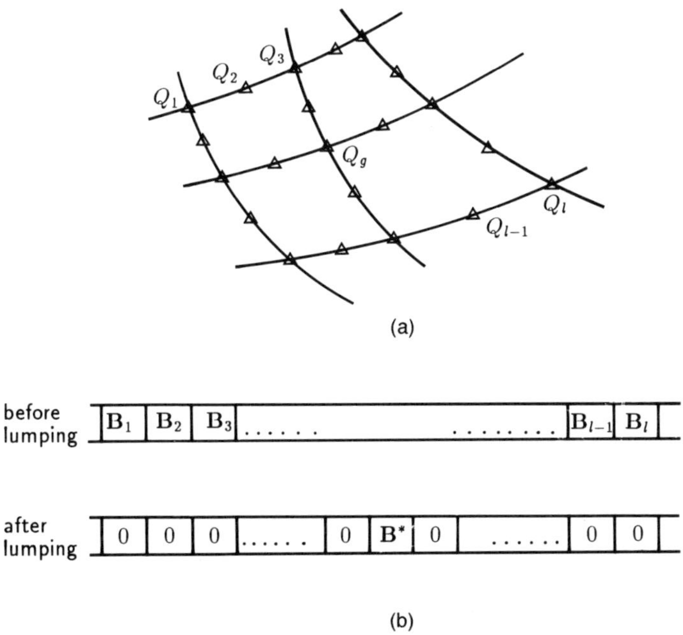

Figure 7.31 Lumping of matrix coefficients. (a) Nodes of lump far from P_a. (b) Unlumped and lumped equations of collocation at P_a

In the resulting modified (lumped) equations for the approximate solution $\{\mathbf{x}^*\}$

$$\mathbf{A}^*\{\mathbf{x}^*\} = \{\mathbf{b}\} \tag{7.140}$$

the submatrix $\mathbf{B}^*$ multiplies the unknowns at node Q_g (see Figure 7.31). Bifactorisation of the lumped matrix $\mathbf{A}^*$ may well take less than 10% of the time required to bifactorise the unlumped matrix $\mathbf{A}$. The techniques of lumping may not be readily applicable if there are constraints on displacement, which give rise to traction unknowns. Care must be taken not to lump together coefficients of displacement with coefficients of traction, or coefficients of components of traction in different directions or acting on parts of the surface with different normal directions. Also, coefficients of traction do not tend to zero with distance from P_a as rapidly as those of displacement. Whether or not there are constraints on displacement, it is impossible to quantify the difference between the solution $\{\mathbf{x}^*\}$ of equation (7.140) and the solution $\{\mathbf{x}\}$ of the unlumped equations.

The error introduced by lumping may be eliminated by a method of successive approximations in which the solution $\{\mathbf{x}^*\}$ is taken to be the first iterated solution. Let $\mathbf{L}^*$ and $\mathbf{U}^*$ be the lower and upper triangular factors of $\mathbf{A}^*$:

$$\mathbf{L}^*\mathbf{U}^* = \mathbf{A}^* \tag{7.141}$$

The first iterated solution is obtained by forward reduction and back substitution on the load vector $\{\mathbf{b}\}$, using the factors $\mathbf{L}^*$ and $\mathbf{U}^*$. In matrix algebra terms, this amounts to solving the equations

$$\mathbf{L}^*\{\mathbf{c}^{(1)}\} = \{\mathbf{b}\} \tag{7.142}$$

$$\mathbf{U}^*\{\mathbf{x}^{(1)}\} = \{\mathbf{c}^{(1)}\} \tag{7.143}$$

in which $\{\mathbf{c}^{(1)}\}$ is the forward reduced load vector, and $\{\mathbf{x}^{(1)}\}$ is the first iterated solution. Then, for $i = 2, 3, 4 \ldots$, we perform the following operations.

1. Compute the residue vector

$$\{\mathbf{b}^{(i)}\} = \{\mathbf{b}\} - \mathbf{A}\{\mathbf{x}^{(i-1)}\} \tag{7.144}$$

2. Solve the equations

$$\mathbf{L}^*\{\mathbf{c}^{(i)}\} = \{\mathbf{b}^{(i)}\} \tag{7.145}$$

$$\mathbf{U}^*\{\delta\mathbf{x}^{(i)}\} = \{\mathbf{c}^{(i)}\} \tag{7.146}$$

3. Take the next iterated solution to be

$$\{\mathbf{x}^{(i)}\} = \{\mathbf{x}^{(i-1)}\} + \{\delta\mathbf{x}^{(i)}\} \tag{7.147}$$

The iteration is continued until coefficients of the residue vector $\{\mathbf{b}^{(i)}\}$ are negligibly small compared with those of the load vector $\{\mathbf{b}\}$, in which case the iterated solution $\{\mathbf{x}^{(i)}\}$ is very nearly the same as the solution $\{\mathbf{x}\}$ of the unlumped equations.

In practice, instead of taking (for three dimensional elastic analysis) the lumped submatrix $\mathbf{B}^*$ shown in Figure 7.31 to be a coefficient of displacement or traction at a specified node Q_g near the centre of the lump, we take it to be a coefficient of average lump displacement or traction. This eliminates the possibility of impairment of convergence of the iteration by poor choice of the node Q_g. Average lump nodal displacements or tractions are introduced as additional unknowns, and constraint equations which state that the additional unknowns are indeed the average values are added to the system as shown in Figure 7.32. Coefficients are lumped if the ratio of lump diameter to minimum distance of the lump from that in which the node P_a is located is less than a threshold value, as illustrated in Figure 7.32 for an infinite domain the

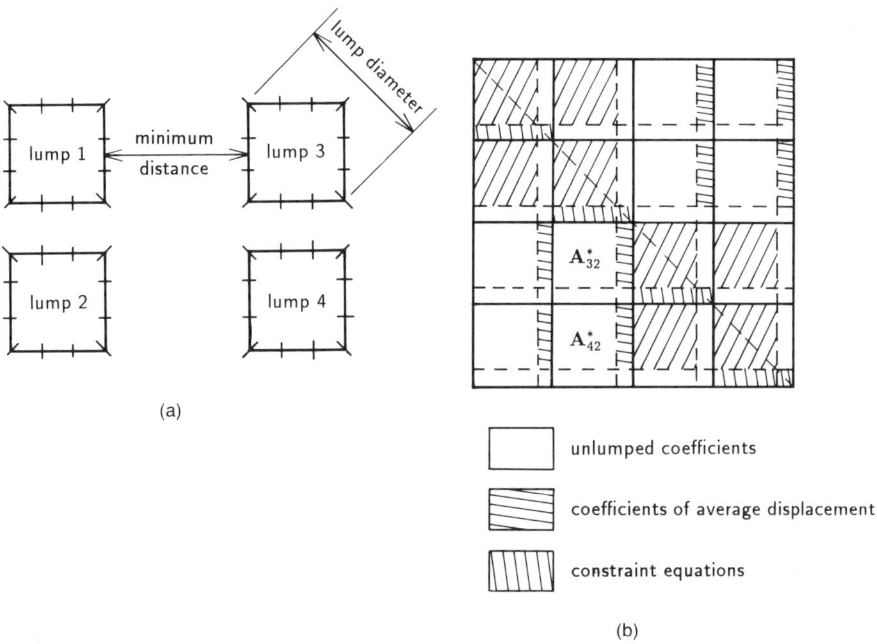

Figure 7.32 Lumping with additional unknowns (infinite domain shown). (a) Boundary elements and lumps. (b) Lumped matrix A*

boundary nodes of which are grouped into four lumps. Note that in principle submatrices such as $\mathbf{A}^*_{32}$ and $\mathbf{A}^*_{42}$ (see Figure 7.32) of the lumped matrix which initially contain zero coefficients are filled with non-zero values during forward reduction. This infill would severely affect the economy of the solution scheme and therefore instead of allowing it to happen we modify the coefficients of average displacement of the lump, on the basis that the average displacement is not very different from displacement at any one of its nodes. Consequently the factors $\mathbf{L}^*$ and $\mathbf{U}^*$ do not exactly satisfy equation (7.141). The effect of this approximation on convergence of the iteration for $\{\mathbf{x}\}$ is negligible.

Finally, if the domain to be analysed is piecewise homogeneous and therefore we write equation (7.136) for each homogeneous subregion, in the matrix of coefficients of the global system of equations there are many zero submatrices. The computer time required for simultaneous equation solution can be greatly reduced by avoiding arithmetic operations on the coefficients of these submatrices. This may conveniently be achieved by the use of a substructuring solver in which forward reduction is carried out in two stages as follows.

1. In equation (7.136) for each of the subregions in turn, eliminate all unknowns not associated with nodes on interfaces between subregions.
2. Perform forward reduction of the resulting system of equations in terms of unknowns associated with nodes on interfaces.

Let there be k subregions. We number the unknowns associated with nodes on interfaces in the following order.

$\{\mathbf{t}^{(1)}\} =$ traction on interfaces between subregion 1 and subregions $2, 3, 4, \ldots, k$.
$\{\mathbf{u}^{(2)}\} =$ displacements or surface tractions at nodes on interfaces with subregion 2, which are not nodes of subregions $3, 4, \ldots, k$.
$\{\mathbf{t}^{(2)}\} =$ tractions on interfaces between subregion 2 and subregions $3, 4, \ldots, k$.
$\{\mathbf{u}^{(3)}\} =$ displacements or surface tractions at nodes on interfaces with subregion 3, which are not nodes of subregions $4, \ldots, k$.

.
.
.

$\{\mathbf{t}^{(k-1)}\} =$tractions on the interface between subregion $k-1$ and subregion k.
$\{\mathbf{u}^{(k)}\} =$ displacements or surface tractions at nodes on interfaces with subregion k.

This numbering system ensures that at no stage of the forward reduction can there be division by zero resulting from an attempt to substitute for the displacements of a subregion in terms of interface tractions alone. Such a substitution is of course impossible, as an arbitrary rigid body displacement field could be added without changing the interface tractions and therefore the displacements are not uniquely defined in terms of those tractions. Equations (7.136) for subregion i are of the form

$$\left[\begin{array}{c|c} \mathbf{A}_{11}^{(i)} & \mathbf{A}_{12}^{(i)} \\ \hline \mathbf{A}_{21}^{(i)} & \mathbf{A}_{22}^{(i)} \end{array}\right]\left[\begin{array}{c} \{\mathbf{v}_1^{(i)}\} \\ \hline \{\mathbf{v}_2^{(i)}\} \end{array}\right] = \left[\begin{array}{c} \mathbf{f}_1^{(i)} \\ \hline \mathbf{f}_2^{(i)} \end{array}\right] \tag{7.148}$$

where $\{\mathbf{v}_1^{(i)}\}$ are the unknowns not associated with nodes on interfaces, and $\{\mathbf{v}_2^{(i)}\}$ is a subset of the groups of unknowns $\{\mathbf{t}^{(1)}\}$ to $\{\mathbf{u}^{(k)}\}$ listed above. In the first stage of forward reduction, the unknowns $\{\mathbf{v}_1^{(i)}\}$ are eliminated to yield the system

$$\hat{\mathbf{A}}_{22}^{(i)}\left\{\mathbf{v}_2^{(i)}\right\} = \left\{\hat{\mathbf{f}}_2^{(i)}\right\} \tag{7.149}$$

In the second stage of forward reduction, a matrix constructed from $\hat{\mathbf{A}}_{22}^{(1)}, \hat{\mathbf{A}}_{22}^{(2)}, \ldots, \hat{\mathbf{A}}_{22}^{(k)}$ is bifactorised.

In Figure 7.33 is shown the example of three subregions, the first and third of which have no nodes or elements in common. In this case, $\{\mathbf{v}_2^{(1)}\} = \{\mathbf{t}^{(1)}, \mathbf{u}^{(2)}\}$, $\{\mathbf{v}_2^{(2)}\} = \{\mathbf{t}^{(1)}, \mathbf{u}^{(2)}, \mathbf{t}^{(2)}, \mathbf{u}^{(3)}\}$ and $\{\mathbf{v}_2^{(3)}\} = \{\mathbf{t}^{(2)}, \mathbf{u}^{(3)}\}$. Note that the matrix to be bifactorised in the second stage of the forward reduction is not fully populated.

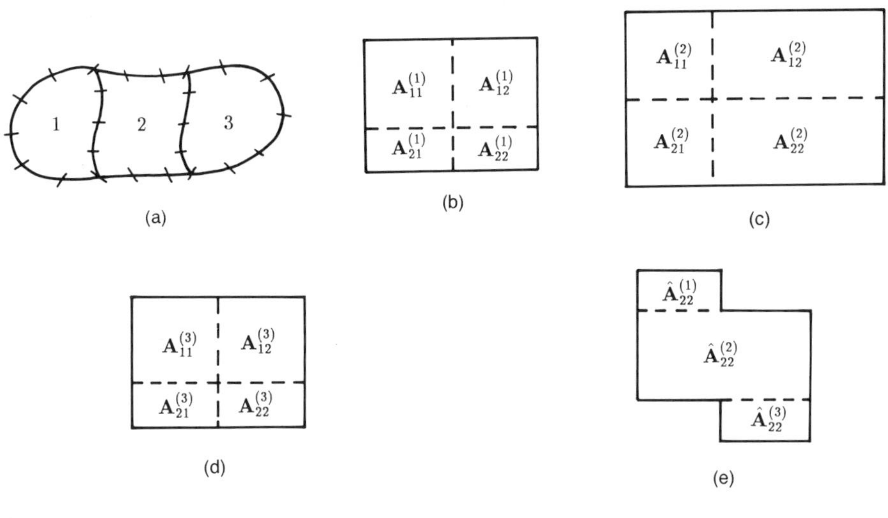

Figure 7.33 Matrices of equation coefficients for the three subregions. (a) Arrangement of subregions. (b) First subregion. (c) Second subregion. (d) Third subregion. (e) Second stage of forward reduction

To summarise: in this section are described a solution scheme in which the matrix of equation coefficients is held on a disc file, techniques of solution by successive approximation, and the solution of simultaneous equations for piecewise homogeneous domains. In each of these developments, account or advantage is taken of special characteristics of the systems of equations obtained by boundary element methods. The techniques described in this section enable more complicated problems to be solved than would otherwise be possible, and reduce the computing time required for solution especially for large three-dimensional analyses.

7.9 POST-PROCESSING

Once the discrete system has been solved for the unknown boundary nodal parameters, additional boundary data and results at internal points of the domain may be computed. For Laplace's equation, the potential u and its normal derivative $\partial u/\partial \mathbf{n}$ are known at every point of the boundary S, and for elasticity the displacement $\mathbf{u}$ and traction $\mathbf{t}$ are known at every point of S. With very little further computational effort it is possible to compute from this data everything else of interest on the boundary.

For Laplace's equation we may wish to know not only the normal derivative $\partial u/\partial \mathbf{n}$ but also tangential derivatives of the potential u, since these derivatives may be important results such as fluid velocity components or electrical current densities depending upon the physical problem being solved. For this purpose we define local cartesian coordinates $(\bar{x}, \bar{y})$ at a point $\xi = \xi_0$ of an element in two dimensions, or $(\bar{x}, \bar{y}, \bar{z})$ at a point $(\xi, \eta) = (\xi_0, \eta_0)$ of an element in three dimensions, as shown in Figure 7.34. The direction $\bar{y}$ in two dimensions, or $\bar{z}$ in three, is the outward normal direction. In two dimensions, the direction $\bar{x}$ is tangential to the element at the point $\xi = \xi_0$. In three dimensions we may arbitrarily take the direction $\bar{x}$ to be tangential to the line $\eta = \eta_0$ in the element, and $\bar{y}$ to be orthogonal to $\bar{x}$ and $\bar{z}$. Note that since the lines $\eta = \eta_0$ and $\xi = \xi_0$ are not necessarily orthogonal, $\bar{y}$ is not in general tangential to the line $\xi = \xi_0$. For the element S_b in two dimensions,

$$\frac{\partial u}{\partial \bar{x}} = \frac{1}{J} \sum_{c=1}^{n(b)} \frac{\mathrm{d}N_c}{\mathrm{d}\xi} u_c^b \tag{7.150}$$

where the Jacobian J and derivatives $\mathrm{d}N_c/\mathrm{d}\xi$ are evaluated for $\xi = \xi_0$. In three dimensions,

$$\left\{ \begin{array}{c} \frac{\partial u}{\partial \bar{x}} \\ \frac{\partial u}{\partial \bar{y}} \end{array} \right\} = \mathbf{J}^{-1} \sum_{c=1}^{n(b)} \left\{ \begin{array}{c} \frac{\partial N_c}{\partial \xi} \\ \frac{\partial N_c}{\partial \eta} \end{array} \right\} u_c^b \tag{7.151}$$

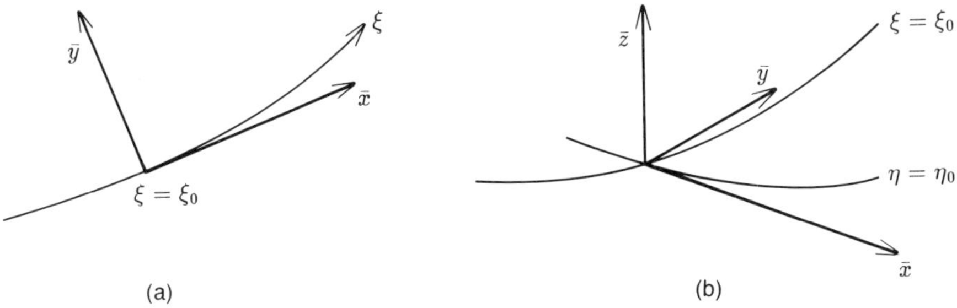

Figure 7.34 Local cartesian coordinates for computation of results on the boundary. (a) Two dimensions. (b) Three dimensions

where the Jacobian matrix $\mathbf{J}$ is given by

$$\mathbf{J} = \begin{bmatrix} \frac{\partial \bar{x}}{\partial \xi} & \frac{\partial \bar{y}}{\partial \xi} \\ \frac{\partial \bar{x}}{\partial \eta} & \frac{\partial \bar{y}}{\partial \eta} \end{bmatrix} \tag{7.152}$$

in which for example $\partial \bar{x}/\partial \xi$ may be computed from the local coordinates $\bar{x}_c^b$ of the nodes of S_b:

$$\frac{\partial \bar{x}}{\partial \xi} = \sum_{c=1}^{n(b)} \frac{\partial N_c}{\partial \xi} \bar{x}_c^b \tag{7.153}$$

In equations (7.151) to (7.153), all functions are evaluated for $(\xi, \eta) = (\xi_0, \eta_0)$. Derivatives of u in the global cartesian directions may be computed from $\partial u/\partial \bar{x}$, $\partial u/\partial \bar{y}$ and $\partial u/\partial \bar{z}$ (i.e. $\partial u/\partial \mathbf{n}$) and the direction cosines of the axes $\bar{x}$, $\bar{y}$ and $\bar{z}$.

For elasticity, tangential stresses are usually the most important results to be obtained from an analysis. In two dimensions, longitudinal strain in the direction $\bar{x}$ is given by

$$\bar{\epsilon}_x = \frac{1}{J} \sum_{c=1}^{n(b)} \frac{\mathrm{d}N_c}{\mathrm{d}\xi} \bar{u}_{xc}^b \tag{7.154}$$

where $\bar{u}_{xc}^b$ is displacement of node c of S_b in the direction $\bar{x}$. For plane strain, the local components of the stress tensor are

$$\begin{aligned} \bar{\sigma}_x &= \frac{1}{(1-\nu)} \left[\frac{E\bar{\epsilon}_x}{(1+\nu)} + \nu \bar{t}_y \right] \\ \bar{\sigma}_y &= \bar{t}_y \\ \bar{\tau}_{xy} &= \bar{t}_x \end{aligned} \tag{7.155}$$

where $(\bar{t}_x, \bar{t}_y)$ are local components of traction. In three dimensions,

$$\left\{ \begin{array}{c} \frac{\partial \mathbf{u}}{\partial \bar{x}} \\ \frac{\partial \mathbf{u}}{\partial \bar{y}} \end{array} \right\} = \mathbf{J}^{-1} \sum_{c=1}^{n(b)} \left\{ \begin{array}{c} \frac{\partial N_c}{\partial \xi} \\ \frac{\partial N_c}{\partial \eta} \end{array} \right\} \mathbf{u}_c^b \tag{7.156}$$

Local components of tangential strain are

$$\begin{aligned} \bar{\epsilon}_x &= \frac{\partial \bar{u}_x}{\partial \bar{x}} \\ \bar{\epsilon}_y &= \frac{\partial \bar{u}_y}{\partial \bar{y}} \\ \bar{\gamma}_{xy} &= \frac{\partial \bar{u}_x}{\partial \bar{y}} + \frac{\partial \bar{u}_y}{\partial \bar{x}} \end{aligned} \tag{7.157}$$

From these strains and local components $(\bar{t}_x, \bar{t}_y, \bar{t}_z)$ of traction, we may compute local components of stress:

$$\begin{aligned} \bar{\sigma}_x &= \frac{E}{(1-\nu^2)}(\bar{\epsilon}_x + \nu\bar{\epsilon}_y) + \frac{\nu\bar{t}_z}{(1-\nu)} \\ \bar{\sigma}_y &= \frac{E}{(1-\nu^2)}(\bar{\epsilon}_y + \nu\bar{\epsilon}_x) + \frac{\nu\bar{t}_z}{(1-\nu)} \\ \bar{\sigma}_z &= \bar{t}_z \\ \bar{\tau}_{xy} &= \frac{E}{2(1+\nu)}\bar{\gamma}_{xy} \\ \bar{\tau}_{yz} &= \bar{t}_y \\ \bar{\tau}_{zx} &= \bar{t}_x \end{aligned} \tag{7.158}$$

Global components of the stress tensor may then be computed from the local components. There is not known to be any advantage in computing potential gradient or stress at the sampling points for Gaussian quadrature, as in the finite element method. Computed potential gradients or stresses at a point common to more than one element (e.g. a corner or side node in three dimensions) are in general different in each of the adjacent elements, and it is the usual practice to take the averages of these values to be the computed values at such a point.

For Laplace's equation, potential at an interior point of the domain may be computed according to the Poisson formula, equation (6.78). Let us denote the interior point by P_a. Then in the notation of Section 7.2,

$$u(P_a) = \sum_{b=1}^{p} \sum_{c=1}^{n(b)} \Delta U_{ac}^b t_c^b - \sum_{b=1}^{p} \sum_{c=1}^{n(b)} \Delta T_{ac}^b u_c^b \tag{7.159}$$

Similarly, for elasticity, displacement at an interior point may be computed according to Somigliana's identity, equation (6.92). In the notation of Section 7.2,

$$\mathbf{u}(P_a) = \sum_{b=1}^{p} \sum_{c=1}^{n(b)} \Delta \mathbf{U}_{ac}^{b} \mathbf{t}_c^b - \sum_{b=1}^{p} \sum_{c=1}^{n(b)} \Delta \mathbf{T}_{ac}^{b} \mathbf{u}_c^b \tag{7.160}$$

It is the usual practice to compute the stress pseudo-vector $\boldsymbol{\sigma}(P_a)$ directly from the boundary data according to the following formula obtained by differentiation of Somigliana's identity and substitution into Hooke's law:

$$\boldsymbol{\sigma}(P_a) = \int_S \left[\mathbf{S}(P_a, Q)\,\mathbf{t}(Q) - \mathbf{R}(P_a, Q)\,\mathbf{u}(Q)\right] \, \mathrm{d}S_Q \tag{7.161}$$

where

$$\mathbf{R}(P_a, Q) = \mathbf{DBT}(P_a, Q) \tag{7.162}$$

and

$$\mathbf{S}(P_a, Q) = \mathbf{DBU}(P_a, Q) \tag{7.163}$$

In equations (7.162) and (7.163), the differentiations implied by the operator $\mathbf{B}$ are with respect to the coordinates of the point P_a. By substituting the parametric interpolates for $\mathbf{u}(Q)$ and $\mathbf{t}(Q)$ we obtain the result

$$\boldsymbol{\sigma}(P_a) = \sum_{b=1}^{p} \sum_{c=1}^{n(b)} \Delta \mathbf{S}_{ac}^{b} \mathbf{t}_c^b - \sum_{b=1}^{p} \sum_{c=1}^{n(b)} \Delta \mathbf{R}_{ac}^{b} \mathbf{u}_c^b \tag{7.164}$$

in which

$$\Delta \mathbf{R}_{ac}^{b} = \int_{S_b} \mathbf{R}(P_a, Q)\, N_c(Q) \, \mathrm{d}S_Q \tag{7.165}$$

$$\Delta \mathbf{S}_{ac}^{b} = \int_{S_b} \mathbf{S}(P_a, Q)\, N_c(Q) \, \mathrm{d}S_Q \tag{7.166}$$

Integration schemes for the computation of results at interior points of the domain differ only in detail from those for matrix construction. There is no integration over elements of which P_a is a node, or over the auxiliary boundary Ω_a. In the solution of problems of elasticity the same integration scheme is used for the calculation of displacement and stress, because in the computation of the kernels $\mathbf{T}$ and $\mathbf{U}$ of equation (7.160) and the kernels $\mathbf{R}$ and

S of equation (7.164) there are many computations in common. The choice of integration formulae is determined by the rapidity of variation of **R**, this being the most rapidly varying kernel. Since for the two dimensional case **R** is of order $1/r^2$, we consider the simplified problem shown in Figure 7.12 and take the function $1/r^2$ to be representative of the kernel **R**. In the notation of Section 7.4, by differentiating $1/r^2$ $2g$ times with respect to β and then substituting $r = R$ we obtain the result analogous to equation (7.53)

$$H_g = \left(\frac{L}{2}\right)^{2g} \frac{(2g+1)!}{R^{2g+2}} \tag{7.167}$$

By substituting into inequality (7.50), we obtain the approximate upper bound

$$\left| \int_{-1}^{1} \frac{1}{r^2}(\beta)\ \mathrm{d}\beta - \sum_{k=1}^{g} B_k^g \frac{1}{r^2}\left(\beta_k^g\right) \right| \le \left(\frac{L}{4}\right)^{2g} \cdot \frac{4(2g+1)}{R^{2g+2}} \tag{7.168}$$

We take as a measure of precision of integration the ratio between the upper bound (7.168) and an upper bound on the integral of $1/r^2$ over the interval. We write

$$\left| \int_{-1}^{1} \frac{1}{r^2}(\beta)\ \mathrm{d}\beta \right| \le \frac{2}{R^2} \tag{7.169}$$

and proceeding as in Section 7.4 find that the number of intervals G into which an element is subdivided must satisfy the inequality

$$G \ge \frac{L_b}{4R_b} \left(\frac{2\left(2g_{\max}+1\right)}{\epsilon} \right)^{\frac{1}{2g_{\max}}} \tag{7.170}$$

The number of Gauss points for each interval is then chosen such that

$$\frac{L}{R} \le 4 \left(\frac{\epsilon}{2\left(2g+1\right)} \right)^{\frac{1}{2g}} \tag{7.171}$$

where L and R are the length and minimum distance to P_a of the interval. The kernel-shape function product integrals $\Delta \mathbf{T}_{ac}^b$ and $\Delta \mathbf{U}_{ac}^b$ are then computed according to equations (7.60) and (7.61), and $\Delta \mathbf{R}_{ac}^b$ and $\Delta \mathbf{S}_{ac}^b$ are computed according to those equations with **T** and **U** replaced by **R** and **S**. The generalisation to infinite elements is similar to that described in Section 7.4 for matrix construction.

In three dimensional elastic analysis, the kernel **R** is of order $1/r^3$. In the notation of Section 7.5, it can be shown that the number of intervals G and H into which a quadrilateral element is subdivided in the ξ and η directions, respectively, must satisfy the inequalities

$$G \geq \frac{L_b^{(\xi)}}{4R_b}\left(\frac{(2g_{\max}+1)\,(2g_{\max}+2)}{\epsilon}\right)^{\frac{1}{2g_{\max}}} \tag{7.172}$$

$$H \geq \frac{L_b^{(\eta)}}{4R_b}\left(\frac{(2g_{\max}+1)\,(2g_{\max}+2)}{\epsilon}\right)^{\frac{1}{2g_{\max}}} \tag{7.173}$$

The number of Gauss points $g(E,F)$ and $h(E,F)$ in the ξ and η directions, respectively, of an interval of integration are chosen such that

$$\frac{L^{(\xi)}}{R} \leq 4\left(\frac{\epsilon}{(2g+1)(2g+2)}\right)^{\frac{1}{2g(E,F)}} \tag{7.174}$$

$$\frac{L^{(\eta)}}{R} \leq 4\left(\frac{\epsilon}{(2h+1)(2h+2)}\right)^{\frac{1}{2h(E,F)}} \tag{7.175}$$

The kernel-shape function product integrals $\Delta\mathbf{T}_{ac}^{b}$ and $\Delta\mathbf{U}_{ac}^{b}$ are then computed according to equations (7.83) and (7.84), and $\Delta\mathbf{R}_{ac}^{b}$ and $\Delta\mathbf{S}_{ac}^{b}$ are computed according to those equations with $\mathbf{T}$ and $\mathbf{U}$ replaced by $\mathbf{R}$ and $\mathbf{S}$. The generalisation to triangular and infinite elements is similar to that described in Section 7.5 for matrix construction.

The integration schemes described in this section ensure adequate precision combined with computational economy. In particular they enable displacements and stresses to be computed accurately at interior points of the domain which are close to the surface compared with the dimensions of the nearest surface elements. The most accurate (but generally least useful) results are those at interior points far from the boundary. The least accurate results are those at points on the boundary, and the accuracy of results computed at interior points near the boundary is impaired by local inaccuracies on the boundary regardless of how precisely the integration is carried out. The computing time required to calculate results at points near the boundary is greater than that for points far from the boundary. If results are required at a number of interior points comparable with the number of nodes of a finite element mesh, then it is usually more economical to carry out an analysis by the finite element method.

7.10 EXAMPLES—1

The purpose of this section is to demonstrate by means of simple examples analysed by program BOUND (see Appendix B) certain characteristics of boundary element methods in general. Program BOUND is a program for the analysis of plane strain of a homogeneous isotropic elastic domain by the direct method and nodal collocation, incorporating quadratic isoparametric and infinite elements. Particular attention is paid to the question of what accuracy may be expected of an analysis. Ignoring machine round-off error which is negligible at IBM double precision, there are two sources of error in a boundary element analysis: discretisation error, and integration error. Discretisation error is error due to assumptions made about geometric and functional variation over each element, as implied by the choice of shape functions. It is expected that if for a given problem a greater number of elements is used to model the boundary then the discretisation error is reduced, and that in the limit as the number of elements tends to infinity the discretisation error tends to zero. Integration error is present independently of degree of mesh refinement, but may be limited by suitable choice of the constant ϵ of equation (7.56). Since for large problems the effect on computing time of reducing ϵ is comparatively small, it is the usual practice to err on the safe side by setting ϵ such that integration error is much smaller than discretisation error. In program BOUND, $\epsilon = 2.7 \times 10^{-5}$ for matrix construction and 5.4×10^{-5} for the computation of results at interior points of the domain. These values have been chosen in the light of experience, taking account of the fact that a solution vector tends to be more sensitive to errors in matrix coefficients than are computed values at an interior point to errors in the integrals of kernel-shape function products referred to in Section 7.9. In the analyses of this section, Poisson's ratio is taken to be zero because for reasons as yet unknown this is generally the value for which boundary element methods are least accurate.

7.10.1 Plate in uniaxial tension

In this example the shape functions are capable of modelling the exact geometry, displacements and traction, so any error in the computed results is due to integration error only. The boundary element mesh consists of eight elements with a total of 16 nodes as shown in Figure 7.35. We take $E = 10\,000$ MPa, and $\nu = 0$ so that there is no difference between plane stress and plane strain. Nodes 9 and 13 are constrained against vertical displacement, and node 11 is constrained against horizontal displacement. Interior nodes at which results are to be computed are spaced at 0.1 m intervals on two straight lines: the series 100 to 109 lies on the inward normal to the boundary at node 15, and the series 200 to 209 lies on the inward normal at a point midway between nodes 15 and 16. BOUND prints computed displacements to four significant figures, and stresses to three significant figures. The exact value of horizontal displacement at nodes 1 and 13 to 16 is 5.0×10^{-4}. From the printed results it is found that the greatest error in computed displacement is less than 5.0×10^{-8}, or one part in 10 000 of the maximum displacement. Likewise from the printed results it is found that the greatest error in computed stress is less

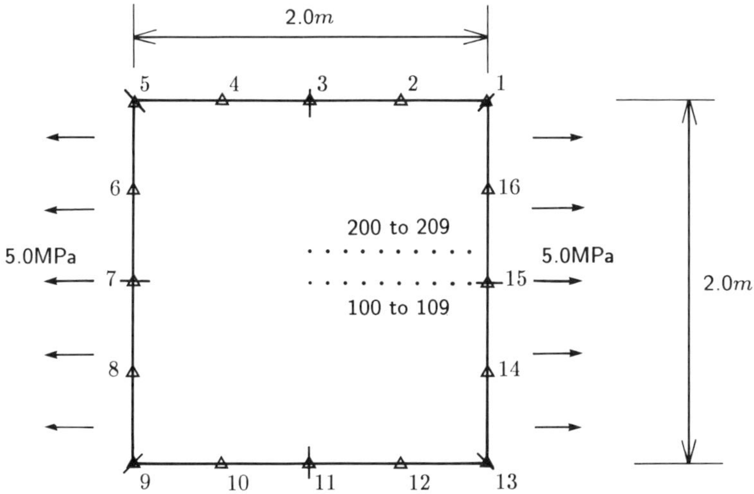

Figure 7.35 Plate in uniaxial tension: boundary element mesh

than 0.005 MPa, or one part in 1000 of the applied uniaxial stress. Due to the use of the integration scheme described in section 7.10, this accuracy is maintained even at nodes 109 and 209 which are only 0.1 of the length of an element from the boundary.

7.10.2 Cantilevers

Problems for which thin plate or thin shell theory is a good approximation can be difficult to solve accurately by boundary element methods. Let us consider a cantilever plate of width from supported side to free side 5 m, with E = 10 000 MPa and $\nu = 0$, subjected to a load of 10 kN/m along its free side as shown in Figure 7.36a. We take the plate to be in a state of plane strain, and consider thickness t = 1.0 m, 0.5 m and 0.2 m. The mesh of 12 boundary elements for these analyses is shown in Figure 7.36b. The element at the root of the cantilever is constrained against horizontal displacement and subjected to a uniformly distributed upwards shear force, whereas the element at the free end or side is subjected to a uniformly distributed downwards shear force, as indicated in Figure 7.36b. Node 2 is constrained against vertical displacement. Computed maximum stresses and displacements are shown in Table 7.3. Computed displacement for the case t=1.0 m is greater than that predicted by beam bending theory because that theory takes no account of displacement due to shear strain, which can be significant for deeper beams. For the lower values of t the beam bending theory is very accurate, and the errors in the boundary element analyses are quite large. Results obtained using the alternative boundary element meshes for t=0.5 m and 0.2 m shown in Figure 7.37 are shown in Table 7.4. It may be seen that the accuracy of results for t=0.5 m and 0.2 m is greatly improved. Possibly the precision of integration set by the choice of ϵ in equation (7.56), whilst adequate for commonly encountered domain geometries, is not adequate for the boundary element mesh of Figure 7.36b with t=0.5 m or 0.2 m.

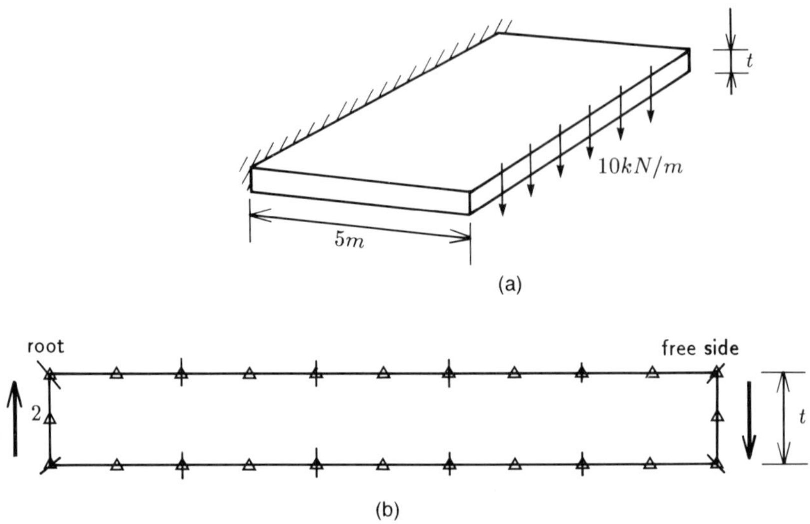

Figure 7.36 Cantilevers. (a) Cantilever plate. (b) Boundary element mesh (12 elements)

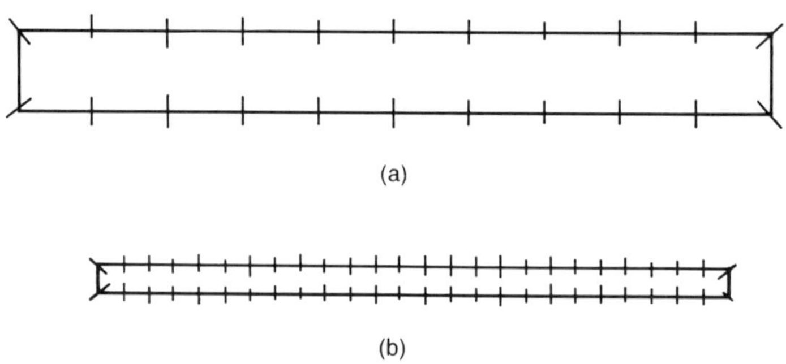

Figure 7.37 Cantilevers: alternative boundary element meshes. (a) t=0.5 m, 22 elements. (b) t=0.2 m, 52 elements

Table 7.3 Displacements of and stresses in cantilevers

t	Maximum deflection, mm		Maximum stress, MPa	
	Computed	Beam theory	Computed	Beam theory
1.0	0.5073	0.5000	0.298	0.300
0.5	3.697	4.000	1.09	1.20
0.2	36.41	62.50	4.23	7.50

Table 7.4 Displacements and stresses computed using alternative meshes

t	Maximum deflection, mm		Maximum stress, MPa	
	Computed	Beam theory	Computed	Beam theory
1.0*	0.5073	0.5000	0.298	0.300
0.5	3.990	4.000	1.19	1.20
0.2	60.45	62.50	7.27	7.50

*same mesh as for Table 7.3

7.10.3 Circular hole in infinite elastic space

For a circular hole, neither the functional variation nor the geometry of the boundary can be modelled exactly by the parametric interpolates over each element. Let us consider a hole of radius 1.0 m in an infinite elastic space for which E=10 000 MPa and $\nu = 0$, with uniaxial stress $\sigma_y = 3.0$ MPa at infinity. We model the boundary by meshes of 4, 8 and 16 elements, as shown in Figure 7.38. In Table 7.5 are shown computed stresses σ_r and σ_θ at two points of the boundary, and at the interior nodes 111 to 115 which are spaced at 0.1 m intervals on the x axis as shown in Figure 7.38. It may be concluded from those results that provided there are no local rapid variations of traction applied to the boundary, an isoparametric quadratic element can model about 45° of boundary arc without significant loss of accuracy of computed stresses.

Table 7.5 Computed stresses around circular hole (in MPa)

Elements			σ_θ					σ_r				
	A	B	111	112	113	114	115	111	112	113	114	115
4	8.69	−2.73	7.25	6.12	5.39	4.88	4.51	0.71	0.98	1.08	1.11	1.09
8	8.99	−2.99	7.31	6.20	5.46	4.93	4.55	0.65	0.95	1.09	1.12	1.11
16	9.00	−3.00	7.31	6.21	5.46	4.94	4.56	0.65	0.95	1.09	1.12	1.11
Exact	9.00	−3.00	7.31	6.21	5.46	4.94	4.56	0.65	0.95	1.09	1.12	1.11

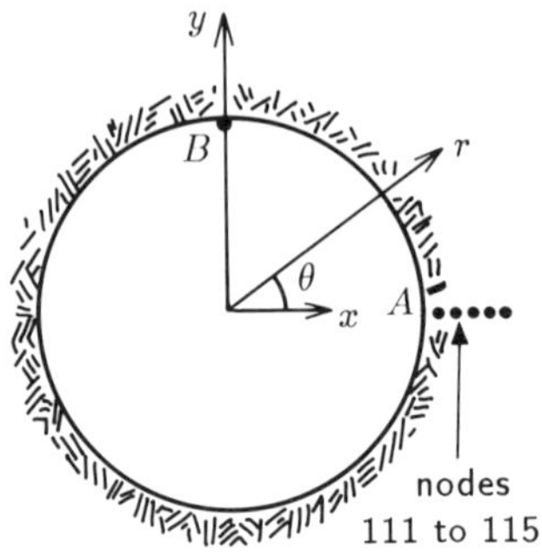

(a)

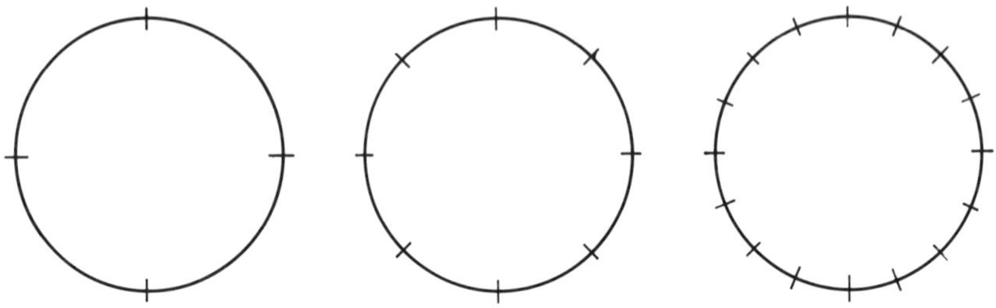

(b)

Figure 7.38 Circular hole in infinite elastic space. (a) Coordinates. (b) Boundary element meshes

7.10.4 Square hole in infinite elastic space

The exact solution for a square hole is not known. According to the theory of elasticity[9] there are components of displacement in the neighbourhood of a corner which tend to zero as $r^{0.544}$ (symmetric notch opening mode) and $r^{0.909}$ (antisymmetric notch opening mode) where r is distance from the corner, and therefore stresses which tend to infinity as $r^{-0.456}$ and $r^{-0.091}$. These components of displacement cannot be modelled accurately by the quadratic shape functions, so there is some loss of accuracy near each corner of the hole. The computed values of stress at a corner are meaningless. Let us consider a hole of side 2 m in an infinite elastic space for which E=10 000 MPa and $\nu = 0$, with uniaxial stress $\sigma_y = 3.0$ MPa at infinity. We model the boundary by meshes of 8, 16 and 32 elements of equal size, and by 32 elements with mesh gradation, as shown in Figure 7.39. In the graded mesh, the elements adjacent to a corner are of length 0.1 m, whereas those adjacent to the points A and B shown in Figure 7.39 are of length 0.4 m. All computed values of boundary stress (tensile from A to C, compressive from C to B)

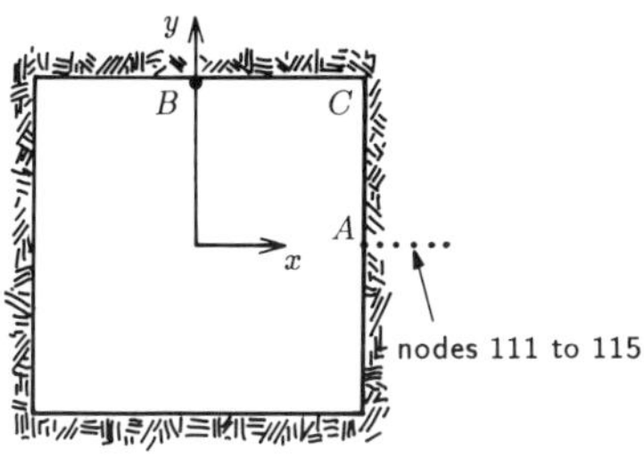

(a)

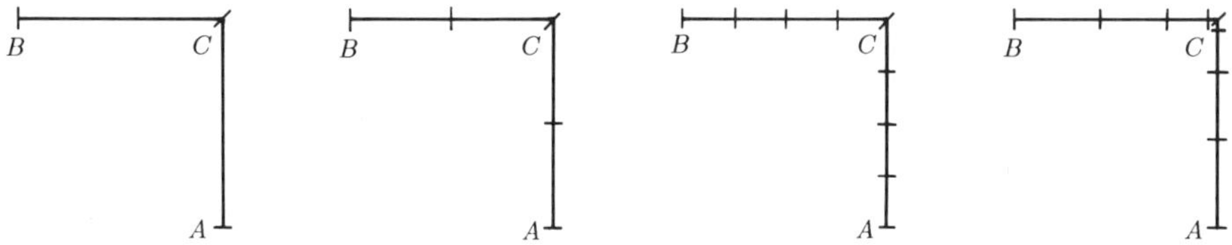

(b)

Figure 7.39 Square hole in infinite elastic space. (a) Coordinates. (b) Boundary element meshes

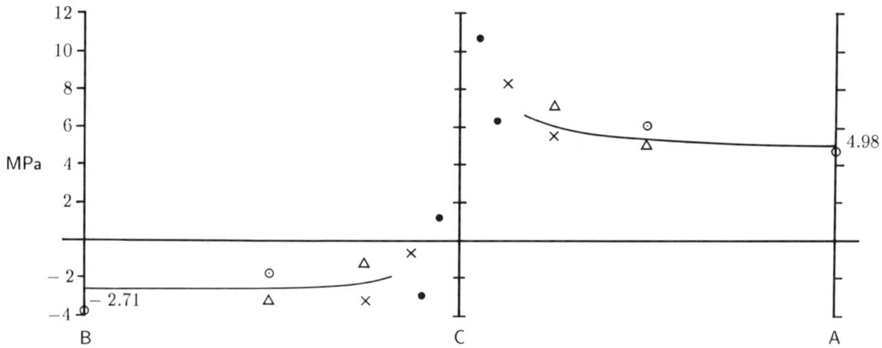

Figure 7.40 Computed variation of boundary stress. (○ 8 elements, △ 16 elements, ×32 elements, • 32 elements graded mesh)

except for those plotted individually lie on the two smooth curves shown in Figure 7.40. These values are presumed accurate. The individually plotted values are always those computed at nodes of the elements adjacent to the corner C, so the rule is that computed values of stress at nodes of elements adjacent to re-entrant corners should be ignored. By grading the mesh it is

Table 7.6 Computed stresses σ_y at point A and interior nodes (in MPa)

Elements	A	111	112	113	114	115
8	4.62	5.09	5.06	4.95	4.79	4.62
16	5.01	5.02	4.96	4.85	4.72	4.58
32	5.01	4.99	4.93	4.83	4.70	4.57
32(graded)	4.98	4.98	4.91	4.81	4.69	4.56

possible, for a given number of elements, to increase the proportion of the boundary for which reliable results are obtained.

In Table 7.6 are given computed values of the stress σ_y at the interior nodes 111 to 115 which are spaced at 0.1 m intervals on the x-axis as shown in Figure 7.39a. The node 111 is 0.1 m from A. Even though the result at A obtained with eight elements is not very accurate, the computed results at the interior nodes are quite good.

7.10.5 Plate with edge crack

Let us consider a long plate for which E=10 000 MPa and $\nu = 0$, with an edge crack of length 0.1 of its width as shown in Figure 7.41a. The applied loading is such that far from the crack there is unit uniaxial stress in the plate. According to the theory of elasticity there exists a component of displacement in the neighbourhood of the crack tip which tends to zero as $r^{0.5}$ where r is distance from the crack tip, and therefore there are stresses which tend to infinity as $r^{-0.5}$. The magnitude of this component of displacement is characterised by the mode 1 stress intensity factor K_1. The purpose of the analyses shown here is to compute K_1, so that for a given material its value can be compared with the critical mode 1 stress intensity factor which is the value of K_1 required to cause the crack to propagate. According to the definition of K_1,[7]

$$K_1 = \frac{E.\Delta u}{8(1+\nu)}\left(\frac{2\pi}{r}\right)^{1/2} \tag{7.176}$$

where (see Figure 7.41b) Δu is the crack opening displacement due to the component of displacement referred to above.

In the boundary element analysis, we model a crack of length 1 mm in a plate of width 10 mm, assuming the stress distribution in the plate to be uniform at a distance 40 mm from the crack. A 40 mm length of plate is modelled as shown in Figure 7.42a. The elements on AB represent one crack surface, and are not subject to any constraint on displacement. The elements on BC are constrained in the horizontal direction but free to displace vertically, and the rest of the boundary is unconstrained. We model the boundary by graded meshes of 29 and 58 elements as shown in Figures 7.42b

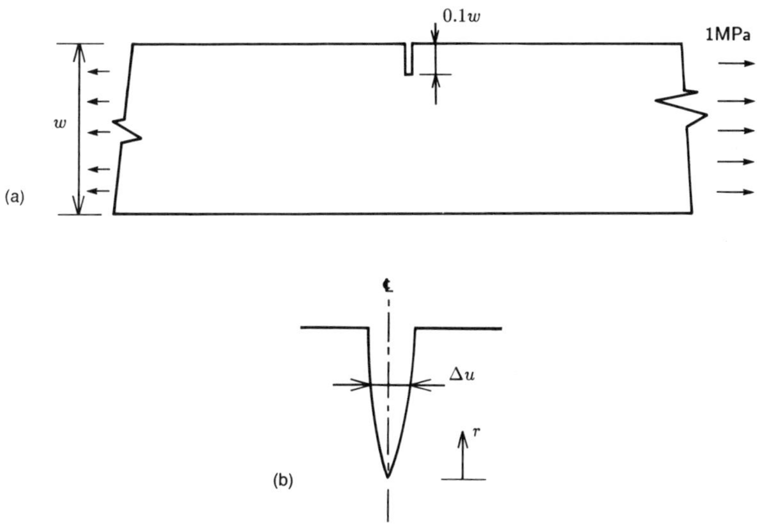

Figure 7.41 Plate with edge crack. (a) Dimensions and loading. (b) Detail of crack

and 7.42c, in which the crack surface AB is modelled by four and eight elements, respectively. In the mesh of 29 elements, the elements adjacent to the crack tip B are of length 0.0625 mm whereas in the mesh of 58 elements, they are of length 0.015 625 mm.

In Figure 7.43 are shown the variations of K_1 obtained by substituting computed boundary nodal values into equation (7.176). There is considerable variation with respect to distance from the crack tip. Far from the crack tip, the apparent value of K_1 is affected by the influence of a component of the displacement field other than that which varies as $\sqrt{r}$. The ratio of the magnitude of this additional component to that of the component which varies as $\sqrt{r}$ tends to zero as the crack tip is approached and therefore in theory the true value of K_1 is the value as r tends to zero. However, near the crack tip the quadratic shape functions do not model the component of displacement which varies as $\sqrt{r}$ very well, so locally there is abnormally high discretisation error. Interpretation of the computed results is therefore difficult. If we suppose that the zone of increased discretisation error extends over two of the elements which model the crack face, then we might take the values of K_1 computed from the nodal displacements of the meshes of 29 and 58 elements to be 2.007 MPa mm$^{1/2}$ and 2.060 MPa mm$^{1/2}$, respectively. Tada[10] gives the value 2.109. The accuracy of the analysis presented here could be improved considerably by moving the centre node of the element adjacent to the crack tip to a point one-quarter of the length of the element from the crack tip, thus enabling the isoparametric interpolate to model accurately the component of displacement which varies as $\sqrt{r}$.[11] There are also other possibly

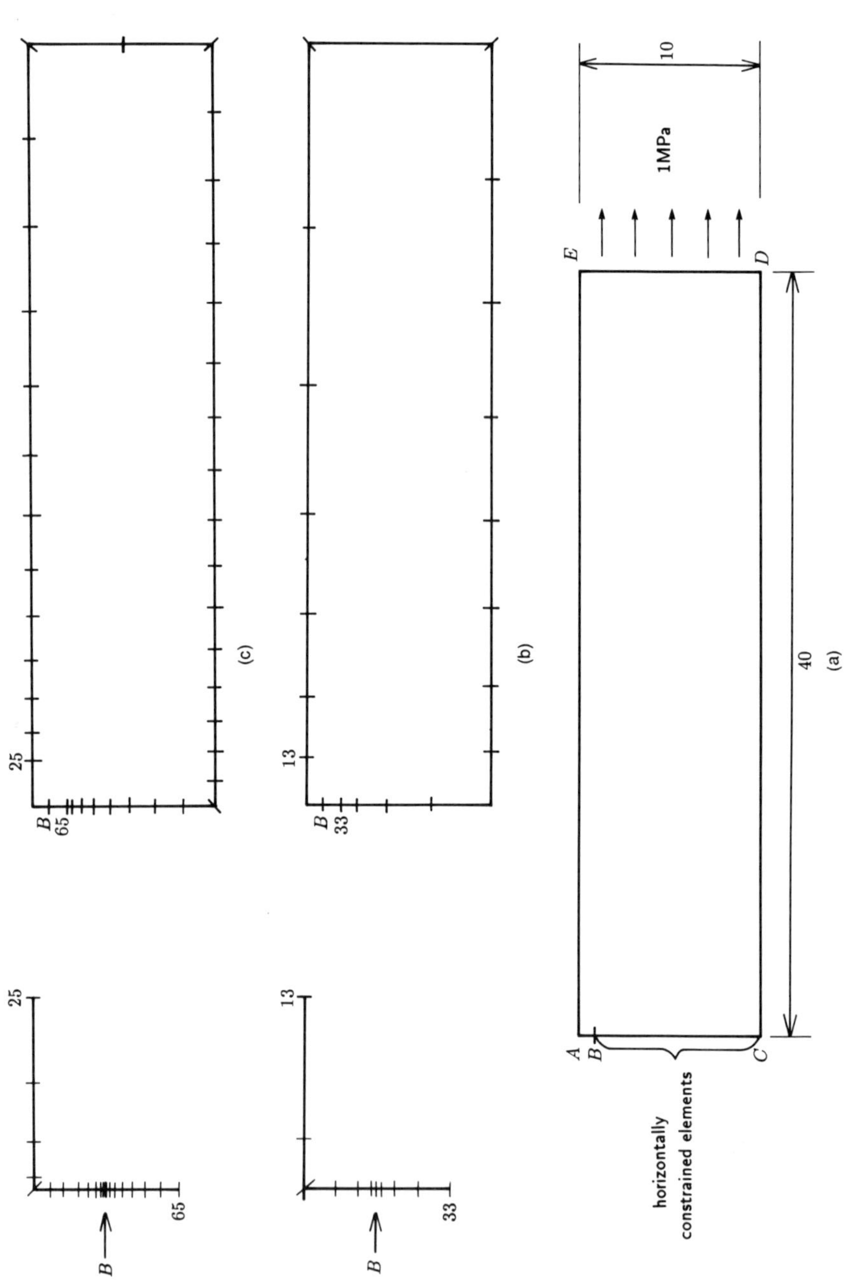

Figure 7.42 Modelling of plate with edge crack. (a) Geometry of domain. (b) 29 element mesh. (c) 58 element mesh

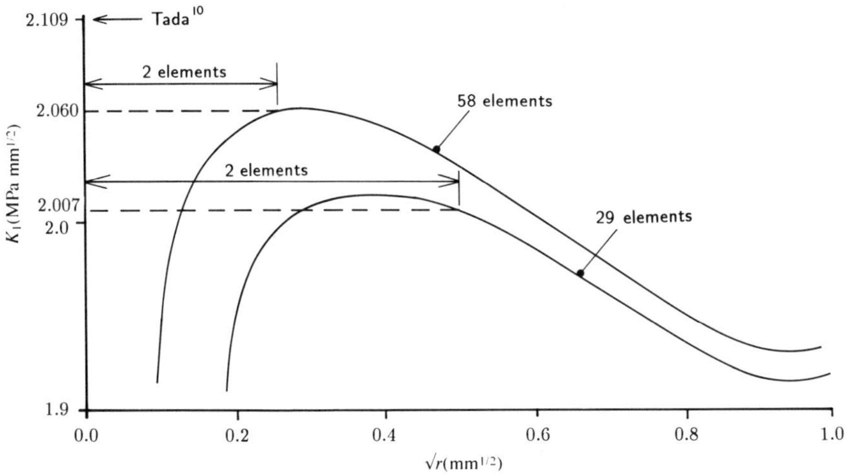

Figure 7.43 Stress intensity factor K_1 calculated from crack opening displacement

more accurate ways of determining K_1 than by consideration of crack opening displacement[7].

7.11 EXAMPLES—2

In this section are described three-dimensional stress analyses of full scale practical problems, by programs THREE and BEFE. Program THREE is a program for the three dimensional elastic analysis of piecewise homogeneous domains by the direct method and nodal collocation, incorporating eight-node and nine-node quadratic isoparametric elements, six-node triangular quadratic isoparametric elements, and infinite elements with asymptotic and plane strain functional variation. Program BEFE is a coupled boundary element–finite element program, in which the implementation of the boundary element method is similar to that of program THREE.

7.11.1 Analysis of stress in a crankshaft

We consider one throw of a scale model marine diesel crankshaft that has been the subject of extensive experimental and theoretical investigation by Lloyd's Register of Shipping[12]. In Figure 7.44 is shown one half of the throw from centreline of a journal to centreline of a crank pin, and in Figures 7.45 and 7.46 are shown an end view and a side view of the throw. Three load cases are considered as shown in Figure 7.47. The stress analysis was carried out using program THREE, with the mesh of eight-node quadrilateral and six-node triangular elements shown in Figure 7.48 to model half of the surface of that part of the crank throw shown in Figure 7.44. Advantage was taken of a facility of program THREE whereby if the elastic body is symmetric about a plane and is subject to constraints and loading which are also symmetric about that plane, then only half the surface need be modelled by boundary elements. The program automatically generates elements on the other half of the surface and integrates over them, but by taking account of the symmetry of the solution constructs a discrete system in terms of unknowns at nodes on one half of the surface only. In the analysis of the crank throw, there are 347 such nodes and therefore 1041 degrees of freedom. No elements are required in the plane of symmetry. Maximum stresses are expected to be on the surfaces of the fillets of radius 0.375 inches between the journal and web and between the web and crank pin. In accordance with the guidelines established in section 7.10 for the modelling of curved surfaces by quadratic boundary elements, these surfaces are modelled by doubly curved elements each of which turns 45° around the fillet and 45° around the journal or crank pin. The length of each of these elements is eight or nine times its width, but this is acceptable because the integration scheme of Section 7.5 is implemented in program THREE. The central processing time and input–output wait time for the analysis on a Cray 1-S were 120 s and 153 s, respectively. Of the central processing time, only 25 s were spent on simultaneous equation solution.

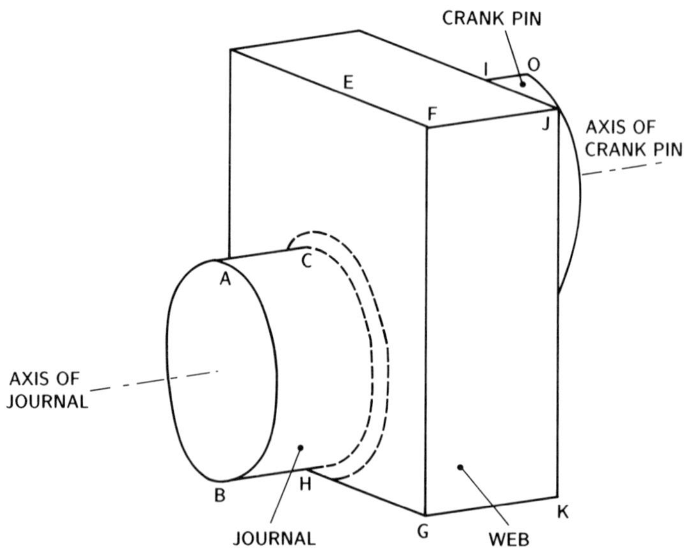

Figure 7.44 One half of crank throw

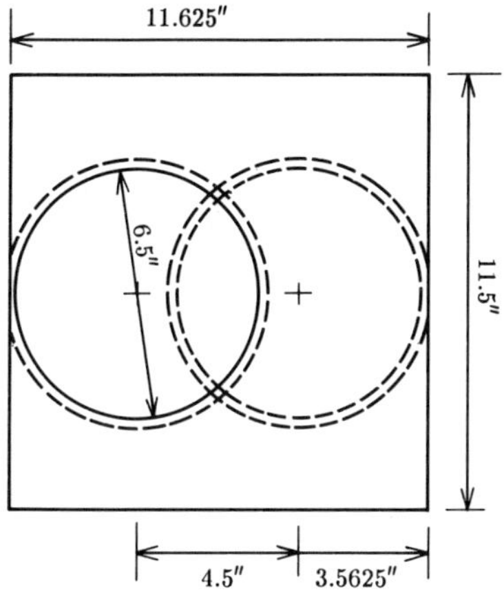

Figure 7.45 End view of crank throw

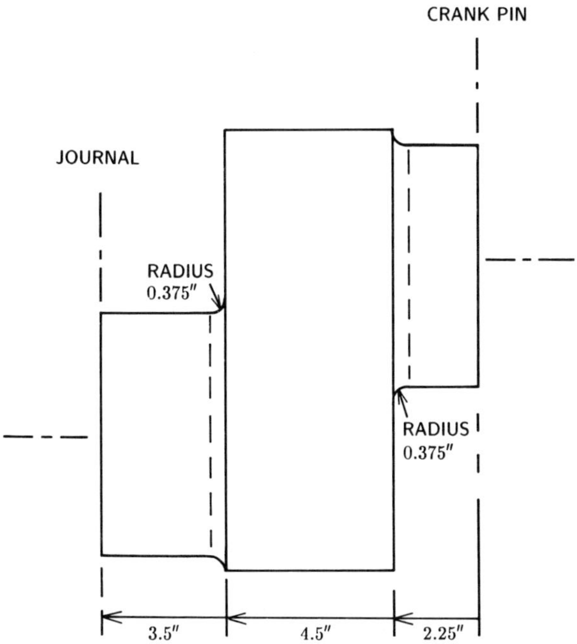

Figure 7.46 Side view of crank throw

In Figures 7.49 and 7.50 are shown the computed meridional stresses in the fillets along lines passing through the points C and N shown in Figure 7.48. The extreme values computed by program THREE are generally in fair agreement with those determined experimentally (in square brackets)

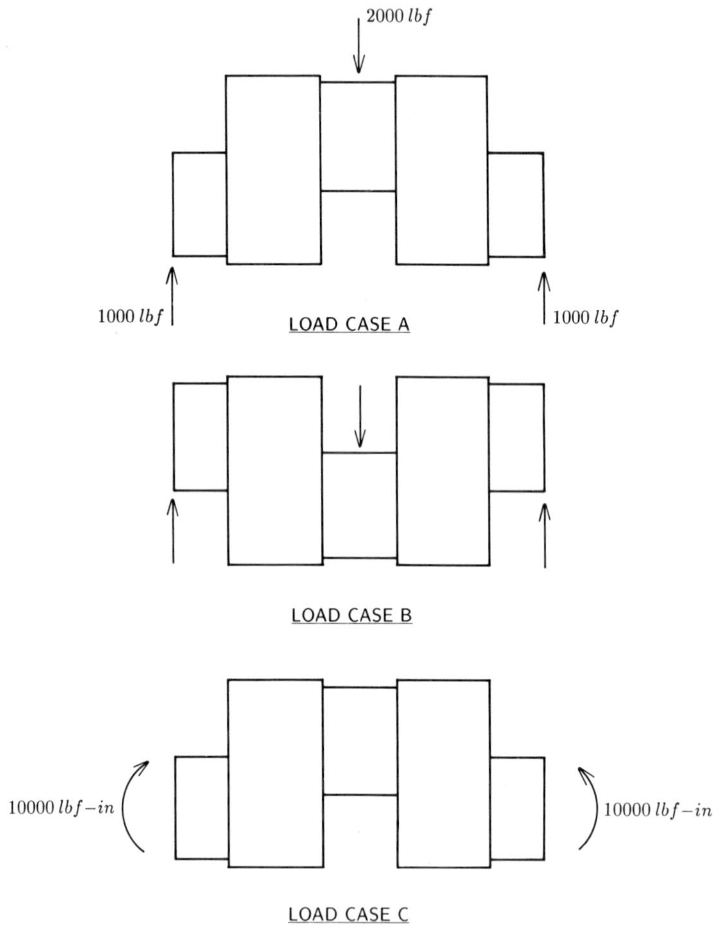

Figure 7.47 Loads applied to crank throw

and extreme values computed by the finite element program NASTRAN (in round brackets). Some of the experimental results were obtained by the photoelastic analysis of stress in epoxy models and so are not strictly comparable with computed values because Poisson's ratio of epoxy resin is somewhat higher than that of steel. In the finite element analysis, 20-node quadratic isoparametric hexahedral elements were used. In the fillet, each exposed face of a finite element turned 22.5° around the fillet and 30° around the journal or crank pin. The finite element mesh was therefore considerably finer than the boundary element mesh shown in Figure 7.48.

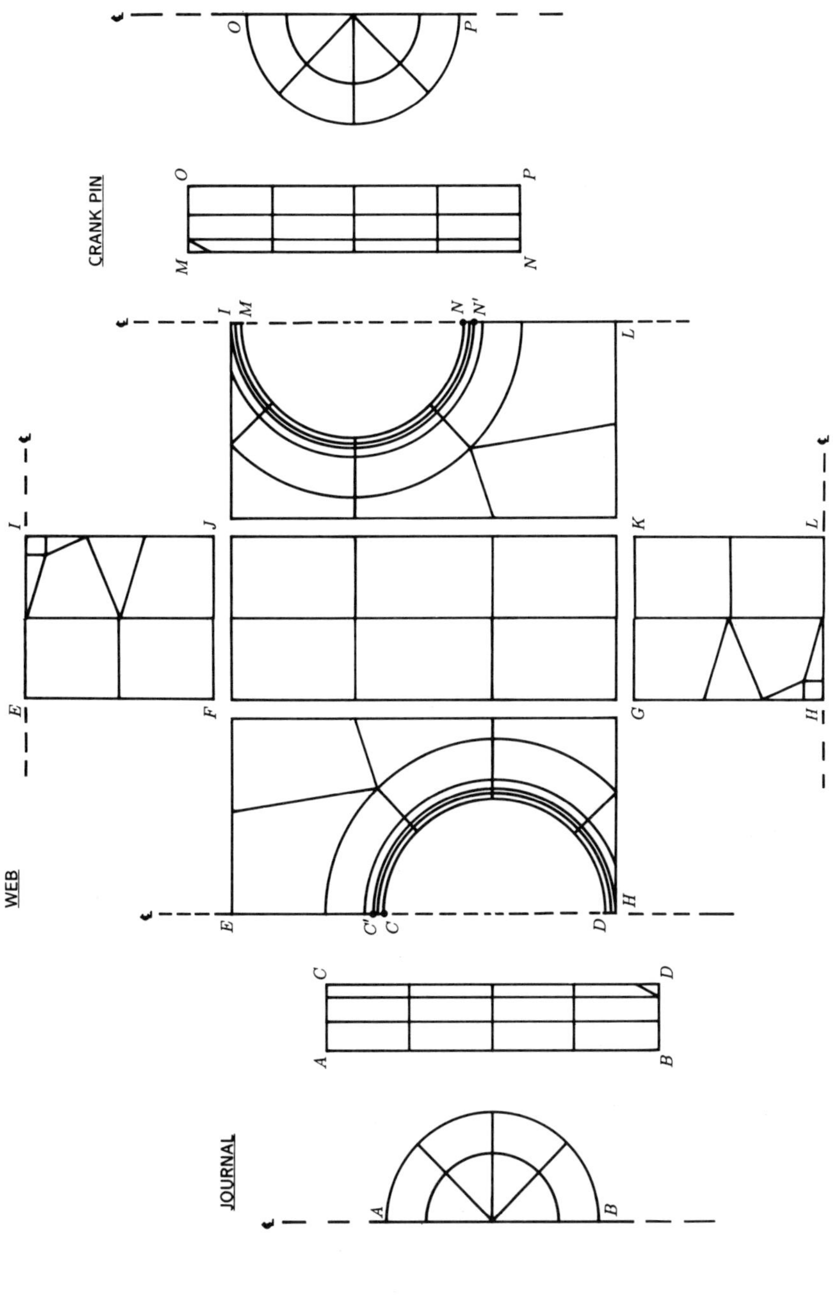

Figure 7.48 Boundary element mesh for stress analysis of crank throw

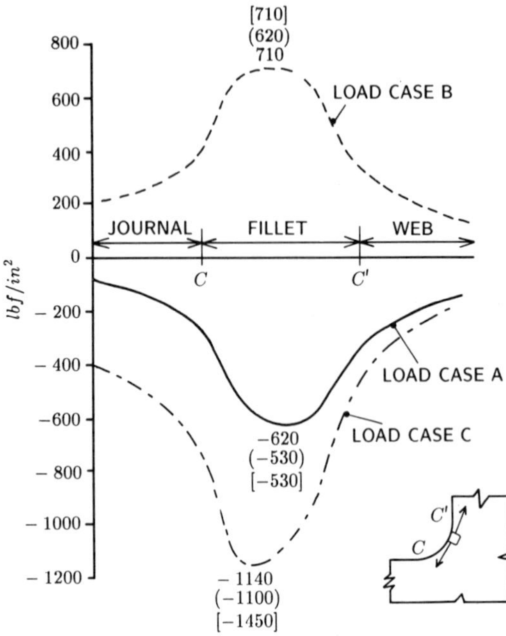

Figure 7.49 Computed meridional stresses in fillet between journal and web. (( ) NASTRAN, [ ] experiment)

7.11.2 Analysis of perforated cased deviated wellbore

Wells drilled from offshore production platforms are deviated from the vertical as shown in Figure 7.51 to recover oil from the outlying parts of the reservoir. Usually the principal directions of *in situ* stress are horizontal and vertical and all three principal stresses (one vertical, two horizontal) are different. In weaker rock it is necessary to case the well, that is to lower into the borehole a steel tube or casing and then fill the space between the casing and the wall of the borehole with cement, as indicated in Figure 7.52.

To enhance oil recovery, we create cracks in the surrounding rock from which oil can flow into the wellbore. This technique is known as hydrofracturing. Where there is no steel casing, the procedure is to pump drilling mud down the wellbore until cracks propagate into the rock. Pumping is continued until the cracks are estimated to have attained the required dimensions. Material known as proppant is carried into the cracks by the mud: the function of proppant is to hold the cracks open after the mud is pumped out and the oil starts to flow. Where there is a casing it is necessary first to form perforations as shown in Figure 7.52. It is expected that as the mud pressure is increased, cracks will propagate from the perforations.

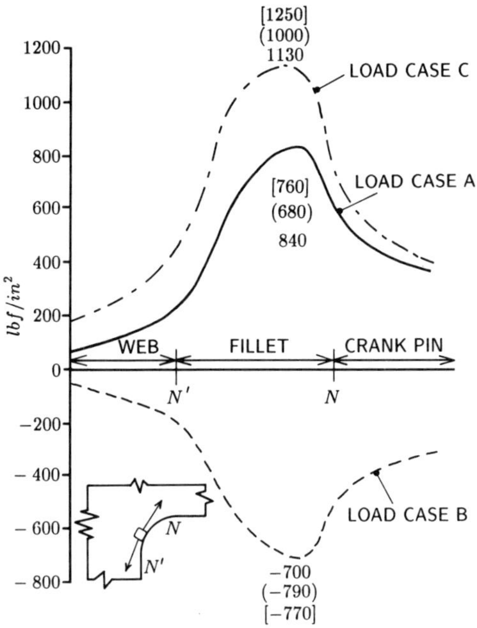

Figure 7.50 Computed meridional stresses in fillet between web and crank pin (( ) NASTRAN, [ ] experiment)

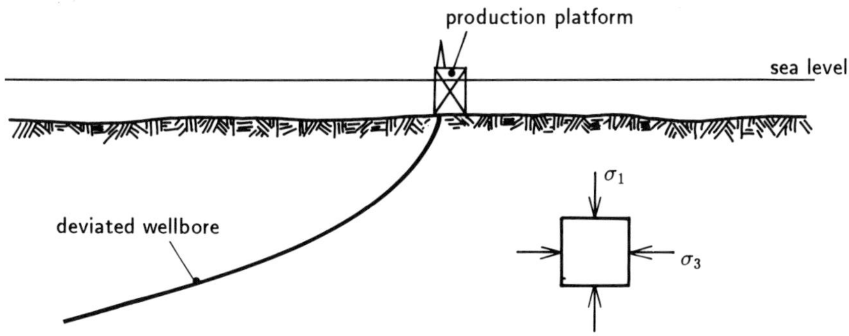

Figure 7.51 Deviated wellbore and *in situ* principal stresses

For the general case of a perforated deviated wellbore, a three-dimensional analysis is required. Analyses have been carried out using program THREE of various configurations proposed by BP Research, to predict the mud pressure at which cracks would be initiated. In Figure 7.52 is shown one of these configurations, in which there are two diametrally opposed perforations. The

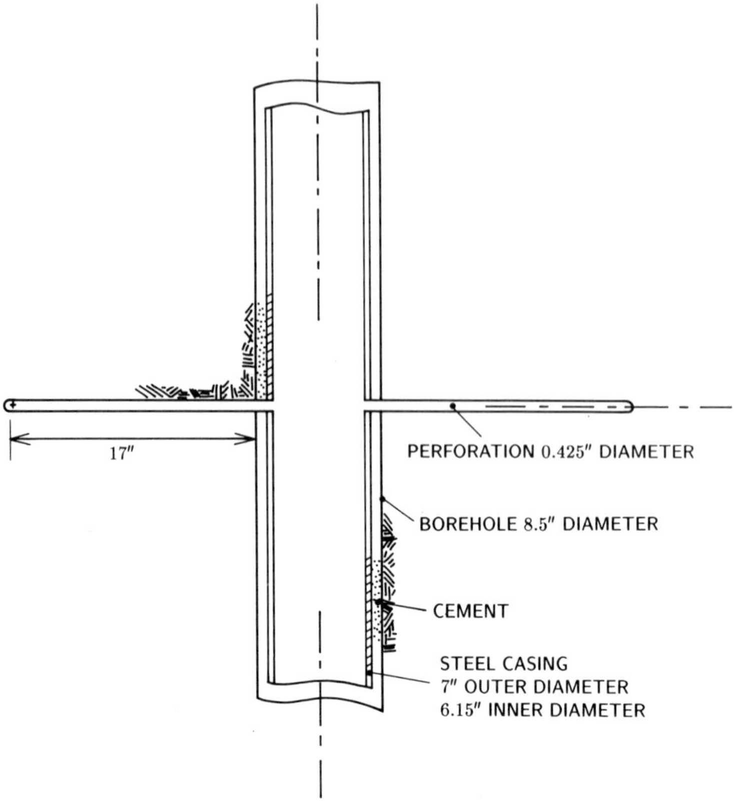

Figure 7.52 Diametral cross section of cased perforated wellbore

analysis presented here is one performed for verification purposes only, in which the *in situ* principal stresses are 6500 lbf/in^2 acting in the axial direction of the perforations, 8500 lbf/in^2 perpendicular to the cross section shown in Figure 7.52, and 10000 lbf/in^2 in the axial direction of the wellbore (all compressive). These data would apply to a vertical wellbore about 8500 feet below the North Sea. In the analysis, account must be taken of the sequence of construction and loading. First the plain borehole is drilled and the *in situ* stresses are redistributed, then the casing is lowered into the hole and wet cement is placed between the casing and the wall of the borehole. According to data supplied by BP Research, the hydrostatic pressure in the cement when placed is 5000 lbf/in^2. This pressure causes stresses to be induced in the casing and surrounding rock, and once the cement has hardened remains as a hydrostatic initial stress field. Stresses are redistributed when the perforations are formed, and change again when the mud pressure is applied.

The structure was divided into seven subregions as shown in Figure 7.53: the infinite subregion A and finite subregions B and C (rock), the infinite subregion D and finite subregions E and F (steel), and the infinite subregion G (cement). Elastic constants are for rock $E = 4.5 \times 10^6$ lbf/in^2 and $\nu = 0.2$, for steel $E = 30.0 \times 10^6$ lbf/in^2 and $\nu = 0.3$, and for cement $E = 2.0 \times 10^6$ lbf/in^2 and $\nu = 0.3$. Three analyses were performed as shown in Figure 7.53, the results of each analysis acting as input to the next. No advantage was taken of symmetries of structure and loading, as analyses were subsequently to be

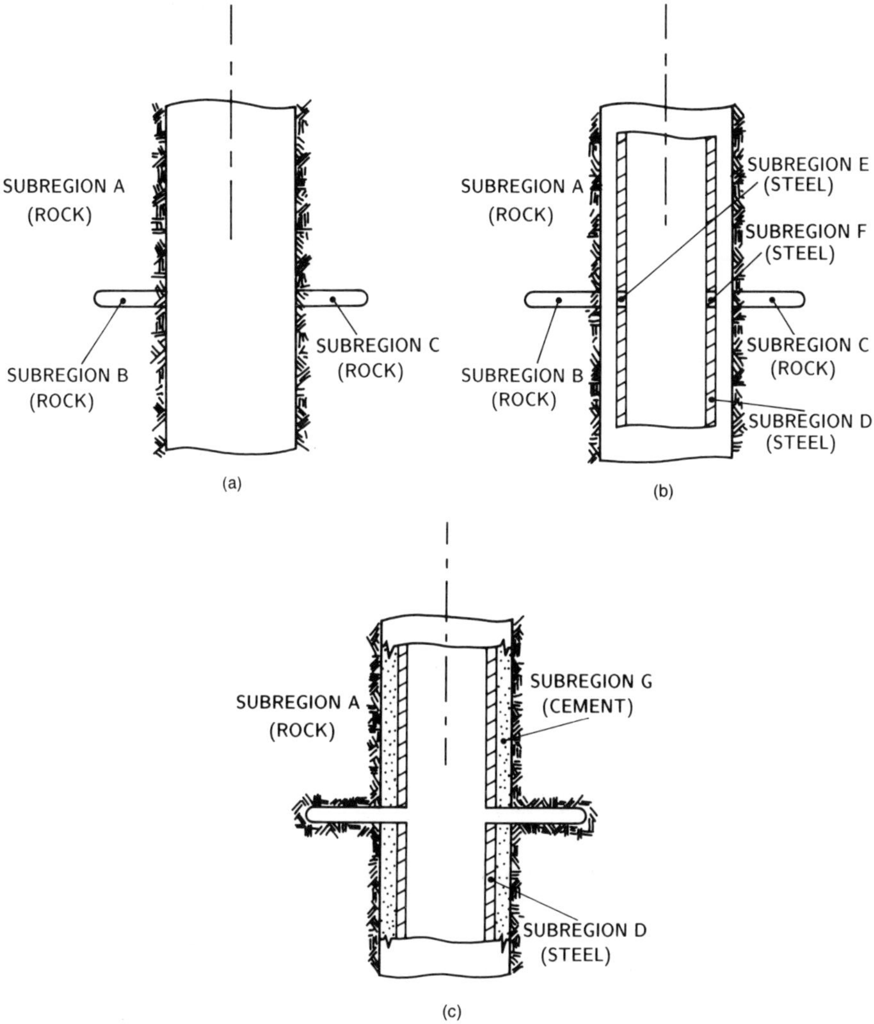

Figure 7.53 Stages of construction (not to scale). (a) Plain borehole. (b) Borehole with steel casing. (c) Perforated borehole

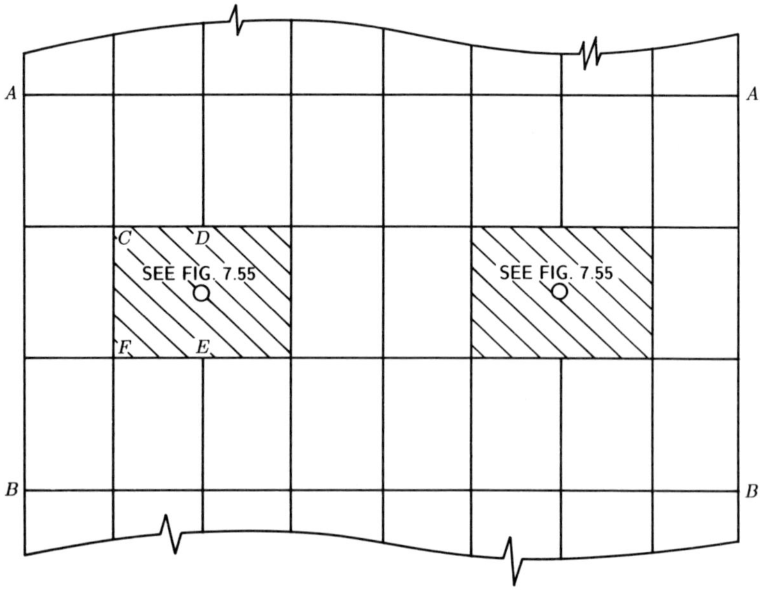

Figure 7.54 Boundary element mesh on rock–cement interface

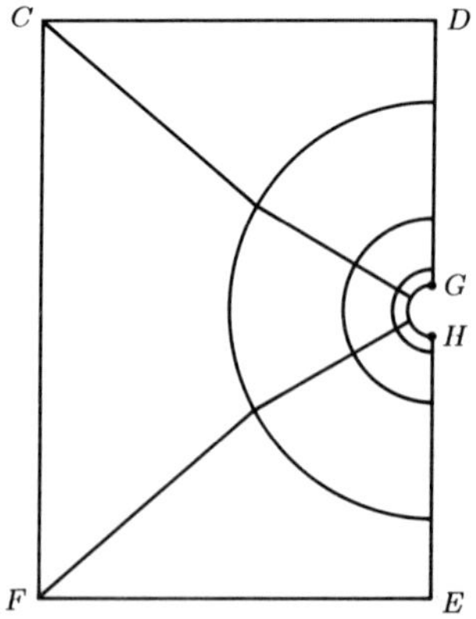

Figure 7.55 Detail of mesh on rock–cement interface

carried out for cases in which the principal directions of *in situ* stress do not coincide with axes of structural symmetry. Surfaces and interfaces were modelled by nine-node quadrilateral elements, six-node triangular elements, and three-node infinite elements with plane strain functional variation. The mesh of elements on the rock–cement interface is shown in Figures 7.54 and 7.55: in Figure 7.54, the elements above AA and below BB are infinite. The meshes on the cement–steel interface and on the inner surface of the casing are similar. These three concentric meshes are separated by distances of only 0.425 in and 0.75 in, compared with a maximum quadrilateral element side length of 5 in. If the same quadrature formula were used for all elements and first arguments of kernels, a greatly increased number of elements would be required. In Figure 7.56 is shown the mesh of elements on the surface of a perforation. The third of the analyses shown in Figure 7.53 is the largest in terms of number of degrees of freedom. In that analysis, there are three subregions arranged as shown in Figure 7.33a, and four systems of equations as described in Section 7.8. The numbers of elements on the surface of each subregion and approximate sizes of matrices of equation coefficients (see Figures 7.33b to 7.33e) are summarised in Table 7.7. The equations were solved iteratively according to equations (7.141) to (7.147), the average number of nodes per lump being about 50. The central processing time and input–output wait time for the analysis on a Cray 1-S were 1787 s and 4260 s, respectively. Of the central processing time, 867 s were spent on simultaneous equation solution.

At a mud pressure of 10 000 lbf/in^2, computed stresses are compressive throughout the rock. In the cement, there is computed to be a small zone of tensile stress close to the wall of a perforation as shown in Figure 7.57. The

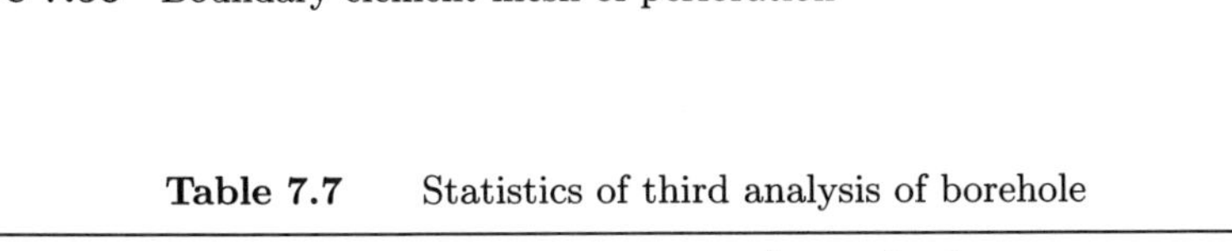

Figure 7.56 Boundary element mesh of perforation

Table 7.7 Statistics of third analysis of borehole

Subregion	Elements			Approximate size of matrix		Remarks
	Surface	Interface	Total	Rows	Columns	
A (rock)	144	84	228	2450	3400	
G (cement)	24	168	192	2100	3900	
D (steel)	108	84	192	2100	3000	
Second stage reduction				3650	3650	matrix not fully populated

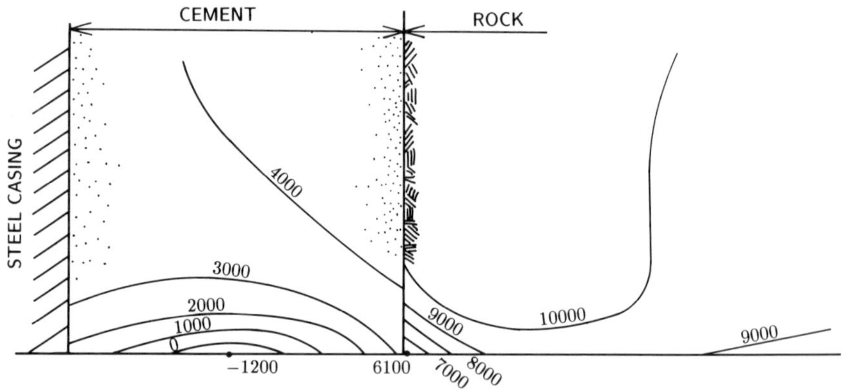

Figure 7.57 Computed normal stresses acting on a vertical plane passing through the axis of a perforation (in lbf/in^2)

computed results indicate that with increasing mud pressure cracks would develop first in the cement, then in the adjacent rock. Shorter perforations would be just as effective in promoting crack propagation as those modelled in the analysis presented here. Also, as might have been predicted without carrying out any computation, there is no significant interaction between diametrally opposed perforations. There may be scope for lowering the mud pressure required to initiate cracks by placing the cement in such a way that the initial stresses in the cement are reduced. To carry out a complete investigation of crack propagation, it would be necessary to model the cracks by displacement discontinuity elements and propose criteria for the prediction of crack propagation.

7.11.3 Analysis of mine excavation

In mining ore, material is sometimes removed by a method known as 'open stoping'. In this method large excavations (stopes) are made underground by mass blasting. These excavations are typically of dimensions $30 \times 30 \times 60\ m^3$. As mining proceeds to deeper levels one ends up with a fairly complex pattern of holes underground. To ensure that the large openings are stable sand, waste rock and mill tailings are used for backfilling. However the elastic properties of fill are an order of magnitude smaller than those of the rock mass.

For the design of the excavations it is sometimes of interest to have knowledge of the stress distribution in the rock mass and the areas of stress concentration. For this purpose the rock mass—although highly jointed and inhomogeneous—can be assumed to be one elastic material and therefore a boundary element analysis can be performed. Because of the low elastic modulus the influence of the fill material can be neglected.

In Figure 7.58 is shown the mesh used for the analysis, consisting of about 800 linear (four node) boundary elements and having about 2500 degrees of freedom. For the purpose of this analysis it was assumed that all excavations were created in one step. The stresses in the rock mass before any excavation takes place (pre-mining stress field) were supposed to vary linearly from the surface. The principal stresses 500 m underground were taken to be as in Table 7.8.

Table 7.8 Pre-mining principal stresses

Stress	Components of direction vector		
	v_x	v_y	v_z
$\sigma_1 = -12.0$ MPa	0.	0.	1.
$\sigma_2 = -18.6$ MPa	0.743	0.669	0.
$\sigma_3 = -31.4$ MPa	0.669	−0.743	0.

For the material properties, $E = 80\ 000$ MPa and $\nu = 0$ were assumed.

The analysis was performed on a VAX station 3000 using the coupled boundary element / finite element program BEFE. Stresses were computed at about 500 interior nodes.

In Figure 7.59 (see colour section) is shown the variation of the minimum principal stress on various 'dummy planes' inside the rock mass. In program BEFE these 'dummy planes' are defined by an assembly of four noded 'dummy elements' each one containing four interior points at which stresses are computed. The stress distribution in the pillar is displayed using the fully interactive postprocessor FEMVUE (see Chapter 11). It can be seen that due to the skew pre-mining stress field there is a stress concentration on one edge of the pillar. In Figure 7.59 red areas indicate unacceptably high stress which could lead to overbreak of rock into the excavation and other problems.

7.12 SUMMARY AND CONCLUSIONS

In the implementation of a boundary element method, it is necessary to propose

1. functional variations over each element in terms of unknown nodal parameters;
2. a system of simultaneous equations in terms of these parameters, approximating to the boundary integral equation;
3. a means of computing the coefficients of that system;
4. a method of simultaneous equation solution.

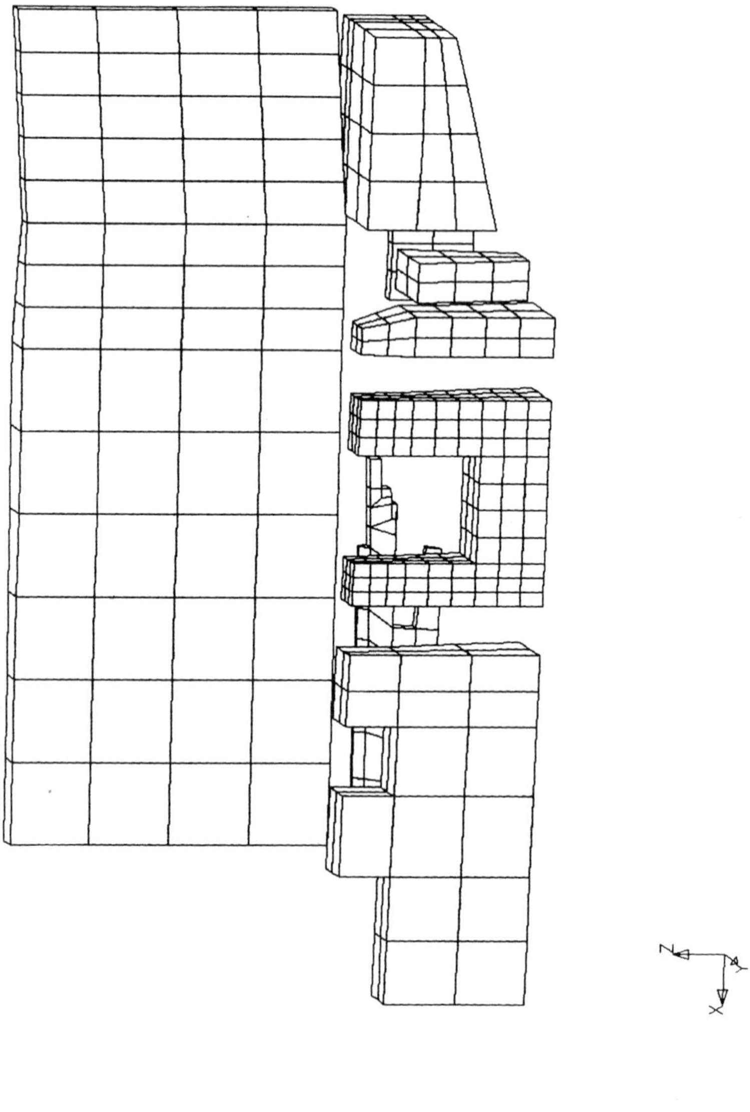

Figure 7.58 Boundary element mesh of mine excavations

The discussion of this chapter is not intended to be a comprehensive account of these issues; rather, it is a description of one way of dealing with them that has been found to work reasonably well. Detailed consideration is given only to the direct method and nodal collocation. The functional variations essentially are borrowed from the finite element method, in which linear and quadratic isoparametric elements have proved very successful. Strategies for the definition of the discrete system have not been given the attention they deserve by developers of boundary element software for use by engineers. It has generally been taken for granted that the system should be obtained by collocation at the nodes with which are associated unknowns, whereas it is possible that a method of moments such as the Galerkin method described in Section 7.2 offers greater computational efficiency. Certainly most mathematicians take the view that a method of moments is preferable to nodal collocation. In defence of the approach adopted by most developers of engineering software to date, it must be pointed out that whereas boundary element methods are more difficult and time-consuming to develop than are finite element methods, the number of competent researchers into boundary elements remains relatively small.

Like functional variation, the means of computing coefficients of the discrete system is borrowed from the finite element method but extensive adaptations have been necessary. The key to development of the first reasonably efficient boundary element computer programs was indeed the accurate evaluation of singular integrals by Gaussian quadrature formulae. In the development of methods of solving the system of simultaneous equations, it has not been possible to draw upon experience with other methods of analysis. Fortunately, the equations obtained show exceptional numerical stability.

The examples are chosen to highlight strengths and weaknesses of the method. Difficulties are evident in the solution of problems such as that of the cantilever, but there is every prospect that the source of inaccuracy will eventually be located and eliminated. The method is seen to have a particular advantage in the solution of problems for infinite domains. The accuracy of boundary element analysis appears to be more severely impaired by the inability of smooth shape functions to model elastic singularities at notch and crack roots than is that of finite element analysis. The solution of this problem is to introduce singular shape functions[3]. Such development of the method of analysis must however be accompanied by education of the user community, within which at present credence is often given to computed results which are actually meaningless.

The techniques described in this chapter are not the most advanced currently available. There now exist Hermitian cubic and singular elements for the analysis of plane strain[3], and soon Hermitian bicubic and singular elements for three dimensional analysis will be implemented. Hermitian cubic functional variation gives interelement continuity not only of displacement but also of strain and stress. Its potential advantage, however, is that it offers greater accuracy for a given number of unknown nodal parameters. If in a boundary element mesh in two dimensions for example quadratic elements are replaced by the same number of Hermitian cubic elements, then the quadratic interpolate over each element is replaced by a cubic interpolate but there are still just four times as many unknowns as elements for plane strain.

One of the most important areas of application of finite and boundary element methods is fracture mechanics. To model cracks which are curved or do not lie in a plane of symmetry of structure and loading, the direct and indirect methods cannot be used in their basic form unless the domain is divided into subregions in such a way that the two surfaces of a crack are not surfaces of the same subregion[11]. This requires the introduction of artificial interfaces extending through intact material ahead of the crack root, which may be inconvenient and in any case increases the number of unknowns. In Chapter 6 the displacement discontinuity method is presented as a means to overcome this problem, and that method is described as an indirect method in which we consider limiting values of traction as the crack is approached from within the material. It is also possible to develop general crack elements in the direct method[13], and in fact these techniques are the same. The principles of the direct method, however, provide a more straightforward means of implementation of higher order elements than do those of the indirect method as enunciated in the discussion of Chapter 6.

7.13 EXERCISES

7.1 The boundary of a circular plate of radius 1 metre is modelled by eight quadratic isoparametric elements:

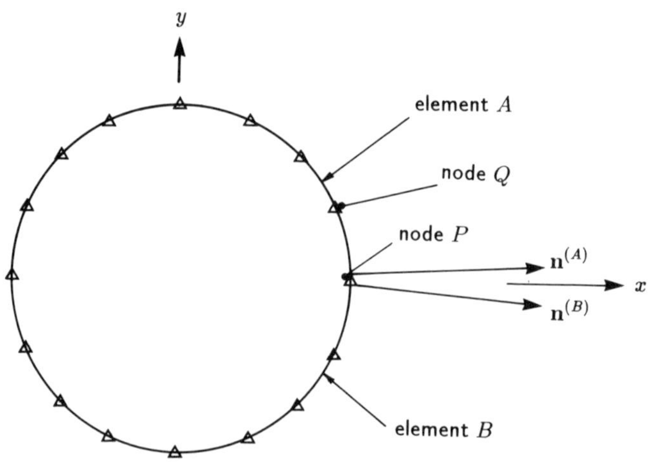

The nodes of these elements are spaced at 22.5° intervals, 1 m from the centre of the plate.

(a) Compute the cartesian components of the unit outward normals $\mathbf{n}^{(A)}$ and $\mathbf{n}^{(B)}$ to elements A and B at their common node P, and the angle between these normals.

(b) Compute the Jacobian at nodes P and Q of element A.

7.2 Compute the integrals of the kernel–shape function products $T(P,Q)M_1(\xi)$ and $T(P,Q)M_2(\xi)$ where $T(P,Q) = \frac{\cos\theta}{2\pi r}$, $M_1(\xi) = \frac{1}{2}(1-\xi)$ and $M_2(\xi) = \frac{1}{2}(1+\xi)$ over the straight line element shown below,

(a) by Simpson's rule
(b) by the three-point Gaussian quadrature formula
(c) by the four-point Gaussian quadrature formula
(d) by the five-point Gaussian quadrature formula

Gauss weights and offsets are as follows.

Formula	Weight	Offset
Three point	0.555 555	0.774 597
	0.888 889	0.000 000
	0.555 555	−0.774 597
Four point	0.347 855	0.861 136
	0.652 145	0.339 981
	0.652 145	−0.339 981
	0.347 855	−0.861 136
Five point	0.236 927	0.906 180
	0.478 629	0.538 469
	0.568 889	0.000 000
	0.478 629	−0.538 469
	0.236 927	−0.906 180

Analytically computed values: 0.112 171 and 0.064 037.

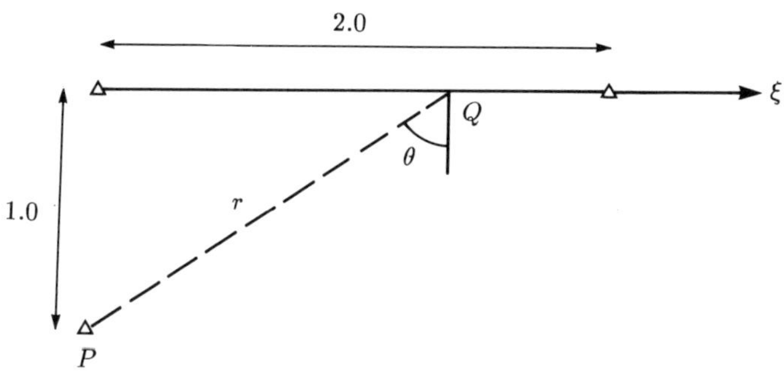

7.3 Stresses and displacements are to be computed at an internal point P of an infinite two-dimensional domain, the boundary of which is modelled by a mesh of 20 elements:

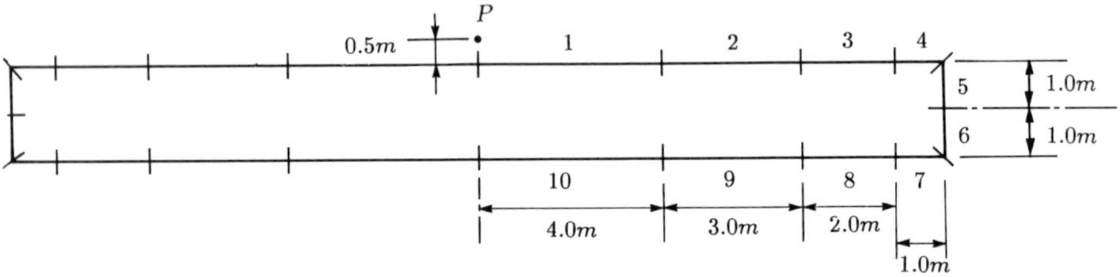

The integration scheme is to be as described in section 7.9, with three, four and five-point Gaussian quadrature formulae and the constant ϵ of equation (7.68) taken equal to 5.0×10^{-5}.

(a) Compute the maximum allowable values of L/R as defined by equation (7.171) for the available Gaussian quadrature formulae.
(b) Determine the numbers of intervals into which elements 1 to 10 should be divided, and the number of Gauss points for each interval.

7.4 A plate of thickness 6 mm with a hole of radius 10 mm is subjected to an axial load of 15 kN, uniformly distributed over its ends:

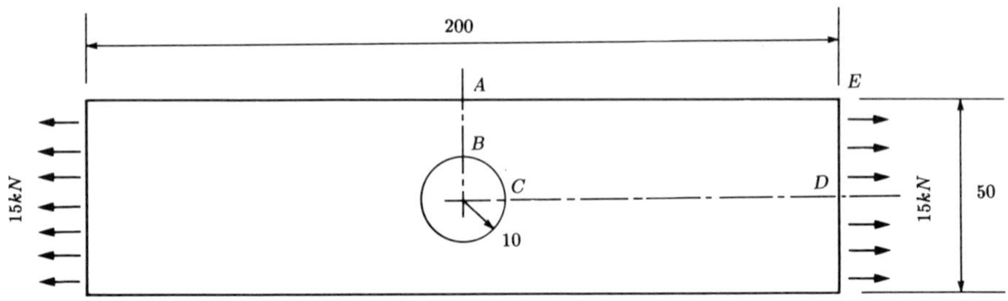

Using program BOUND, compute the horizontal direct stress at the points A and B on the vertical centreline of the plate. Compare the results with the values that would be obtained by taking stress to be uniform on the vertical centreline, and by taking the plate to be of infinite extent and referring to the exact solution for stress around a circular hole.
Suggested modelling strategy: use the SYMX and SYMY facilities of program BOUND. Try the following meshes.

(a) Two elements on BC, one on DE and four on EA preferably with mesh gradation on EA.
(b) Four elements on BC, two on DE and eight on EA preferably with mesh gradation on EA.

The plate is in a state of plane stress and in BOUND a state of plane strain is analysed, but for this problem the stress distributions for plane stress and plane strain are the same except for σ_z. Arbitrarily take E=10 000 MPa, $\nu = 0$.

7.5 Ore is to be extracted from a stope of height 50 m and width 10 m, the vertical centreline of which is located 20 m from the centreline of an existing drive:

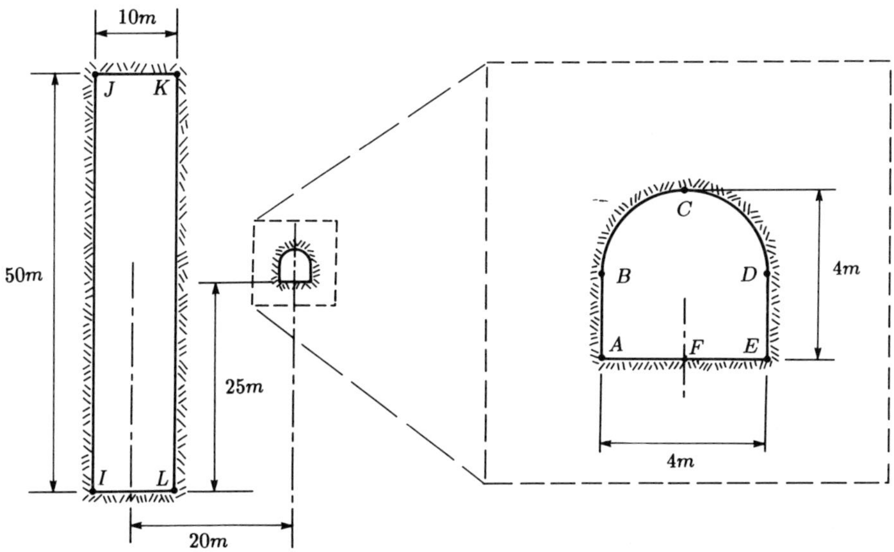

The pre-mining state of stress is 20 MPa hydrostatic, and both the drive and the stope continue a considerable distance perpendicular to the cross-section above; therefore, a state of plane strain may be supposed to exist. Young's modulus and Poisson's ratio are 40 000 MPa and 0.25, respectively. Compute the principal stresses at each of the four points B, C, D and F before and after extraction of the ore.
Suggested modelling strategy: try the following meshes.

(a) Two elements on AB, four on BCD, two on DE, four on EFA, four on IJ, two on JK, four on KL and two on LI.
(b) Four elements on AB, eight on BCD, four on DE, eight on EFA, eight on IJ, four on JK, eight on KL and four on LI.

Mesh on IJ and KL should preferably be graded so that the first and last elements are approximately 1.5 times longer than the elements on JK and LI. If stresses in the neighbourhood of the stope were of interest, then it would be desirable to increase the numbers of elements on IJ, JK, KL and LI.

7.14 REFERENCES

1. Lachat, J.C. and Watson, J.O. (1976) Effective numerical treatment of boundary integral equations. *Int. J. Num. Meth. Eng.*, **10**, 991–1005.
2. Watson, J.O. (1979) Advanced implementation of the boundary element method for two and three dimensional elastostatics. *Developments in Boundary Element Methods–1* (eds P.K. Banerjee and R. Butterfield), Applied Science Publishers, London, pp. 31–63.
3. Watson, J.O. (1982) Hermitian cubic boundary elements for plane problems of fracture mechanics. *Res Mechanica*, **4**, 23–42.
4. Lean, M.H., Friedman, M. and Wexler, A. (1979) Application of the boundary element method in electrical engineering problems. *Developments in Boundary Element Methods–1* (eds P.K. Banerjee and R.Butterfield), Applied Science Publishers, London, pp. 207–250.
5. Stroud, A.H. and Secrest, D. (1966) *Gaussian Quadrature Formulas.* Prentice-Hall, Englewood Cliffs, New Jersey.
6. Beer, G. and Watson, J.O. (1989) Infinite boundary elements. *Int. J. Num. Meth. Eng.*, **28**, 1233–1247.
7. Cruse, T.A. (1979) Two and three dimensional problems of fracture mechanics. *Developments in Boundary Element Methods–1* (eds P.K. Banerjee and R. Butterfield), Applied Science Publishers, London, pp. 97–120.
8. Lachat, J.C. and Watson, J.O. (1975) A second generation boundary integral equation program for three dimensional elastic analysis. *Boundary Integral Equation Method: Computational Applications in Applied Mechanics* (eds T.A. Cruse and F.J. Rizzo), ASME, New York, pp. 85–100.
9. Williams, M.L. (1952) Stress singularities resulting from various boundary conditions in angular corners of plates in extension. *J. Appl. Mech.*, **74**, 526–528.
10. Tada, H., Paris, P.C. and Irwin, G.R. (1973) *The Stress Analysis of Cracks Handbook.* Del Research Corporation, Hellertown, Pennsylvania.
11. Blandford, G.E., Ingraffea, A.R. and Liggett, J.A. (1981) Two dimensional stress intensity factor computations using the boundary element method. *Int. J. Num. Meth. Eng.*, **17**, 387–404.
12. (1980) *The Calculation of Crankshaft Stress.* Lloyds Register of Shipping, London.
13. Watson, J.O. (1986) Hermitian cubic and singular elements for plane strain. *Developments in Boundary Element Methods–4* (eds P.K. Banerjee and J.O. Watson), Elsevier Applied Science Publishers, London, pp. 1–28.

8

Coupled Boundary Element/Finite Element Analysis

Marriage à la mode

O. C. Zienkiewicz

8.1 INTRODUCTION

During the 'journey' through the last chapters the reader should have become fairly familiar with both the finite and the boundary element methods. It is obvious that both methods have their own range of application where they are most efficient. The finite element method is most suitable for problems:

- with a high ratio of boundary surface to volume;
- where boundary stresses are not of primary importance;
- where the material is nonhomogeneous, behaves nonlinearly and/or contains joints and cracks.

The boundary element method, on the other hand, is most suitable for applications involving:

- a low ratio of boundary surface to volume;
- a high accuracy of boundary stresses;
- homogeneous and linear elastic materials.

Typical applications with a high surface to volume ratio are plate and shell structures in civil engineering. Extremely low surface to volume ratios occur in geomechanics problems where the domain is in most cases assumed to be infinite. Cases where boundary stresses are of primary importance occur mainly in fracture mechanics and many applications in mechanical engineering.

Materials in geomechanics, especially rock mechanics, usually are non-homogeneous, anisotropic, non-linear and contain joints and cracks. On the other hand, materials used in mechanical engineering and civil engineering can usually be assumed to be homogeneous and behave linear elastic.

There are problems, therefore, which are not ideally suited for either the boundary or the finite element method. For example, problems in geomechanics not only have a low surface to volume ratio but also involve nonhomogeneous and jointed material. For such applications a combination of finite element and boundary element methods seem to be the most efficient way of analysis.

This chapter then, is about the coupling of both methods: a 'marriage à la mode'.† In a coupled analysis the user can have the best of both worlds, i.e. use the advantages of both methods. We shall see that this results in an efficient analysis of many engineering problems.

8.2 COUPLING THEORY

There are basically two approaches to coupling boundary element and finite element regions. In the first approach, the BE region is treated as a super finite element and its stiffness matrix is computed and assembled into the global stiffness matrix using the assembly procedure of Chapter 4.

In the second approach, the FE region is treated as an equivalent BE region and its coefficient matrix is determined and assembled into the global coefficient matrix using the procedures described in Chapter 7.

The choice of the coupling method depends mainly on the software available for the implementation of coupling, i.e. if boundary element capabilities are to be added to a finite element program or finite element capabilities to a boundary element program.

In the following we will discuss the theoretical basis and the implementation of each approach in detail. The coupling theory is discussed on problems in elasticity. However, as demonstrated throughout this book, potential problems can be considered in the same way by substituting temperature/potential for displacement and flux for stress.

† This term has been used by O.C. Zienkiewicz in his now classical paper presented at the International Conference on Innovative Numerical Analyses[1]. It means a marriage of convenience, not love. Judging from comments made by BE and FE researchers there is still not much love lost between the two.

8.2.1 Coupling to finite elements

The FEM leads to a system of simultaneous equations which relate the displacements at all the nodes to *nodal forces*. In the BEM on the other hand a relationship between nodal displacements and *nodal tractions* is established.

Consider the two regions in Figure 8.1 which are connected at a common interface. Region A is a finite element region and region B is a boundary element region. For the FE region A the following system of equations is obtained:

$$[\mathbf{K}]^{\mathrm{A}}\{\mathbf{u}\}^{\mathrm{A}} = \{\mathbf{F}\}^{\mathrm{A}} \tag{8.1}$$

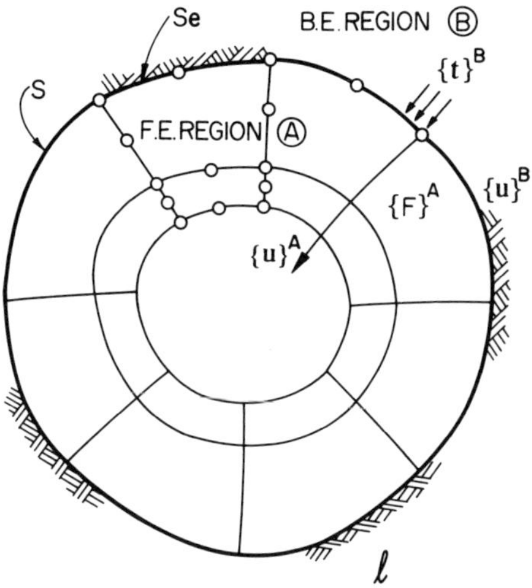

Figure 8.1 An excavation is surrounded by finite elements which are connected to a boundary element region: fully coupled analysis

where $\{\mathbf{u}\}^{\mathrm{A}}$ and $\{\mathbf{F}\}^{\mathrm{A}}$ are vectors of displacement and nodal forces at the nodes of region A and $[\mathbf{K}]^{\mathrm{A}}$ is the stiffness matrix of the region. For the BE region B the system of equations is written for the direct method

$$[\Delta\mathbf{T}]^{\mathrm{B}}\{\mathbf{u}\}^{\mathrm{B}} = [\Delta\mathbf{U}]^{\mathrm{B}}\{\mathbf{t}\}^{\mathrm{B}} \tag{8.2}$$

where $\{\mathbf{u}\}^{\mathrm{B}}$ and $\{\mathbf{t}\}^{\mathrm{B}}$ are displacements and tractions at the interface of region B and $[\Delta\mathbf{T}]^{\mathrm{B}}$, $[\Delta\mathbf{U}]^{\mathrm{B}}$ are coefficient matrices obtained using nodal collocation. Our approach is to rewrite equation (8.2) into the same form as equation (8.1) i.e. compute a stiffness matrix for the BE region. This requires the conversion of tractions into nodal point forces.

A relationship between nodal boundary traction values and nodal point forces can be established using the principle of virtual work. The equation for the force $\mathbf{F}_l^e$ at node l of an element e which is subjected to a distributed traction $\mathbf{t}$ is given by:

$$\mathbf{F}_l^e = \int_{S_e} N_l \, \mathbf{t} \, \mathrm{d}S \tag{8.3}$$

where the integration is carried out over the length or area of a boundary element (S_e). Using the usual interpolation of tractions from J nodal values gives

$$\mathbf{t} = \sum_{j=1}^{J} N_j \mathbf{t}_j^e \tag{8.4}$$

Substituting into equation (8.3) and considering that $\mathbf{t}_j^e$, being traction values at the nodes of element e, can be taken outside the integral we obtain

$$\mathbf{F}_l^e = \sum_{j=1}^{J} \left(\int N_l N_j \, \mathrm{d}S \right) \mathbf{t}_j^e \tag{8.5}$$

The relationship between nodal tractions $\mathbf{t}^e$ and nodal forces $\mathbf{F}^e$ on all nodes of element e is conveniently written in matrix form

$$\mathbf{F}^e = \mathbf{M}^e \mathbf{t}^e \tag{8.6}$$

where

$$\mathbf{M}^e = \begin{bmatrix} N_{11}^e \mathbf{I} & . & . & . & . & . & N_{1j}^e \mathbf{I} \\ . & & & & & & . \\ . & & & & & & . \\ . & & & & & & . \\ N_{j1}^e \mathbf{I} & . & . & . & . & . & N_{jj}^e \mathbf{I} \end{bmatrix} \tag{8.7}$$

In the above $\mathbf{I}$ is a 2×2 or 3×3 unit matrix depending if two-dimensional or three-dimensional problems are considered and

$$N_{ij}^e = \int_{S_e} N_i N_j \, \mathrm{d}S \tag{8.8}$$

To explain how to obtain the stiffness matrix of the boundary element region, consider that all degrees of freedom of the boundary element region are on the interface (fully coupled analysis) and that no loading is applied at the interface.

Equation (8.2) can be solved for the tractions at the interface nodes due to displacements $\{\mathbf{u}\}^{\mathrm{B}}$:

$$\{\mathbf{t}\}^{\mathrm{B}} = \mathbf{Q}\{\mathbf{u}\}^{\mathrm{B}} \tag{8.9}$$

where $\mathbf{Q}$ is a matrix of solutions due to unit displacements i.e. in matrix notation omitting the superscript B

$$\mathbf{Q} = [\Delta\mathbf{U}]^{-1}[\Delta\mathbf{T}] \tag{8.10}$$

although as we will see later the inverse of $[\Delta\mathbf{U}]$ need not be computed.

Next supposing the interface is smooth, the nodal traction values are converted into global nodal point forces using equation (8.6).

$$\{\mathbf{F}\}^{\mathrm{B}} = \mathbf{M}\{\mathbf{t}\}^{\mathrm{B}} = \mathbf{MQ}\{\mathbf{u}\}^{\mathrm{B}} \tag{8.11}$$

where $\mathbf{M}$ is a diagonally dominant matrix assembled from element contributions $\mathbf{M}^e$. The stiffness matrix of the boundary element region is thus given by

$$[\mathbf{K}]^{\mathrm{B}} = \mathbf{MQ} \tag{8.12}$$

This stiffness matrix will, in general, not be symmetric as was the case with finite elements because although $\mathbf{M}$ is symmetric since $N_{ij} = N_{ji}$, $\mathbf{Q}$ is in most cases not. One of the reasons lies in the way the boundary element equations are set up, i.e. for example by using a finite number of collocation points. It can be observed that the finer the boundary element mesh is, the more symmetric the resulting matrix becomes.

Let us look at an alternative way of computing the stiffness matrix using the principle of minimum potential energy. Consider the coupled mesh in Figure 8.1 and replace the BE region by the tractions $\hat{\mathbf{t}}$ at the interface acting on the FE region.

The potential energy of the FE region is then given by

$$\pi = \frac{1}{2}\int_V \boldsymbol{\sigma}^{\mathrm{T}}\boldsymbol{\epsilon}\, \mathrm{d}V - \frac{1}{2}\int_S \mathbf{u}^{\mathrm{T}}\hat{\mathbf{t}}\, \mathrm{d}S \tag{8.13}$$

where the last integral is taken over the interface only.

Using the standard discretisation procedure for the FE region the first integral and the energy of nodal point forces is written as:

$$\frac{1}{2}\left(\{\mathbf{u}\}^{\mathrm{A}}\right)^{\mathrm{T}}[\mathbf{K}]^{\mathrm{A}}\{\mathbf{u}\}^{\mathrm{A}} - (\{\mathbf{u}\}^{\mathrm{A}})^{\mathrm{T}}\{\mathbf{F}\}^{\mathrm{A}} \tag{8.14}$$

The second (surface) integral can be discretised by

$$\frac{1}{2}\sum_{e=1}^{E}\int_{S_e}\left(\mathbf{N}^e\mathbf{u}^e\right)^{\mathrm{T}}\mathbf{N}^e\hat{\mathbf{t}}^e\, \mathrm{d}\mathbf{S} \tag{8.15}$$

where E is the number of elements on the boundary and

$$\mathbf{N}^e = [\mathbf{I}N_1 \ldots \mathbf{I}N_j] \tag{8.16}$$

($\mathbf{I}$ is a unit matrix as explained previously.) Next, we substitute for vectors $\mathbf{u}^e$ and $\mathbf{t}^e$ in equation (8.15), which contain element displacements and tractions, the pseudo-vectors $\{\mathbf{u}\}^{\mathrm{B}}$ and $\{\hat{\mathbf{t}}\}$ which contain displacements and tractions at all the nodes of the interface. We obtain for (8.15)

$$\frac{1}{2}\left(\{\mathbf{u}\}^{\mathrm{B}}\right)^{\mathrm{T}}\mathbf{M}\{\hat{\mathbf{t}}\} \tag{8.17}$$

where $\mathbf{M}$ is the same matrix as in equation (8.11).

The global traction vector $\{\hat{\mathbf{t}}\}$ is given by equation (8.9) as

$$\{\hat{\mathbf{t}}\} = -\{\mathbf{t}\}^{\mathrm{B}} = -\mathbf{Q}\{\mathbf{u}\}^{\mathrm{B}} \tag{8.18}$$

The discretised form of equation (8.13) is therefore:

$$\pi = \frac{1}{2}\left(\{\mathbf{u}\}^{\mathrm{A}}\right)^{\mathrm{T}}[\mathbf{K}]^{\mathrm{A}}\{\mathbf{u}\}^{\mathrm{A}} - (\{\mathbf{u}\}^{\mathrm{A}})^{\mathrm{T}}\{\mathbf{F}\}^{\mathrm{A}} + \frac{1}{2}\left(\{\mathbf{u}\}^{\mathrm{B}}\right)^{\mathrm{T}}\mathbf{MQ}\{\mathbf{u}\}^{\mathrm{B}} \tag{8.19}$$

Minimizing the potential energy with respect to all displacements $\{\mathbf{u}\}$ we obtain the equilibrium equation as:

$$\frac{\partial \pi}{\partial \{\mathbf{u}\}} = [\mathbf{K}]^{\mathrm{A}}\{\mathbf{u}\}^{\mathrm{A}} - \{\mathbf{F}\}^{\mathrm{A}} + [\mathbf{K}]^{\mathrm{B}}\{\mathbf{u}\}^{\mathrm{B}} = \mathbf{0} \tag{8.20}$$

where

$$[\mathbf{K}]^{\mathrm{B}} = \frac{1}{2}\left[(\mathbf{MQ})^{\mathrm{T}} + \mathbf{MQ}\right] \tag{8.21}$$

is the stiffness of the interface. Although $\mathbf{Q}$ may not be symmetric it follows from equation (8.21) that $[\mathbf{K}]^{\mathrm{B}}$ is.

Next we divide the displacement vector $\{\mathbf{u}\}^{\mathrm{A}}$ into displacements at nodes inside the FE region $\{\mathbf{u}\}^{\mathrm{F}}$ and on the interface $\{\mathbf{u}\}^{\mathrm{I}}$. Then equation (8.20) can be written as

$$[\mathbf{K}]^{\mathrm{F}}\{\mathbf{u}\}^{\mathrm{F}} + [\mathbf{K}]^{\mathrm{I}}\{\mathbf{u}\}^{\mathrm{I}} - \{\mathbf{F}\}^{\mathrm{A}} + [\mathbf{K}]^{\mathrm{B}}\{\mathbf{u}\}^{\mathrm{B}} = \mathbf{0} \tag{8.22}$$

where $[\mathbf{K}]^{\mathrm{F}}$ and $[\mathbf{K}]^{\mathrm{I}}$ are submatrices of $[\mathbf{K}]^{\mathrm{A}}$ containing stiffness coefficients for the nodes inside the FE region and on the interface. Ensuring compatibility at the interface i.e.

$$\{\mathbf{u}\}^{\mathrm{I}} = \{\mathbf{u}\}^{\mathrm{B}} \tag{8.23}$$

we obtain

$$[[\mathbf{K}]^{\mathrm{F}}, [\mathbf{K}]^{\mathrm{I}} + [\mathbf{K}]^{\mathrm{B}}] \left\{ \begin{array}{c} \{\mathbf{u}\}^{\mathrm{F}} \\ \{\mathbf{u}\}^{\mathrm{I}} \end{array} \right\} - \{\mathbf{F}\}^{\mathrm{A}} = \mathbf{0} \tag{8.24}$$

The summation $[\mathbf{K}]^{\mathrm{I}} + [\mathbf{K}]^{\mathrm{B}}$ implies that the boundary element region can be treated as a super 'finite element' and assembled with the FE mesh in the usual way.

This derivation of the stiffness matrix $[\mathbf{K}]^{\mathrm{B}}$ is not quite correct because in equation (8.15) we have assumed a distribution of both the tractions and displacements on the interface. However, if the displacements on the boundary of the BE region are prescribed then the traction distribution follows from the boundary element solution. By assuming such a distribution of tractions we are therefore applying additional (artificial) constraints which are a result of the discretisation process. Again, as the number of elements on the boundary increases the error will become smaller.

The coupling process requires the inversion of matrix $[\Delta\mathbf{U}]$ (equation (8.10)). However, it is not necessary to do this inversion and the matrix multiplication.

As explained previously, the matrix $\mathbf{Q}$ actually contains solutions due to unit displacement. In an efficient implementation we proceed as follows:

1. Do a Gauss reduction of the matrix $[\Delta\mathbf{U}]$ as explained in section 4.5 treating each column of matrix $[\Delta\mathbf{T}]$ in turn as a right hand side.
2. Obtain a matrix of solutions $\mathbf{Q}$ by back substitution.

For an analysis where only some of the nodes are connected as shown in Figure 8.2 special care has to be taken since two different traction values may exist at one node. For example, node n has a traction of t^l on the left (which is known) and a traction of t^r (which is unknown).

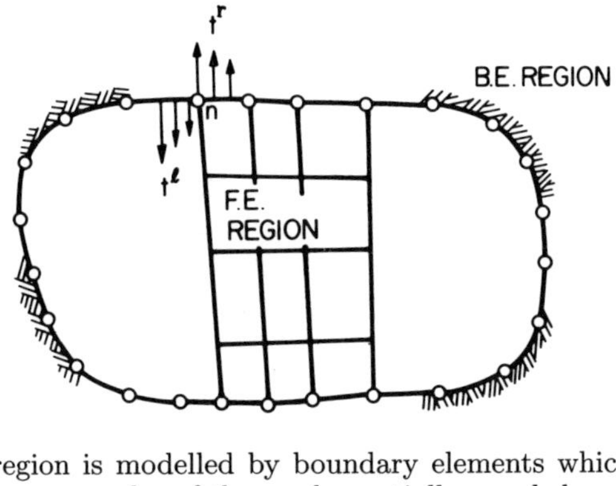

Figure 8.2 A region is modelled by boundary elements which are connected to finite elements on some nodes of the mesh: partially coupled analysis

This has to be considered when assembling the boundary element equations. To explain this let us define the global tractions at the interface nodes by $\{\mathbf{t}\}^{\mathrm{B}}$, the applied tractions at each boundary element not at the interface by $\tilde{\mathbf{t}}^e$, and the corresponding global displacement at the interface nodes by $\{\mathbf{u}\}^{\mathrm{B}}$ and non-interface nodes by $\{\mathbf{u}\}^{\mathrm{C}}$; then the discretised boundary element equation (8.2) can be written as:

$$\left[[\Delta\mathbf{T}]^{\mathrm{B}}, [\Delta\mathbf{T}]^{\mathrm{C}}\right]\begin{Bmatrix}\{\mathbf{u}\}^{\mathrm{B}}\\ \{\mathbf{u}\}^{\mathrm{C}}\end{Bmatrix} = \sum_{e=1}^{E}\Delta\mathbf{U}^e\tilde{\mathbf{t}}^e + [\Delta\mathbf{U}]^{\mathrm{B}}\{\mathbf{t}\}^{\mathrm{B}} \tag{8.25}$$

where E is the number of boundary elements not on an interface. Note that for the applied tractions at the non-interface nodes a summation instead of a matrix multiplication has been used because of the local effect of the traction at the node n. Rearrangement of equation (8.25) gives

$$\left[-[\Delta\mathbf{U}]^{\mathrm{B}}, [\Delta\mathbf{T}]^{\mathrm{C}}\right]\begin{Bmatrix}\{\mathbf{t}\}^{\mathrm{B}}\\ \{\mathbf{u}\}^{\mathrm{C}}\end{Bmatrix} = \mathbf{f} - [\Delta\mathbf{T}]^{\mathrm{B}}\{\mathbf{u}\}^{\mathrm{B}} \tag{8.26}$$

where $\mathbf{f} = \sum_{e=1}^{E}\Delta\mathbf{U}^e\tilde{\mathbf{t}}^e$.

We proceed as before with the solution. This time, however, we have one additional right-hand side vector, namely $\mathbf{f}$, and therefore two sets of solutions, $\{\hat{\mathbf{t}}\}^{\mathrm{B}}$ and $\hat{\mathbf{Q}}$ are obtained on the interface, i.e. the tractions at the interface are given by

$$\{\mathbf{t}\}^{\mathrm{B}} = \{\hat{\mathbf{t}}\}^{\mathrm{B}} + \hat{\mathbf{Q}}\{\mathbf{u}\}^{\mathrm{B}} \tag{8.27}$$

where $\{\hat{\mathbf{t}}\}^{\mathrm{B}}$ are nodal tractions at the interface for zero displacements and $\hat{\mathbf{Q}}$ are solutions for the tractions due to unit displacements of the interface nodes (Figure 8.3).

In addition two sets of solutions are obtained for the displacements at the non interface nodes

$$\{\mathbf{u}\}^{\mathrm{C}} = \{\hat{\mathbf{u}}\}^{\mathrm{C}} + \mathbf{G}\{\mathbf{u}\}^{\mathrm{B}} \tag{8.28}$$

where $\{\hat{\mathbf{u}}\}^{\mathrm{C}}$ are displacements with fixed interface nodes and $\mathbf{G}$ are solutions due to unit displacements of the interface.

The derivation of the stiffness matrix of the interface nodes proceeds as before with the exception that the surface integral in equation (8.15) extends only over the interface (i.e. where connection to finite elements occurs) and expression (8.17) is replaced by

$$\frac{1}{2}\left(\{\mathbf{u}\}^{\mathrm{B}}\right)^{\mathrm{T}}\mathbf{M}\left(\{\hat{\mathbf{t}}\}^{\mathrm{B}} + \hat{\mathbf{Q}}\{\mathbf{u}\}^{\mathrm{B}}\right) \tag{8.29}$$

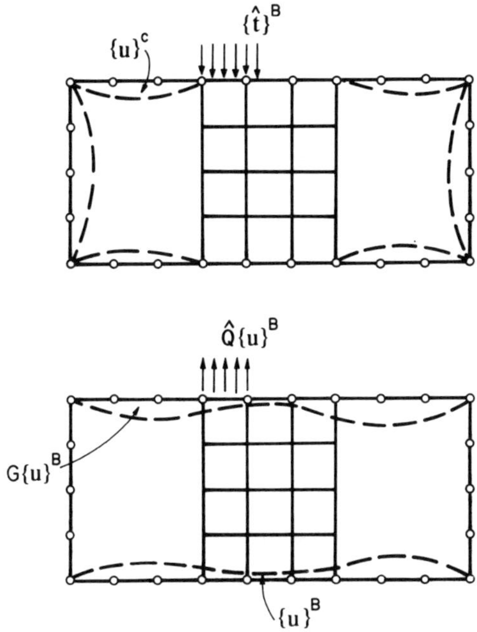

Figure 8.3 Partially coupled analysis: the analysis proceeds in two steps. (a) The interface nodes are fixed and load is applied at the non-interface nodes. (b) The load is removed and the nodes on the interface are allowed to displace

Equation (8.19) becomes now

$$\pi = \frac{1}{2}\left(\{\mathbf{u}\}^{\mathrm{A}}\right)^{\mathrm{T}}[\mathbf{K}]^{\mathrm{A}}\{\mathbf{u}\}^{\mathrm{A}} - (\{\mathbf{u}\}^{\mathrm{A}})^{\mathrm{T}}\{\mathbf{F}\}^{\mathrm{A}} + \frac{1}{2}\left(\{\mathbf{u}\}^{\mathrm{B}}\right)^{\mathrm{T}}\mathbf{M}\left(\{\hat{\mathbf{t}}\}^{\mathrm{B}} + \hat{\mathbf{Q}}\{\mathbf{u}\}^{\mathrm{B}}\right) \tag{8.30}$$

The minimum potential energy is then given by

$$\frac{\partial\pi}{\partial\{\mathbf{u}\}} = [\mathbf{K}]^{\mathrm{A}}\{\mathbf{u}\}^{\mathrm{A}} - \{\mathbf{F}\}^{\mathrm{A}} + [\mathbf{K}]^{\mathrm{B}}\{\mathbf{u}\}^{\mathrm{B}} + \{\mathbf{F}\}^{\mathrm{B}} = \mathbf{0} \tag{8.31}$$

where

$$\{\mathbf{F}\}^{\mathrm{B}} = \mathbf{M}\{\hat{\mathbf{t}}\}^{\mathrm{B}}$$
$$[\mathbf{K}]^{\mathrm{B}} = \frac{1}{2}\left[(\mathbf{M}\hat{\mathbf{Q}})^{\mathrm{T}} + \mathbf{M}\hat{\mathbf{Q}}\right]$$

By separating the displacements at nodes inside the FE region and at the interface we obtain for the partially coupled analysis analogous to equation (8.24)

$$[[\mathbf{K}]^{\mathrm{F}}, [\mathbf{K}]^{\mathrm{I}} - [\mathbf{K}]^{\mathrm{B}}] \left\{ \begin{matrix} \{\mathbf{u}\}^{\mathrm{F}} \\ \{\mathbf{u}\}^{\mathrm{I}} \end{matrix} \right\} - \{\mathbf{F}\}^{\mathrm{A}} + \{\mathbf{F}\}^{\mathrm{B}} = \mathbf{0} \tag{8.32}$$

The difference of a partially coupled and fully coupled analysis is that $[\mathbf{K}]^{\mathrm{B}}$ now only involves the interface nodes and not all the nodes of the BE region and that there is an additional force term $\{\mathbf{F}\}^{\mathrm{B}}$.

Solution of equation (8.32) gives the displacements at the interface and inside the FE region. To obtain the displacements and tractions at the non-interface nodes we back-substitute into equations (8.27) and (8.28).

The advantage of using the principle of minimum potential energy is that the resulting stiffness matrix of the boundary element region is symmetric and after assembly the system of equations to be solved remains symmetric. In the solution of such equations we can therefore use finite element solvers which exploit all the characteristics of these equations (i.e. symmetry and sparseness) resulting in economical use of storage space.

8.2.2 Coupling to boundary elements

Consider again the fully coupled problem in Figure 8.1. On the interface between the BE and FE region now both displacements and tractions are unknown. Rearranging all unknowns to the left hand side of equation (8.2) we obtain

$$[[\Delta\mathbf{T}]^{\mathrm{B}}, -[\Delta\mathbf{U}]^{\mathrm{B}}] \left\{ \begin{matrix} \{\mathbf{u}\}^{\mathrm{B}} \\ \{\mathbf{t}\}^{\mathrm{B}} \end{matrix} \right\} = \mathbf{0} \tag{8.33}$$

There are too many unknowns and not enough equations for a solution. The additional equations come from the finite element region. For the finite element region we have after partitioning $[\mathbf{K}]^{\mathrm{A}}$ of equation (8.1):

$$\begin{bmatrix} [\mathbf{K}]^{\mathrm{II}} & [\mathbf{K}]^{\mathrm{IF}} \\ [\mathbf{K}]^{\mathrm{FI}} & [\mathbf{K}]^{\mathrm{FF}} \end{bmatrix} \left\{ \begin{matrix} \{\mathbf{u}\}^{\mathrm{I}} \\ \{\mathbf{u}\}^{\mathrm{F}} \end{matrix} \right\} = \left\{ \begin{matrix} \{\mathbf{F}\}^{\mathrm{I}} \\ \{\mathbf{F}\}^{\mathrm{F}} \end{matrix} \right\} \tag{8.34}$$

where the superscripts I and F denote interface nodes and nodes inside the FE region, respectively.

Using the relationship between forces at the interface $\{\mathbf{F}\}^{\mathrm{I}}$ and tractions $\{\mathbf{t}\}^{\mathrm{I}}$ we can write:

$$\begin{bmatrix} [\mathbf{K}]^{\mathrm{II}} & [\mathbf{K}]^{\mathrm{IF}} \\ [\mathbf{K}]^{\mathrm{FI}} & [\mathbf{K}]^{\mathrm{FF}} \end{bmatrix} \left\{ \begin{matrix} \{\mathbf{u}\}^{\mathrm{I}} \\ \{\mathbf{u}\}^{\mathrm{F}} \end{matrix} \right\} = \left\{ \begin{matrix} \mathbf{M}\{\mathbf{t}\}^{\mathrm{I}} \\ \{\mathbf{F}\}^{\mathrm{F}} \end{matrix} \right\} \tag{8.35}$$

Compatibility on the interface stipulates for a fully coupled analyses:

$$\{\mathbf{u}\}^{\mathrm{I}} = \{\mathbf{u}\}^{\mathrm{B}} \tag{8.36}$$

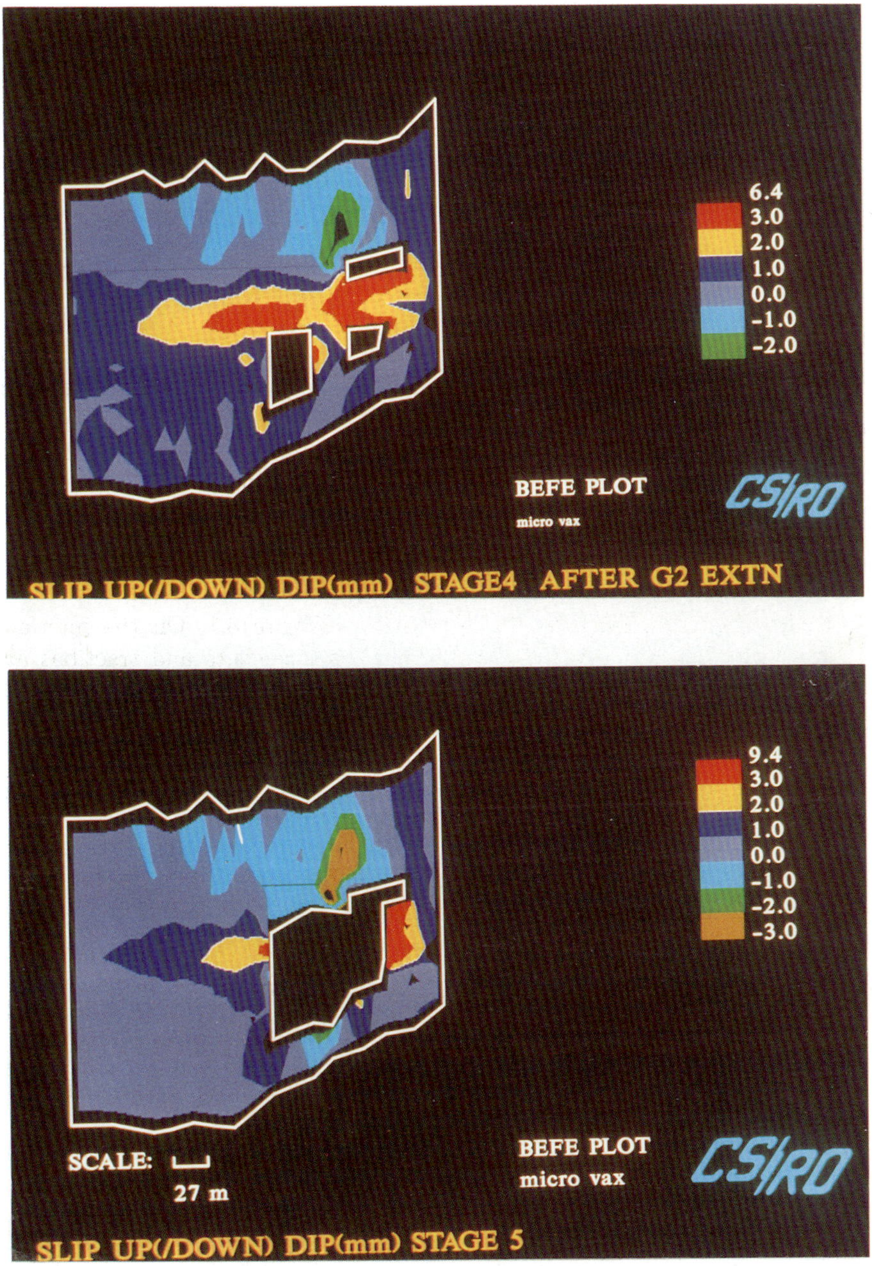

(c)

Figure 1.2 Mining excavations: prediction of slip on major fault (c) contours of slip before and after mining step 5

Figure 4.20 (b) Perspective view of finite element mesh used for analysis

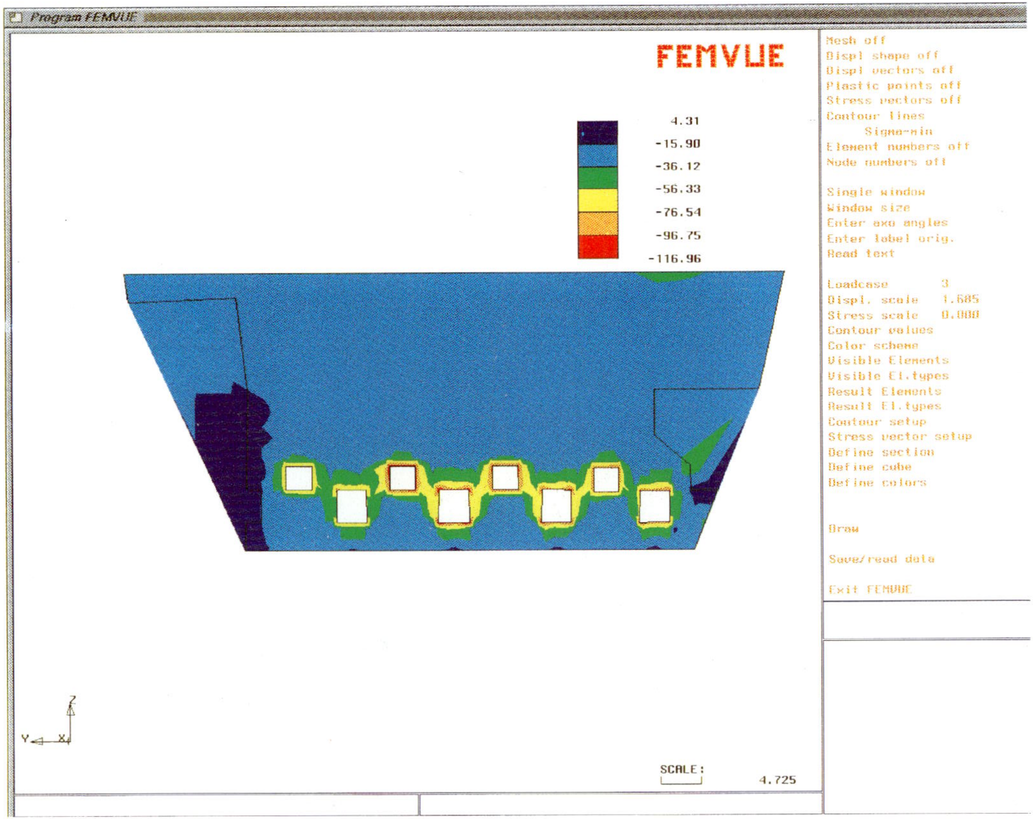

Figure 4.21 Contours of max. compressive stress on section for mining step 3

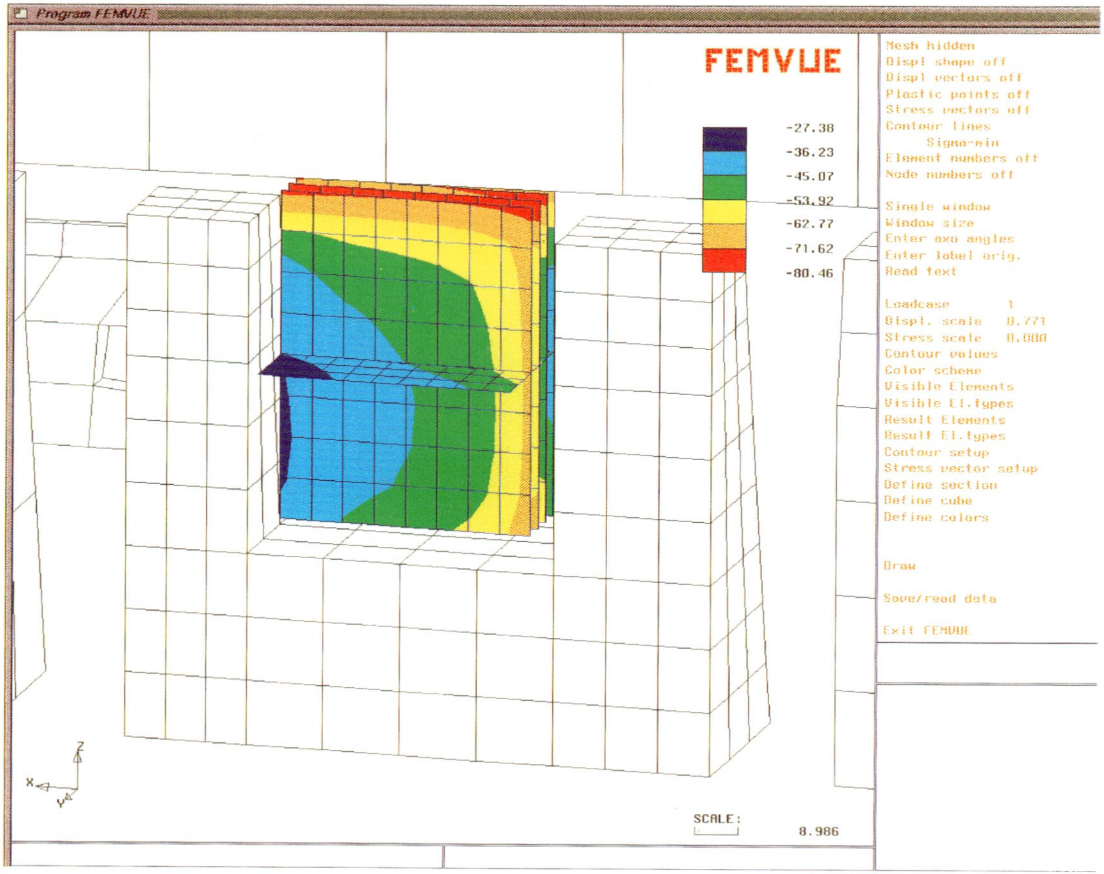

Figure 7.59 Maximum compressive stress plotted on dummy planes

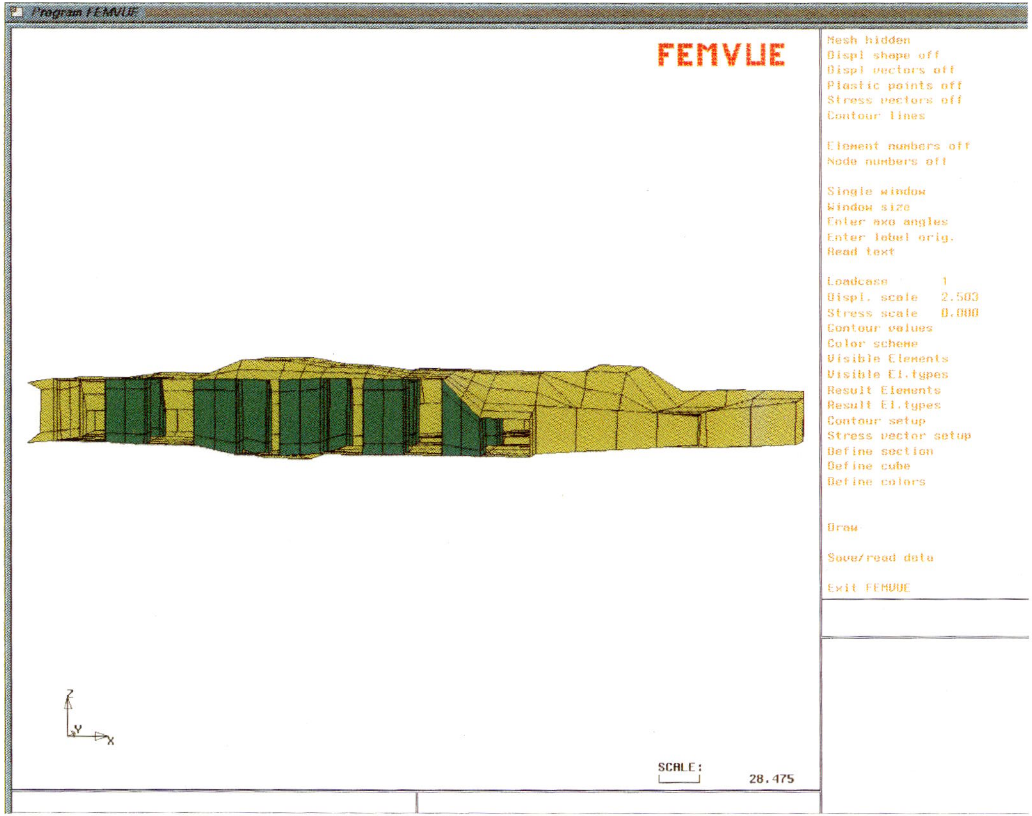

Figure 8.12 Partially coupled mesh of mine excavation with pillars. (In this view the mesh has been cut open to reveal the rock pillars which are discretised into finite elements)

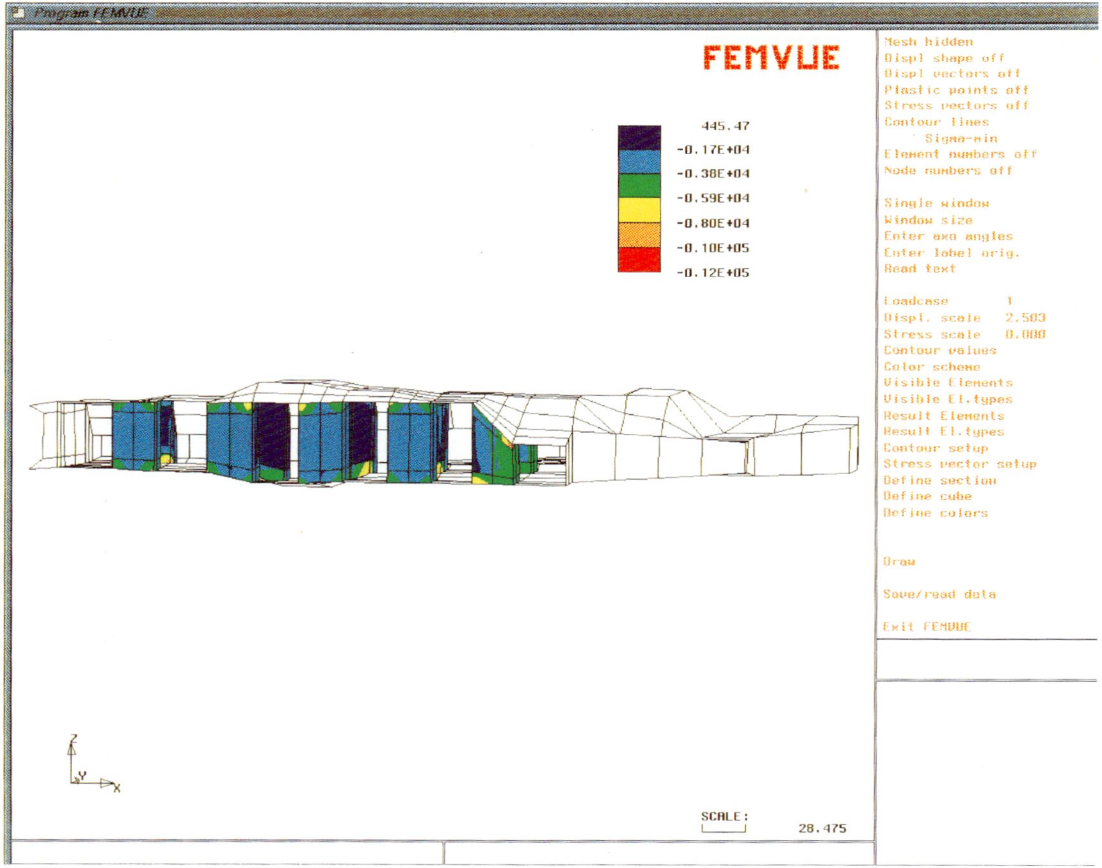

Figure 8.13 Maximum compressive stress (psi) in pillars for excavation stage 1

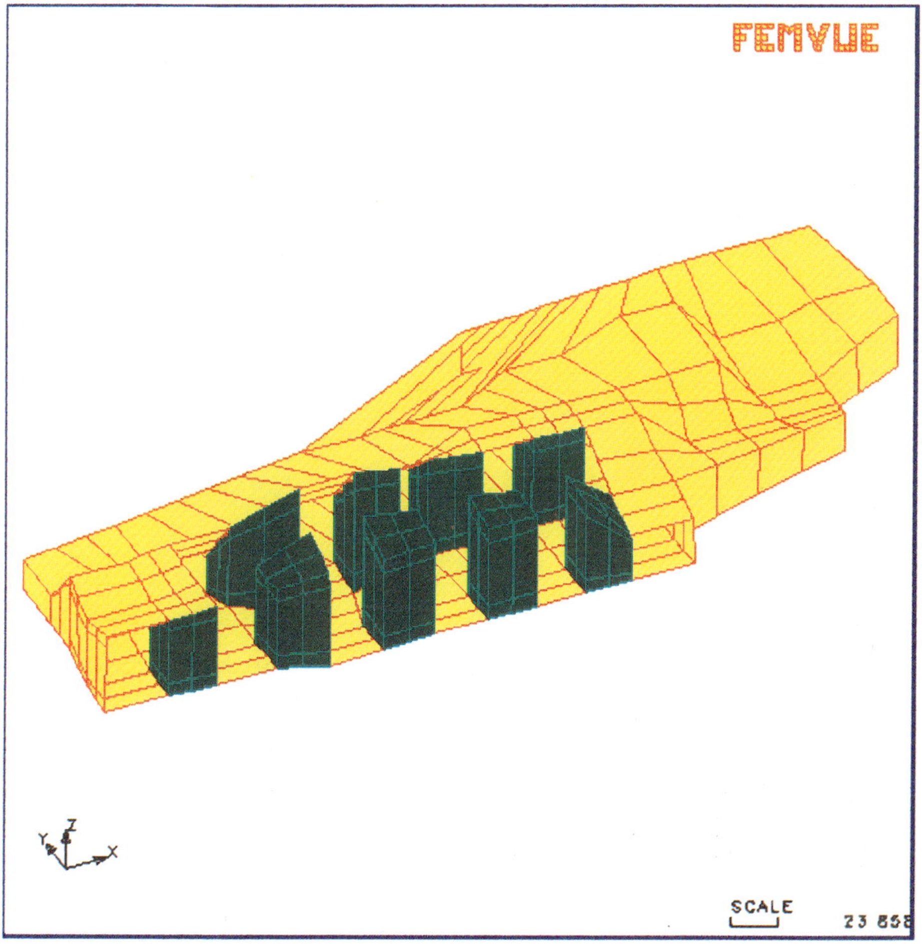

Figure 11.10 Example of simple hidden line algorithem

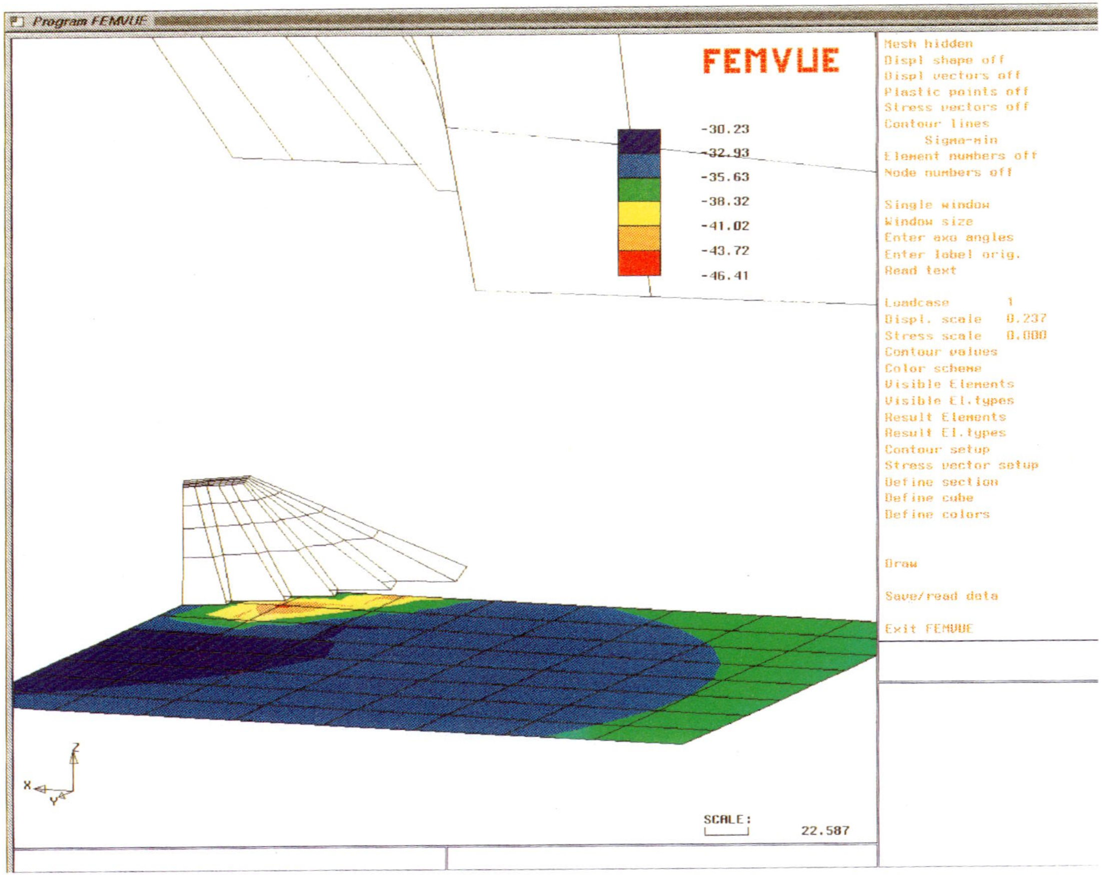

Figure 11.17 Example of colour contouring

and equilibrium

$$\{\mathbf{t}\}^{\mathrm{I}} = -\{\mathbf{t}\}^{\mathrm{B}} \tag{8.37}$$

Combining equation (8.33) and equation (8.35) we obtain the following system of equations

$$\begin{bmatrix} [\Delta\mathbf{T}]^{\mathrm{B}}, & -[\Delta\mathbf{U}]^{\mathrm{B}}, & \mathbf{0} \\ [\mathbf{K}]^{\mathrm{II}}, & \mathbf{M}, & [\mathbf{K}]^{\mathrm{IF}} \\ [\mathbf{K}]^{\mathrm{FI}}, & \mathbf{0}, & [\mathbf{K}]^{\mathrm{FF}} \end{bmatrix} \begin{Bmatrix} \{\mathbf{u}\}^{\mathrm{B}} \\ \{\mathbf{t}\}^{\mathrm{B}} \\ \{\mathbf{u}\}^{\mathrm{F}} \end{Bmatrix} = \begin{Bmatrix} \mathbf{0} \\ \mathbf{0} \\ \{\mathbf{F}\}^{\mathrm{F}} \end{Bmatrix} \tag{8.38}$$

which now can be solved for all unknowns.

8.2.3 Non-smooth interfaces

The above coupling theory is only valid for interfaces that are smooth i.e. do not have edges or corners. This restriction occurs because we have assumed implicitly that there is only one value of traction at a point on the interface. For example this is assumed in equation (8.11) because $\{\mathbf{t}\}^{\mathrm{B}}$ is defined as a vector containing tractions $\mathbf{t}_k$ at nodes, k, of the interface. Indeed, the assembly process of matrix $\mathbf{M}$ implies this because contributions of N^e_{ij} for adjacent elements are added.

Where there are sharp corners of the interface the coupling theory has to be modified to account for different traction values to the 'left' and to the 'right' of a node.

Consider the coupled mesh in Figure 8.4. At node k two different values of traction exist, $\mathbf{t}^l$ and $\mathbf{t}^r$. Both are unknown. This means that the vector of nodal tractions $\{\mathbf{t}\}^{\mathrm{B}}$ will become larger than for the case with a smooth interface because there are additional traction components. This also means that we need additional equations (or rows in the $[\Delta\mathbf{U}]^{\mathrm{B}}$ matrix) to be able to solve for $\mathbf{Q}$.

Consider the corner point of the two-dimensional problem in Figure 8.5 where two elements join. One way to get additional equations is from the condition that provided the interface occurs inside a material where properties are the same on both sides at that point there must be a unique value of stress. That is, the stress computed from element l must be equal to the stresses computed from element r:

$$\boldsymbol{\sigma}^l = \boldsymbol{\sigma}^r \tag{8.39}$$

These stresses are computed according to Chapter 7, for the element l

$$\boldsymbol{\sigma}^l = \begin{Bmatrix} C_1\epsilon^l_\theta + C_2 t^l_n \\ t^l_n \\ t^l_s \end{Bmatrix} \tag{8.40}$$

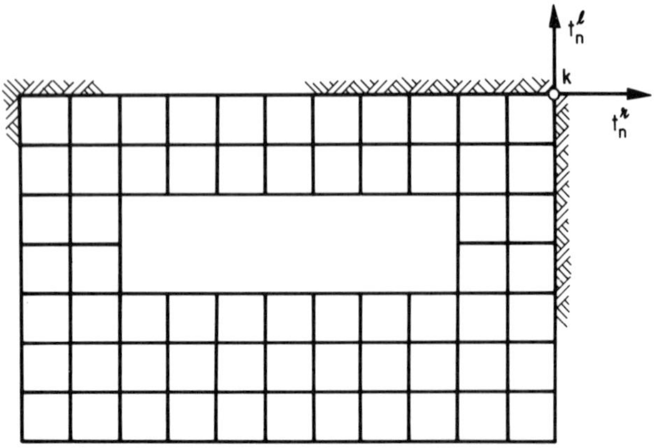

Figure 8.4 Coupled problem with corners at the interface

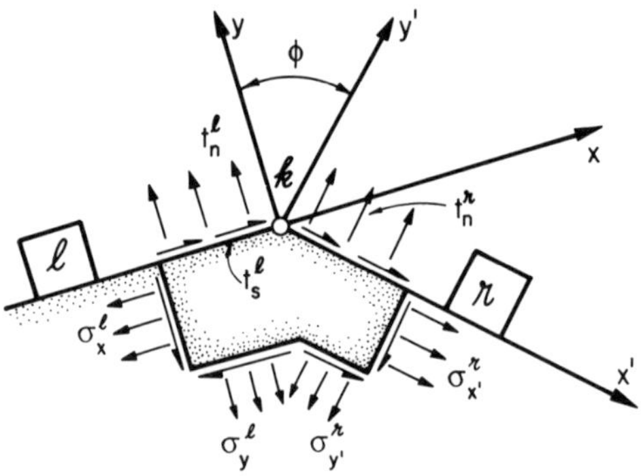

Figure 8.5 Tractions and strains at a corner point

where for plane strain $C_1 = \frac{E}{(1-\nu^2)}$ and $C_2 = \frac{\nu}{1-\nu}$, ϵ_θ^l is the strain in tangential direction at node k computed from element l, t_n^l and t_s^l are values of normal and tangential traction at point k.

For the stress computed from element r

$$\boldsymbol{\sigma}'^r = \begin{Bmatrix} C_1\epsilon_\theta^r + C_2 t_n^r \\ t_n^r \\ t_s^r \end{Bmatrix} \tag{8.41}$$

These stresses are however in the local x', y' coordinate system and have to be transformed into the x, y system. Equation (8.41) is written as

$$\boldsymbol{\sigma}^l = \mathbf{T}_\sigma \boldsymbol{\sigma}'^r \tag{8.42}$$

The reader may verify that this gives the following system of equations:

$$\begin{aligned} C_1\epsilon_\theta^l + C_2 t_n^l &= (C_1\epsilon_\theta^r + C_2 t_n^r)\cos^2\phi + t_n^r \sin^2\phi - 2t_s^r \sin\phi\cos\phi \\ t_n^l &= (C_1\epsilon_\theta^r + C_2 t_n^r)\sin^2\phi + t_n^r \cos^2\phi + 2t_s^r \sin\phi\cos\phi \\ t_s^l &= (C_1\epsilon_\theta^r + C_2 t_n^r)\sin\phi\cos\phi - t_n^r \sin\phi\cos\phi + t_s^r(\cos^2\phi - \sin^2\phi) \end{aligned} \tag{8.43}$$

Adding the first and the second equations we obtain the first of the additional equations for point k:

$$\epsilon_\theta^l - \epsilon_\theta^r = \frac{1+\nu}{E}(t_n^r - t_n^l) \tag{8.44}$$

whereas subtracting from the third equation $\frac{\cos\phi}{\sin\phi}$ times the second gives

$$(t_n^l - t_n^r)\cos\phi = (t_s^r + t_s^l)\sin\phi \tag{8.45}$$

These are the two additional equations needed at point k to obtain the two additional traction components.

The tangential strain is given by

$$\epsilon_\theta = \frac{\partial u_\theta}{\partial s} = \frac{\partial u_\theta}{\partial \xi}\left(\frac{\mathrm{d}s}{\mathrm{d}\xi}\right)^{-1} \tag{8.46}$$

where u_θ is the displacement in tangential direction and $\frac{\mathrm{d}s}{\mathrm{d}\xi} = J$ is defined in Chapter 2.

Expressed in terms of nodal displacement $u_{\theta j}$ the tangential strain is given by

$$\epsilon_\theta = \sum_{j=1}^{J} \frac{\partial N_j}{\partial \xi} u_{\theta j} \frac{1}{J} \tag{8.47}$$

The tangential displacements are

$$u_{\theta j} = \mathbf{v}_1^e \mathbf{u}_j^e \tag{8.48}$$

where

$$\mathbf{v}_1^e = \frac{1}{J}\left[\frac{\partial x}{\partial \xi}, \frac{\partial y}{\partial \xi}\right]$$

Equation (8.47) can be conveniently written in matrix form as

$$\epsilon_\theta = \mathbf{N}^e \mathbf{u}^e \tag{8.49}$$

where

$$\mathbf{N}^e = \frac{1}{J}\left[\frac{\partial N_1}{\partial \xi}\mathbf{v}_1^e,\ \frac{\partial N_2}{\partial \xi}\mathbf{v}_1^e,\ \ldots,\ \frac{\partial N_J}{\partial \xi}\mathbf{v}_1^e\right] \tag{8.50}$$

Substitution of equation (8.49) gives for equation (8.44)

$$\mathbf{N}^l\mathbf{u}^l - \mathbf{N}^r\mathbf{u}^r = \frac{1+\nu}{E}(t_n^r - t_n^l) \tag{8.51}$$

Finally substituting for the normal and tangential tractions in terms of global components of traction we have for the left-hand side, assuming k to be the jth node of element l

$$t_n^l = \mathbf{v}_1^l\mathbf{t}_j^l \quad \text{and} \quad t_s^l = \mathbf{v}_2^l\mathbf{t}_j^l$$

Similarly for the right-hand side (assuming k to be the ith node of element r)

$$t_n^r = \mathbf{v}_1^r\mathbf{t}_i^r \quad \text{and} \quad t_s^r = \mathbf{v}_2^r\mathbf{t}_i^r$$

where

$$\mathbf{v}_2 = \frac{1}{J}\left[-\frac{\partial y}{\partial \xi},\ \frac{\partial x}{\partial \xi}\right]$$

The two additional equations become

$$\begin{aligned}\mathbf{N}^l\mathbf{u}^l - \mathbf{N}^r\mathbf{u}^r &= \frac{1+\nu}{E}\left(\mathbf{v}_1^r\mathbf{t}_i^r - \mathbf{v}_1^l\mathbf{t}_j^l\right)\\ (\mathbf{v}_1^l\mathbf{t}_j^l - \mathbf{v}_1^r\mathbf{t}_i^r)\cos\phi &= (\mathbf{v}_2^r\mathbf{t}_i^r + \mathbf{v}_2^l\mathbf{t}_j^l)\sin\phi\end{aligned} \tag{8.52}$$

For the special case where $\phi = 90°$ the second equation above gives

$$t_s^r + t_s^l = 0 \tag{8.53}$$

which means that shear tractions must be equal and opposite.

The two additional equations (8.52) are assembled into the global system of equations for points where the outward normal computed for adjacent elements at this point is not the same. In a computer implementation one may check outward normals at all points where interface elements join and implement the above algorithm where they differ within a certain tolerance.

The work involved in implementing capabilities for dealing with nonsmooth interfaces may be quite significant and in some instances it may be more economical to avoid occurrences of discontinuous interfaces. Indeed, a procedure, where by the computer program automatically constructs smooth transition surfaces at sharp corners may be optimal.

Another simpler way of treating discontinuities of tractions at the interface is to increase the number of collocation points at corners when establishing matrix $[\Delta\mathbf{U}]^B$. As we know from Chapter 7 this will give additional rows in the matrix $[\Delta\mathbf{U}]^B$. One approach would be to have two collocation points at a corner, one slightly to the left, the other slightly to the right. Care has to be taken that the points are not too near to each other because the two equations obtained will be very similar. Details of such double node schemes are given for example in reference 9.

Finally it should be pointed out that the above procedure does not have to be applied to the case where corners occur at the end or begining of an interface as shown in Figure 8.2 because here the traction on one side is known.

8.3 EXAMPLES—1

Here some examples are presented which demonstrate the accuracy of combined finite and boundary element analyses.

Figure 8.6 shows the discretisation of one quarter of a circular cavity in an infinite solid subjected to a uniaxial stress field. The mesh consists of

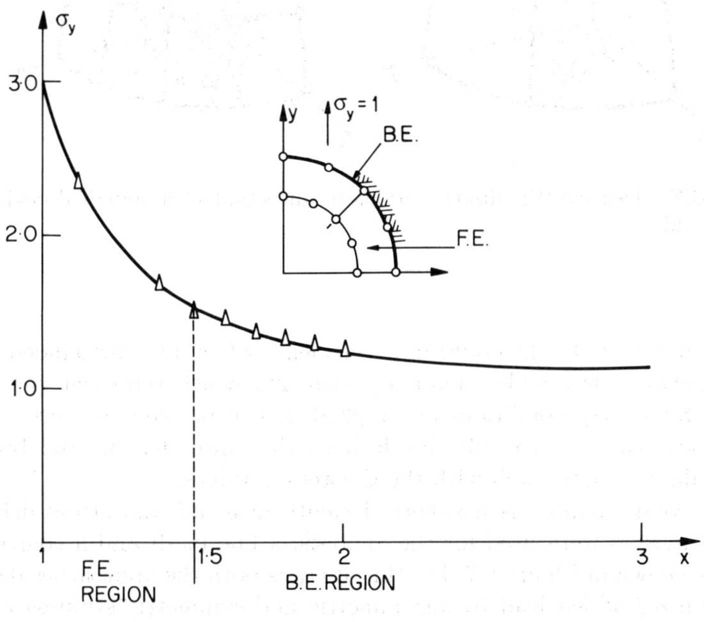

Figure 8.6 Circular excavation in infinite domain coupled mesh and stress results. (△ computed, — theory)

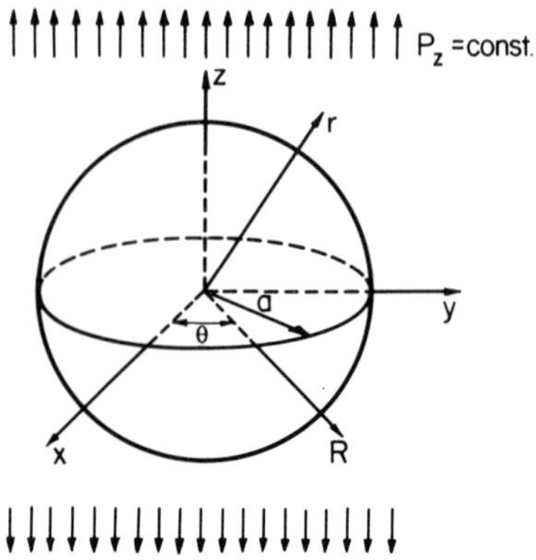

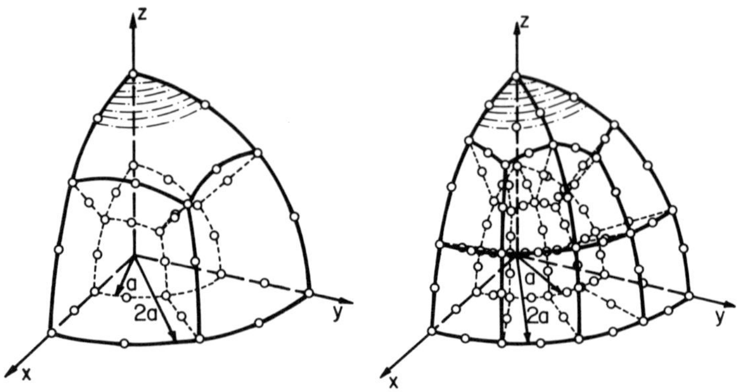

Figure 8.7 Two BE/FE discretisations of an octant of a spherical cavity in an infinite solid

two isoparametric (eight-node) finite elements which are surrounded by two isoparametric (three-node) boundary elements which represent the infinite domain. Symmetry conditions are applied. It can be seen that even for this coarse mesh the stress results inside both the finite element and boundary element domain agree well with the theoretical values.

The second example is a spherical cavity in a uniaxial stress field. Two different meshes were used for the analysis: a fine mesh and a coarse mesh. They are shown in Figure 8.7. For this analysis both the approaches discussed in Section 8.2 which lead to unsymmetric and symmetric stiffness matrices i.e. the one using virtual work and minimum potential energy have been used. The results are summarised in Table 8.1.

Table 8.1 Comparison of results for normalised stresses σ_z/p_z in the plane $z = 0$ as a function of R/a (after reference 6)

R/a	Subregion	Virtual Work		Potential Energy		Exact solution
		Coarse	Fine	Coarse	Fine	
1.0	FE	1.1136	1.9691	1.1153	1.9700	2.0455
1.50		1.8235	1.1211	1.8170	1.1216	1.1751
2.00		1.0597	1.0763	1.0656	1.0773	1.0540
2.00	BE	1.0000	1.0203	1.0018	1.0136	1.0540
2.50		1.0183	1.0232	1.0186	1.0233	1.0229
3.00		1.0111	1.0120	1.0112	1.0121	1.0118
4.00		1.0042	1.0044	1.0042	1.0044	1.0044
6.00		1.0011	1.0012	1.0011	1.0012	1.0012
10.00		1.0002	1.0002	1.0002	1.0002	1.0002

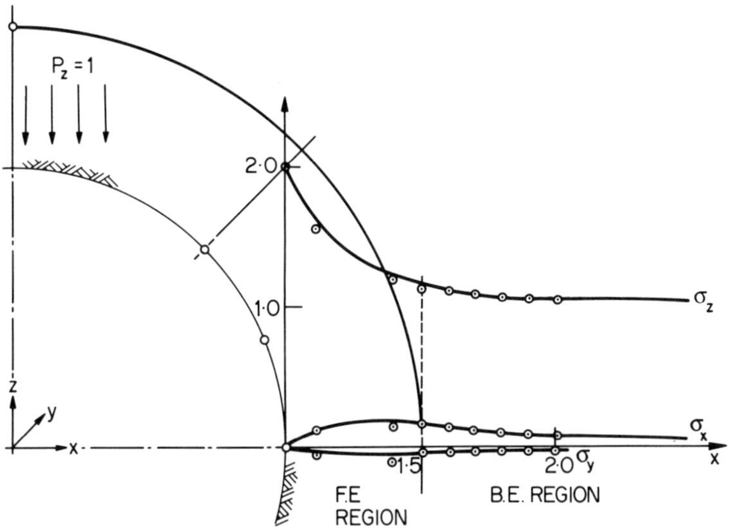

Figure 8.8 Distribution of vertical stress inside FE and BE region for spherical excavation. (△ computed, — theory)

The results for all stress components for coupling method 2 (minimum potential energy) and the coarse mesh are plotted in Figure 8.8. Both symmetric and unsymmetric methods give good results for this case.

The third example is a cantilever beam modelled in three dimensions using quadratic finite element for one half of the beam and quadratic boundary elements for the other. The beam is subject to a shear traction distributed parabolically at the end.

Figure 8.9 shows the coupled mesh used for the analysis and the results using the approach with minimum potential energy and virtual work. It can be seen that although the displaced shape is approximated well by both methods, there is a local error in the distribution σ_x affecting the first row of boundary elements from the interface for the minimum potential energy method. It has been found for this example that, as the mesh is refined and more elements are used for the boundary element region this local error vanishes.

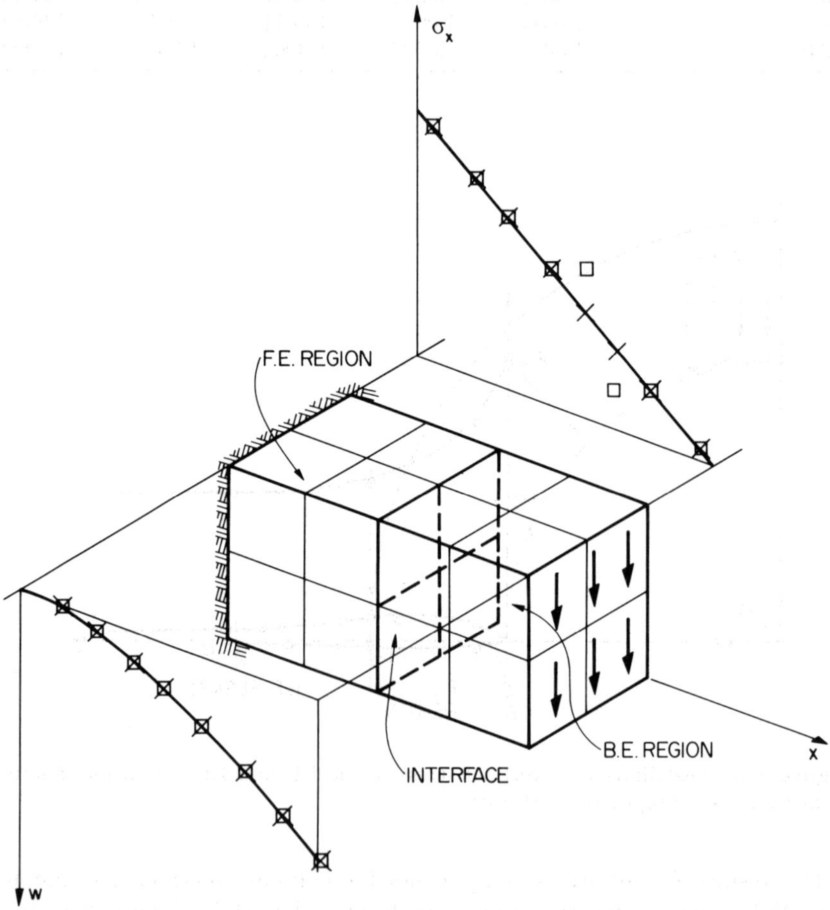

Figure 8.9 Coupled analysis of cantilever beam: comparison of vertical displacement and σ_x for symmetric and non-symmetric coupling. (x virtual work, □ minimum potential energy)

There has been much discussion about the two approaches of obtaining the stiffness matrix for the BE region[6,9]. There is no doubt to the advantages of having a symmetric stiffness matrix in terms of storage requirements and equation solution. Finite element software is geared towards symmetric equations and very efficent solvers have been developed over the last decades. As we have seen, one may construct examples where symmetric coupling leads to localised error however it is questionable whether one would sacrifice the advantages of having to store only half the number of coefficients in order to obtain higher accuracy.

There is—in terms of computational efficency—a compromise solution. As detailed in reference 10 with the virtual work approach one may obtain the solution for the symmetric stiffness matrix first and then iterate with the unsymmetric part. The algorithm for the nth iteration is

$$[\mathbf{K}]_{\mathrm{s}}\{\mathbf{u}\}^{n} = \{\mathbf{F}\} - [\mathbf{K}]_{\mathrm{a}}\{\mathbf{u}\}^{n-1} \tag{8.54}$$

where

$$\begin{aligned} [\mathbf{K}]_{\mathrm{s}} &= \frac{1}{2}\left([\mathbf{K}]^{\mathrm{T}} + [\mathbf{K}]\right) \\ [\mathbf{K}]_{\mathrm{a}} &= \frac{1}{2}\left([\mathbf{K}] - [\mathbf{K}]^{\mathrm{T}}\right) \\ \{\mathbf{u}\}^{0} &= \{\mathbf{0}\} \end{aligned} \tag{8.55}$$

In an efficient implementation we first do a Gauss reduction with the symmetric part and at the second and subsequent iterations we solve the equations for a new right-hand side involving the unsymmetric part.

This is more efficient than solving for the unsymmetric matrices because only 1/2 of the matrix is operated upon during Gauss elimination. This method may be used as a 'safety net'. If the symmetric method works well very few applications of (8.54) will be needed.

8.4 EXAMPLES—2

Two applications of the coupled FEM/BEM method are shown. Both are related to underground excavations in rock and are analysed with the coupled program BEFE.

8.4.1 Tunnel

In tunnel construction the stress state near the tunnel face is of importance. In particular it is of interest to calibrate approximate plane models which have been used to model the change in stress and displacement in a cross section near the tunnel face.

The three-dimensional model of a tunnel[11] is shown in Figure 8.10 and consists of four zones:

Zone A.	Excavation region
Zone B.	Shell elements to simulate shotcrete lining
Zone C.	Non-linear near field region
Zone D.	Elastic far field region

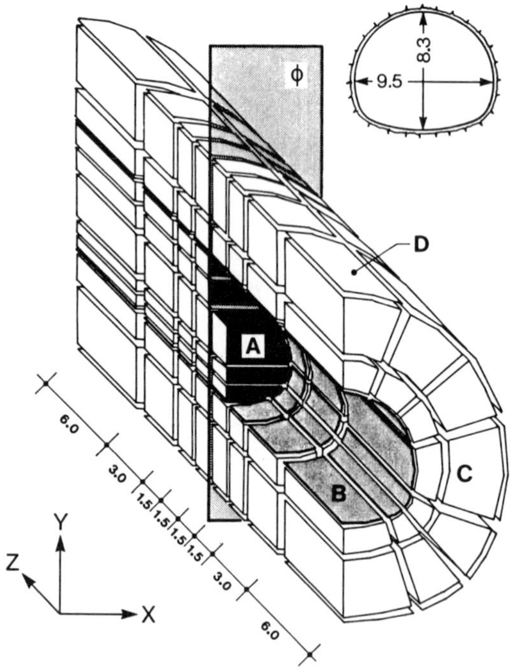

Figure 8.10 Three-dimensional coupled mesh for a tunnel excavation problem. (A excavation region, B shotcrete lining, C non-linear immediate region, D elastic far region)

Zones A to C are described by finite elements (they consist of 20-node isoparametric quadratic solid elements) and Zone B is described by 8-node shell elements. Zone D, which is assumed to be of infinite extent, is modelled by eight-node boundary elements. Symmetry conditions are applied to the boundary element mesh as well as to the finite element mesh (x-displacement restrained).

Comparison of the three-dimensional results is made with approximate two-dimensional results using the 'stiffness reduction method'[12] in Figure 8.11. Here the membrane forces in the tunnel lining are shown for both three- and

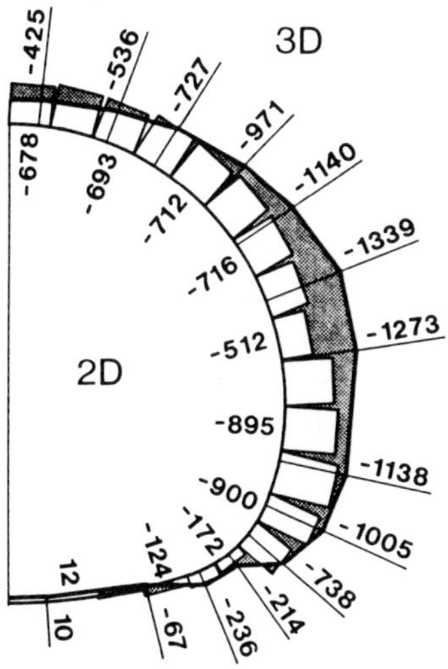

Figure 8.11 Comparison of simplified plane and three-dimensional analyses

two-dimensional analysis. It can be seen that the approximate plane analysis underestimates the lining forces as computed by the more realistic three-dimensional analysis. The coupled model used here results in substantial reduction in effort because the far field can be modelled accurately with the boundary elements and there is no need to extend the mesh a large distance away from the opening.

The analysis proceeds as follows.

1. The stiffness matrix of the boundary element region is computed.
2. The stiffness matrices of the finite elements are computed.
3. The finite elements for each excavation stage are assembled and the system of equations is solved.

Thus, for each excavation stage, only step 3 needs to be repeated whereas step 1—which takes most computation time—and 2 need only be done once.

8.4.2 Mining excavations

In mining excavations often pillars of ore are left to ensure stability of the mine workings. These pillars contain valuable ore and it is of economic interest to recover them. Care has to be taken in the recovery to ensure that no catastrophic failure occurs in other areas of the mine which would endanger lives and/or prevent futher pillar recovery. In many cases it is required to use mine backfill to provide additional support.

The partially coupled BE/FE analysis reported here has been carried out for the Doe Run lead mine in Missouri, U.S.A. Figure 8.12 (see colour section) shows a perspective view of the coupled BE/FE mesh which has been cut open to show the discretisation of some of the pillars into eight-node linear 'brick' finite elements. The reason these pillars have been discretised into finite elements is because this allows greater flexibility in simulating the sequential excavation and also allows non-linear material behaviour to be considered. The boundary element region is represented by four-node linear boundary elements. A total of 1045 elements was used of which 630 were boundary elements and 415 finite elements. As can be seen from Figure 8.12 the finite elements are only connected to parts of the BE region. In the analysis the stiffness matrix is only computed for nodes at the interface between pillars and the boundary elements representing the infinite elastic rock mass.

The analysis was performed on a 486 PC with a WEITEK coprocessor and 16Mb of RAM at Golder Associates, Vancouver. Table 8.2 shows some statistics of the problem.

Table 8.2

	No. of degrees of freedom (BE region)	1872
	No. of degrees of freedom (FE region)	1926
	Maximum front width	1143
	CPU times (seconds)	
1.	Coefficient matrices	964
2.	BE solution	7030
3.	BE stiffness matrix computation	233
4.	Front solution	1568
5.	Making results	980
	Total	10775 (approx 3h)

A total of five excavation stages were computed. Each of the additional excavation stages which involved the removal of pillar elements and the installation of backfill elements did not require the computation of the BE stiffness matrix and only involved the frontal solution part.

Figure 8.13 (see colour section) shows one of the results obtained namely the distribution of the maximum compressive stress in the pillars for excavation stage 1.

8.5 CONCLUDING REMARKS

It is significant that both practical examples of a coupled analysis shown here are in geomechanics because it is in this area that most benefits are gained. There we are dealing with both an unbounded (infinite or semi infinite) domain and inelastic materials. In an analysis using finite elements only we would have to extend the mesh to a large distance away from the area of interest and then apply some artificial boundary conditions. In the coupled meshes presented here we let the boundary elements 'do the work' of modelling the rock mass surrounding the excavations and hence provide a natural boundary condition. On the other hand an analysis with boundary elements only—although possible in this case—would have provided us with less flexibility with regard to material non-linearity, and inhomogeneity of the rock mass and would not have allowed us to remove elements to simulate excavations. Other applications of coupled analysis can be found in fracture mechanics where we concentrate on a very small region of the analysed domain.

In these and other applications the user can have the 'best of both worlds' in a skillful application of the two numerical modelling techniques discussed in this book.

Indeed, the purpose of this book is to show that the finite and boundary element method, allthough conceptually quite different can be applied together.

8.6 EXERCISES

8.1 Using the principle of virtual work derive equation (8.3) for a linear isoparametric boundary element

(a) for a plane problem
(b) for a three-dimensional problem

8.2 On an example show how matrix $\mathbf{M}$ is assembled from boundary element contributions $\mathbf{M}^e$.

8.3 In the derivation of the stiffness matrix of the boundary element region using minimum potential energy a distribution of both tractions and displacements along the interface was assumed. This is—as pointed out in the text—not quite correct as for a particular distribution of displacement the actual traction distribution must follow from a boundary element solution. Some researchers have suggested that the interpolation of the tractions should be one order less than the interpolation of the displacements (i.e. linear variation of tractions—parabolic variation of displacements).
Using the Kelvin fundamental solutions investigate if there is any theoretical basis to this suggestion.

8.4 Based on the derivations in Section 8.2.3, derive the auxiliary equations for a sharp corner in the interface of a three-dimensional problem.

8.5 Derive the coupled equation (8.22) for heat flow/potential problems.

8.6 As explained in Section 8.2.3, if the interface boundary is not smooth then, because the traction distribution is discontinuous, additional equations have to be considered. Comment on the error introduced by assuming that the tractions are continuous if the corner angle ϕ is $90°$. Will the error be localised?

8.7 Derive coupled equations (8.38) for the case where only part of the boundary element mesh is coupled to finite elements.

8.7 REFERENCES

1. Zienkiewicz, O.C., Kelly, D.W. and Bettess, P. (1977). *Marriage a la mode—The best of both worlds.* Int. Symp. on Innovative Numerical Analysis in Applied Engineering Science, Versailles, France, May 1977.
2. Zienkiewicz, O.C., Kelly, D.W. and Bettess, P. (1977). The coupling of the finite element method and boundary solution procedures. *Int. J. Num. Meth. Eng.* **11**, 355–375.
3. Kelly, D.W., Mustoe, G.G.W. and Zienkiewicz, O.C. (1979). Coupling boundary element methods with other numerical methods. In *Developments in Boundary Element Methods-1* (eds P.K. Banerjee and R. Butterfield), Applied Science, London.
4. Beer, G. (1983). Finite element, boundary element and coupled analysis of unbounded problems in elastostatics. *Int. J. Num. Meth. Eng.*, **19**, 567–580.
5. Beer, G. (1986). Implementation of combined boundary element-finite element analysis with applications in Geomechanics. *Developments in Boundary Elements-4* (eds P.K. Banerjee and J.O. Watson), Applied Science, London. pp. 191–226.
6. Li, H.B., Han, G.M., Mang, H.A. and Torzicky, P. (1986). A new method for the coupling of finite element and boundary element discretized subdomains of elastic bodies. *Comp. Meth. Appl. Science and Eng.* **54**, 161–185.
7. Brebbia, C.A. and Georgiou, P. (1979). Combination of boundary and finite element methods. *Appl. Math. Modelling*, **3**, 212–220.
8. Rudolphi, T.J. (1983). An implementation of the boundary element method for zoned media with stress discontinuities. *Int. J. Num. Meth. Eng.* **19**, 1–15.
9. Mustoe G.G.W., Volait F. and Zienkiewicz O.C. (1982) A symmetric direct boundary integral equation for two dimensional elastostatics, *Res Mechanica*, **4**, 57–82.
10. Ma Jingming and Le Meifeng (1985) *The symmetrization—iteration method for coupling of BEM and FEM in two dimensional elastostatics.* 1st Conf. of BEM in Eng., Chongqing, China.
11. Beer, G., Watson, J.O. and Swoboda, G. (1987). Three-dimensional analysis of tunnels using infinite boundary elements. *Comput. Geotechn.* **3**, 37–58.
12. Swoboda, G. (1979). *Finite element analysis of the New Austrian Tunnelling Method (NATM).* Proc. 3rd Int. Conf. on Numer. Meth. in Geomechanics (ed. A.A. Balkema). pp. 581–586.

9

Non-linear Problems

$\Pi\alpha\nu\tau\alpha \quad \rho\epsilon\iota$

Aristotle

9.1 INTRODUCTION

So far we have discussed problems where there is a linear relationship between applied loading and displacement. For example the finite element equations

$$[\mathbf{K}]\{\mathbf{u}\} = \{\mathbf{F}\} \tag{9.1}$$

correspond to a linear analysis because the displacements $\{\mathbf{u}\}$ are a linear function of the loads $\{\mathbf{F}\}$; i.e. if loads $\lambda\{\mathbf{F}\}$ are applied instead of $\{\mathbf{F}\}$; the corresponding displacements are $\lambda\{\mathbf{u}\}$. Here λ is a constant, also known as load factor.

The linearity of equation (9.1) arises because of certain assumptions we have made when deriving the finite element stiffness matrix or the boundary element coefficient matrices. These assumptions are:

1. The relationship between stress and strain (or thermal gradient and flux) is linear.
2. All differentiations and integrations are with respect to the original geometry unaffected by displacements which occur as a result of applied loads.
3. The boundary conditions do not change during the loading.

If the above assumptions are not satisfied then the relationship between forces and displacements will become non-linear and we have to perform a non-linear analysis. In a non-linear analysis basically the stiffness matrix or coefficient matrix is dependent on the displacements (temperatures). Since these are not known until after a solution is obtained we can not solve the system of equations directly. Instead the solution must proceed in increments whereby displacements are updated after each solution step.

Therefore non-linear problems add a further complexity to the solution process. Now not only the convergence of the solution in terms of mesh size has to be considered but also the convergence with respect to the nonlinear iteration procedure. In addition, there is the added possibility that computed results may not converge to the right solution if the increments used are too large.

In the following we will discuss general solution procedures for non-linear analyses. Next, the non-linearity introduced by the first assumption is discussed in detail. To deal with the second and the third type of non-linearity, namely geometrical and boundary condition non-linearity is beyond the scope of this introductory text. The reader is referred to the many excellent texts on this subject cited at the end of this chapter.

9.2 GENERAL SOLUTION PROCEDURES

The equilibrium equation of a system can be expressed quite generally as

$$ {}^{t}\{\mathbf{F}\} - {}^{t}\{\mathbf{R}\} = \mathbf{0} \tag{9.2} $$

where ${}^{t}\{\mathbf{F}\}$ is a vector containing all externally applied nodal point forces at time t and ${}^{t}\{\mathbf{R}\}$ is a vector of nodal point forces corresponding to the internal stresses at time t.

For a finite element mesh for example we have

$$ {}^{t}\{\mathbf{R}\} = \sum \int_V ({}^{t}\mathbf{B}^e)^{\mathrm{T}}\, {}^{t}\boldsymbol{\sigma}\, \mathrm{d}V \tag{9.3} $$

where $\sum$ implies the summation of all element contributions and ${}^{t}\mathbf{B}^e$ and ${}^{t}\boldsymbol{\sigma}$ are the strain–displacement matrix and stress vector at time t.

Equation (9.2) must be satisfied for any time or load step t and must express the equilibrium in the current (deformed) geometry. Furthermore, when the analysis includes path-dependent or time-dependent response characteristics equation (9.2) need to be solved for the complete time range of interest. For example, in many cases the total load is applied in increments and a solution obtained for every increment in load.

Let us assume that the solution at time t is known and the solution at time $t + \Delta t$ is required. At time $t + \Delta t$ we have

$$ {}^{t+\Delta t}\{\mathbf{F}\} - {}^{t+\Delta t}\{\mathbf{R}\} = \mathbf{0} \tag{9.4} $$

where

$$ {}^{t+\Delta t}\{\mathbf{R}\} = {}^{t}\{\mathbf{R}\} + {}^{\Delta t}\{\mathbf{R}\} \tag{9.5} $$

and ${}^{\Delta t}\{\mathbf{R}\}$ is the increment in nodal point forces corresponding to the increment of stresses during interval Δt. For example for a finite element analysis it can be computed from

$$ {}^{\Delta t}\{\mathbf{R}\} = {}^{t}[\mathbf{K}]\ {}^{\Delta t}\{\mathbf{u}\} \tag{9.6} $$

where ${}^{t}[\mathbf{K}]$ is a tangential stiffness matrix at time t and ${}^{\Delta t}\{\mathbf{u}\}$ is an increment in displacements. Substituting equation (9.6) and equation (9.5) into equation (9.4) we obtain

$$ {}^{t}[\mathbf{K}]\ {}^{\Delta t}\{\mathbf{u}\} = {}^{t+\Delta t}\{\mathbf{F}\} - {}^{t}\{\mathbf{R}\} \tag{9.7} $$

and solving for ${}^{\Delta t}\{\mathbf{u}\}$ an approximation to the displacements at time $t + \Delta t$ is

$$ {}^{t+\Delta t}\{\mathbf{u}\} = {}^{t}\{\mathbf{u}\} + {}^{\Delta t}\{\mathbf{u}\} \tag{9.8} $$

In a non-linear analysis ${}^{t}[\mathbf{K}]$ changes throughout the time step Δt. Because of the assumption of a constant stiffness matrix ${}^{t}[\mathbf{K}]$ throughout the time increment Δt in equation (9.6) a solution may be subject to significant errors of Δt is not very small. It is therefore necessary to iterate within each time step.

The recursive steps for iteration i are

$$ {}^{t}[\mathbf{K}]^{i-1}\Delta\{\mathbf{u}\}^{i} = \Delta\{\mathbf{R}\}^{i-1} \tag{9.9} $$

where the increment in the residual force vector is given by:

$$ \Delta\{\mathbf{R}\}^{i-1} = {}^{t+\Delta t}\{\mathbf{F}\} - {}^{t+\Delta t}\{\mathbf{R}\}^{i-1} $$

The total displacement at iteration i is:

$$ {}^{t+\Delta t}\{\mathbf{u}\}^{i} = {}^{t+\Delta t}\{\mathbf{u}\}^{i-1} + \Delta\{\mathbf{u}\}^{i} \tag{9.10} $$

Here ${}^{t}[\mathbf{K}]^{i}$ is the stiffness matrix and $\Delta\{\mathbf{u}\}^{i}$ is the increment in displacement at the ith iteration. The above iteration procedure is also called the *Newton–Raphson* procedure. If the stiffness matrix is kept constant throughout the time step then this method is termed *modified Newton–Raphson*. For this method the recursion formula (9.9) is given by

$$ {}^{t}[\mathbf{K}]\,\Delta\{\mathbf{u}\}^{i} = \Delta\{\mathbf{R}\}^{i-1} \tag{9.11} $$

i.e. the stiffness matrix ${}^{t}[\mathbf{K}]$ is not changed during iteration.

In the following we shall discuss one type of non-linearity in detail.

9.3 NON-LINEAR MATERIAL PROBLEMS

Here we discuss problems where the non-linearity of the response stems from the nonlinearity in the stress–strain relationship. Consider for example the system consisting of two bars in Figure 9.1. The force F acts on a rigid end member which can not rotate. Bar A is assumed to have a lower yield strength, σ_{YA}, than bar B, σ_{YB}.

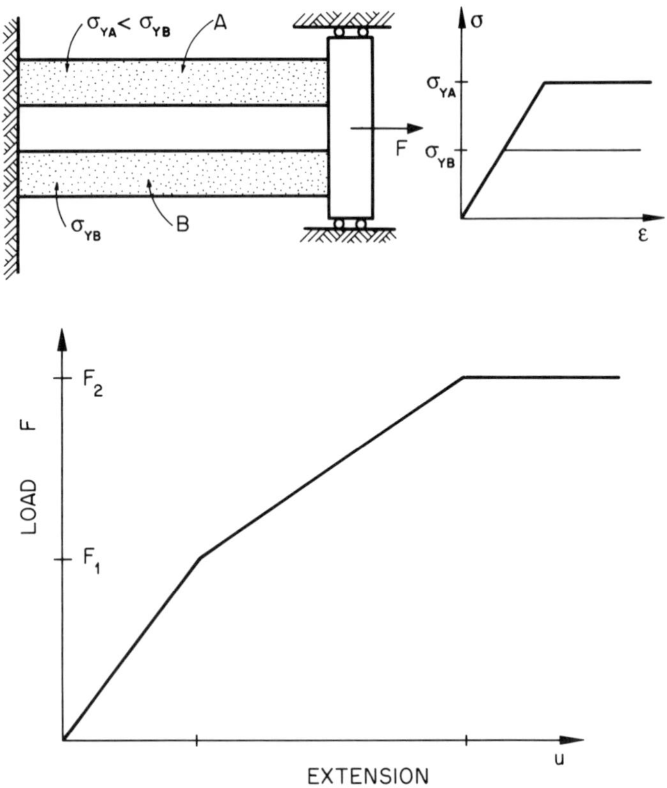

Figure 9.1 A two-bar system where the force acts on a member which can not rotate

Assuming that both bars have the same modulus of elasticity E, and cross sectional area A, then the stiffness of the two-bar structure is

$$^{0}k = \frac{EA}{l} + \frac{EA}{l} = \frac{2EA}{l} \tag{9.12}$$

We now plot the load–deflection curve for the structure and find that up to a load $F_1 = 2A\,\sigma_{Y\mathrm{A}}$ there is a linear relationship between displacement u and applied force F, i.e.

$$u = F\,\frac{l}{2EA} \tag{9.13}$$

If the force is increased, then the stress for bar A—as the stress–strain curve indicates—can not increase regardless of displacement. This means that all the additional load will be taken by bar B. The tangential stiffness due to an increment in force ΔF will be

$$^1k = \frac{EA}{l} \tag{9.14}$$

Again the load-displacement curve is linear. When the bar A has yielded the displacement increment Δu is given by

$$\Delta u = {}^1k\,\Delta F \tag{9.15}$$

Finally, after load $F_2 = A(\sigma_{Y\mathrm{A}} + \sigma_{Y\mathrm{B}})$ has been reached, there can be no increase in stress in either bar and the tangential stiffness is

$$^2k = 0 \tag{9.16}$$

This is reflected by the horizontal part of the load–deflection curve in Figure 9.1. This is a trivial example and solution has been obtained without iteration. In a general two- or three-dimensional situation, however, the solution becomes more complicated. For example, it is no longer possible to calculate the exact load where the weakest member or element has reached the yield point. Also, the load–displacement curve will no longer be piecewise linear but curved.

To demonstrate the Newton–Raphson procedure on this simple problem assume that our first loading step has taken us to ^0F instead of F_1. The first estimate of displacement is computed from

$$^0u^0 = {}^0F\,\frac{l}{2EA} \tag{9.17}$$

According to Figure 9.2 this means that we have assumed that the system is loaded elastically up to point B which is off the actual load displacement curve. This essentially means that the bar A has taken more stress than it should have. We must therefore reduce the stress in the bar from $^0\sigma = {}^0F/2A$ to $\sigma_{Y\mathrm{A}}$ and the resultant force due to the reduced internal stress in bar A and the internal stress in bar B is

$$^0R^0 = (\sigma_{Y\mathrm{A}} + {}^0\sigma)A \tag{9.18}$$

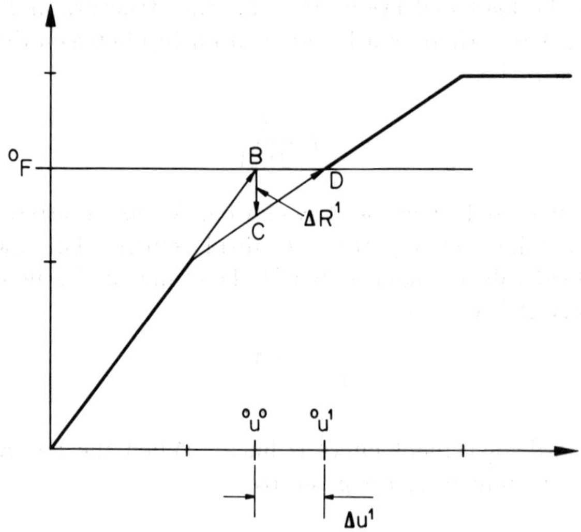

Figure 9.2 Solution of the two-bar problem by the Newton–Raphson method

whereas the applied force is still

$$^0F = 2\,{}^0\sigma A \tag{9.19}$$

According to equation (9.9) we have for the first iteration

$$\Delta u^1 = \frac{\Delta R^1}{{}^0k^1} \tag{9.20}$$

where

$$\Delta R^1 = A({}^0\sigma - \sigma_{YA}) \tag{9.21}$$

and ${}^0k^1 = \frac{EA}{l}$ is the tangential stiffness at point C.

The total displacement is computed from

$$^0u^1 = {}^0u^0 + \Delta u^1 \tag{9.22}$$

Since point D now lies on the curve the computation has converged with only one iteration. This is because the load–displacement relationship has been piecewise linear in this example and because we have used an updated stiffness ${}^0k^1$ during the iteration.

Consider what happens if the stiffness matrix is not updated during the iteration, i.e. instead of equation (9.20) we use

$$\Delta u^1 = \frac{\Delta R^1}{{}^0k^0} \tag{9.23}$$

where ${}^0k^0$ is the initial stiffness $\left(\frac{2EA}{l}\right)$

This then means that instead of going straight from C to D on the curve a point is reached which is again off the curve. A new resultant due to internal stresses has to be computed.

$${}^0R^1 = \frac{3}{2}\,{}^0\sigma A \tag{9.24}$$

and the resultant of the applied force is

$${}^0F = 2\,{}^0\sigma\ A \tag{9.25}$$

and we find that

$$\Delta R^2 = A\frac{1}{2}\,{}^0\sigma \neq 0 \tag{9.26}$$

The iteration proceeds until ΔR is sufficiently small. Figure 9.3 shows graphically the iteration steps needed.

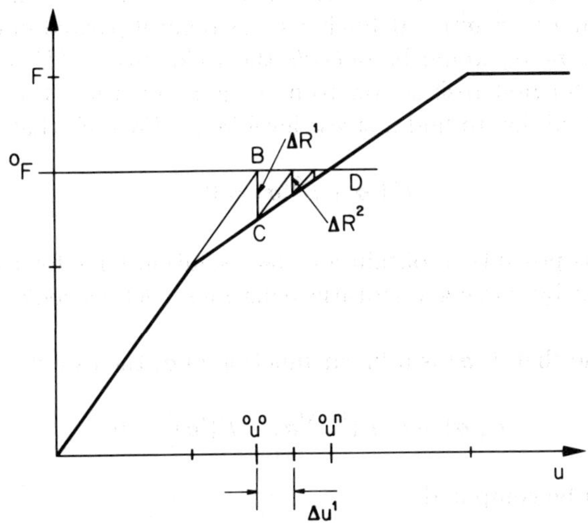

Figure 9.3 Solution of the two-bar problem by the modified Newton–Raphson method

9.3.1 Elasto-plastic problems

Consider now a three-dimensional problem subjected to a stress state $\boldsymbol{\sigma}$. As explained in Chapter 3 a material may behave elastically until a limiting stress state is reached; i.e. the linear stress–strain relationship is given by

$$\boldsymbol{\sigma} = \mathbf{D}\boldsymbol{\epsilon} \tag{9.27}$$

The limiting state is given by the yield condition

$$F(\boldsymbol{\sigma}, C_1, C_2, \dots) = 0 \tag{9.28}$$

where $C_1, C_2, \dots$ are plastic material properties. The incremental stress–strain relationship at yield is given by

$$\mathrm{d}\boldsymbol{\sigma} = \mathbf{D}_{ep}\, \mathrm{d}\boldsymbol{\epsilon} \tag{9.29}$$

where $\mathbf{D}_{ep}$ is the elasto-plastic constitutive matrix i.e.

$$\mathbf{D}_{ep} = \mathbf{D} - \mathbf{D}_p \tag{9.30}$$

The incremental relationship in equation (9.29) is only valid when $F = 0$ and the elasto-plastic theory stipulates that $F > 0$ is not permitted.

Let us consider what happens in an elasto-plastic analysis. Initially the element response matrices are computed using the elastic material properties. Thus it may occur that at time $t + \Delta t$ the state of stress ${}^{t+\Delta t}\boldsymbol{\sigma}^0 = {}^{t}\boldsymbol{\sigma} + {}^{\Delta t}\boldsymbol{\sigma}$ will be such that $F({}^{t+\Delta t}\boldsymbol{\sigma}^0) > 0$. Such a stress point if plotted in the principal stress space σ_1, σ_2, σ_3 would lie *outside* the yield surface ($F = 0$) as shown in Figure 9.4. Our first task is now to find the point where the yield surface has been traversed. i.e. to find a stress level ${}^{t}\boldsymbol{\sigma} + r\,{}^{\Delta t}\boldsymbol{\sigma}$ such that

$$F\left({}^{t}\boldsymbol{\sigma} + r\,{}^{\Delta t}\boldsymbol{\sigma}\right) = 0 \tag{9.31}$$

Although it is possible to obtain an exact solution for r for most common yield conditions by solving quadratic equations[2], let us look at a simpler alternative.

Let us assume that $F(\boldsymbol{\sigma})$ is a linear function of $\boldsymbol{\sigma}$, then we can write

$$F({}^{t}\boldsymbol{\sigma}) + r\left[F({}^{t+\Delta t}\boldsymbol{\sigma}) - F({}^{t}\boldsymbol{\sigma})\right] = 0 \tag{9.32}$$

from which can be computed

$$r = \frac{-F({}^{t}\boldsymbol{\sigma})}{F({}^{t+\Delta t}\boldsymbol{\sigma}) - F({}^{t}\boldsymbol{\sigma})} \tag{9.33}$$

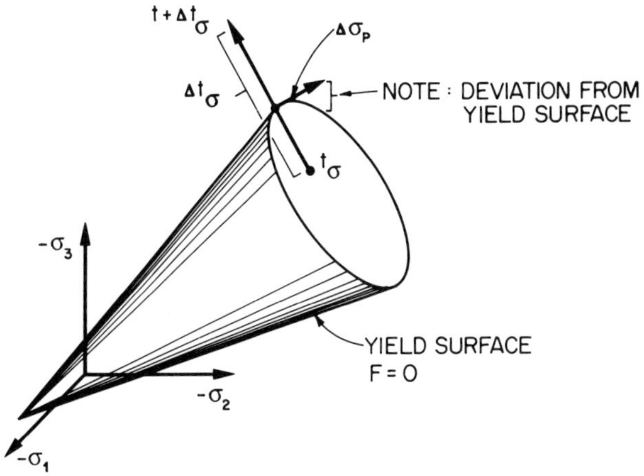

Figure 9.4 Elasto-plastic analysis: case where stress point is outside the yield surface

From the discussion of yield surfaces in Chapter 3 it is evident that $F(\boldsymbol{\sigma})$ is far from linear for most yield surfaces, thus relationship (9.33) is an approximation only. It is possible to improve this approximation by iteration[3] but this will not be discussed in detail here.

Any subsequent increase in strain from the point on the yield surface causes an increase in stress according to equation (9.29). This means that the stress path if plotted in the $\sigma_1, \sigma_2, \sigma_3$ space is tangential to the yield surface (Figure 9.4).

The equivalent increment of elastic strain which corresponds to the portion of stress increment $\left[(1-r)\,{}^{\Delta t}\boldsymbol{\sigma}\right]$ which lies outside the yield surface is

$$\Delta\boldsymbol{\epsilon}_p = \mathbf{D}^{-1}(1-r)\,{}^{\Delta t}\boldsymbol{\sigma} \tag{9.34}$$

The stress increment corresponding to this strain increment which was wrongly computed according to the elastic relationship (9.27) has to be computed according to the incremental relationship (9.29). The total stress at $t+\Delta t$ is therefore given by

$$^{t+\Delta t}\boldsymbol{\sigma}^1 = {}^{t}\boldsymbol{\sigma} + r\,{}^{\Delta t}\boldsymbol{\sigma} + \Delta\boldsymbol{\sigma}_p \tag{9.35}$$

where $\Delta\boldsymbol{\sigma}_p$ can be approximated by

$$\Delta\boldsymbol{\sigma}_p = \mathbf{D}_{ep}\mathbf{D}^{-1}(1-r)\,{}^{\Delta t}\boldsymbol{\sigma} \tag{9.36}$$

The approximation occurs because equation (9.29) is only valid for infinitesimally small increments of stress/strain. Therefore, if the yield condition is again checked for the corrected stresses ${}^{t+\Delta t}\boldsymbol{\sigma}^1$ then we may have $F({}^{t+\Delta t}\boldsymbol{\sigma}^1) > 0$. Various methods of making sure that the yield condition is not violated during an iteration exist. The simplest one is to split $\Delta\boldsymbol{\epsilon}_p$ into increments and apply equation (9.36) several times, each time recomputing the $\mathbf{D}_{ep}$ matrix after ensuring that the stress state is on the yield surface. Further, more sophisticated 'return' algorithms are discussed in reference 1.

Because we have computed our displacements during time step Δt using the elastic stress–strain relationship equation (9.27) there is now a discrepancy between the internal stresses ${}^{t+\Delta t}\boldsymbol{\sigma}^0$ and ${}^{t+\Delta t}\boldsymbol{\sigma}^1$ i.e. the results of the first iteration. This difference is computed by

$$\Delta\boldsymbol{\sigma}^1 = {}^{t+\Delta t}\boldsymbol{\sigma}^0 - {}^{t+\Delta t}\boldsymbol{\sigma}^1 \tag{9.37}$$

and contributes to the residual force vector $\Delta\{\mathbf{R}\}'$ in equation (9.9).

Equation (9.35) is quite general and can be used for computing the stresses at time $t + \Delta t$ in all cases. Table 9.1 gives the value of r in equations (9.35) and (9.36) dependent on the status of stress at time t and $t + \Delta t$. If $r = 1$ then we have obviously $\Delta\boldsymbol{\sigma}_p = \mathbf{0}$ and $\Delta\boldsymbol{\sigma}^i = \mathbf{0}$ i.e. there is no contribution to the right-hand side vector.

Table 9.1 Values of r dependent on value of F at time t and $t + \Delta t$

	t	$t + \Delta t$	r
Case 1	Elastic $F < 0$	Elastic $F < 0$	1
Case 2	Elastic $F < 0$	Plastic $F > 0$	1 to 0
Case 3	Plastic $F = 0$	Plastic $F > 0$	0
Case 4	Plastic $F = 0$	Elastic $F < 0$	1

Case 1 is where no plastic deformation occurs. Case 2 is the case just discussed when the stress state changes from elastic to plastic. Case 3 is where the point is already on the yield surface at time t and continues into the plastic range with all the stress change occurring on the yield surface. Case 4 finally is the case where the point unloads elastically from the yield surface.

9.3.2 Time dependent plasticity: visco-plasticity and creep

In the previous section we have assumed that plastic strains develop instantaneously. In reality, however, such strains take time to develop. Visco-plasticity takes this time dependence into account. The main difference to elasto-plasticity is that the yield surface is assumed to be a *plastic potential.* In visco-plasticity we postulate that the rate of plastic straining is zero if $F \leq 0$ and increases as $F > 0$. Recapitulating from Section 3.5.4 we have for the plastic strain rate:

$$\dot{\boldsymbol{\epsilon}}_p = \frac{1}{\eta}\phi(F)\frac{\partial Q}{\partial \boldsymbol{\sigma}} \tag{9.38}$$

If a power law is used then we have

$$\begin{aligned} \phi(F) &= 0 && \text{for } F < 0 \\ \phi(F) &= F^n && \text{for } F > 0 \end{aligned}$$

The procedure for a visco-plastic analysis is quite different to that of an elasto-plastic analysis as equation (9.38) is valid regardless of the value of F. There is no need to compute the point where the stresses cross the yield surface $F = 0$. Instead an infinite number of equipotential surfaces exist as illustrated in Figure 9.5. Thus a particular point may reach a higher plastic potential, $F > 0$. Instead of an iterative scheme we use a time stepping scheme. Here it is assumed that no plastic strain occurs at the instant when the point reaches the point $F > 0$. The plastic strains develop with time. Proceeding in time steps Δt, the plastic strain increment at time t is computed by

$${}^{\Delta t}\boldsymbol{\epsilon}_p = {}^{t}\dot{\boldsymbol{\epsilon}}_p \, \Delta t \tag{9.39}$$

and the total stress at $t + \Delta t$ is given by

$${}^{t+\Delta t}\boldsymbol{\sigma} = {}^{t}\boldsymbol{\sigma} + \mathbf{D}\left({}^{\Delta t}\boldsymbol{\epsilon} - {}^{\Delta t}\boldsymbol{\epsilon}_p\right) \tag{9.40}$$

where ${}^{\Delta t}\boldsymbol{\epsilon}$ is the increment in total strain for an increment in time of Δt.

The residual stress $\Delta\boldsymbol{\sigma}$ is computed from

$$\Delta\boldsymbol{\sigma} = \mathbf{D}\, {}^{\Delta t}\boldsymbol{\epsilon}_p \tag{9.41}$$

Unlike the elasto-plastic method discussed previously the time increment Δt can no longer be chosen freely and certain stability criteria dependent on the material parameter η (viscosity) are applied. If the time step is chosen too large then oscillations will occur in the iteration procedure i.e. the results become erratic. Further details can be found in reference 4.

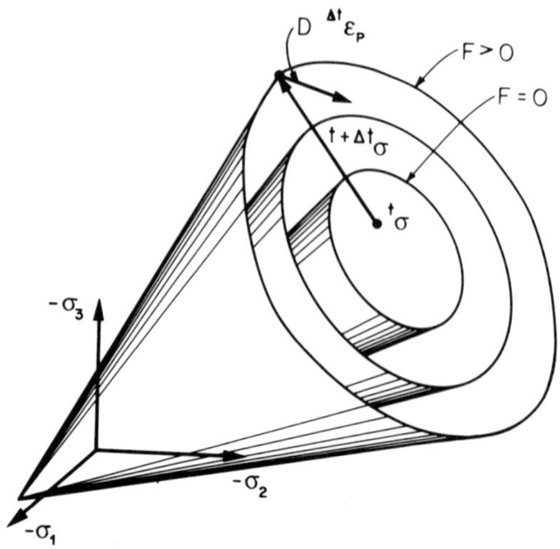

Figure 9.5 Visco-plastic analysis: state of stress at times t and $t + \Delta t$

9.3.3 Computational steps—finite element method

In the numerical implementation of the elasto-plastic and visco-plastic analysis the stress level and yield condition is checked at a finite number of points. For a finite element analysis, for example, it is convenient to use the Gaussian integration points for this purpose. The reasons for this are two-fold: firstly, these are the points where we usually compute the stresses within the element; secondly, this will allow us to compute the residual force vector by simply adding the excess stresses at the Gauss points. The computational procedure is to compute the stress and the yield condition at each Gauss point and using Sections 9.3.1 or 9.3.2 compute the reduced stresses and the stresses $\Delta\boldsymbol{\sigma}^i$ to be redistributed. These stresses contribute to the residual force vector. For a two-dimensional problem the residual force vector for element e is given by

$$\Delta\mathbf{R}^e = \int_V \mathbf{B}^{e\mathrm{T}}\,\Delta\boldsymbol{\sigma}^i\,\mathrm{d}V = \sum_{l=1}^{L}\sum_{m=1}^{M}\left(\mathbf{B}^{e\mathrm{T}}\left(\xi_l,\eta_m\right)\right)\,\Delta\boldsymbol{\sigma}^i\,\left(\xi_l,\eta_m\right)W_l W_m J \tag{9.42}$$

The computational steps for each iteration (time step) of an elasto-plastic or visco-plastic analysis are:

For each element **DO**
    For each Gauss point **DO**
        Compute the stress $\boldsymbol{\sigma}$.
        Compute yield function F.

IF $(F > 0)$ **THEN**
Compute excess stress $\Delta\boldsymbol{\sigma}^i$.
Compute $\mathbf{B}^e$ matrix.
Compute contribution to $\Delta\mathbf{R}^e =$
$(\mathbf{B}^e)^{\mathrm{T}}\,\Delta\boldsymbol{\sigma}^i\,|J|W_m W_l$
Add to other contributions
END IF
END DO
Assemble element force matrices
END DO
Assemble and solve with new right hand side

Depending on the method of analysis the global stiffness matrix is either kept constant or updated at certain intervals. The updated stiffness matrix is computed for elements which have at least one Gauss point yielded by substituting $\mathbf{D}_{ep}$ instead of $\mathbf{D}$ for yielded Gauss points.

The advantage of using a constant stiffness matrix is that the stiffness matrix once computed and reduced can be stored away so that each subsequent iteration/time step only involves a resolution which requires much less computation than an initial solution. On the other hand for such a solution many more iterations may be needed until convergence is reached.

9.3.4 Computational steps—boundary element method

In contrast to the finite element method in the boundary element method we have to first establish an additional term in the boundary element equations which deals with an initial stress inside the boundary element region. Using Betti's reciprocal theorem as explained in Chapter 6, there is additional work done by the stress increments $\Delta\boldsymbol{\sigma}$ of the real system (load case (1)) multiplied by strains $\mathbf{E}(P,Q) = \mathbf{B}\mathbf{U}(P,Q)$ which are due to the unit force (load case (2)):

$$\int_{V_p} \Delta\boldsymbol{\sigma}^{\mathrm{T}}\,(P,Q)\,\mathbf{E}(P,Q)\,\mathrm{d}V \tag{9.43}$$

The integration in the term (9.43) is taken over the plastic subregion V_p of the total region V, i.e. that part of V where plastic strains develop.

The volume/area may be divided into 'cells' as shown in Figure 9.6 which are very similar to triangular finite elements.

The procedure is then as follows:

- Values of $\Delta\boldsymbol{\sigma}$ are computed at each node of the internal mesh. We denote $\Delta\boldsymbol{\sigma}_j^c$ as the stress at the jth node of cell c.

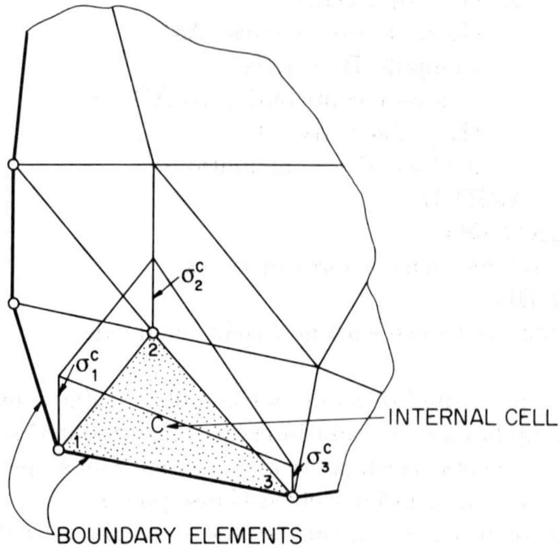

Figure 9.6 Internal cell discretisation

- Inside each cell the stresses are interpolated from the nodal values using linear interpolation functions. For example, for a point with the local coordinates $\boldsymbol{\xi}$ we have

$$\Delta\boldsymbol{\sigma}(\boldsymbol{\xi}) = \sum N_j(\boldsymbol{\xi})\Delta\boldsymbol{\sigma}_j^c \tag{9.44}$$

With this approximation it is possible to write for the term (9.43)

$$\sum_{c=1}^{n_c}\sum_{j=1}^{3}(\Delta\boldsymbol{\sigma}_j^c)^{\mathrm{T}}\ \Delta\mathbf{E}_{aj}^c \tag{9.45}$$

where $\Delta\mathbf{E}_{aj}^c = \int_{V_c}\mathbf{E}\,(P_a, Q)\,N_j\ \mathrm{d}V_c$ and n_c is the number of cells.

The discretised boundary element equations in Chapter 7 now have an additional term, i.e. equation (7.5) becomes

$$\mathbf{A}(P_a)\mathbf{u}(P_a) + \sum_{\substack{b=1\\ d(b,c)\neq a}}^{p}\sum_{c=1}^{n(b)}\Delta\mathbf{T}_{ac}^b\mathbf{u}_c^b = \sum_{b=1}^{p}\sum_{c=1}^{n(b)}\Delta\mathbf{U}_{ac}^b\mathbf{t}_c^b + \sum_{c=1}^{n_c}\sum_{j=1}^{3}\left(\Delta\boldsymbol{\sigma}_j^c\right)^{\mathrm{T}}\Delta\mathbf{E}_{aj}^b \tag{9.46}$$

The integration can be carried out analytically as explained in detail in reference 3.

Analogous to the finite element method, increments of stress inside the region will change the right-hand side of the system of equations only and require a resolution of the system of equations.

The computational steps for each iteration are

```
For each cell DO
    For each cell node DO
        Compute σ
        Compute yield condition F
        IF (F > 0) THEN
            Compute excess stress Δσ^i
            Compute contribution to right hand side
            Add contributions
        END IF
    END DO
END DO
```

Solve equations with new right-hand side.

Note that the coefficients $\Delta\mathbf{E}^{b}_{aj}$, once computed, can be stored and need not be computed at each iteration step.

9.4 EXAMPLES

A few examples of elasto-plastic and visco-plastic analyses are shown to illustrate the theory presented in this chapter.

Example 1 is that of a circular opening in an infinite domain subjected to a stress at infinity[4]. The material (properties $E = 206$ GPa, $\nu = 0.45$) was assumed to follow the von Mises yield condition (σ_Y = yield stress = 245 MPa). This example was chosen because the results can be compared with a closed form solution by Savin[5]. Figure 9.7a shows the coupled BE/FE mesh used for the analysis. The non-linear material behaviour is confined to the finite element mesh. Figure 9.7b shows the distribution of tangential stress along horizontal and vertical lines as predicted by theory and this is compared with stress values at the Gauss points. It can be seen that the tangential stress decreases sharply in the plastic region, i.e. the region where the yield condition has been reached.

The *second example* relates to a circular tunnel near the surface[6]. Two meshes are used. Figure 9.8a shows a coupled BE/FE mesh and the initial stress field in the ground before the tunnel is excavated whereas Figure 9.8b shows a mesh with boundary elements only with internal cell discretisation. The following material properties were assumed:

$$E = 1000$$
$$\nu = 0.2$$

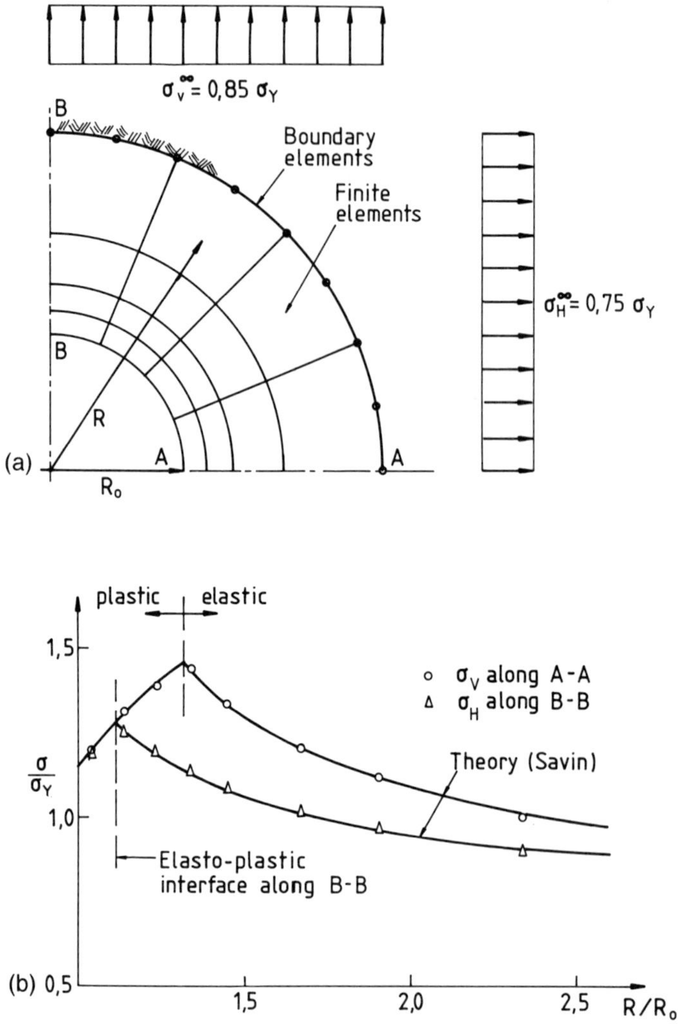

Figure 9.7 Example 1: circular opening in infinite domain (a) mesh and loading, (b) tangential stress distribution

Drucker–Prager yield condition with

$$\phi = 30°$$
$$c = 0.28$$

In Figure 9.8c is compared the extent of the zone where plastic straining has taken place and we find that both methods of analysis give similar results.

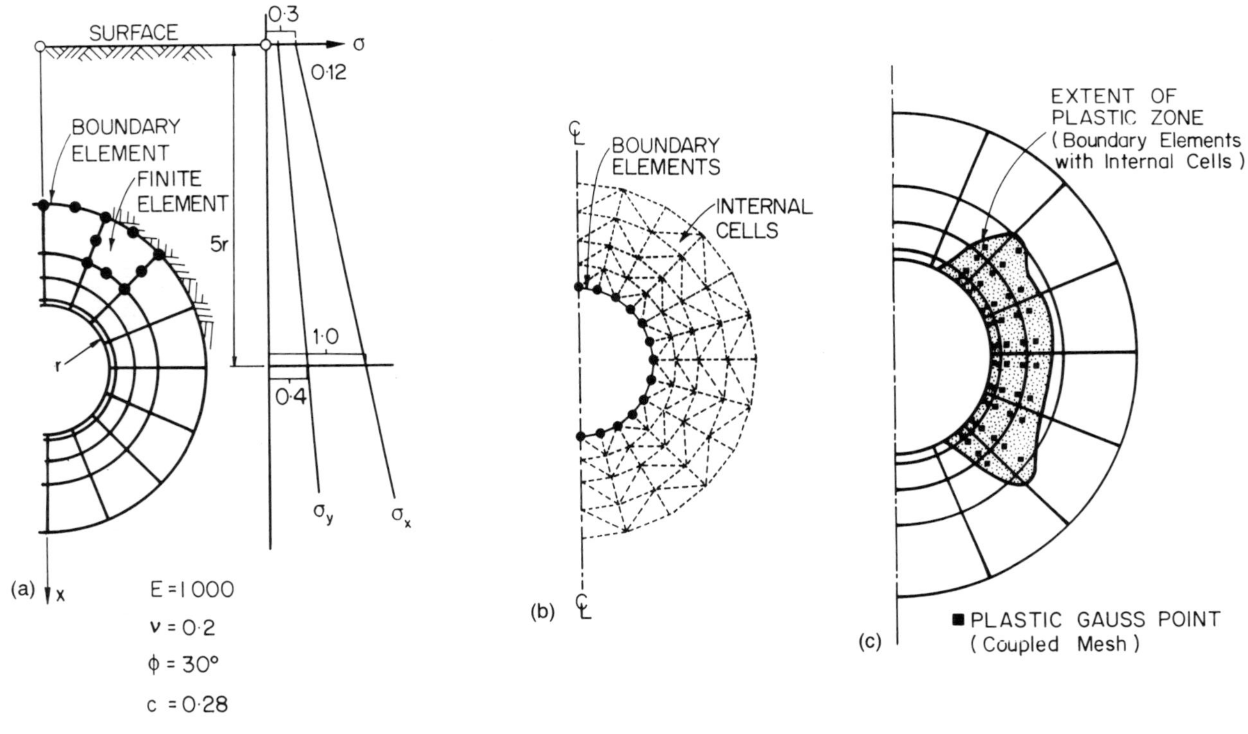

Figure 9.8 Example 2: Circular excavation near the surface. (a) Coupled mesh and virgin stress conditions. (b) Boundary element mesh and cell discretisation. (c) Comparison of plastic zones

The *third example* is taken from mining and is concerned with the behaviour of a pillar of rock left between excavations[7] in an infinite rock mass. The pre-mining stress field was assumed to be 30 MPa compression in the vertical direction. The properties of the rock mass were

$$E = 80 \text{ GPa}$$
$$\nu = 0.34$$

Two cases were considered: the first with a Drucker–Prager yield condition and

$$\phi = 50°$$
$$c = 20 \text{ MPa}$$

and a non-associative flow law ($\psi = 0°$), the second with a multi-laminate model with one plane of weakness at 30° to the horizontal and

$$\phi = 30°$$
$$c = 1 \text{ MPa}$$

and a non-associative flow law ($\psi = 0°$). A visco-plastic algorithm was used with $n = 1$ and $\eta = 6000$ d MPa.

Figure 9.9a shows the coupled BE/FE mesh used for the analysis. Here non-linear behaviour is confined to the pillar which is modelled by finite elements. Figure 9.9b shows the results obtained for the homogeneous Drucker–Prager material for the first three time steps after the excavations have been made. It can be seen that the stress decreases at the pillar surfaces as a consequence of visco–plastic deformation. Figure 9.10 shows the same results for the case where planes of weakness are assumed. Note that because the material is anisotropic the resulting stress distribution is no longer symmetric.

9.5 CONCLUDING REMARKS

In this chapter we have discussed problems which require special solution procedures because the relationship between load and displacement is no longer linear. General solution procedures for such problems were discussed.

We have discussed in detail only one type of non-linearity (where the relationship between stress and strain is non-linear) and in particular elasto–plastic problems.

A detailed discussion of the other types of non-linearity where the force–displacement relationship is affected by the displacements (large displacement problems) has not been included because they are beyond the scope of this introductory text. For such problems we have to write down the equilibrium

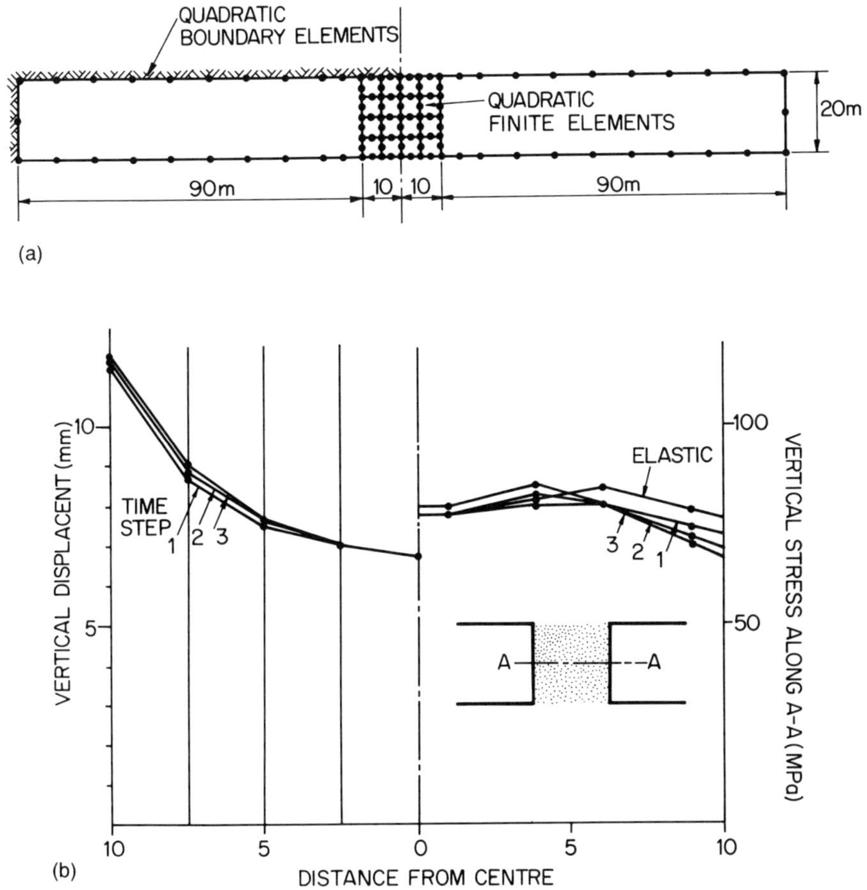

Figure 9.9 Example 3: Mine pillar problem. (a) mesh, (b) results, isotropic rock properties

equations in the displaced shape and worry about the definition of stress for excessive distortions of an element. The reader is referred to the text books cited at the end of this chapter for further in-depth study of how the residual force vectors and the tangential stiffness matrix required for the general solution procedures of Section 9.2 are obtained.

9.6 EXERCISES

9.1 Consider the two-bar problem in Figure 9.1. Assume that both bars have a cross sectional area of 1 and a Young's modulus of 10^4. Assume

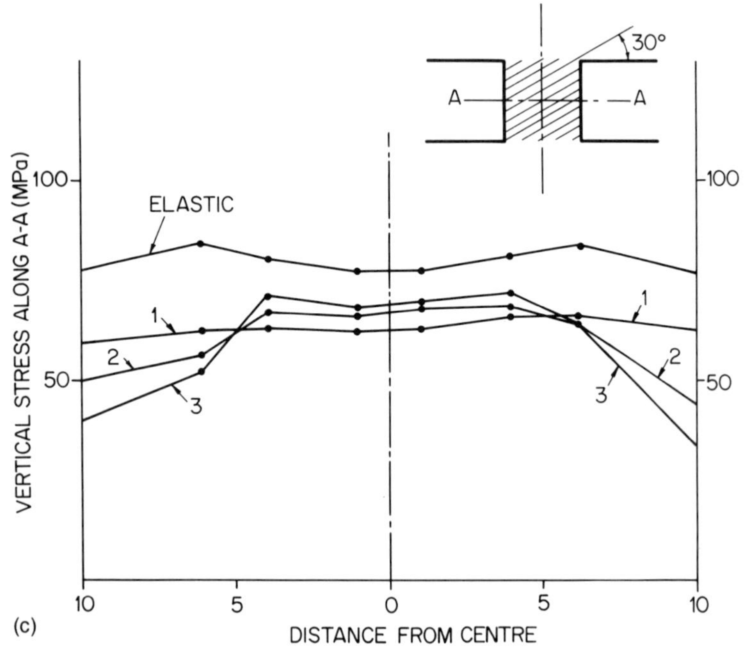

Figure 9.10 Example 3: results, anisotropic rock properties

that the yield stress in bar A is 10 and that in bar B is 20. Plot the load displacement curve for this problem

(a) Using common sense
(b) Using Newton–Raphson method
(c) Using modified Newton–Raphson.

9.2 Simulate a triaxial test of soil material. Assume the confining stress to be 1 and increase the axial stress from 1 until yielding occurs. The material properties are

$$
\begin{aligned}
&\mathrm{E} = 1000 \text{ MPa}\\
&\nu = 0.3\\
&\phi = 30^\circ\\
&C = 0\\
&\text{Drucker–Prager yield condition}
\end{aligned}
$$

Draw the stress path of the test in $\sigma_1, \sigma_2, \sigma_3$ stress space. Comment on the drift from the yield surface depending on increment of axial load used. Neglect scale effects of the specimen.

9.3 Do example 9.2 using visco-plasticity with the material values

$$\phi(F) = F^n \quad \text{for} \quad F > 0$$
$$n = 2 \quad \text{and} \quad \eta = 6000 \text{ d MPa}$$

9.4 Comment on the suitability of the Gauss points for sampling the stresses, computing excess stresses and for sampling contributions to the residual force vectors. Consider the distribution of strain—for example—in an element where only one Gauss point has yielded and the others have not.

9.5 Comment on the suitability of the boundary element method to analyse nonlinear material problems. What asumptions are made with respect to the distribution of plastic strain inside a cell[2]. Will cell size affect the results and how?

9.6 In an elasto-plastic analysis using the boundary element method are there additional degrees of freedom introduced by the internal cells?

9.7 In the finite element method it is possible to update the global coefficient matrix during nonlinear iterations. Can the same be done for the boundary element method? Give reasons for your answer.

9.7 REFERENCES

1. Owen, O.R.J. and Hinton, C. (1980) *Finite Elements in Plasticity.* Pineridge Press, Swansea.
2. Bicancic, N.P. (1989) Exact evaluation of contact stress state in computational elasto-plasticty. *Eng. Comput.* **6**, 67–73.
3. Margues, J.M.M. (1984) Stress computation in Elasto-plasticity. *Eng. Comput.* **1**, 42–51.
4. Zienkiewicz, O.C. and Cormeau, I.C. (1979) Visco-plasticity, plasticity and creep in elastic solids–a unified numerical solution approach. *Int. J. Num. Meth. Eng.*, **8**, 821–45.
5. Telles, J.C.F. (1983) *The Boundary Element Method Applied to Inelastic Problems (Lecture Notes in Engineering)*, Springer Verlag, Berlin.
6. Beer, G. (1986) Implementation of combined boundary element—finite element analysis with applications in geomechanics. *Developments in Boundary Element Methods—4, Applied Science.*
7. Savin, G.N. (1961) *Stress Concentration Around Holes*, Pergamon Press, Oxford.
8. Beer, G. and Swoboda, G. (1985) On the efficient analysis of shallow tunnels. *Comput. and Geotech.* **1**, 15–31.
9. Beer, G., Meek, J.L. and Friday, R.G. (1982) *Efficient analysis of problems in geomechanics.* Proc. 4th Int. Conf. on Num. Meth. in Geomechanics, Edmonton (ed. A.A. Balkema). pp. 5–13.

9.8. SUGGESTED FURTHER READING

Zienkiewicz, O.C. (1977) *The Finite Element Method—Third Edition*, McGraw-Hill. Chapter 19.
Bathe, K.-J. (1982) *Finite Element Procedures in Engineering Analysis*, Prentice-Hall. Chapter 6.
Venturini, W.S. (1983) *Boundary Element Method in Geomechanics (Lecture Notes in Engineering)*. Springer Verlag, Berlin.

10

Mesh Design and Error Analysis

The truth is experienced not taught

Herman Hesse

10.1 INTRODUCTION

In the previous chapters we have discussed the background to the numerical models and the techniques used to implement them. Here we shall discuss the use of the methods, in particular the design of finite, boundary element and coupled meshes.

Mesh design is an art which requires experience. A well designed mesh will give good results and be economical in the use of both human and computer resources. A badly designed mesh may give inaccurate results at a high cost and in extreme cases may give the wrong result.

In the 1970s the meshes were designed by putting transparent paper over an engineering drawing and by sketching the mesh using pencil and eraser. Nowadays it is more likely that meshes are generated using an electronic digitizer, workstation and mouse.

However, in the foreseeable future the computer is unlikely to completely take over the task of designing a mesh. Therefore this chapter will give an introduction to mesh design.

During the design of meshes the designer has to make important decisions which will affect the accuracy of the results and the economy of the analysis (i.e. the amount of computer resources needed to obtain a result).

One of the first decisions will be regarding the type of analysis used, i.e. if finite elements, boundary elements or both are used. The second decision will be regarding the type of elements used, i.e. of linear, parabolic or shell elements are used. The third decision is with respect to the fineness of the mesh, i.e. how many of the elements are used.

It is often necessary to obtain—after designing a mesh—an indication of the accuracy of the result. There are several sources of errors but the most related to mesh design is the discretisation error. We will show how estimates of the discretisation error can be obtained in a fairly simple way.

Finally there have been efforts to develop computer codes which automatically—based on error indicators—change the mesh, i.e. refine the mesh in certain areas. We will give an introduction into these adaptive mesh refinement techniques but we will see that these procedures may not be as efficient as a well designed mesh.

10.2 FINITE ELEMENT, BOUNDARY ELEMENTS OR BOTH?

The first decision which has to be made when designing a mesh is which analysis method will be most optimal.

The decision of which analysis method to use will be influenced by:

(1) The nature of the problem i.e. if it is bounded or unbounded.
(2) The nature of the material, i.e. if the material is elastic or inelastic, homogenous or inhomogeneous, isotropic or anisotropic.
(3) The types of results expected: i.e. surface stresses or stresses inside the material.

It should have become clear from the previous chapters that the BEM is best suited for unbounded problems such as excavations in rock or soil but that the FEM is best for inelastic, non-homogeneous, anisotropic material. Also, as explained previously the BEM usually gives better results on the boundary and is therefore better suited for problems where boundary stresses are important.

Figure 10.1 shows two typical examples where the BEM would be applied:

(a) an excavation in an infinite or semi-infinite elastic rock mass; and (b) a machine component with a stress concentration.

Figure 10.2 shows examples where the FEM would be typically applied:

(a) a thin plate or shell structure; and (b) a machine component where the material is non-linear and/or anisotropic.

Finally Figure 10.3 shows two typical applications of the coupled method:

(a) an excavation in an infinite domain were the material near the excavation is expected to behave in an inelastic manner; and (b) a machine component where the non-linear behaviour is localised.

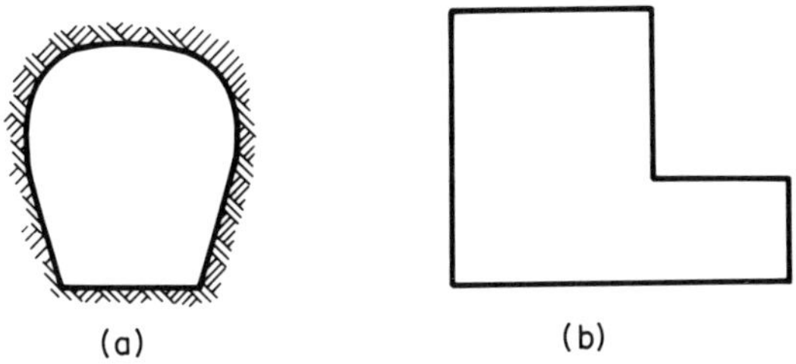

Figure 10.1 Ideal application for the BEM. (a) Excavation in infinite domain. (b) Mechanical engineering problem with surface stress concentration

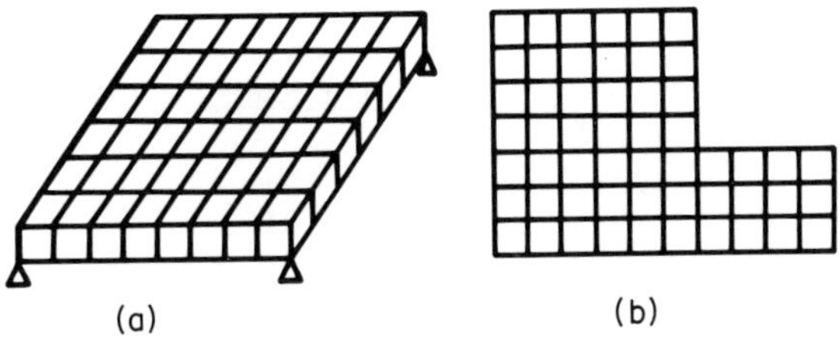

Figure 10.2 Ideal application for the FEM. (a) A plate structure. (b) A machine component with non-linear material

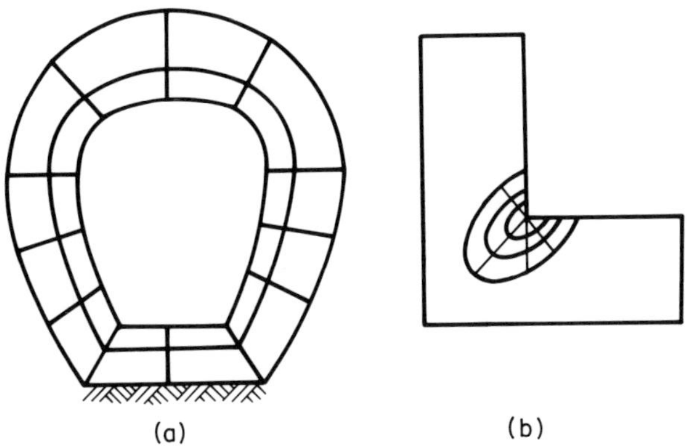

Figure 10.3 Ideal applications of the coupled method. (a) Excavation with localised material non-linearity. (b) Machine component with localised non-linear behaviour

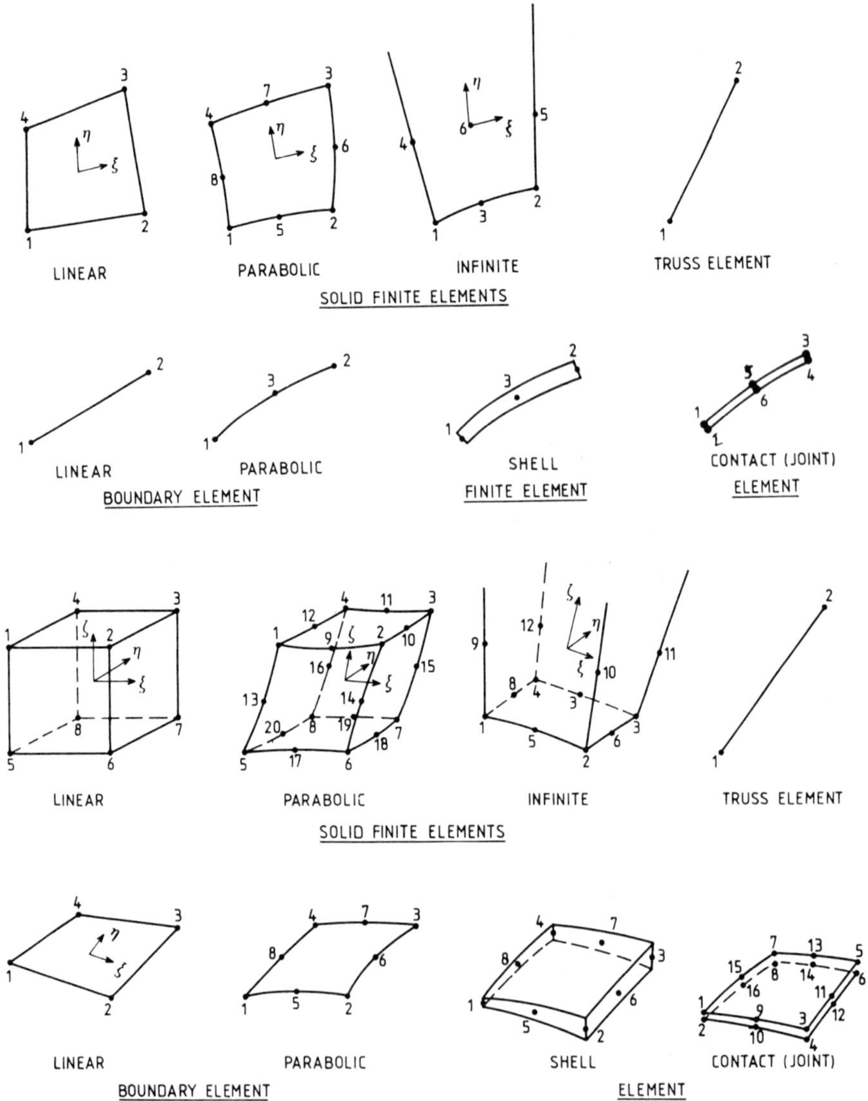

Figure 10.4 Element library of the coupled BE/FE program BEFE

A computer program with an element library such as the one shown in Figure 10.4 will allow the user to select linear and parabolic finite and boundary elements, truss elements, shell elements and joint elements.

10.3 DESIGN OF MESHES

Once we have selected the most appropriate method of analysis the next task is to design an appropriate mesh.

The design will be influenced by:

(a) shape of the boundary
(b) material interfaces
(c) zones of rapid variation of the unknown
(d) available resources, program, cost limitations
(e) expected accuracy of the results.

An example of constraints (a) and (b) is given in Figure 10.5 which shows the mesh for analysing a layered embankment. Note the awkward shape of some elements because of the need to follow material interfaces. The use of triangular elements would have made things easier.

An example of constraint (c) is given in Figure 10.6 where the mesh of a lined tunnel has been refined near the lining because of the more rapid variation in stress and displacement there.

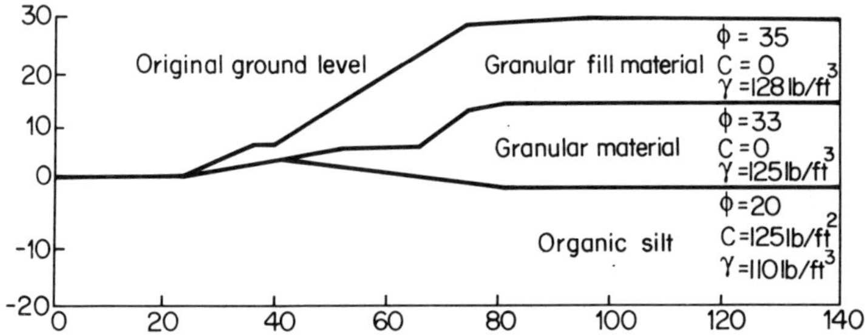

(a)

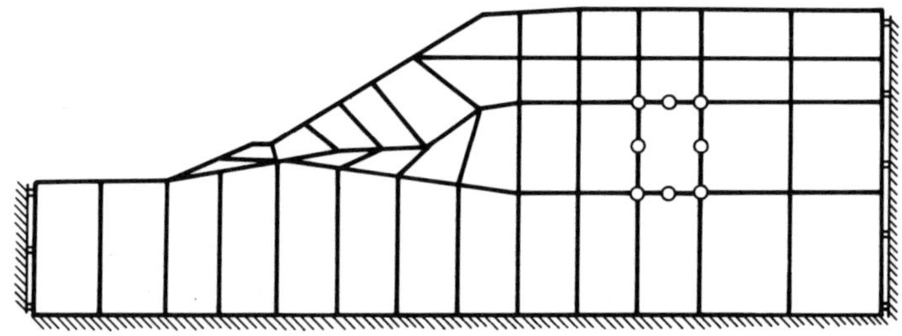

(b)

Figure 10.5 Layered embankment problem. (a) Geometry and material properties. (b) finite element mesh

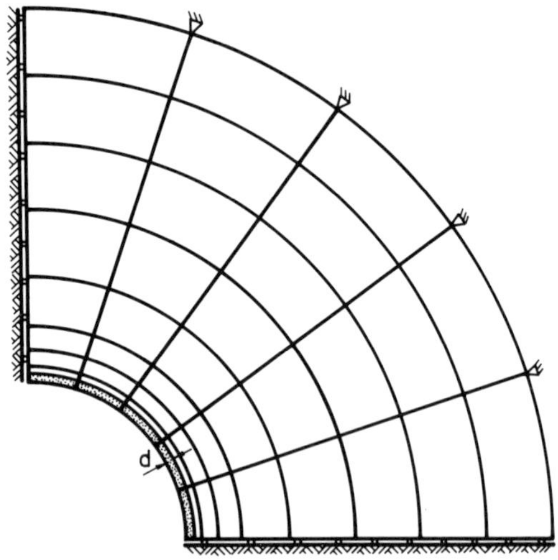

Figure 10.6 Lined tunnel—finite element mesh

Any mesh must be a compromise between available resources, both in computer and man hours and required accuracy. Ideally we would like a very fine mesh but program and resource limitations may make this impossible. So we must always aim for an optimal mesh. This means that we use a coarse mesh where the variation of the unknown is less rapid or where detailed results are not required and refine the mesh in other areas.

Figure 10.7a and b show methods of refining a conventional finite or boundary element mesh, i.e. ways of increasing the element density from one to two, three and more elements.

Another way of refining a mesh is to increase the element order by using the mixed linear/parabolic elements discussed in the chapter on solid modelling and shown in Figure 10.7c.

These mesh refinement types allow us to rapidly increase and decrease mesh density in areas where the unknown varies more rapidly.

10.4 PITFALLS

Here we shall discuss common errors made in the design of meshes.

One common mistake made by novices to the method is to design meshes which are not properly connected. We know from the previous chapters that elements must be connected to each other by the nodes as for example shown in Figure 10.8a. A subdivision as in Figure 10.8b is not usually allowed unless special elements (such as, for example, the hierarchic elements discussed in Section 10.7) are used.

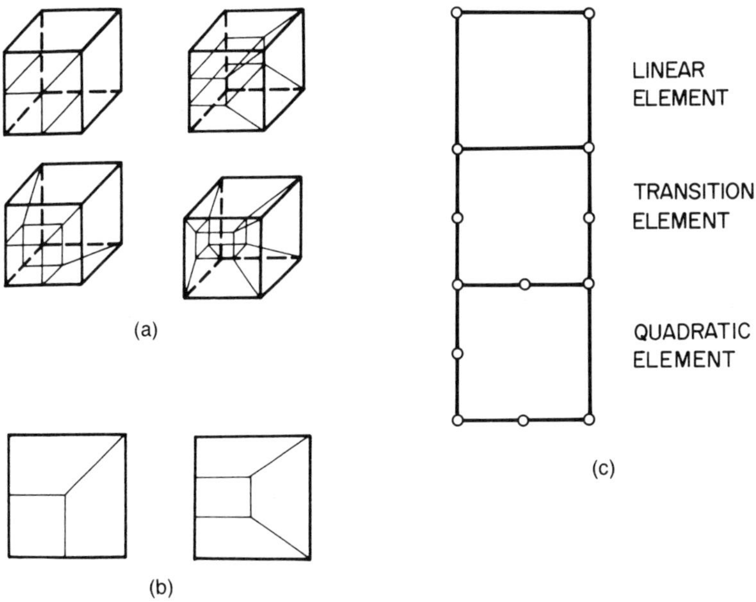

Figure 10.7 Possible refinement of meshes. (a) For three-dimensional finite elements. (b) For boundary elements and two-dimensional finite elements. (c) By increasing the degree of polynomal

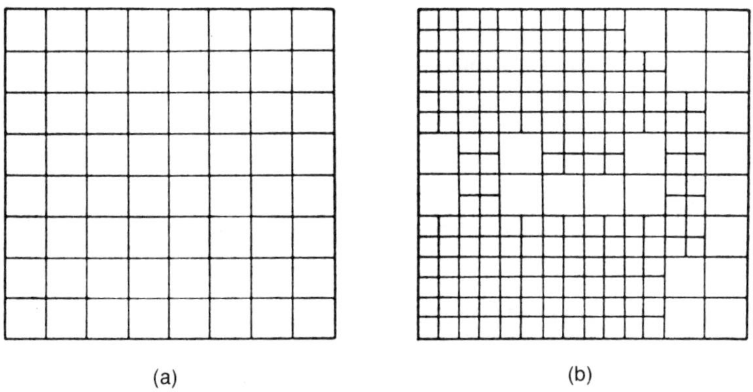

Figure 10.8 Elements have to be connected to each other by their nodes as shown in (a). The mesh in (b) is not allowed unless constraint equations are used on the interfaces between elements

A second common mistake is to allow highly distorted elements in the mesh. Figure 10.9 shows an example of distorted shapes of a quadilateral element. We know from Chapter 2 that the element in (a) will not work because the Jacobian is negative. The element in Figure 10.9b is a borderline case and not recommended because inaccurate results will be obtained near node n. The element in Figure 10.9c is adequate. Figure 10.9d shows in more detail the non-unique mapping obtained for the element in Figure 10.9a.

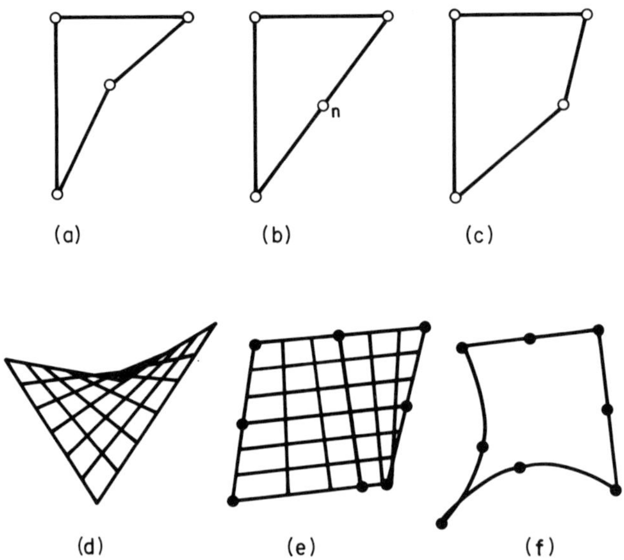

Figure 10.9 Examples of distorted elements

If the midside node is severely off-centre then this will have an undesired effect on the mapping as shown in Figure 10.9e. Finally if two sides of a quadratic element are too curved there will be an overlap as shown in Figure 10.9f.

It is fairly easy to implement coding in the computer program which at an early stage—before the analysis is started—computes the Jacobian at each Gauss point and prints out an error message if it is negative.

10.5 SOURCES OF ERROR

In a BE or FE analysis we try to approximate the behaviour of the prototype (rock mass, machine component etc) as closely as possible. The degree of accurary with which we can model the behaviour will depend on the assumptions introduced in the modelling process.

In 1982 E. L. Wilson wrote 'The majority of finite element analyses being conducted are in large error and, in many cases, are not as good as approximate solutions...'. Whilst this statement sounds rather extreme there is probably some element of truth in it.

There are several sources of error which can lead to poor approximation of the real situation.

(1) Poor input data because of lack of information.
(2) Simplified assumptions.
(3) Error in describing the boundary of problem.
(4) Numerical round-off (in solving the simultaneous equations) in the computer.
(5) Discretisation error.

Each of the sources of error in the analysis can be significant. In the extreme case they can lead to analysis results which are completely different to the actual prototype behaviour.

The first three sources of error relate to the input into the model. Here the old saying of *'garbage in, garbage out'* holds true. The fourth source of error has to do with the accuracy with which the computer stores numbers. The final source of error is related to the approximation introduced by the numerical modelling technique, i.e. errors introduced by the use of shape functions in the finite element method etc. This source of error will be discussed in more detail in Section 10.6.

10.5.1 Poor input data

There are some examples of modelling where we have exact information about the prototype to be modelled. For example the geometry of a machine component will be exactly defined to the last hundredth of a millimetre and the material properties will be known with an accuracy of $\pm 1\%$. There are other examples however where the prototype is poorly defined. For example the geometry of a mine excavation is in many cases not exactly defined but depends on the blasting pattern and the way the rock mass will fragment. Real excavation outlines may often differ by meters from the design outline. Also, the properties of the material rock are not accurately known and there may be up to 30% error. One can not do much except, maybe, be aware that these errors exist. There seems, for example, to be no sense in trying to achieve a very high accuracy in the model if we know that the input data is of poor quality.

10.5.2 Simplified modelling assumptions

In modelling we sometimes try to simplify the model so as to save effort in data preparation and computing time. One of the most common simplifying assumptions is to assume plane strain or plane stress conditions for a real three-dimensional state of stress. For example the stress distribution around a very long tunnel may be assumed to be plane strain but this will be in error near the end of the tunnel where it becomes three dimensional. On the other hand we may assume that the material is linearly elastic and homogeneous where in reality it is not. Such simplified assumptions can give answers which are sufficiently accurate for some purposes but care must be taken, because in some cases oversimplifications may lead to completely wrong answers.

10.5.3 Error in describing boundary of problems

In the model we have to approximate the boundary and the boundary conditions in some way. For example linear elements are a poor approximation for a circular tunnel unless many elements are used. Some problems are more sensitive to errors in boundary shape. For example a shell structure loaded in compression (see Chapter 5) may be much more sensitive to boundary shape than, for example, an excavation. This sort of error can be minimised by using a mesh of sufficient density and order.

10.5.4 Numerical round-off

Numbers are stored in the computer in an internal (binary) format with a limited number of bits. For example in normal (single) precision 32 bits are allocated to store the mantissa and exponent. In practice this means that only up to eight significant digits can be stored.

Consider for example the case where during assembly, an element is added whose response coefficient is very much bigger than the already assembled coefficient (e.g. if we add a very stiff spring to a structure):

$$\begin{array}{rl} 898.00 & \\ & + \\ 1234\,567\,000.00 & \\ \hline 1234\,567\,898.00 & \end{array}$$

Then, since the result of the addition can only be stored as eight digits, the computer—depending on make—either truncates the number after the eighth digit i.e. to 1234 567 800 or rounds it off to 1234 567 900. Both cases are loosely known as 'round-off'. In any case we can see that the newly introduced coefficient will completely 'swamp' the existing coefficient. This may occur not only when we add a much stiffer element but on other less obvious occasions. Consider for example, the long thin cantilever in Figure 10.10. As we assemble elements at the end, round-off may occur because the bending stiffness of the whole cantilever is very much lower than the stiffness of an individual (small) element.

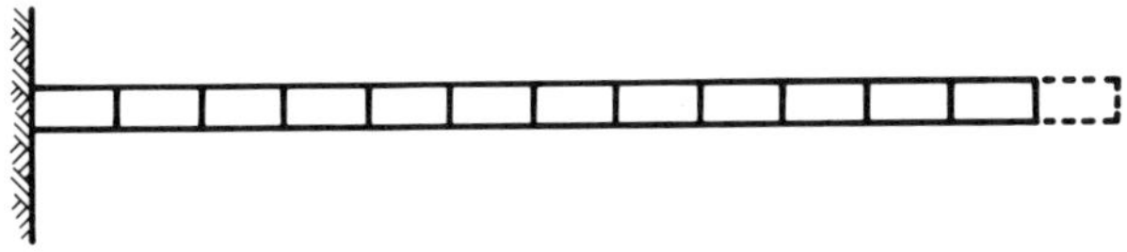

Figure 10.10 A slender cantilever beam. Round-off may occur as more and more elements are added to the ends because the stiffness of the cantilever is much less than the stiffness of a small element

Round-off is a potentially serious problem because answers may be obtained from the analysis which look reasonable but are completely wrong. In most cases this problem can be solved by declaring the variables in the solution to be double precision with the effect that we double the number of digits which can be stored.

A relatively simple and effective check can be made of the round-off error and this is strongly recommended. One simply multiplies the result vector $\{\mathbf{u}\}$ with the global stiffness or coefficient matrix and subtracts the force vector. Then we have

$$[\mathbf{K}]\{\mathbf{u}\} - \{\mathbf{F}\} = \{\mathbf{R}\} \tag{10.1}$$

where $\{\mathbf{R}\}$ is a residual force vector which should be zero, but is not zero if round-off errors have occurred. It gives an indication of the round-off error introduced in the solution. Usually we take a norm of $\{\mathbf{R}\}$ to be the measure of error. If this error is too great we can refine the solution—i.e. make it more accurate by resolving the system with $\{\mathbf{R}\}$ as the new right-hand side until the residual becomes sufficiently small. For example, a correction for the displacements $\delta\{\mathbf{u}\}$ may be obtained by

$$\delta\{\mathbf{u}\} = [\mathbf{K}]^{-1}\{\mathbf{R}\} \tag{10.2}$$

The computational overhead is not too great because each application of (10.2) only involves a re-solution of the system of equations.

10.5.5 Discretisation errors

Errors are also introduced by the approximation made to the actual variation of the unknown inside a finite element or boundary element because either the mesh is not sufficiently fine or the order of the shape function is not sufficiently high.

Finally errors are introduced into the results due to lack of satisfaction of the integral equations in a boundary element analysis. For example if point collocation is used then satisfaction of Green's or Betti's theorem is only guaranteed at the collocation points and if these are too widely spaced because the mesh is too coarse then there will be a lack of satisfaction between these points.

The last two sources of errors are known as discretisation errors and will be discussed in detail in the next section.

10.6 ESTIMATES OF DISCRETISATION ERROR

Discretisation errors may be introduced by the fact that the actual variation of the unknown is not approximated well by the element shape functions.

To explain this, consider the problem in Figure 10.11 consisting of an elastic string[2] of length 1 and cross sectional area A which is subjected to a body force b. The differential equation is

$$E\frac{\mathrm{d}^2u}{\mathrm{d}x^2} + b = 0 \tag{10.3}$$

subject to boundary conditions $u = 0$ at $x = 0$ and $\mathrm{d}u/\mathrm{d}x = 0$ at $x = 1$, and the exact solution for the displacement is given by

$$u = \frac{b}{E}\left(x - \frac{x^2}{2}\right) \tag{10.4}$$

Let us try to solve this problem with one linear element only (mesh 1). Referring to Chapter 4 the solution at the end of the string ($x = 1$) is

$$u_2 = \frac{b}{2E} \tag{10.5}$$

which is exactly the same as the theoretical solution and since $u_1 = 0$ also the error is zero at both ends.

However, since the assumed distribution of displacement is linear, errors will occur between the nodes. The assumed displacement field is

$$u_h = \frac{1}{2}\frac{b}{E}x \tag{10.6}$$

and the error

$$e = u - u_h = \frac{1}{2}\frac{b}{E}(x - x^2) \tag{10.7}$$

As a global measure of error an energy norm is usually used. This is expressed by the strain energy of the error:

$$\|e\|^2 = \frac{1}{2}\int_\Omega (\boldsymbol{\epsilon} - \boldsymbol{\epsilon}_h)^{\mathrm{T}}\, \mathbf{D}\, (\boldsymbol{\epsilon} - \boldsymbol{\epsilon}_h)\ \mathrm{d}\Omega \tag{10.8}$$

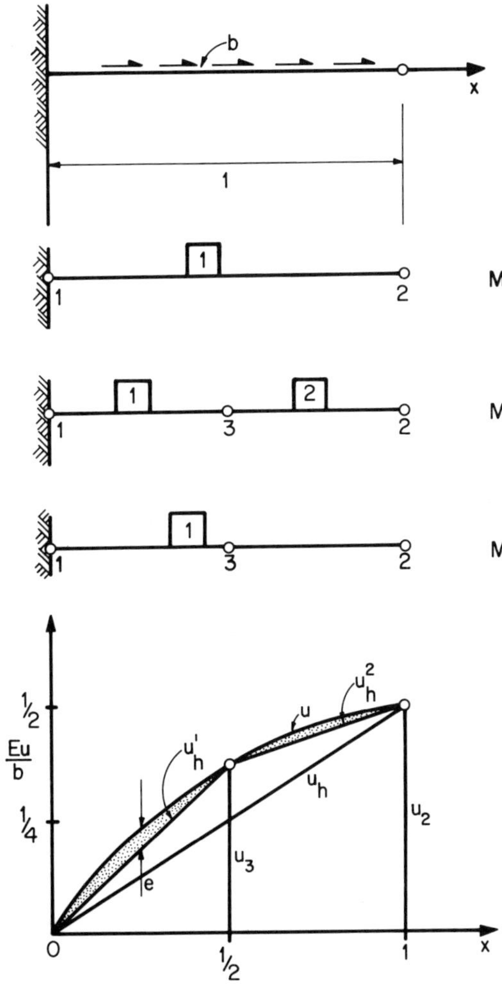

Figure 10.11 Simple example to explain discretisation error

For the one dimensional problem this is given by

$$\|e\|^2 = \frac{K}{2} \int_0^1 \left(\frac{\mathrm{d}e}{\mathrm{d}x}\right)^2 \mathrm{d}x \tag{10.9}$$

where $K = EA$. For our particular example with one linear element,

$$\|e\| = \frac{b}{\sqrt{24K}} \tag{10.10}$$

We find that although the error in the displacement is zero at the end of the string an error in the stress occurs, i.e. the stress is constant,

$$\sigma = E\frac{\mathrm{d}u_h}{\mathrm{d}x} = \frac{b}{2} \tag{10.11}$$

whereas it should be increasing from zero towards the fixed end of the string.

Consider now mesh 2 with two linear elements. The finite element solution gives

$$u_3 = \frac{b}{E}\frac{3}{8}$$

$$u_2 = \frac{b}{E}\frac{1}{2} \tag{10.12}$$

and again this is equal to the exact solution at these points.

However the assumed displacement field is linear in element 1,

$$u_h^1 = \frac{3}{4}\frac{b}{E}x \tag{10.13}$$

and element 2,

$$u_h^2 = \frac{b}{E}\left(\frac{1}{4} + \frac{x}{4}\right) \tag{10.14}$$

The error in element 1 is given by

$$e^1 = u - u_h^1 = \frac{b}{E}\left(\frac{x}{4} - \frac{x^2}{2}\right) \tag{10.15}$$

and in element 2 by

$$e^2 = u - u_h^2 = \frac{b}{E}\left(\frac{3x}{4} - \frac{x^2}{2} - \frac{1}{4}\right) \tag{10.16}$$

The energy norm of the error is given by the sum of the error norms in elements 1 and 2.

$$\|e\|^2 = \frac{K}{2}\left(\int_0^{\frac{1}{2}}\left(\frac{\mathrm{d}e^1}{\mathrm{d}x}\right)^2 \mathrm{d}x + \int_{\frac{1}{2}}^{1}\left(\frac{\mathrm{d}e^2}{\mathrm{d}x}\right)^2 \mathrm{d}x\right) \tag{10.17}$$

$$\|e\| = \frac{b}{\sqrt{96K}} \tag{10.18}$$

Again, although the error in displacement is zero at $x = 1/2$ there is an error in the stress values i.e. the stresses in element 1 are not in equilibrium with the stresses in element 2.

We have

$$\sigma^1 = E\frac{\mathrm{d}u^1}{\mathrm{d}x} = \frac{3}{4}b$$

$$\sigma^2 = E\frac{\mathrm{d}u^2}{\mathrm{d}x} = \frac{1}{4}b \tag{10.19}$$

and the difference in stress is

$$\Delta\sigma = \sigma^2 - \sigma^1 = -\frac{1}{2}b \tag{10.20}$$

If we use mesh 3, i.e. one parabolic element then it can be easily shown that the exact solution is obtained i.e. the error in displacements and stress is zero. This is because the solution can be exactly represented by a parabola.

The results of the analysis so far are summarised in Table 10.1 and we can see that if the mesh is increased from one to two elements the error norm is exactly halved i.e. the error can be related to the size of the element by

$$\|e\| = Ch \tag{10.21}$$

where h is the size of the element and C is a constant.

Table 10.1 Summary of results: one-dimensional example

No. of elements n	Degree of polynomial p	Error norm $\|e\|\sqrt{K}/b$
1	1	0.204
2	1	0.102
1	2	0

It can be seen that the most efficient way of increasing the accuracy of the results in this case is by increasing the degree of polynomial rather than increasing the number of elements. However, this is not always the case.

In most cases when we do a finite element analysis, we do not know the exact solution of the problem (if we do, there is no need for an approximate solution). Therefore we can not compute the error norm exactly as shown above but have to estimate it.

There are basically two types of error which occur in a finite element analysis:

(1) The error in the satisfaction of the differential equation due to the assumed variation of the unknown.
(2) The error in the satisfaction of the interelement equilibrium due to the fact that we use only piecewise continuous elements.

To determine the error of type (1) we simply substitute the computed displacement field into the governing differential equation. In general this equation will then not be satisfied and we can work out a residual r for each element such that

$$r = Lu_h + b \tag{10.22}$$

where for our simple one-dimensional case

$$L = E\frac{\partial^2}{\partial x^2}$$

To determine the second type of error we may integrate over the surfaces of a two- or three-dimensional element the discrepancy in the stress $\Delta\boldsymbol{\sigma}$ as computed for example in (10.20) for the one-dimensional case. A possible estimate of the energy norm is given by[5]

$$\|e\|^2 = C_1 \int_\Omega r^2 \,\mathrm{d}\Omega + C_2 \int_S \Delta\boldsymbol{\sigma}^2 \,\mathrm{d}S \tag{10.23}$$

Details of such estimates are given in reference 5.

For boundary element analysis the situation is somewhat simpler because the differential equation is always satisfied inside the BE region. Therefore errors only occur due to the satisfaction of the boundary conditions and the satisfaction of Green's and Betti's theorems. If point collocation is used these theorems are only satisfied at the collocation points. Formal error estimates have been worked out for the boundary element method but they are restricted to regular meshes using the Galerkin method.

There are simpler ways of checking that the results of an analysis are sufficiently accurate. One way would be to simply double the number of elements in the mesh, for example by subdividing each element in the mesh, and then comparing the results. We can then estimate the error as the norm of differences between fine and coarse meshes. An example of this method is shown in Figure 4.15.

Another way of estimating the error is by increasing the order of the shape functions i.e. from linear to parabolic, parabolic to cubic etc. The reader is referred to an example of this shown in Section 4.6.1.

Obviously such methods of determining the errors are very wasteful in resources. Simply increasing the number of elements for the whole mesh and not only in areas where needed, is not very efficient and can be very costly for large meshes.

10.7 HIERARCHICAL ELEMENTS AND ADAPTIVE MESH REFINEMENT

There are computer techniques available which allow us to automatically change the mesh locally during or after an analysis based on local error estimates. Here we will briefly introduce one such method: i.e. the *hierarchical finite element method.* There are basically two ways in which we can refine the mesh:

(1) by subdividing elements (h-refinement)
(2) by increasing the degree of polynomial (p-refinement)

The concept of the method is that the element shape functions are changed in such a way that additional nodes (degrees of freedom) do not change the original shape functions but introduce new ones which are added to the original ones.

Consider for example the one dimensional linear shape function in Figure 10.12. The original shape functions are known to be

$$\begin{aligned} N_1 &= \frac{1}{2}(1-\xi) \\ N_2 &= \frac{1}{2}(1+\xi) \end{aligned} \qquad (10.24)$$

and the variation of some quantity a inside the element is given by

$$a = N_1 a_1 + N_2 a_2 \qquad (10.25)$$

In a normal refinement we simply divide the element into two elements and have

$$\begin{aligned} N_1^1 &= \frac{1}{2}(1-\xi^{(1)}) \qquad & N_1^2 &= \frac{1}{2}(1-\xi^{(2)}) \\ N_2^1 &= \frac{1}{2}(1+\xi^{(1)}) \qquad & N_2^2 &= \frac{1}{2}(1+\xi^{(2)}) \end{aligned} \qquad (10.26)$$

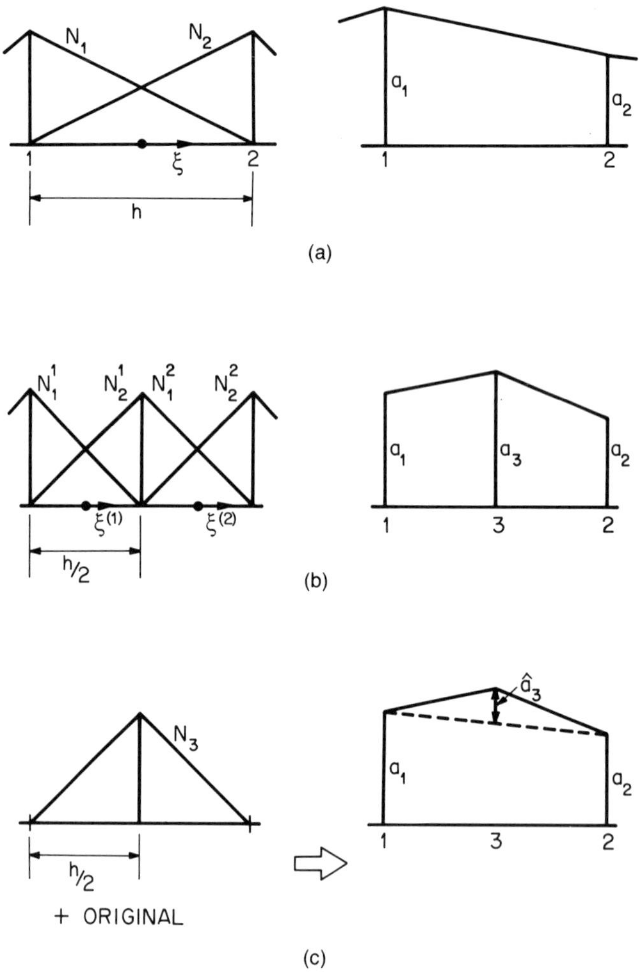

Figure 10.12 One-dimensional shape functions. (a) original, (b) normal h-refinement, (c) hierarchical h-refinement

and

$$\begin{aligned} a &= N_1^1 a_1 + N_2^1 a_3 \quad \text{for element 1} \\ a &= N_1^2 a_3 + N_2^2 a_2 \quad \text{for element 2} \end{aligned} \tag{10.27}$$

In a hierarchic refinement we introduce

$$N_3 = \begin{cases} 1 + \xi, & \text{if } \xi < 0 \\ 1 - \xi, & \text{if } \xi > 0 \end{cases} \tag{10.28}$$

and compute the distribution of a by

$$a = N_1 a_1 + N_2 a_2 + N_3 \hat{a}_3 \tag{10.29}$$

where N_1 and N_2 are the original shape functions and $\hat{a}_3$ is the difference to the linear distribution at point 3. We can immediately see the advantage of the hierarchical approach as the refined interpolation is obtained by simply adding shape functions. In a sense this technique is similar to the well known Fourier series approach to approximating functions.

In the p-refinement (Figure 10.13) we have the normal refinement as

$$\begin{aligned} \bar{N}_1 &= \frac{1}{2}\left((1-\xi) - (1-\xi^2)\right) \\ \bar{N}_2 &= \frac{1}{2}\left((1+\xi) - (1-\xi^2)\right) \\ \bar{N}_3 &= 1-\xi^2 \end{aligned} \tag{10.30}$$

and

$$a = \bar{N}_1 a_1 + \bar{N}_2 a_2 + \bar{N}_3 a_3 \tag{10.31}$$

In the hierarchical refinement we simply add,

$$N_3 = 1 - \xi^2 \tag{10.32}$$

to the linear functions and have

$$a = N_1 a_1 + N_2 a_2 + N_3 \hat{a}_3 \tag{10.33}$$

where $\hat{a}_3$ is again the deviation of a from a linear distribution.

The use of hierarchical shape functions for two-dimensional problems is completely analogous. Figures 10.14 and 10.15 show the hierarchical form of shape functions for the h- and p-refinement.

The hierarchical forms of the shape function have one major advantage: as we introduce refinement (i.e. new nodes) the original shape functions are retained and new ones added. This means that the original stiffness matrix can be retained and the effect of the hierarchical nodes simply added on.

Let us apply the hierarchical h-refinement to the problem of the loaded string in Figure 10.11 and let us introduce a new hierarchic node in mesh 1. Then we have for the displacement distribution,

$$u = N_1 u_1 + N_1 u_2 + N_3 \hat{u}_3 \tag{10.34}$$

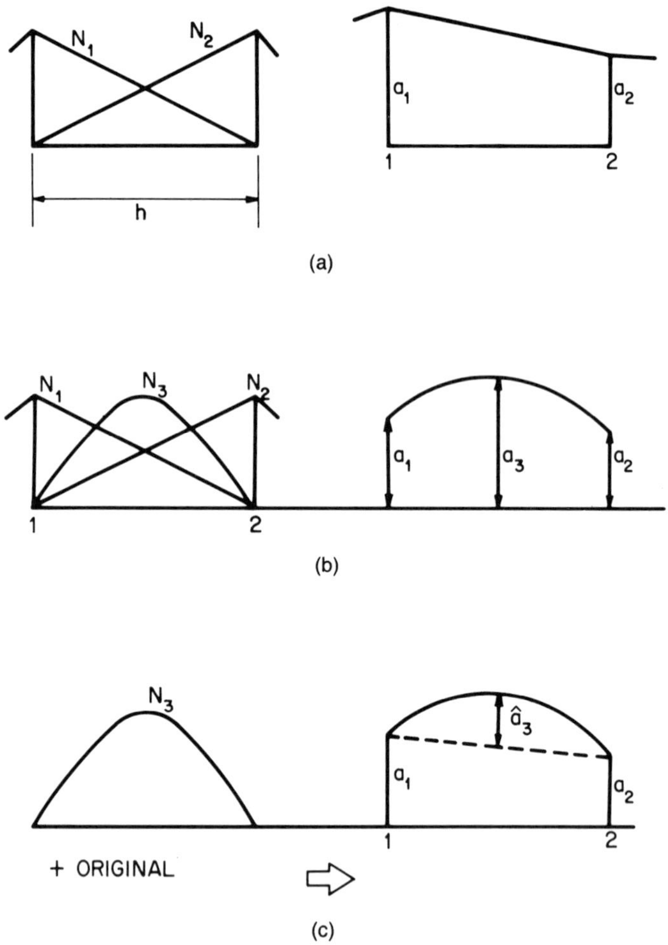

Figure 10.13 One-dimensional shape functions. (a) original (unrefined), (b) normal p-refinement, (c) hierarchical p-refinement

For this particular example $\hat{u}_3$, being the difference between the linear and the bilinear distribution is actually equal to the error of the solution at point 3.

Following Chapter 4 we have for the $\mathbf{B}^e$ matrix,

$$\mathbf{B}^e = \left[\frac{\partial N_1}{\partial x}, \frac{\partial N_2}{\partial x}, \frac{\partial N_3}{\partial x} \right] \tag{10.35}$$

where the first two terms are the same as for the linear element and the global derivatives of N_3 as defined in equation (10.28) are given by

$$\begin{aligned}\frac{\mathrm{d}N_3}{\mathrm{d}x} &= 2 \quad \text{for} \quad \xi < 0\\ \frac{\mathrm{d}N_3}{\mathrm{d}x} &= -2 \quad \text{for} \quad \xi > 0\end{aligned} \tag{10.36}$$

Hence the $\mathbf{B}^e$ matrix is given by

$$\begin{aligned}\mathbf{B}^e &= [-1, 1,\ 2] \quad \text{for} \quad \xi < 0\\ \mathbf{B}^e &= [-1, 1, -2] \quad \text{for} \quad \xi > 0\end{aligned} \tag{10.37}$$

Because $\mathbf{B}^e$ is discontinuous over the integration range the strain energy is given by the sum of two integrals,

$$U = \frac{A}{2}\left(\int_{x=0}^{\frac{1}{2}} \varepsilon\sigma \ \mathrm{d}x + \int_{x=\frac{1}{2}}^{1} \varepsilon\sigma \ \mathrm{d}x\right) \tag{10.38}$$

Using the standard process of Chapter 4 the stiffness matrix is computed as

$$\mathbf{K}^e = \frac{A}{2}(\mathbf{B}^{e\mathrm{T}} E\, \mathbf{B}^e + \mathbf{B}^{e\mathrm{T}} E\, \mathbf{B}^e) \tag{10.39}$$

which gives

$$\mathbf{K}^e = EA\begin{bmatrix} 1 & -1 & 0\\ -1 & 1 & 0\\ 0 & 0 & 4\end{bmatrix} \tag{10.40}$$

Note that this stiffness matrix is the same as for the linear element except that a row and a column have been added. Note also that the new off-diagonal terms are zero but unfortunately this is only true for one-dimensional problems.

The right hand side vector due to a body force b is given by

$$\mathbf{F}^e = Ab\int_{x=0}^{1} \mathbf{N}^e \ \mathrm{d}x \tag{10.41}$$

that is,

$$\mathbf{F}^e = \begin{Bmatrix} \frac{1}{2}\\ \frac{1}{2}\\ \frac{1}{2}\end{Bmatrix} Ab \tag{10.42}$$

where A is the cross sectional area.

Without solving the system of equations we can obtain immediately by back-substitution into the last equation,

$$\hat{u}_3 = \frac{b}{8E} = e_3 \tag{10.43}$$

The error corrected by the new degree of freedom is given by

$$e = N_3 e_3 \tag{10.44}$$

The derivative of the error is given by

$$\frac{\mathrm{d}e}{\mathrm{d}x} = \frac{\partial N_3}{\partial x} e_3 = \pm \frac{b}{4E} \tag{10.45}$$

and an estimate of the error now is obtained after substitution in equation (10.17) and integration

$$\begin{aligned} \|e\|^2 &= \frac{K}{2}\left[\left(\frac{b}{4K}\right)^2 \frac{1}{2} + \left(\frac{b}{4K}\right)^2 \frac{1}{2}\right] \\ &= \frac{b^2}{32K} \end{aligned} \tag{10.46}$$

or,

$$\|e\| = \frac{1}{\sqrt{32}} \frac{b}{\sqrt{K}} = 0.176 \frac{b}{\sqrt{K}} \tag{10.47}$$

which underestimates the exact global error as calculated under equation (10.8) by 14%.

The example demonstrates the following.

(1) The introduction of an hierarchic node gives just one additional term in the stiffness matrix with the remainder unchanged.
(2) A simple error estimate which estimates the improvement in the solution accuracy due to introducing the additional degree of freedom is obtained without having to solve the entire system of equations.

The concept may be generalised to two- and three-dimensional finite and boundary elements.

For the finite element method consider the system of equations for the original (unrefined) mesh,

$$[\mathbf{K}]\{\mathbf{u}\} = \{\mathbf{F}\} \tag{10.48}$$

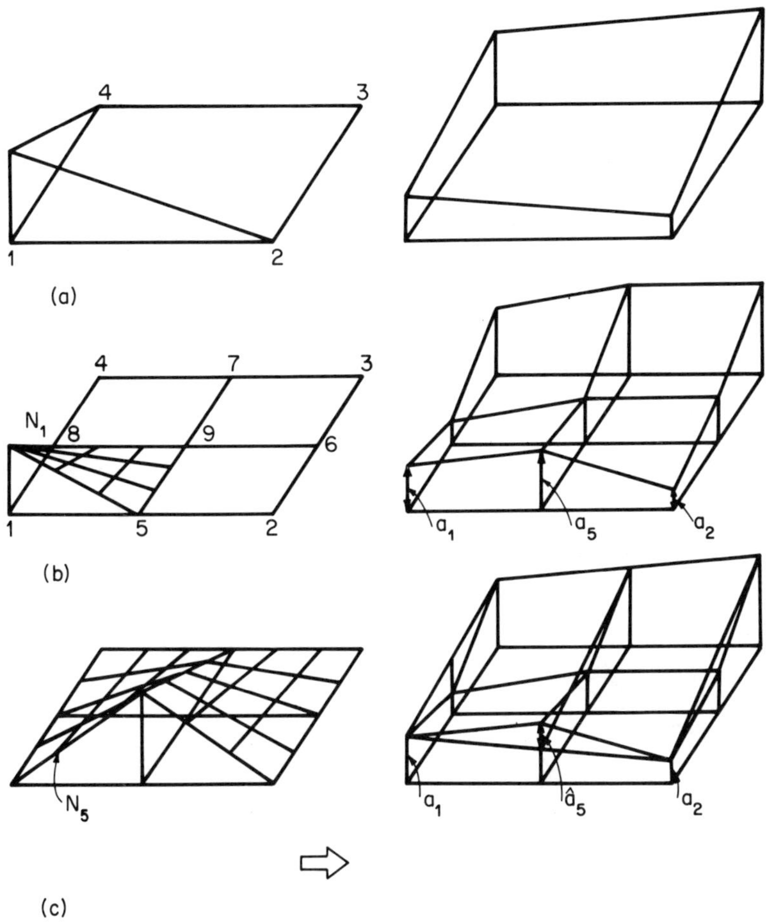

Figure 10.14 Two-dimensional shape functions. (a) original, (b) normal h-refinement, (c) hierarchical p-refinement

If we introduce m new hierarchical degrees of freedom then we have:

$$\begin{bmatrix} [\mathbf{K}] & [\mathbf{K}]^{1,m} \\ [\mathbf{K}]^{m,1} & [\mathbf{K}]^{m} \end{bmatrix} \left\{ \begin{matrix} \{\mathbf{u}\}' \\ \{\hat{\mathbf{u}}\}^{m} \end{matrix} \right\} = \left\{ \begin{matrix} \{\mathbf{f}\} \\ \{\hat{\mathbf{f}}\}^{m} \end{matrix} \right\} \tag{10.49}$$

where the original stiffness matrix $[\mathbf{K}]$ is retained and there are m additional columns and rows. Note that for the two- and three-dimensional case the off-diagonal terms $[\mathbf{K}]^{m,1}$ and $[\mathbf{K}]^{1,m}$ are no longer zero.

We can now solve for the unknowns at the newly introduced hierarchical nodes by

$$\{\hat{\mathbf{u}}\}^{m} = ([\mathbf{K}]^{m})^{-1}(\{\hat{\mathbf{f}}\}^{m} - [\mathbf{K}]^{m,1}\{\mathbf{u}\}') \tag{10.50}$$

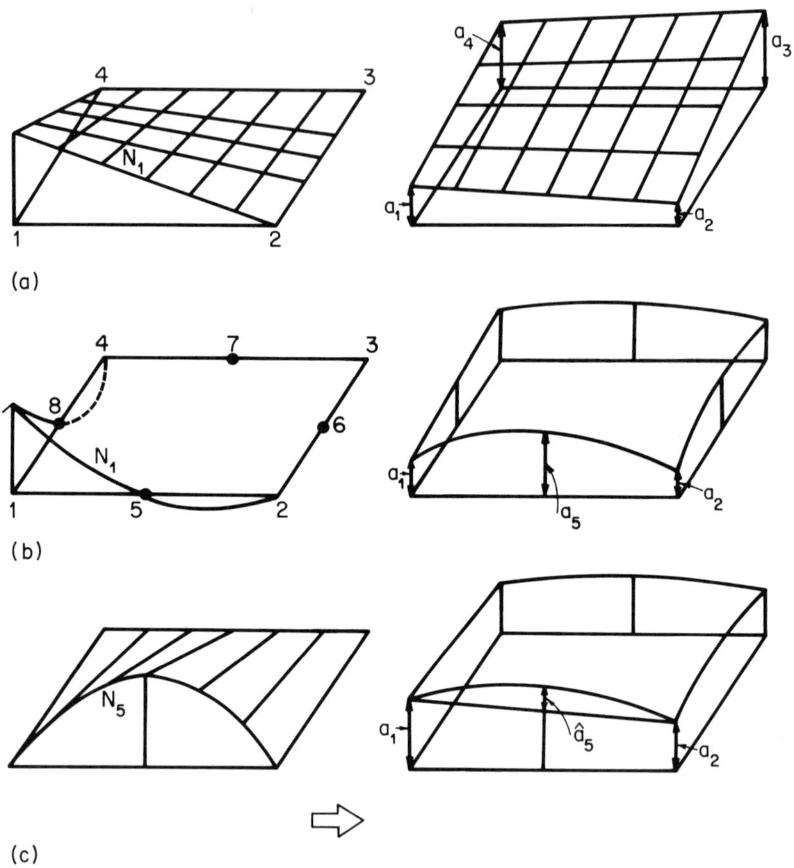

Figure 10.15 Two-dimensional shape functions. (a) original, (b) normal p-refinement, (c) hierarchical p-refinement

Let us now introduce one degree of freedom only. Then we have

$$\hat{u}^m = \frac{1}{K^m}(\hat{f}^m - [\mathbf{K}]^{m,1}\{\mathbf{u}\}') \tag{10.51}$$

where K^m and $\hat{f}^m$ are now scalars. $\hat{u}^m$ is equal to the error corrected by the introduction of the new degree of freedom at the new hierarchical node. The distribution of the error across the element is given by

$$e^m = \hat{u}^m N_m \tag{10.52}$$

where N_m is the shape function corresponding to the newly introduced node.

The energy norm of the error is given by

$$\|e\|^2 = (\hat{u})^2 \; \|N_m\|^2 \tag{10.53}$$

where the energy norm of N_m is given by

$$\|N_m\|^2 = K^m \tag{10.54}$$

To be able to solve equation (10.49) without having to solve the whole system of equations we make the following approximation,

$$\{\mathbf{u}\}' = \{\mathbf{u}\} \tag{10.55}$$

where $\{\mathbf{u}\}$ are the displacements obtained for the unrefined mesh.

For the boundary element method we may apply similar reasoning. Before the introduction of the hierarchical refinement we have

$$[\Delta\mathbf{T}]\,\{\mathbf{u}\} = [\Delta\mathbf{U}]\,\{\mathbf{t}\} \tag{10.56}$$

If we introduce m new hierarchical degrees of freedom equation (10.56) becomes:

$$\begin{bmatrix} [\Delta\mathbf{T}] & [\Delta\mathbf{T}]^{1,m} \\ [\Delta\mathbf{T}]^{m,1} & [\Delta\mathbf{T}]^{m} \end{bmatrix} \begin{Bmatrix} \{\mathbf{u}\}' \\ \{\hat{\mathbf{u}}\}^m \end{Bmatrix} = \begin{bmatrix} [\Delta\mathbf{U}] & [\Delta\mathbf{U}]^{1,m} \\ [\Delta\mathbf{U}]^{m,1} & [\Delta\mathbf{U}]^{m} \end{bmatrix} \begin{Bmatrix} \{\mathbf{t}\}' \\ \{\hat{\mathbf{t}}\}^m \end{Bmatrix} \tag{10.57}$$

In this case we not only introduce m new degrees of freedom but also add m collocation points, i.e. we will improve the satisfaction of Green's or Betti's theorem as well.

10.7.1 Adaptive strategies

Automatic mesh refinement uses the approximate error or rather improvement indicators computed previously to decide if a particular degree of freedom is to be introduced or not. Generally if the improvement (change) in the solution is expected to be large then the degree of freedom is introduced, if it is small then it is not.

The difficulty is in determining what is a small and what is a big improvement. Here it is convenient to work with a relative energy norm error in percent:

$$\eta = \frac{\|e\|}{\|u\|} \times 100 \;\; (\%) \tag{10.58}$$

where

$$\|u\|^2 = \int_{\Omega} u^2 \, \mathrm{d}\Omega \tag{10.59}$$

One way would be to decide *a priori* on a value for η and add all degrees of freedom which have an η above this value. Another way would be to first compute the relative errors for all added nodes, determine the maximum error $\eta_{\max}$ and then add only the degrees of freedom where

$$\eta > j \; \eta_{\max}$$

The parameter j controls the speed of convergence. If j is zero, then all possible degrees of freedom are added if it is one, none are added.

The automatic mesh refinement procedure is as follows:

(1) Possible new degrees of freedom are introduced and tested, using the approximate error indicator, if they improve the solution sufficiently or not.
(2) All degrees of freedom which improve the solution significantly are added to the mesh and the whole system of equations is solved again.
(3) A second and subsequent iteration may be necessary to further improve the solution.

10.8 CONCLUSIONS

In this chapter we have tried to convey to the reader the art of mesh design but this is something which will come out of experience and can not be taught. A well designed mesh will not only minimise errors but also lead to a more efficient analysis and save computer time. This depends on the good judgement of the analyst and the aim is to obtain results to the accuracy required with the minimum of effort.

However, even with automatic refinement procedures there is a need for skill and experience in designing a good mesh which is both accurate and economic.

Furthermore it is a good strategy to assume that answers from an analysis are wrong until proven otherwise. There have been many cases in the authors' experience where simple 'back of the envelope' computations have proved that the results obtained from an analysis could not possibly be right.

10.9 EXERCISES

10.1 In the figure a part of a mesh is shown with two different node numbering schemes. Show why

(a) for boundary elements the numbering will have no effect on solution time and storage.
(b) for finite elements where the system of equations is solved by a frontal solution scheme there will also be no effect.
(c) for finite elements, when the system of equations is solved by a band solver there will be significant effect on the storage solution time.

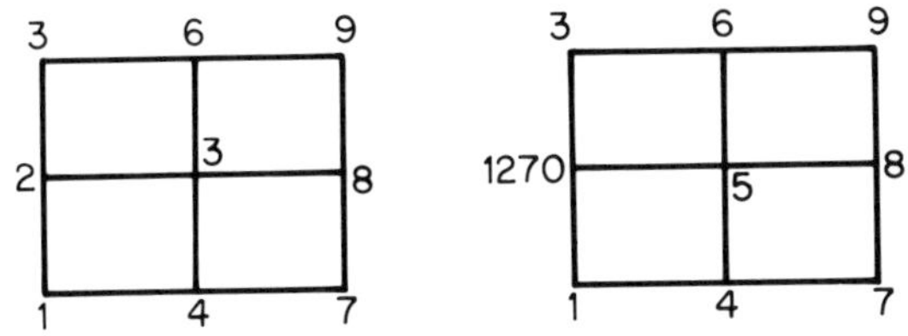

10.2 For the finite element mesh below how would you number the nodes and elements to reduce

(a) the bandwidth,
(b) the front width,

of the system of equations?

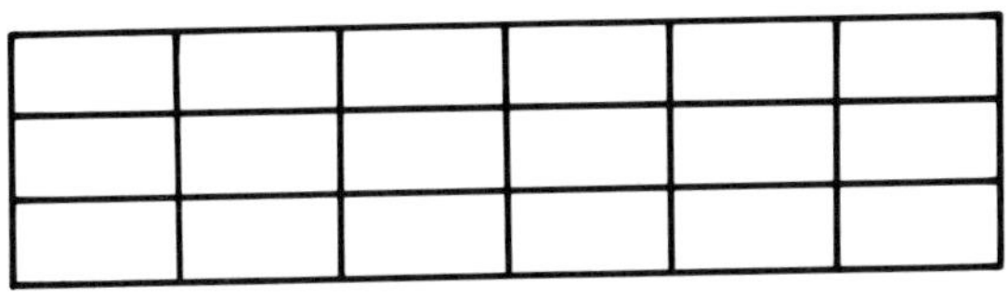

10.3 Solve the string problem in Figure 10.11 using the finite element method and

(a) one linear element,
(b) two linear elements,
(c) one parabolic element.

10.4 What is the significance of approximations made by equation (10.55) in the computation of error estimates?

10.5 Derive equation (10.54) using the definition of the energy norm given under section 10.6 and the derivation of the stiffness matrix in Chapter 4.

10.10 REFERENCES

1. Zienkiewicz, O.C., Humpheson, C. and Lewis, R. (1975) Associated and non-associated visco-plasticity and plasticity in soil mechanics. *Geotechnique*, **25**, 671–689.
2. Kelly, D.W. and Isels, J.D. (1992) A procedure for a posteriori error analysis for the finite element method which contains a bounding measure. *Computers & Structures* (in press).
3. Timoshenko, S. and Goodier, O. (1974) *Theory of Elasticity*, John Wiley, New York.
4. Zienkiewicz, O.C. and Craig, A. (1986) Adaptive refinement, error estimates, multigrid solution and hierarchic finite element method concepts. *Accuracy Estimates and Adaptive Refinements in Finite Element Computations.* (eds I. Babuska *et al.*) John Wiley, New York.
5. Zienkiewicz, O.C., Kelly, D.W., Gago, J. and Babuska, I. (1982) Hierarchial finite element approaches, error estimates and adaptive refinement, *The Mathematics of Finite Elements and Applications IV* (eds J.R. Whiteman), Academic Press, London. p313–346.

10.11 SUGGESTED FURTHER READING

Babuska, I., Zienkiewicz, O.C., Gago, J. and A. de Oliviera, E.R. (1986) *Accuracy Estimates and Adaptive Refinements in Finite Element Computations.* John Wiley New York.

Irons, B. and Ahmad, S. (1980) *Techniques of Finite Elements.* Ellis Horwood Chapter 10.

11

Computer Aided Pre- and Post-processing

A picture says more than 1000 words

Anon

11.1 INTRODUCTION

Entering data for a BE or FE analysis used to be a painful process. Firstly a mesh had to be sketched and all the nodes and elements numbered. Then the coordinates of each node had to be measured and typed in. Finally the connectivity of each element had to be specified. The first author remembers many hours/days/weeks spent on inputting the mesh. During the typing in of hundreds of numbers obviously errors were made and thus the second phase in the specification of input data was the debugging of data by plotting the mesh on a plotter. For three-dimensional analyses perspective views of the mesh had to be plotted and this necessitated the need for hidden line removal as it was difficult to see the mesh and errors if all lines were plotted. This required the mesh to be viewed from various angles and many errors were missed because they were in locations obscured from view. When the mesh data was finally in the computer, debugged and—hopefully—error free then the problem was analysed and usually a large amount of printout obtained. The user then had to go through all the printout and extract the values of interest and plot them by hand so that the design engineer could understand them.

Fortunately nowadays with the development of graphics workstations and powerful PCs we do not have to go through this painful process any more. In this chapter we will show how the computer can assist us in all phases of pre-processing to such a degree that almost no numbers have to be typed in. We will find that this not only dramatically reduces the effort spent in defining meshes but also leads to reduced possibility of errors.

Apart from assisting in the input phase the computer can also help in the presentation of the output of results from an analysis. Instead of looking through tons of output the user can see a picture of the results—in colour—on the graphics screen.

In summary this chapter is about making numerical modelling fun.

11.2 PRE-PROCESSING

11.2.1 Mesh generation

The first aspect of pre-processing is the generation of meshes from as little information as possible.

One—fairly obvious—possibility is to use the finite element shape functions to generate a mesh. In such a scheme we define large elements (superelements) and subdivide them as shown in Figure 11.1. The coordinates of the generated element nodes $\mathbf{X}(\xi,\eta)$ are then computed as follows,

$$
\begin{aligned}
&\textbf{DO for } \eta = -1, 1, \ \Delta\eta \\
&\qquad \textbf{DO for } \xi = -1, 1, \ \Delta\xi \\
&\qquad\qquad \mathbf{X} = \sum N_j(\xi,\eta)\mathbf{X}_j \\
&\qquad \textbf{END DO} \\
&\textbf{END DO}
\end{aligned}
$$

where $\Delta\xi$ and $\Delta\eta$ are increments in ξ and η directions and $\mathbf{X}_j$ are the coordinates of the nodes of the superelement.

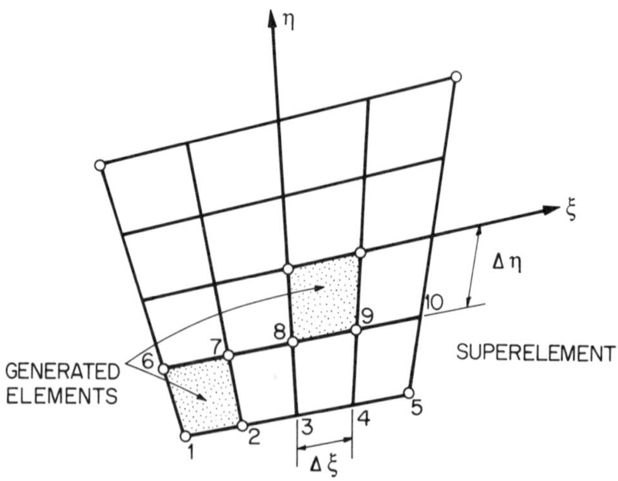

Figure 11.1 Superelement subdivided into smaller elements

Each node of the generated mesh can be numbered automatically by incrementing from the last number used. Here we have to take care to account for the case where more than one superelement is used. That is we have to check that the node numbers are the same at the interface where two superelements meet. A simple way to achieve this is to check the coordinates of any newly generated node against the coordinates of existing nodes. If identical nodes exist then the newly generated node is given the number of the already existing node. Usually a 'search square' is used to capture nodes whose co-ordinates are not exactly the same within the computed accuracy.

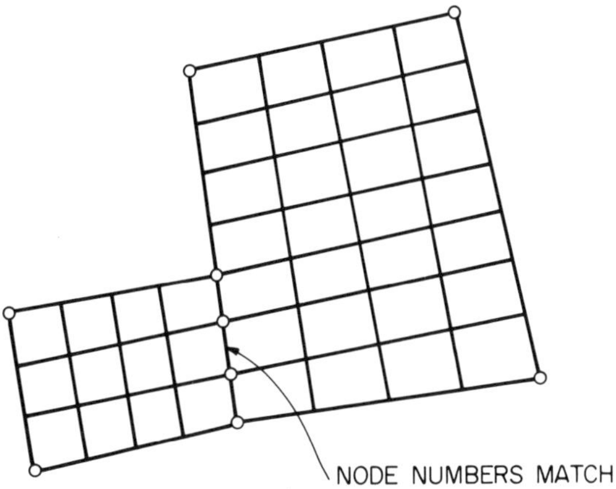

Figure 11.2 Two superelements connected to each other

Newly generated elements can be automatically assigned numbers by incrementing from the last number or from a number specified by the user. Element connectivities are determined in a logical way. For example the connectivity of the first element generated in Figure 11.1 is NODE (1), NODE (2), NODE (NXSi+2), NODE (NXSi+1) where NODE (i) is the node number assigned to the ith generated node and NXSi is the number of nodes in the ξ direction.

The mesh generation just discussed allows us to not only generate the coordinates of all nodes within the superelement but also element connectivities by specifying the nodes and connectivity of the superelement only.

The mesh generation described above works well if the shape of the boundary curve is smooth, i.e. can be described by a parabola or straight line. In many cases however boundary lines can not be approximated in such a way. An alternative mesh generation uses polygons to approximate the boundary. To determine the coordinates of generated element nodes we can no longer use the usual finite element shape functions but use a different method, also called *transfinite mapping.*

Consider the surface bounded by a polygon in Figure 11.3. First we 'map' the polygon into a unit square with 'local' coordinates u, v.

The local coordinates u, v of points located on the sides are given by

$$
\begin{aligned}
&\textit{Side 1}\\
&u_{i,1} = \frac{\sum_{n=2}^{i} \Delta l_{n,1}}{l_{g,1}}\\
&v_{i,1} = 0\\
&\textit{Side 2}\\
&u_{i,j_{\max}} = \frac{\sum_{n=2}^{i} \Delta l_{n,j_{\max}}}{l_{g,2}}\\
&v_{i,j_{\max}} = 1\\
&\textit{Side 3}\\
&u_{1,j} = 0\\
&v_{1,j} = \frac{\sum_{n=2}^{j} \Delta l_{1,n}}{l_{g,3}}\\
&\textit{Side 4}\\
&u_{i_{\max},j} = 1\\
&v_{i_{\max},j} = \frac{\sum_{n=2}^{j} \Delta l_{i_{\max},n}}{l_{g,4}}
\end{aligned}
\tag{11.1}
$$

where

$$\Delta l_{n,1} = \left((x_{n,1} - x_{n-1,1})^2 + (y_{n,1} - y_{n-1,1})^2 + (z_{n,1} - z_{n-1,1})^2 \right)^{\frac{1}{2}}$$

and

$$l_{g,1} = \sum_{n=2}^{i_{\max}} \Delta l_{n,1}$$

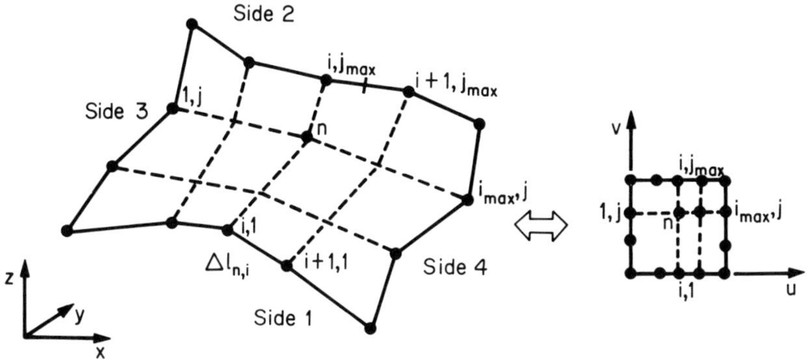

Figure 11.3 A surface bounded by polygon, and logical square

The co-ordinates $\mathbf{x}$ of the internal mesh points of the surface are computed by the linearly blended interpolation formula of Coons[4]:

$$\begin{aligned} \mathbf{x}(u,v) =&(1-v)\cdot\mathbf{x}(u,0)+v\cdot\mathbf{x}(u,1) \\ &+(1-u)\cdot\mathbf{x}(0,v)+u\cdot\mathbf{x}(1,v) \\ &-(1-u)\cdot(1-v)\cdot\mathbf{x}(0,0)-u\cdot(1-v)\cdot\mathbf{x}(1,0) \\ &-(1-u)\cdot v\cdot\mathbf{x}(0,1)-u\cdot v\cdot\mathbf{x}(1,1) \end{aligned} \tag{11.2}$$

where $\mathbf{x}$ is a vector of cartesian (x, y, z) coordinates. The coordinates $\mathbf{x}(0,0)$, $\mathbf{x}(0,1)$, $\mathbf{x}(1,0)$ and $\mathbf{x}(1,1)$ stand for the corner nodes of the *logical square*. The coordinates $\mathbf{x}(u,0)$,$\mathbf{x}(u,1)$, $\mathbf{x}(0,v)$ and $\mathbf{x}(1,v)$ are the coordinates of points on the boundary sides 1,2,3 and 4.

During mesh generation the boundary points are used for determining the number of nodes and elements to be generated in a particular direction.

For example to work out the coordinates of the generated mesh point n in Figure 11.3 we substitute,

$$\begin{aligned} \mathbf{x}(u,0) =&\mathbf{x}(u_{i,1},0) \\ \mathbf{x}(u,1) =&\mathbf{x}(u_{i,j_{\max}},1) \\ \mathbf{x}(0,v) =&\mathbf{x}(0,v_{1,j}) \\ \mathbf{x}(1,v) =&\mathbf{x}(0,v_{i_{\max},j}) \end{aligned} \tag{11.3}$$

The mesh generation for a two-dimensional mesh proceeds as follows:

(1) The boundary polygon is mapped into the unit square and values of local coordinates u, v corresponding to the polygon nodes are obtained. As explained previously the number of polygon points will determine the number of subdivisions in the generated mesh.

(2) The coordinates of internal points are determined by applying equation (11.2).
(3) Node numbers and element connectivity are generated.

The method for numbering the nodes and elements and for generating connectivity informaton is the same as previously described.

For three-dimensional meshes where we have a logical cube (Figure 11.4) instead of a logical square we proceed as follows:

(1) For each surface of the cube we determine the generated mesh nodes as explained previously.
(2) Mesh nodes inside the cube are determined by the three-dimensional version of the Coons formula.

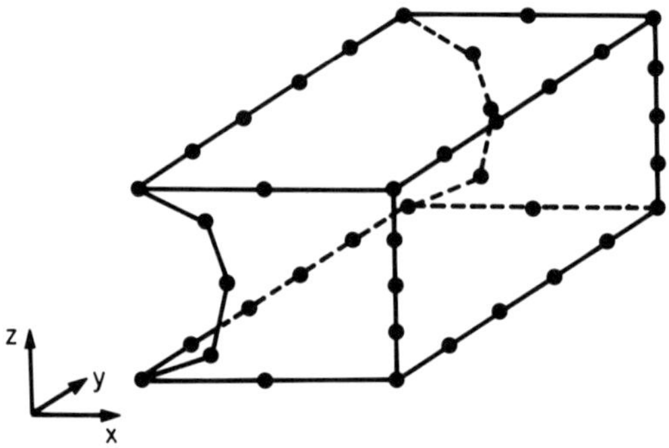

Figure 11.4 A logical cube

For three-dimensional meshes the coordinates of a point inside the logical cube with local coordinates u, v, w are obtained by

$$\begin{aligned}\mathbf{x}(u,v,w) = \frac{1}{2}\cdot\big[&(1-v)\cdot\mathbf{x}(u,0,w)+v\cdot\mathbf{x}(u,1,w)\\ &+(1-u)\cdot\mathbf{x}(0,v,w)+u\cdot\mathbf{x}(1,v,w)\\ &+(1-w)\cdot\mathbf{x}(u,v,0)+w\cdot\mathbf{x}(u,v,1)\big]\\ -\frac{1}{2}\cdot\big[&(1-u)\cdot(1-v)\cdot(1-w)\cdot\mathbf{x}(0,0,0)\\ &+(1-u)\cdot(1-v)\cdot w\cdot\mathbf{x}(0,0,1)\\ &+(1-u)\cdot v\cdot(1-w)\cdot\mathbf{x}(0,1,0)\\ &+(1-u)\cdot v\cdot w\cdot\mathbf{x}(0,1,1)\\ &+u\cdot(1-v)\cdot(1-w)\cdot\mathbf{x}(1,0,0)\\ &+u\cdot(1-v)\cdot w\cdot\mathbf{x}(1,0,1)\\ &+u\cdot v\cdot(1-w)\cdot\mathbf{x}(1,1,0)\\ &+u\cdot v\cdot w\cdot\mathbf{x}(1,1,1)\big]\end{aligned} \tag{11.4}$$

Here, $\mathbf{x}(u, 0, w)$ etc. stands for the coordinates of the points on one of the boundary surfaces and $\mathbf{x}(0, 0, 0)$ etc. for the coordinates of the corner nodes.

With this type of mesh generation there is no restriction on the shape of a boundary, i.e. boundaries do not have to be smooth and may contain corners.

11.2.2 Computer graphics—hardware and software

To interactively generate and to display meshes and the results a graphics workstation is needed. A graphics workstation is characterised by the following:

- Fast CPU with graphics processor
- Large colour screen (15 in to 19 in, 37 to 47 cm)
- Graphics input devices
    - Digitiser
    - Mouse
- Hard copy device (inkjet plotter, thermal plotter, laser printer)

The role of the CPU is to do the basic 'number crunching' i.e. the computations needed for the mesh generation and the display of meshes (computation of plot coordinates, hidden line algorithm etc.). The graphics processor interprets the draw commands given by the program to generate a picture on the monitor.

The screen is divided into a matrix of dots where each dot (also called a pixel) can be given a certain light intensity and colour. The number of pixels will determine the resolution of the screen. Typical screen resolutions for graphics work are in the order of 640 × 480 for low to about 1000 × 1000 for medium and 3000 × 3000 for high resolution displays. For a 1000 × 1000 pixel display the graphics processor has to store several million bits of information about the screen. If colour is used then we have at least three bit planes, each bit plane handling one of the three prime colours (red, blue, green).

Digitising tablets have a electrostatically sensitive surface which allows them to pass the location, in device coordinates, of the cross hair cursor to the computer via a serial port.

The mouse is used to control the location of a cursor (small square, arrow etc.) on the screen by moving it on a surface. As the mouse is moved electronic signals are sent to the computer which in turn cause the cursor to move on the screen.

All graphics functions are invoked by using native (hardware dependent) graphic commands (graphics primitives).

In the last decade standard graphics languages have been developed which allow to write programs which are hardware independent. One of these languages is GKS[5] (Graphics Kernel System) which basically is a set of

FORTRAN or C subroutines which allow to draw polygons, fill areas inside polygons with colour, write text, set colour attributes etc. For example to draw a polygon the call to GKS is

CALL GPL (N, PX,PY)

where N is the number of polygon points and PX(N), PY(N) are the x and y coordinates of points.

Figure 11.5 shows a flow diagram of how the FORTRAN applications program communicates with various graphics devices.

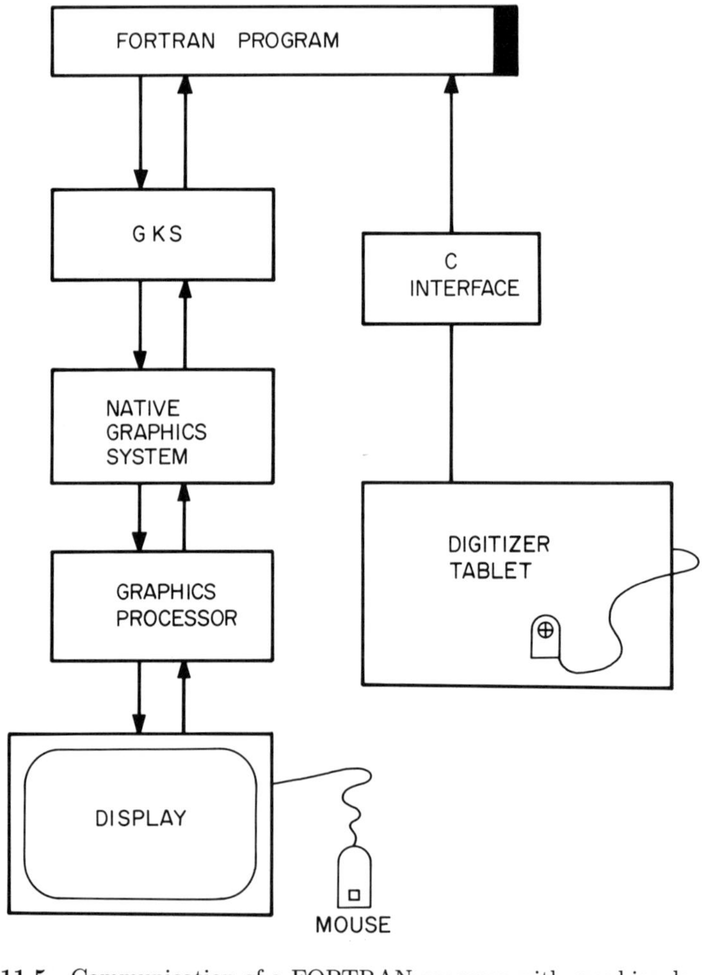

Figure 11.5 Communication of a FORTRAN program with graphics devices

If a GKS subroutine is called the GKS driver will translate the call into native graphics language. The graphics processor then changes the intensity of pixels on the screen such that the desired effect is achieved, for example a line appears on the screen. If a GKS call is made to interrogate the location of the cursor on the screen then the reverse occurs and the coordinates of the cursor are returned.

A digitiser does not usually operate through GKS and transfers the local coordinates of the cross hair device as a string of numbers which can be read with a READ statement in the application program.

11.2.3 User friendly interfaces

The automatic mesh generation schemes described in Section 11.2.1 assist greatly in reducing the effort in specifying the input data for numerical models. However, to make the data preparation really fun requires a user friendly interface between the human and the computer.

Let us investigate the main difference between humans and computers. Computers understand only numbers. Humans on the other hand understand graphics. An engineer may communicate with another by sketches, engineering drawings etc. A set of coordinates and element connectivity on the other hand would be meaningless to an engineer.

A user friendly interface between humans and computers would therefore have to use graphics. The best way to communicate with the computer is by using graphical input and output devices.

As explained previously one way of communicating with the computer is via a mouse which moves a cursor on the screen. We may use the mouse to select items on a menu displayed on the screen much in the same way as we select an item from a menu in a restaurant. Figure 11.6 shows an example of a menu from the FEMCAD preprocessor[6].

A user friendly preprocessor has the following features:

(1) Easy to read, clear menu system with submenus if necessary.
(2) Help messages which instruct the user what to do.
(3) Error recovery procedures which allow the user to quit out of a selected function at any time.
(4) Safety procedures to safeguard against lost data. For example when the user wants to exit the program it is checked whether data has been saved etc.
(5) Fast screen refresh.

Whereas the computer program requires numbers as input (element numbers, node numbers, connectivity etc) graphical preprocessors work completely without numbers as such numbers are not of interest to the design engineer. In some cases (for example when defining material properties,

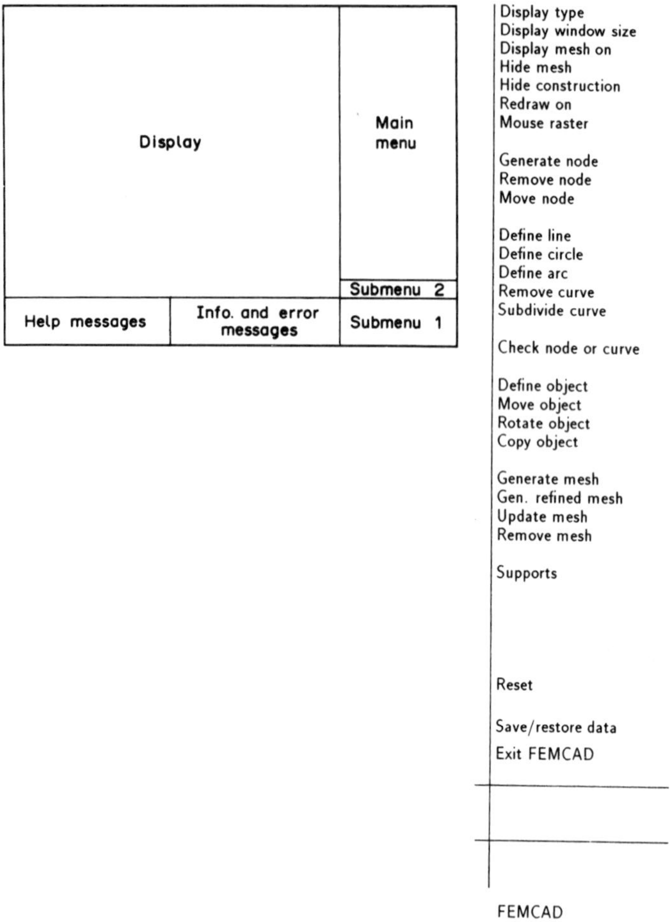

Figure 11.6 Example of a menu

boundary loading, etc) such numbers are needed and the preprocessor should have the capability of extracting these numbers.

11.2.4 Display of meshes

One of the most important aspects of computer aided pre- and post-processing is the display of BE or FE meshes. Meshes are displayed by plotting the boundaries of elements. This is done by drawing a polygon where the number of polygon points is determined by the order of shape function used and the deviation from linearity of the curve (Figure 11.7). The number of polygon points n may be determined by

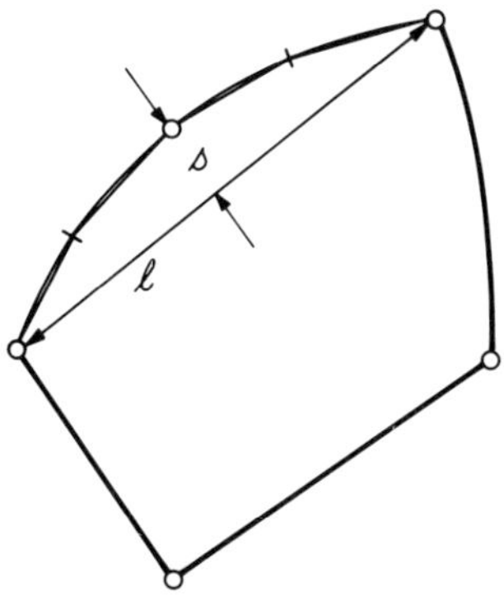

Figure 11.7 Plotting of element boundaries using polygons

$$n = INTEG\left(\frac{s}{lr}\right) + 2 \tag{11.5}$$

where s is the 'sag' of the curve, or its deviation from linearity, l is the length and r is the resolution of the graphics device. $INTEG(\cdot)$ means a truncation i.e. all the values after the decimal point are removed. Equation (11.5) reverts to two points for a linear shape function. In an efficient display program algorithms have to be devised which prevent the drawing of polygons twice at the interface where two elements meet.

For three-dimensional meshes, we have to plot an axonometric or perspective view of the mesh. First we have to select the position of the viewpoint or eye (Figure 11.8). This can be specified by two angles: one horizontal, the other vertical or by the position of the eye point $(\bar{x}, \bar{y}, \bar{z})$. Next we have to work out picture plane coordinates of the mesh polygons, x', y' and these are the coordinates of points plotted on the screen.

For an axonometric view the transformation from x, y, z coordinates to x', y', z' coordinates, where z' denotes distance to the eye, is given by

$$\begin{Bmatrix} z' \\ x' \\ y' \end{Bmatrix} = \begin{bmatrix} \cos\beta\cos\theta & \cos\beta\sin\theta & \sin\beta \\ -\sin\theta & \cos\theta & 0 \\ -\sin\beta\cos\theta & -\sin\beta\sin\theta & \cos\beta \end{bmatrix} \begin{Bmatrix} x \\ y \\ z \end{Bmatrix} \tag{11.6}$$

For a perspective view we have:

$$\begin{Bmatrix} x'_p \\ y'_p \end{Bmatrix} = \begin{bmatrix} -\sin\theta & \cos\theta & 0 \\ -\sin\beta\cos\theta & -\sin\beta\sin\theta & \cos\beta \end{bmatrix} \left(\begin{Bmatrix} \bar{x} \\ \bar{y} \\ \bar{z} \end{Bmatrix} - c \begin{Bmatrix} x \\ y \\ z \end{Bmatrix} \right) \tag{11.7}$$

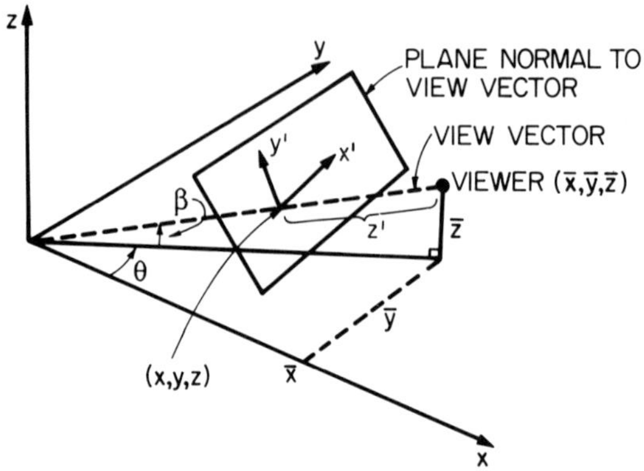

Figure 11.8 Definition of angles for axonometric display of meshes

where

$$c = \frac{D_o}{\cos\beta\,\cos\theta(\bar{x}-x)+\cos\beta\,\sin\theta(\bar{y}-y)+\sin\beta(\bar{z}-z)}$$

and D_o is the distance from the observer to the projection plane.

Once we have obtained the picture plane coordinates of a point we have to scale them so that all points fit within the display area.

Basically we have to scale the picture plane coordinates which range from $x'_{\text{min}}, y'_{\text{min}}$ to $x'_{\text{max}}, y'_{\text{max}}$ such that they fit into the range of screen coordinates $x_{s,\text{min}}, y_{s,\text{min}}$ and $x_{s,\text{max}}, y_{s,\text{max}}$ (Figure 11.9).

The scaled screen coordinates are given by

$$x_s = (x' - x'_{\text{min}})S_x + x_{s,\text{min}}$$
$$x_s = (y' - y'_{\text{min}})S_y + y_{s,\text{min}}$$

where S_x and S_y are scale factors given by

$$S_x = \frac{x_{s,\text{max}} - x_{s,\text{min}}}{x'_{\text{max}} - x'_{\text{min}}}$$

and

$$S_y = \frac{y_{s,\text{max}} - y_{s,\text{min}}}{y'_{\text{max}} - y'_{\text{min}}}$$

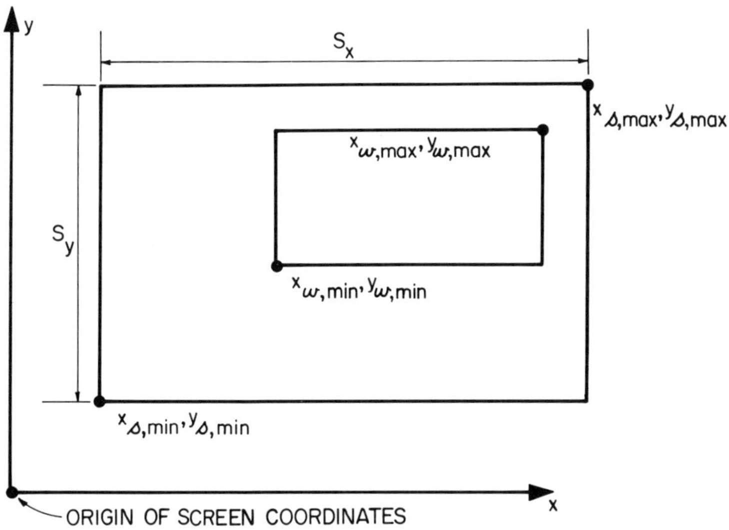

Figure 11.9 Definition of screen coordinates

To avoid distortion of the image we assume that S_x, S_y are replaced by one scale factor S which is defined by

$$S = \min(S_x, S_y)$$

Sometimes it is necessary to window in on a certain area of the screen and enlarge this area.

Referring to Figure 11.9 the screen coordinates of a window are defined by

$$X_w = (X_s - x_{w,\min})/S_x^w$$
$$Y_w = (Y_s - y_{w,\min})/S_y^w$$

where

$$S_x^w = \frac{x_{w,\max} - x_{w,\min}}{x_{s,\max} - x_{s,\min}}$$

The coordinates of points which do not lie within the window area will be outside the range of screen coordinates. Usually the device driver will ensure that such points are clipped i.e. no attempt is made at plotting them.

For three-dimensional meshes it is confusing if all mesh lines are plotted, i.e. those which are visible and those which are not. It is necessary in this case to only plot the visible portions of the mesh. Various algorithms exist for determining which portions of the mesh are hidden. In one such algorithm[7] each line is checked against each surface to determine if it is fully or partially hidden and if the latter applies the intersection of the line and the boundary of the surface is found. Obviously such algorithms involve a significant amount of numerical work but such procedures are necessary if the meshes are displayed on a plotter.

If on the other hand the meshes are displayed on a graphics screen a much simpler faster algorithm can be used. This algorithm is based on the fact that on a graphics screen we can fill the area bounded by a polygon with a colour and that any filled area will overwrite any existing information on the screen.

The simple hidden line algorithm works as follows.

(1) Create polygons for each element surface i.e. one surface for a boundary element and six surfaces for each finite element.
(2) Eliminate duplicate surfaces i.e. surfaces which are common to two finite elements.
(3) Sort polygons in order of distance to the eye point, i.e. the furthest away first, then the nearest. Sorting is done by comparing distances from the eye to the centroid of the surface.
(4) Plot the polygons with colour infill as shown previously.

This algorithm is very fast and has the added advantage that we can actually see the hidden surfaces as the mesh is drawn on the screen.

An example of a mesh drawn using this algorithm as shown in Figure 11.10 (see colour section).

11.3 POST-PROCESSING

11.3.1 Display of results

The output from an analysis basically consists of numbers. These are displacements at nodes, stresses inside elements or displacements and stresses inside a BE region. Alternatively temperatures and fluxes are obtained. The most efficient way of presenting the results of analysis is to display them graphically. Basically there are three types of display:

(1) displaced shape of a mesh
(2) vectors
(3) contours.

For displaying deformed meshes we simply plot the original mesh then add the displacements (multiplied with a magnification factor) to the coordinates of each node and plot the mesh again in a different colour (Figure 11.11).

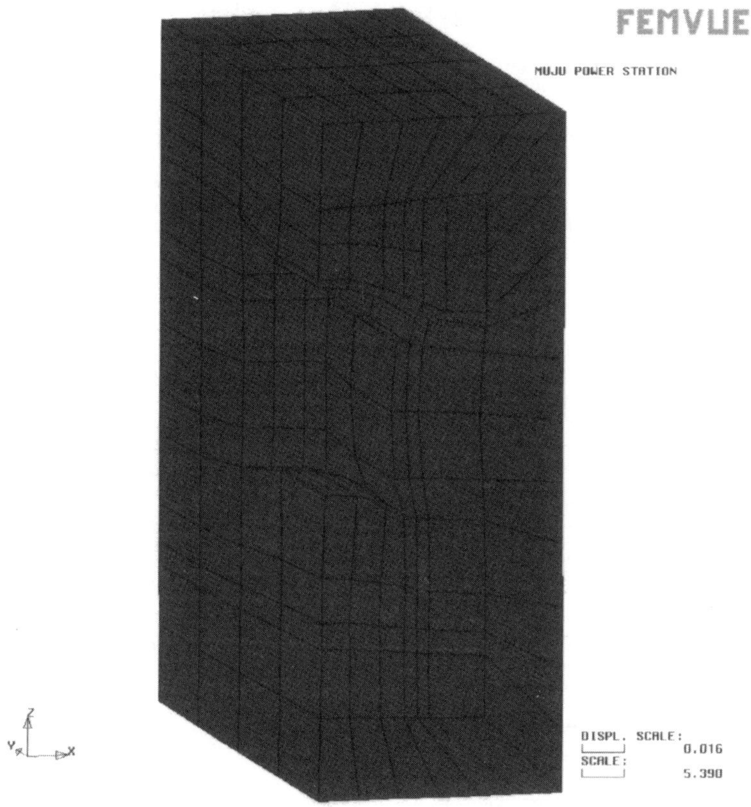

Figure 11.11 Plot of displaced shape

Most results we obtain from a boundary element or a finite element analysis are vector quantities, i.e. have more than one component. An obvious way is therefore to plot the results as vectors. For example a displacement vector plot would have arrows which indicate the magnitude and direction of the displacement. In a stress vector plot we plot arrows for each of the principal stress components. It is convenient to plot principal stress vectors at the Gauss points because, as we have seen these are the optimal result points. Figure 11.12 shows an example of a stress vector plot.

Contours are lines along which the value to be displayed (such as, for example, stress) is constant. Contour lines are usually drawn at certain intervals. In many cases however the engineer is interested in ranges of stress or flux

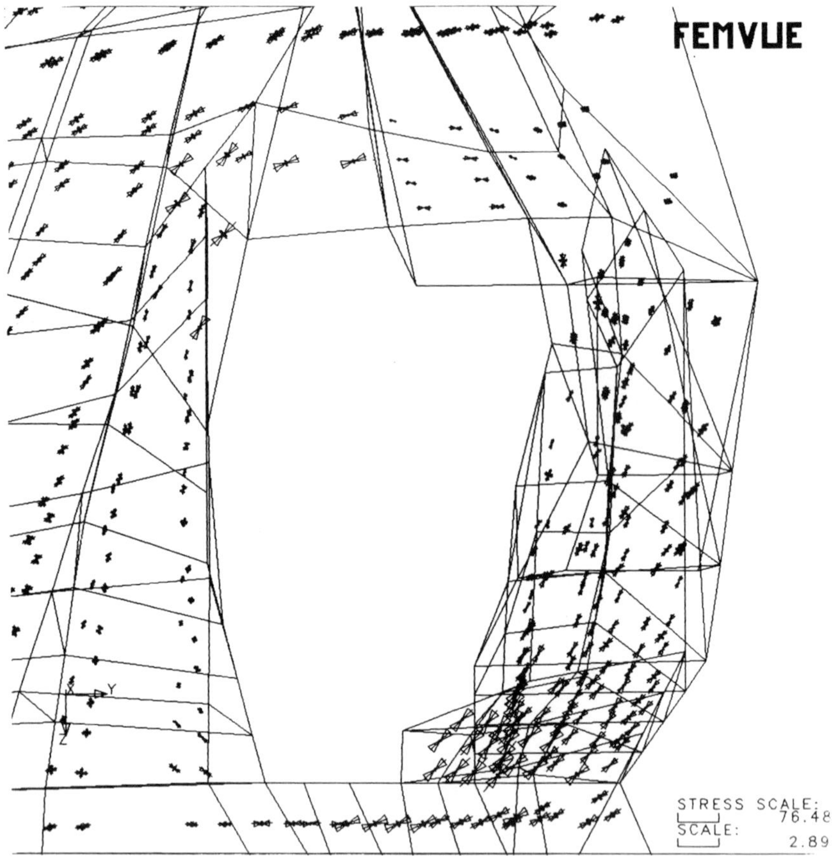

Figure 11.12 Plot of stress vectors

values in particular areas rather than lines of constant stress. A very effective way of displaying stress results is by painting areas which fall within a certain range in a particular colour. Apart from being a much better way of conveying information, colour also allows to put some interpretation into the results. For example, we may choose to paint in red colours, zones which we presume to be at a critical stress level and blue or green colours for the zones which we perceive to be safe.

We explain here how the colour painting algorithm is implemented.

The basic principle is to break up the area to be colour painted into triangles. On the three sides of a triangle we find the intersection points of the two stress levels of the stress range to be painted. Once these points have been found they are connected by a polygon and plotted using the colour fill option described previously. This proceeds until all triangles have been colour painted for all possible stress ranges.

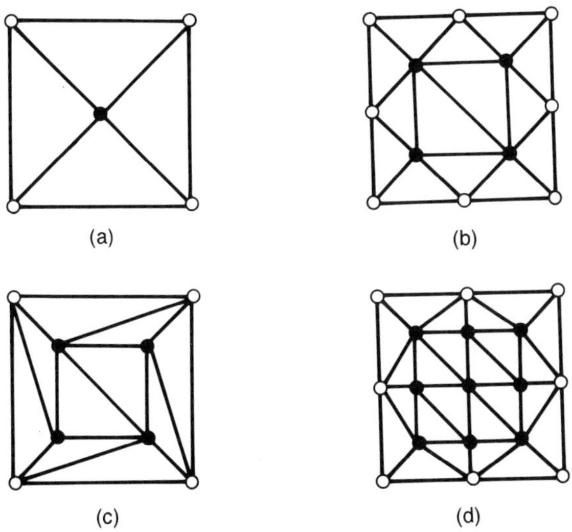

Figure 11.13 Examples of element triangularisation, (● Gauss points, ○ nodes). (a) Linear, one Gauss point. (b) Quadratic, four Gauss points. (c) Linear, four Gauss points. (d) Quadratic, 9 Gauss points

Figure 11.13 shows examples of how elements can be triangularised depending upon the number of nodes and Gauss points used. Here the filled circles represent Gauss points and the hollow circles node points. The results of an analysis are usually available at either the Gauss points or the nodes.

If only the results σ_g at Gauss points with coordinates ξ_g and η_g are available, then the nodal values $\sigma(\xi_n, \eta_n)$ can be computed by linear extrapolation (Figure 11.14).

The value of stress at node n for example is given by

$$\sigma(\xi_n,\ \eta_n) = \sum_{g=1}^{4} N_g(\xi_n,\ \eta_n)\ \sigma_g$$

where

$$N_g = 2.25\left(\frac{1}{3} + \xi_g\xi_n\right)\left(\frac{1}{3} + \eta_g\eta_n\right)$$

For three-dimensional finite elements where the Gauss points are actually inside the element they have to be extrapolated to each of the six boundary surfaces first. This is done by (Figure 11.15)

$$\sigma_g = \sigma_{II} - 1.366(\sigma_{II} - \sigma_I)$$

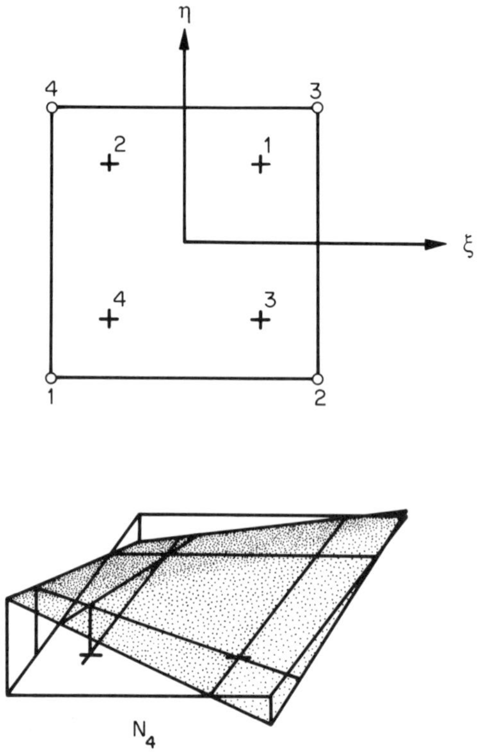

Figure 11.14 Extrapolation of Gauss point stress to the node

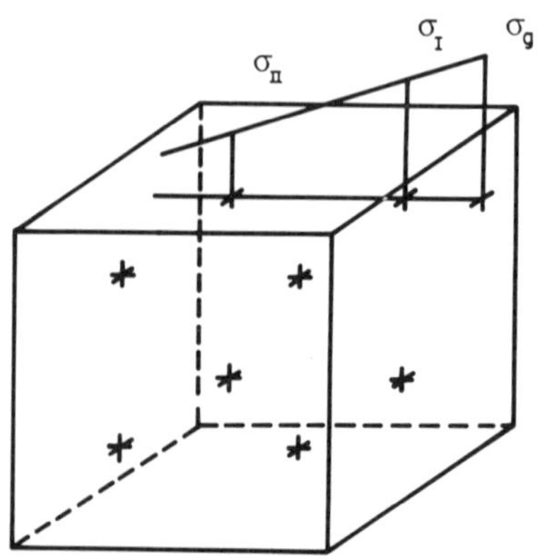

Figure 11.15 Extrapolation of Gauss point values to the boundary surfaces

To obtain smooth contours stress values have to be averaged at node points if the stresses can be assumed to be continuous, for example for smooth surfaces. In the averaging procedure the stress values at the node for all elements connected to the node are added and then divided by the number of elements which connect to this node. As explained before this should only be done if the stresses can be assumed to be continuous. One case where this can not be done in three-dimensional analysis is when the surface is not smooth. A way of checking for surface smoothness is to compare the outward normals of connecting surfaces and not to average values if the outward normals differ by more than a certain amount.

Another case where stress discontinuities can occur is at material interfaces. In such a case we only average the stresses of elements which have the same material number.

After all the values at the Gauss and nodal points have been determined and the elements have been subdivided into triangles we next explain how the triangles can be colour painted. For each colour we have a maximum and a minimum stress level. First we find the number of intersection points between these minimum and maximum stress levels on the three sides of the triangle assuming a linear variation of the variable along each side. Depending on the values at the triangle corner nodes we may get zero to four intersection points (Figure 11.16) and 0–5 nodes of the polygon whose inside is to be colour painted. To paint the contour we simply plot the polygon with the desired colour fill option switched on. We repeat the above procedure for all stress ranges or colours, then for all triangles within one element and then for all elements. For a three-dimensional mesh which is plotted in perspective view elements must be painted in a special order beginning with the one furthest away from the observer so as to preserve the three-dimensional appearance of the mesh.

When we want to plot contours of stresses etc. inside a BE region then we have to provide a mesh of so-called 'dummy' elements i.e. elements which are not used for the analysis but whose sole purpose is to enable the plotting of results. It is convenient from the programming point of view to calculate the stresses at the Gauss points of such dummy elements and treat them in the same way as the other elements. Figure 11.17 (see colour section) shows an example of contour painting on 'dummy' elements inside a three-dimensional BE region.

During the plotting of results it may be useful to be able to interrogate a point on the mesh—by pointing to it with a cursor and to inquire the level of displacement and stress there. This facility is implemented by checking all surfaces bounded by polygons to determine if the required location is inside the surface. Next we determine the local (ξ, η) coordinates of the point and the value of stress there by interpolation (or extrapolation) from the Gauss point values.

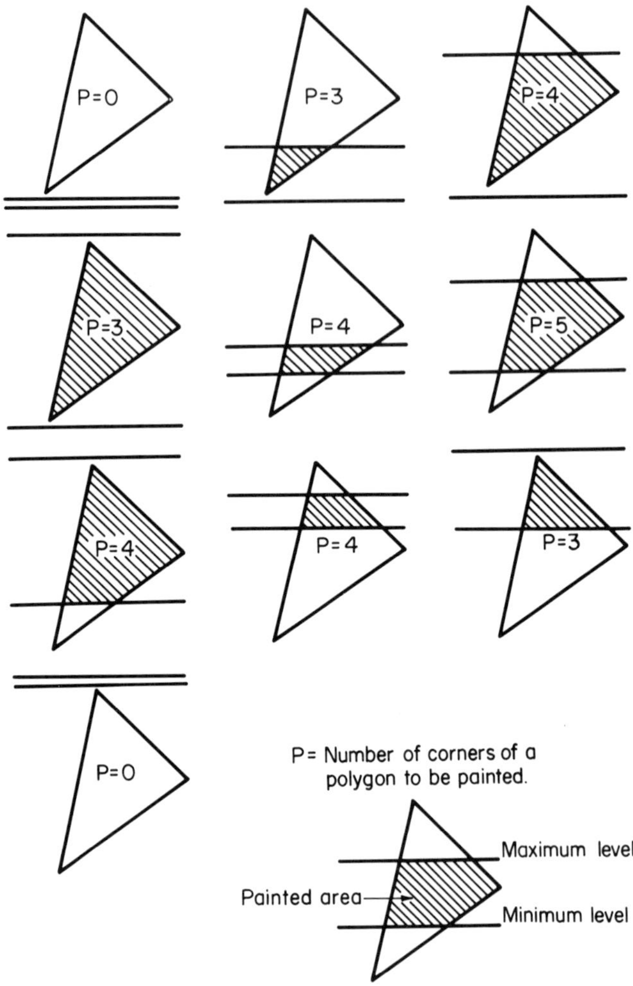

Figure 11.16 Possible shapes of polygons for the contour painting algorithm

11.4 CONCLUSIONS

We have shown here how the computer can be used to aid in the task of communicating with the numerical model. With the emergence of low cost, high performance workstations the communication with the computer will be humanised to a large degree, i.e. engineers will be able to communicate with the computer in much the same way as with each other using graphics. Therefore the communication with the computer model will be on a higher level and the tedious task of preparing and inputting mesh data will be a thing of the past. This then should make such models more readily available to engineers and no longer make them the domain of computer enthusiasts.

11.5 REFERENCES

1. Zienkiewicz, O.C. and Phillips, D.V. (1971) An automatic mesh generation scheme for plane and curved surfaces by 'isoparametric' coordinates. *Int. J. Num. Meth. Eng.*, **3**, 519–528.
2. Cook, W.A. (1974) Body oriented (natural) coordinates for generating three dimensional meshes. *Int. J. Num. Meth. Eng.*, **8**, 27–43.
3. Haber, R., Shephard, M.S., Abel, J.F., Gallagher, R.H. and Greenberg, D.P. (1981) A general two-dimensional, graphical finite element preprocessor using discrete transfinite mappings. *Int. J. Num. Meth. Eng.*, **17**, 1015–1044.
4. Coons, S.A. (1967) *Surfaces for computer-aided design of space forms.* Report MAC-TR-44, M.I.T., Cambridge, Mass.
5. Endele, *et al* (1990) *Computer Graphics Programming*, Springer, Berlin.
6. Beer, G. and Mertz, W. (1990) A user friendly interface for numerical modelling in mining. *Int. J. Rod Medr. Min. Sci. Geomech. Abotr.*, **27**, 541–552.
7. Janssen, T.L. (1983) A simple efficient hidden line algorithm. *Computers Structures*, **17**, 563–571.

11.6 SUGGESTED FURTHER READING

Baily, B.C, Haggar, F. and Abel J.F. (1986) Towards effective interactive three dimensional colour post processing, *Eng. Comput.* **3**, 90–98.
Hoffmeister, L.D. (1978) A drafting program for isoparametric finite elements. *Int. J. Num. Meth. Eng.*, **12**, 505–529.
Peruchio, R., Ingraffea, A.R. and Abel, J. (1982) Interactive computer graphics processing for three dimensional finite element analysis. *Int. J. Num. Meth. Eng.*, **18**, 909–926.

Appendix A
Program DEMON

A1 INTRODUCTION

Program DEMON is a program for the analysis of plane strain by the finite element method. Many programs capable of such analysis already exist, but program DEMON is specifically written for the following purposes.

1. To familiarise intending users of the finite element method with the practical details of using finite element programs, and with the results that may be expected from finite element analysis.
2. For those interested in developing their own software, to demonstrate typical programming techniques of pre-processing, equation construction and solution, and post-processing. Like nearly all finite and boundary element software, program DEMON is written in FORTRAN. To the best of the authors' knowledge no language has yet been devised which is better suited to the task: probably no other high-level language enables the generation of faster executable code, and in FORTRAN the syntax of file handling operations is well established and standardised. Whilst it is true that the syntactical rules of FORTRAN allow badly structured code to be written by the novice, an experienced programmer can adopt structures as rigid as those imposed by languages with stricter syntactical rules where appropriate, without being obliged to obey restrictive rules in all circumstances.

In program DEMON, eight-node isoparametric quadrilateral and six-node isoparametric triangular elements are used. In these elements, displacement is taken to vary quadratically with respect to intrinsic coordinates. The shape functions used to interpolate displacement are also used to interpolate cartesian coordinates, with the result that if the midside node of a side of an element does not lie on the straight line joining the ends of the side, then the side is curved. This allows curved boundaries to be modelled accurately using relatively few elements, and thereby improves efficiency of analysis.

Sign convention for stress

In the sign convention of program DEMON, tensile stress is regarded as positive.

Problem size limits

The following limits may not be exceeded.

- Number of material types 10
- Number of nodes 1600
- Highest node number 4000
- Number of elements 400
- Highest element number 800
- Maximum internal node numbering difference 95
- Number of load cases 5

Note that the user is allowed to leave gaps in the node and element numbering, so that for example the highest node number is not necessarily equal to the number of nodes. This possibility enables certain automatic data generation facilities to be used, and also simplifies modifications of the mesh to take account of design changes for example. The program automatically generates an internal numbering system by renumbering nodes in ascending order of number assigned by the user, from 1 to the number of nodes. The internal node numbering difference of an element is the greatest difference between the internal numbers of any of its nodes. In the interest of economy of solution it is desirable to minimise the maximum internal node numbering difference. It is expected that for all but very unusual problem geometries, the limit on number of elements will be exceeded before the other limits are attained.

A load case is some distribution of applied load, for a given structural geometry and set of constraints on displacement. Generally, it takes little more computer time to analyse several load cases than it does to analyse a single load case.

Modes of execution of the program

If the input data ends with the keyword `GO`, then the program carries out the requested analysis if it has found no syntactical errors in the input data. If errors were found then the program stops after reading and checking as fully as possible all the input data but before embarking on any computation requiring significant amounts of computer time. If the input data ends with the keyword `NOGO`, then the program stops after reading and checking the input data, regardless of whether or not syntactical errors are found.

It is advisable to carry out a `NOGO` run before attempting a full analysis and indeed to obtain a plot of the mesh, since errors in specification of the problem to be solved do not necessarily manifest themselves as syntactical errors detectable by the preprocessing routines. Note that program DEMON will always read the input data to the end and provide as full an account as possible of errors, no matter how many errors exist in the data.

Input and output files

Program DEMON reads input data from the file DEMON.DAT, writes printable output (including any error messages) to the file DEMON.OUT, and writes plot data if requested to the file DPLOT.PLT:

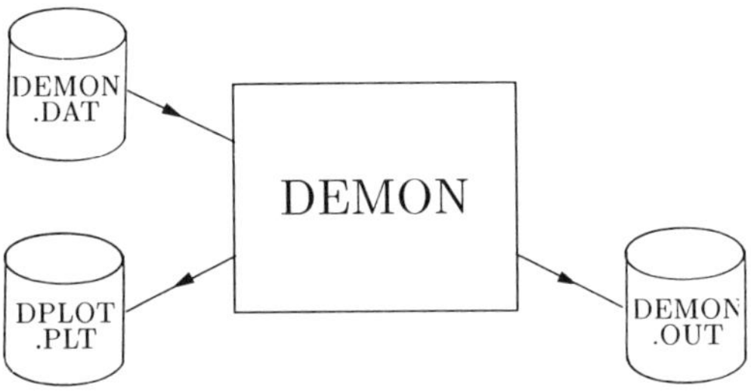

A2 INPUT DATA (FILE DEMON.DAT)

The first four characters of each line of input data are reserved for a keyword which identifies the type of data on the rest of the line.

Keywords are from one to four characters in length, and must always begin at the first character of the line. The remaining 76 characters may be used for numeric data or, for problem and load case titles, for text. Items of numeric data must be separated from one another by at least one space. Data must be presented in the following order.

(A) Problem title, plot control data, material properties
(B) Coordinates of nodes
(C) Nodes of elements
(D) Constraints on displacement
(E) Load case title
(F) Applied loads
(G) `GO` or `NOGO`

(E) and (F): repeat for each load case

The use of keywords to identify data allows the user some flexibility in the order in which lines of data are presented within each of the groups above; indeed, the order is only of any significance within Group B, and then only if use is made of the keyword `ORIG` to move the origin of coordinates. The specification of data within each group is as follows.

(A) PROBLEM TITLE, PLOT CONTROL DATA, MATERIAL PROPERTIES

Problem title

`T` *text*

text = title of problem to appear on printout and plots if any.

Plot of mesh

`MESH` *xlo* *xhi* *ylo* *yhi*

xlo = minimum x coordinate *xhi* = maximum x coordinate *ylo* = minimum y coordinate *yhi* = maximum y coordinate	all elements which lie in the window defined by these limits are plotted during preprocessing to enable geometry to be checked.

Nodes are represented by triangles. The program chooses the largest convenient scale that will allow the elements within the window to be plotted in the plot area defined by the keyword `SIZE` (see below).

Plot of displacement

`DISP` *xlo* *xhi* *ylo* *yhi*

Numeric data are as for `MESH`. A plot of the deformed mesh with suitably magnified displacements, superimposed on the undeformed mesh, is produced for each load case.

Plot of principal stresses

`PRST` *xlo* *xhi* *ylo* *yhi*

Numeric data are as for `MESH`. A plot is produced for each load case and material type, showing the magnitudes and orientations of principal stresses at nodes.

Size of plot

`SIZE` a b

a = first dimension of plot in millimetres
b = second dimension of plot in millimetres

Default values are $a = 290,\ b = 205$ (just smaller than A4).

Annotation of plot

`NUMS`

Node and element numbers are drawn on the plot specified by `MESH`. By default, these numbers are not plotted.

Definition of isotropic material

`MAT` $n \quad E \quad \nu \quad w$

n = material number ($1 \leq n \leq 10$)
E = Young's modulus
ν = Poisson's ratio
w = unit weight (operative in conjunction with load type `GRAV`).

Definition of transversely isotropic material

`MAN` $n \quad E_1 \quad E_2 \quad \nu_1 \quad \nu_2 \quad G \quad \beta \quad w$

n = material number ($1 \leq n \leq 10$)
E_1 = elastic modulus in direction 1
E_2 = elastic modulus in direction 2
ν_1 = Poisson's ratio in plane $1z$
ν_2 = Poisson's ratio in planes $2z$ and 12
G = shear modulus in planes $2z$ and 12
β = angle measured anticlockwise from direction x to direction 1 in degrees
w = unit weight (operative in conjunction with load type `GRAV`)

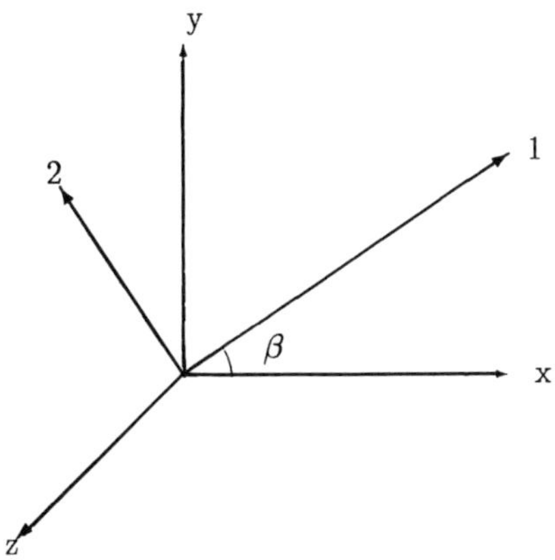

(B) COORDINATES OF NODES

The positions of nodes may be defined by reference to either the cartesian system (x, y) or the associated polar system (r, θ) where $x = r\cos\theta$, $y = r\sin\theta$. Furthermore, the common origin of these systems may be moved from its initial position O, by means of the `ORIG` keyword. Prior to reading of an `ORIG` line of data, $x_0 = y_0 = 0$. Note that if isoparametric patches of elements are specified (see next group of data) it is usually sufficient to specify the positions of a very small proportion of the nodes of the mesh at this stage.

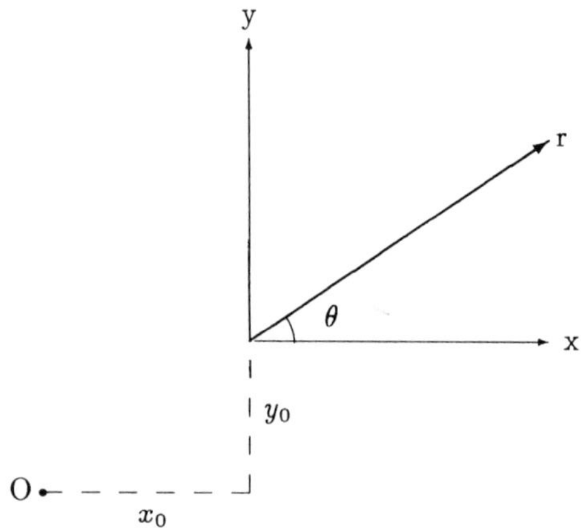

Relocation of the origin

`ORIG` x_0 y_0

x_0
 y_0 as shown above

Cartesian coordinates of single node

`CAR` i x_i y_i

$i =$ node number $1 \leq i \leq 4000$
$x_i = x$ coordinate of node i
$y_i = y$ coordinate of node i

Cartesian coordinates of series of nodes

`CARS` i j k x_i y_i x_j y_j

$i =$ first node number $1 \leq i \leq 4000$
$j =$ last node number $1 \leq j \leq 4000$
$k =$ node number increment
$x_i = x$ coordinate of node i
$y_i = y$ coordinate of node i
$x_j = x$ coordinate of node j
$y_j = y$ coordinate of node j

The program generates the nodes $i,\ i+k,\ i+2k, \ldots,\ j$ equally spaced along a straight line.

Polar coordinates of single node

`POL` i r_i θ_i

$i =$ node number $1 \leq i \leq 4000$
$r_i = r$ coordinate of node i
$\theta_i = \theta$ coordinate of node i in degrees

Polar coordinates of series of nodes

POLS $\quad i \quad j \quad k \quad r_i \quad \theta_i \quad r_j \quad \theta_j$

$i =$ first node number $1 \leq i \leq 4000$

$j =$ last node number $1 \leq j \leq 4000$

$k =$ node number increment

$r_i = r$ coordinate of node i

$\theta_i = \theta$ coordinate of node i in degrees

$r_j = r$ coordinate of node j

$\theta_j = \theta$ coordinate of node j in degrees

The program generates the nodes $i,\ i+k,\ i+2k, \dots,\ j$ at equal intervals of r and θ, so that for example if $r_i = r_j$ then the nodes will lie on an arc of a circle.

(C) NODES OF ELEMENTS

The nodes of triangular and quadrilateral elements are numbered in the orders shown below, anticlockwise starting from any corner node.

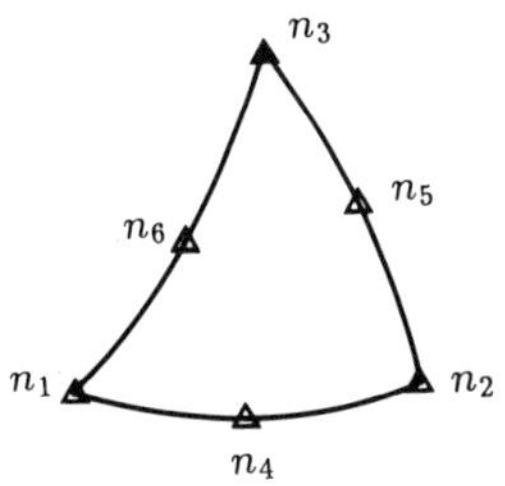

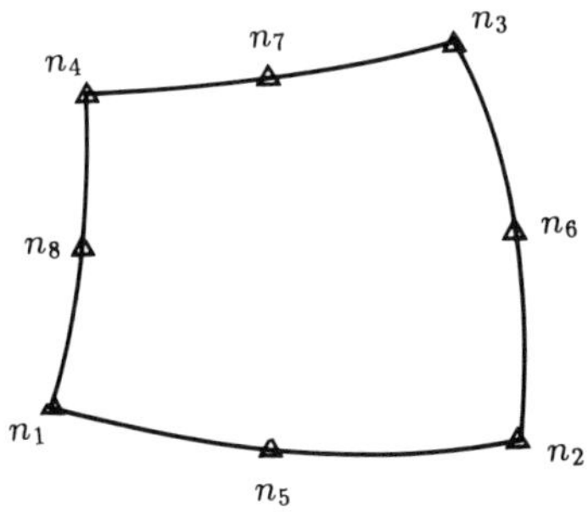

Elements may be specified individually, and also in the case of quadrilateral elements by isoparametric patch (the case $i = 3,\ j = 4$ shown: m_1 etc. are nodes of the patch, n_1 etc. are nodes of element e_{11}) on p. 386.

Coordinates of nodes of elements of an isoparametric patch are computed from the coordinates of the corner and side nodes of the patch, using quadratic shape functions as if the patch itself were a single isoparametric element. If a side node of the patch is nearer to one end of the side than to the other, then there is mesh gradation along that side of the patch, with smaller elements towards the end to which the side node is closer. Elements of an isoparametric patch are automatically numbered consecutively in the order $e_{11},\ e_{12},\ \dots,\ e_{1i},\ e_{21},\ e_{22},\ \dots,\ e_{2i}$ etc. (see p. 386) and nodes of elements are numbered with constant increment per element in the i and j directions of the patch. The corner and side nodes of the patch are also nodes of elements, and must be numbered in a manner consistent with this scheme.

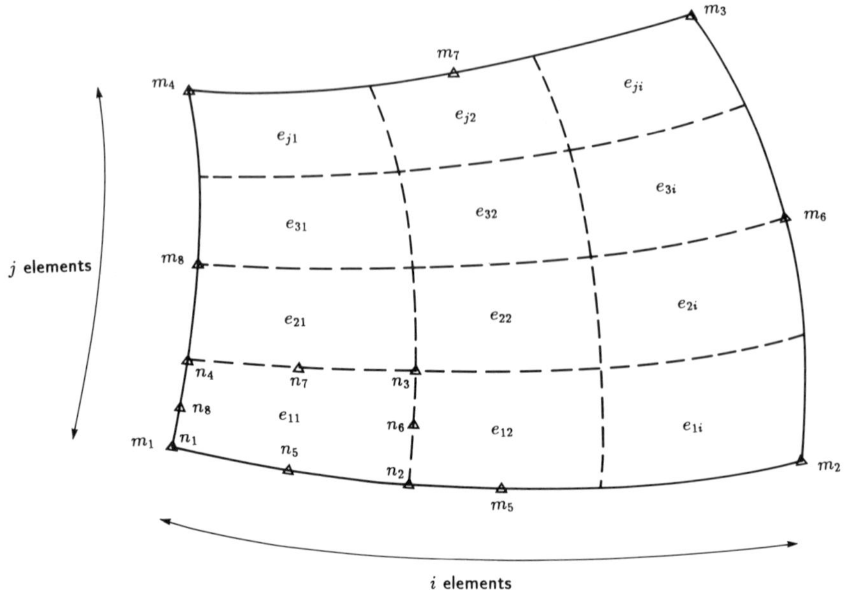

The coordinates of at least the corner nodes of the patch must have been specified in Data Group B. A side node of a patch is taken to be located halfway along the straight line joining the ends of the side unless its coordinates are specified in Data Group B.

When specifying elements individually it is advisable to locate side nodes roughly equidistant from the ends of sides. When generating elements of an isoparametric patch, the program automatically imposes this rule whether or not there is mesh gradation.

Single triangular element

ELT e k n_1 n_2 n_3 n_4 n_5 n_6

$e =$ element number $1 \leq e \leq 800$

$k =$ material number

n_1 to $n_6 =$ numbers of nodes

Single quadrilateral element

ELQ e k n_1 n_2 n_3 n_4 n_5 n_6 n_7 n_8

$e =$ element number $1 \leq e \leq 800$

$k =$ material number

n_1 to $n_8 =$ numbers of nodes

Patch of quadrilateral elements

`ELQS` e_{11} i j k m_1 m_2 m_3 m_4

e_{11} = first element number $1 \leq e \leq 800$
i = how many elements along sides m_1 m_2 and m_4 m_3
j = how many elements along sides m_1 m_4 and m_2 m_3
k = material number
m_1 to m_4 = numbers of corner nodes of patch

Note that if i or j are odd and there is mesh gradation, then midside nodes of the patch will be displaced from the locations specified in Data Group B in order to satisfy the requirement that midside nodes of elements be roughly equidistant from ends of sides.

(D) CONSTRAINTS ON DISPLACEMENT

Nodal displacements may be specified to be zero in either or both cartesian directions.

Constraint in x direction of single node

`FX` i

i = number of node to be constrained.

Constraint in x direction of series of nodes

`FXS` i j k

i = number of first node
j = number of last node
k = node number increment

The program constrains the nodes $i,\ i+k,\ i+2k,\ \ldots,\ j$.

Constraint in y direction of single node

`FY` i

i = number of node to be constrained

Constraint in y direction of series of nodes

`FYS` i j k

i = number of first node
j = number of last node
k = node number increment

The program constrains the nodes $i,\ i+k,\ i+2k,\ \ldots,\ j$.

(E) LOAD CASE TITLE

`L` *text*

text = title of load case, to appear on printout and plot if any.

(F) APPLIED LOADS

Within a load case, applied loads may be specified in any order and are additive in effect. The following side codes are used to identify sides of elements subject to load.

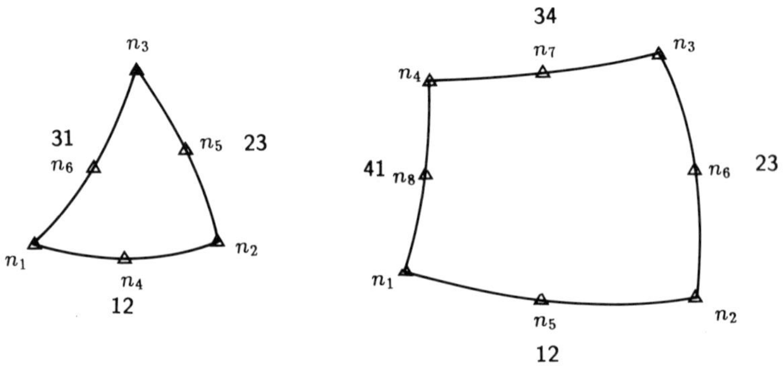

Self weight

GRAV

Self weight of w per unit volume (see data group A, definition of material type) is taken to act in the $-y$ direction.

Traction on side of single element

TR $\quad t_x \quad t_y \quad s \quad i$

$\left.\begin{matrix} t_x \\ t_y \end{matrix}\right\}$ cartesian components of traction

$s =$ side code (see above)
$i =$ element subject to load

Traction on sides of series of elements

TRS $\quad t_x \quad t_y \quad s \quad i \quad j \quad k$

$\left.\begin{matrix} t_x \\ t_y \end{matrix}\right\}$ cartesian components of traction

$s =$ side code (see above)
$i =$ number of first element subject to load
$j =$ number of last element subject to load
$k =$ element number increment

The traction (t_x, t_y) is applied to side s (see above for side codes) of elements $i,\ i+k,\ i+2k,\ \ldots,\ j$.

Pressure on side of single element

`PR` $p \quad s \quad i$

$p =$ magnitude of pressure (compression positive)
$s =$ side code (see above)
$i =$ element subject to load

Pressure on sides of series of elements

`PRS` $p \quad s \quad i \quad j \quad k$

$p =$ magnitude of pressure (compression positive)
$s =$ side code (see above)
$i =$ number of first element subject to load
$j =$ number of last element subject to load
$k =$ element number increment

The pressure p is applied to side s (see above for side codes) of elements $i,\ i+k,\ i+2k,\ \ldots,\ j$.

Initial stress field

`INF` $\sigma_x \quad \sigma_y \quad \tau_{xy} \quad \partial\sigma_x/\partial y$

$\left.\begin{array}{l}\sigma_x\\ \sigma_y\\ \tau_{xy}\end{array}\right\}$ components of initial stress, acting in all elements (Note: σ_x is the value at $y = 0$).

$\partial\sigma_x/\partial y =$ rate of change of σ_x with respect to y

(G) `GO` OR `NOGO`

The last line of input data is one of the following keywords.

Keyword	Action taken by program
`GO`	Analysis carried out if no syntactical errors found.
`NOGO`	Data check only carried out, including any plot requested by keyword `MESH`.

Examples of input data

The following input data files are on the disc obtainable by reply card.

```
T       CANTILEVER - COARSE MESH
MESH     0.0   5.0     0.0   1.0
DISP     0.0   5.0     0.0   1.0
NUMS
MAT    1     10000   0.0    1.0
CAR      1     0.0   0.0
CAR      5     0.0   1.0
CAR    201     5.0   0.0
CAR    205     5.0   1.0
ELQS   1   5   1   1     205     5     1  201
FXS    1   5   2
FYS    1   5   2
L     SELF WEIGHT
GRAV
L     UNIT LOAD AT END
TR     0.0  -1.0    41    1
GO

T       CANTILEVER - FINE MESH
MESH     0.0   5.0     0.0   1.0
DISP     0.0   5.0     0.0   1.0
NUMS
MAT    1     10000   0.0    1.0
CAR      1     0.0   0.0
CAR      5     0.0   1.0
CAR    201     5.0   0.0
CAR    205     5.0   1.0
ELQS   1  10   2   1     205     5     1  201
FXS    1   5   1
FYS    1   5   1
L     SELF WEIGHT
GRAV
L     UNIT LOAD AT END
TRS    0.0  -1.0    41    1    11    10
GO
```

A3 PRINTABLE OUTPUT (FILE DEMON.OUT)

DEMON.OUT file records are up to 120 characters in length. Data is written to the file in the following order.

(a) The input data read from the file DEMON.DAT. Note that a blank line is inserted after every five records to improve appearance, and each record of input data is preceded by three asterisks so that a blank line of input data can readily be identified as such.
(b) Title of problem, plot control data and material properties. In the table of material properties, all materials are treated as transversely isotropic.
(c) Coordinates of nodes as specified in Group B of the input data. The coordinates tabulated are cartesian coordinates with respect to the initial position of the origin, before the reading of any `ORIG` data.
(d) Nodes of elements including elements of isoparametric patches.
(e) A complete table of coordinates of nodes, including any node generated by `ELQS` data. At the end of this table is given the greatest internal node numbering difference.

```
T     HORSESHOE SECTION TUNNEL
PRST   0.0    5.0    -5.0    5.0
MAT     1    10000    0.2    0.0
CAR      1      0.0    40.0
CAR    121     40.0    40.0
CAR    241     40.0     0.0
CAR    361     40.0   -40.0
CAR    481      0.0   -40.0
POLS    13   253   60      2.0   90.0     2.0    0.0
CAR    373      2.0    -2.0
CAR    493      0.0    -2.0
POL      7     11.8    90.0
POL    127     15.8    45.0
POL    247     11.8     0.0
CAR    367     11.8   -11.8
CAR    487      0.0   -11.8
ELQS     1    3    6    1     121     1    13   133
ELQS    21    3    6    1     241   121   133   253
ELQS    41    3    6    1     361   241   253   373
ELQS    61    3    6    1     481   361   373   493
FXS      1    13    1
FXS    481   493    1
FXS    121   361   20
FYS      1   121   20
FYS    361   481   20
L     HYDROSTATIC FIELD STRESS
INF    -1.0   -1.0    0.0    0.0
L     HORIZONTAL STRESS = 0.5 TIMES VERTICAL
INF    -0.5   -1.0    0.0    0.0
GO

T      BEAM WITH HOLE
MAT     1      10000   0.25    0.0
POLS       1   801   100    125    90      125   -90
CAR        9                  0   250
CAR      809                  0  -250
CARS     209   609   200    250   250      250  -250
CARS     217   617   200    500   250      500  -250
CARS     225   625   200   1000   250     1000  -250
ELQS       1    2     2     1      209    9    1  201
ELQS       5    2     2     1      409  209  201  401
ELQS       9    2     2     1      609  409  401  601
ELQS      13    2     2     1      809  609  601  801
ELQS      17    4     2     1      225  209  409  425
ELQS      25    4     2     1      425  409  609  625
FXS    1    9    2
FXS   801  809   2
FY    425
L     UNIFORMLY DISTRIBUTED LOAD
TRS    0    -1    12    1    2    1
TRS    0    -1    12   17   20    1
TRS    0     2    41   17   29    4
GO
```

(f) Constraints on nodes.
(g) Applied loads.
(h) Computed displacements of, and stresses at, nodes. The following results are written to the file for each node.

(i) Cartesian components of displacement.
(ii) Cartesian components of the stress tensor. The values written to file are the averages of the values computed in each adjacent element by linear extrapolation with respect to intrinsic coordinates, from stresses computed at the sampling points of the 2×2 Gaussian quadrature formula.
(iii) Principal stresses and angles measured anticlockwise from the x axis to the directions in which they act, computed from the averaged cartesian components of the stress tensor.

(i) Time taken for execution of the program, and for various parts of the program.

A4 PLOT FILE (DPLOT.PLT)

The data on DPLOT.PLT defines the plots requested by keywords `MESH`, `DISP` and `PRST`. To make implementation of plotting as simple as possible and to avoid dependency on any advanced features of plotting packages, only four plotting operations are recognised. These are as follows.

1. *Initialisation of plot*: specify the dimensions of a plot window and (notionally) position a pen at the origin of plotter coordinates, this being located at the bottom left corner of the window. The unit of plotter coordinates is the millimetre.
2. *Move pen, with pen down*: move the pen in a straight line from its current position to a new position, drawing a line between the two positions.
3. *Move pen, with pen up*: move the pen from its current position to a new position, without drawing a line between the two positions.
4. *Plot text*: draw text of a specified height in a specified direction, beginning from a given position.

The primary consideration affecting the specification of the plot file is that it should be as small as possible. The most common operations are 2 and 3 above, because each side of an element is drawn as either ten straight lines or as five straight dashes. The least efficient mode of storage of the coordinates of the point to which the pen is to be moved is as formatted data. Unformatted floating point representation is better, but the most efficient technique is to multiply the coordinates by some factor and store them in unformatted integer representation. In subroutine DPLOT of program DEMON, coordinates (and also angles in degrees) are multiplied by a factor which depends upon plot size as defined by keyword SIZE, and written to the file DPLOT.PLT as two byte integers (INTEGER*2). For an A4 size plot (the default), this gives a plotting resolution of better than ± 0.01 mm.

In subroutine DPLOT, plotting operations are blocked into records each of 500 two byte integers in length:

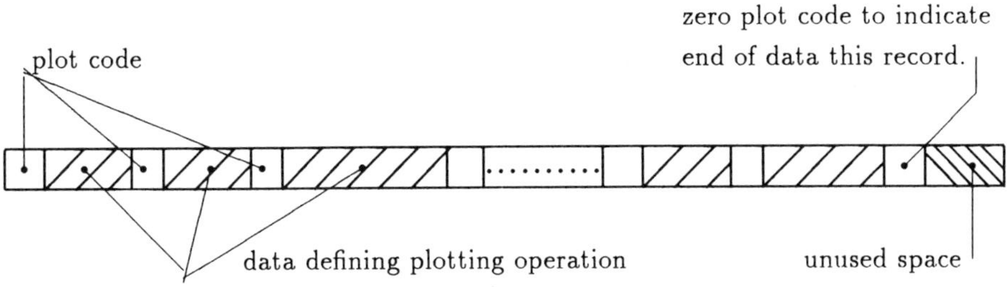

Plot codes are as follows:

1. initialisation of plot.
2. end of last plot.
3. move pen, with pen down.
4. move pen, with pen up.
5. plot text.

Program READP

To facilitate the implementation of plotting, there is a copy of program READP on the disc obtainable by reply card. Program READP reads the file DPLOT.PLT, and may be used as the framework for a program to draw or display the plots defined on that file. The source code of the program is as follows.

```
      PROGRAM READP
C     READ PLOT FILE WRITTEN BY SUBROUTINE DPLOT
      INTEGER*2 ICO,JBUF
      CHARACTER JCHA*1,ICHA*80
      DIMENSION SIZ(2),X(2),JBUF(500),JCHA(80)
      DATA LPL/9/
      OPEN (UNIT=LPL,STATUS='OLD',FILE='DPLOT.PLT',
     1 FORM='UNFORMATTED',IOSTAT=IERR,ERR=999)
      READ (LPL,IOSTAT=IERR,ERR=999) JBUF
      JZ=1
C     DATA TYPE CODE
    5 ICO=JBUF(JZ)
      IF (ICO.NE.0) GOTO 6
C     REFILL BUFFER OF PLOT DATA
      READ (LPL,IOSTAT=IERR,ERR=999) JBUF
      JZ=1
      ICO=JBUF(JZ)
    6 CONTINUE
      GOTO (1,2,3,3,4),ICO
```

```
C     INITIALISATION OF PLOT - INITIAL POSITION OF PEN IS (0,0)
C     X RANGE IS (0,SIZ(1)), Y RANGE IS (0,SIZ(2))
C     SIZ(1) AND SIZ(2) ARE IN MILLIMETRES
C     SCALE FACTOR AS SET IN SUBROUTINE DPLOT
    1 ECH=JBUF(JZ+1)
      SIZ(1)=JBUF(JZ+2)/ECH
      SIZ(2)=JBUF(JZ+3)/ECH
C     *** INSERT CODE TO INITIALISE PLOT ***
      JZ=JZ+4
      GOTO 5
C     STRAIGHT LINE FROM CURRENT POSITION OF PEN TO (X(1),X(2))
C     PEN DOWN IF IPEN=0, UP IF IPEN=1
C     X(1) AND X(2) ARE IN MILLIMETRES
    3 IPEN=ICO-3
      X(1)=JBUF(JZ+1)/ECH
      X(2)=JBUF(JZ+2)/ECH
C     *** INSERT CODE TO MOVE PEN ***
      JZ=JZ+3
      GOTO 5
C     TEXT - HT MILLIMETRES HIGH, LEN CHARACTERS AT TH DEGREES
C     ANTICLOCKWISE FROM X AXIS, STARTING AT THE POINT (X(1),X(2))
C     X(1) AND X(2) ARE IN MILLIMETRES
    4 X(1)=JBUF(JZ+1)/ECH
      X(2)=JBUF(JZ+2)/ECH
      HT=JBUF(JZ+3)/ECH
      TH=JBUF(JZ+4)/ECH
      LEN=JBUF(JZ+5)
      JD=(LEN-1)/2+1
      JA=JZ+6
      JB=JA+JD-1
      WRITE (ICHA,102) (JBUF(IO),IO=JA,JB)
      READ (ICHA,103) (JCHA(IO),IO=1,LEN)
  102 FORMAT (40A2)
  103 FORMAT (80A1)
C     *** INSERT CODE TO PLOT TEXT ***
      JZ=JB+1
      GOTO 5
  999 WRITE (*,104) IERR
  104 FORMAT (8H IERR = ,I10)
    2 CONTINUE
      END
```

Note that data to be processed are placed in the variables SIZ, IPEN, X, HT, TH, LEN and JCHA.

A5 PROGRAM DEMON SOURCE CODE

On the disc obtainable by reply card there are the files DROOT.FOR, DOVLA.FOR, DOVLB.FOR, DOVLC.FOR, DOVLD.FOR, DOVLE.FOR and DOVLF.FOR; these files comprise the source code of program DEMON.

DROOT.FOR contains the main program and a number of general purpose subroutines, which form the root segment if the program is overlain. The functions of the subroutines on the other files are as follows.

DOVLA.FOR DOVLB.FOR DOVLC.FOR	preprocessing and equivalent nodal force vector construction.
DOVLD.FOR	stiffness matrix construction.
DOVLE.FOR	simultaneous equation solution.
DOVLF.FOR	postprocessing.

Communication between routines is via parameter lists, COMMON blocks and scratch files. General purpose subroutines such as those which zeroise and copy arrays for example almost always communicate by parameter lists, to allow complete flexibility of use. Such flexibility is not required of subroutines with specialised purpose that are only called at one place in the program, so those subroutines usually communicate entirely by means of COMMON blocks or files. There are eight small COMMON blocks, and one large block named /CHANGE/. The small blocks contain data which largely remain unchanged during program execution, such as material properties or number of nodes. The large block /CHANGE/ is used for different purposes at different stages of execution, so that for example during preprocessing is contains such data as coordinates of nodes and nodes of elements, but during simultaneous equation solution it contains only equation coefficients. Generally, data are held in a file if (like the global stiffness matrix) they cannot be accommodated in /CHANGE/, or if (like coordinates of nodes during simultaneous equation solution) they occupy space that is temporarily required for other purposes.

COMMON blocks are declared in the main program to lengths which are at least as long as required in any subroutine. Different lengths of a COMMON block (especially /CHANGE/) may be required in different subroutines but to satisfy the requirements of certain FORTRAN compilers, in all references to a given block the list of variables is padded out to the length declared in the main program by the addition of a trailing integer array. To meet the requirements of yet other FORTRAN compilers, padding in the form of unsubscripted integers or additional coefficients of integer arrays is inserted where necessary to ensure that the offsets within COMMON blocks of double precision variables are always exact multiples of eight bytes.

Note that where it simplifies the program, use is made of the convention in FORTRAN whereby for multiply dimensioned arrays, the order in which coefficients are stored in memory is such that the first subscript varies most rapidly with respect to address, the second subscript the next most rapidly and so on. Very often therefore, whereas an actual parameter in a subroutine call is a two-dimensional array, the corresponding formal parameter is only a one dimensional array.

Unit numbers of files are held as variables in COMMON blocks. The contents of files used by program DEMON are shown against the variable names in the following table.

File	Contents	Organisation
LIN	the file DEMON.DAT (q.v.)	formatted, sequential
LUO	the file DEMON.OUT (q.v.)	formatted, sequential
LPL	the file DPLOT.PLT (q.v.)	unformatted, sequential
LUI	a copy of the file LIN	formatted, sequential
LUT	a single record of the file LIN, less the keyword	formatted, sequential
LUB	equivalent nodal forces and basic geometric data	unformatted, sequential
LUA	global stiffness matrix	unformatted, sequential
LUU	triangular factor of stiffness matrix	unformatted, direct access
LUS	solution vectors	unformatted, sequential

Data (apart from error messages which are always written to LUO) are written to and read from files as shown on p. 397.

Description of variables in COMMON blocks

Block	Variable	Description
/CLOCK/	SEC(10)	times in seconds spent in various segments of code
/FORM/	LUO	
	LIN	unit numbers of
	LUI	formatted files
	LUT	
	IREC(20)	single record buffer to hold data from LUI
	LULL	length of IREC
/PLOT/	ECH	factor by which plot coordinates in millimetres are multiplied before being written to LPL as integers
	SIZ(2)	dimensions of plot in millimetres
	SCA(3)	plotting scale factors for `MESH`, `DISP` and `PRST`
	CENP(2)	plot coordinates of centre of area available for plot
	CEN(2,3)	user's coordinates of centre of plot window for `MESH`, `DISP` and `PRST`
	DCO(4,3)	direction cosines for transformation from user's coordinates to plot coordinates
	LOT(3)	= 1 if plot type (`MESH`, `DISP` or `PRST`) requested, zero otherwise
	XLO(2,3)	user's coordinates of lower left corner of plot window for `MESH`, `DISP` and `PRST`

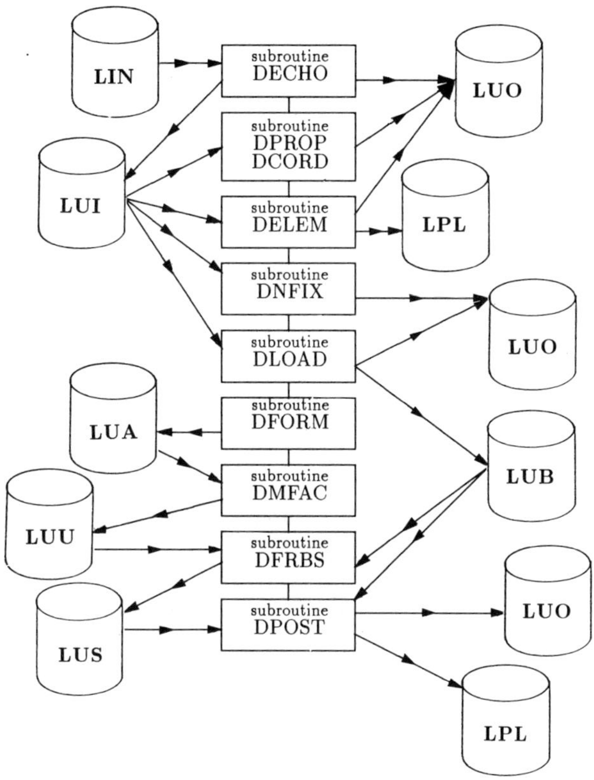

Block	Variable	Description
/PLOT/ (cont'd)	XHI(2,3)	user's coordinates of upper right corner of plot window for `MESH, DISP` and `PRST`
	NUM	= 1 if node and element numbers to be plotted for plot type `MESH`, zero otherwise
	LOS(10)	LOS(I) = 1 if there is to be a `PRST` plot for material type I, zero otherwise
/FLAG/	IGER	= 1 if error found in specification of geometry or material properties, = 0 otherwise.
	ILER	= 1 if error found in specification of constraints or applied loads, = 0 otherwise. Note: this applies to preprocessing routines; elsewhere in the program IGER=1 indicates an error condition of any type, and ILER is unused.

Block	Variable	Description
/TITLE/	IDENT(20)	title of problem
	JDENT(20,5)	titles of load cases
/PROP/	MMAT	maximum allowable material number
	IMAT(10)	IMAT(I) = 1 if material number I is defined, zero otherwise
	YMX(10)	elastic modulus in direction 1 of anisotropy
	YMY(10)	elastic modulus in direction 2 of anisotropy
	PRX(10)	Poisson's ratio in plane *1z*
	PRY(10)	Poisson's ratio in planes *2z* and *12*
	GEE(10)	shear modulus in planes *2z* and *12*
	BET(10)	angle measured anticlockwise from direction x to direction 1 in radians **Note:** an isotropic material is treated as an anisotropic material with YMY = YMX, PRY = PRX, and GEE computed from $G = \frac{E}{2(1+\nu)}$
	DEN(10)	unit weight of material
/GEOM/	MXA	maximum allowable node number
	NXB	number of nodes
	MEA	maximum allowable element number
	NEB	number of elements
	NDIF	greatest internal node numbering difference within an element
	LUA	unit number of global stiffness matrix file
	LUU	unit number of triangular factor file
	LUUL	triangular factor file record length
	LUUM	number of records in triangular factor file
/LOAD/	NCC	number of load cases
	LUB	unit number of file of equivalent nodal forces and basic geometric data
	LUS	unit number of file of solution vectors
	STR(4,5)	for each load case, the initial stresses at y=0 and rate of change of σ_x with respect to y
/CHANGE/ in subroutine DELEM before call of DPGEO	X(2,4000)	(X(1,I),X(2,I)) = coordinates of node with external number I
	IXY(4000)	IXY(I) = 1 if node with external number I is defined, zero otherwise
	IEL(800)	IEL(I) = 1 if element with external number I is defined, zero otherwise

Block	Variable	Description
/CHANGE/ (cont'd)	JEN(800)	JEN(I) = number of nodes of element with external number I
	JEM(800)	JEM(I) = material number of element with external number I
	IPT(800)	IPT(I) = i+2j+4k where i=1 if element with external number I appears in MESH plot, zero otherwise j=1 if element with external number I appears in DISP plot, zero otherwise k=1 if element with external number I appears in PRST plot, zero otherwise
	KEL(8,800)	KEL(I,J) = external number of node I of element with external number J **Note:** external node and element numbers are those referred to in the input data and understood by the user of the program. In subsequent documentation of /CHANGE/, 1. node and element numbers are internal node and element numbers unless otherwise stated. 2. data in subscripted variables are indexed on internal node and element numbers, unless otherwise stated.
/CHANGE/ in subroutine	X(2,1600)	coordinates of nodes
DLOAD	IXY(4000)	internal node numbers indexed on external node numbers
	JXY(1600)	external node numbers
	IEL(800)	internal element numbers indexed on external element numbers
	JEL(400)	external element numbers
	JEN(400)	JEN(I) = number of nodes of element I
	JEM(400)	JEM(I) = material number of element I
	IPT(400)	IPT(I) = i+2j+4k as described above but for element with internal number I
	KEL(8,400)	KEL(I,J) = number of node I of element J
	IFX(1600)	IFX(I) = i+2j where i=1 if node I is constrained against displacement in the x direction, zero otherwise j=1 if node I is constrained against displacement in the y direction, zero otherwise
	ENF(2,1600)	(ENF(1,I),ENF(2,I)) = components of equivalent nodal force at node I, for load case currently being read.

Block	Variable	Description
/CHANGE/ upon return from subroutine DCSTR called by subroutine DPOST	X(2,1600) JXY(1600) JEL(400) JEN(400) JEM(400) IPT(400) KEL(8,400)	same as for /CHANGE/ in DLOAD
	DN(2,1600)	(DN(1,I),DN(2,I)) = components of displacement at node I, for load case defined by the loop control variable IA of subroutine DPOST.
	SN(3,1600)	(SN(1,I),SN(2,I),SN(3,I)) = components of stress tensor at node I, for load case and material number defined by the loop control variables IA and IB respectively of subroutine DPOST.

A6 SELECTED LISTINGS

The complete listing of over 3000 lines of source code may be obtained by printing the files DROOT.FOR etc. referred to earlier. The listings reproduced here are of the routines held on files DROOT.FOR, DOVLD.FOR and DOVLE.FOR. The following notes are to be read in conjunction with the listings.

FILE DROOT.FOR

DEMON

Except for /FLAG/, COMMON blocks are declared to contain single subscripted variables dimensioned to at least the greatest length required at any stage of execution. Generally there is trailing unused space which may be used in future program development. The names of the arrays have no significance elsewhere in the program.

PTIME

There is provision for timing up to 10 segments of code. GETTIM is a library routine which returns the current system clock reading: NH = hours, NM = minutes, NS = seconds and ND = hundredths of a second. For a system on which a different syntax is used to read the clock, PTIME must be modified accordingly. If there exists no method of reading the system clock then the calls of GETTIM and the two lines immediately following each of those calls should be commented out or deleted.

DFERR

The error number printed by DFERR is as specified in the users' manual for the FORTRAN compiler. Normally the only reason for termination of execution in DFERR is insufficient disk space for scratch files.

DSHTD

Note that in this subroutine the variables ELA, ELB and ELC which are commented as area coordinates vary between -1.0 and 1.0, rather than between 0.0 and 1.0 as area coordinates are usually defined in the literature.

DDMAT

This routine is written for an anisotropic material only, because when subroutine DPROP reads the properties of an isotropic material it converts them to those of an anisotropic material. The computer time that would be saved by treating an isotropic material as a special case is negligible.

DPLOT

All plotting operations are effected by means of calls to entry points in this subroutine.

DCSCA

This routine chooses the scales for plotting of the mesh, displacements and principal stresses. To facilitate scaling off plots, the scales are chosen from the series ...... $2:1,\ 1:1,\ 1:2,\ 1:2.5,\ 1:5,\ 1:10,\ 1:20,$ ....

FILE DOVLD.FOR

DFORM

This matrix is assembled by JC blocks, each (except the last) consisting of JB pairs of rows. Only coefficients below and on the leading diagonal of the stiffness matrix are computed. Subscripts of the arrays A and AE are as follows.

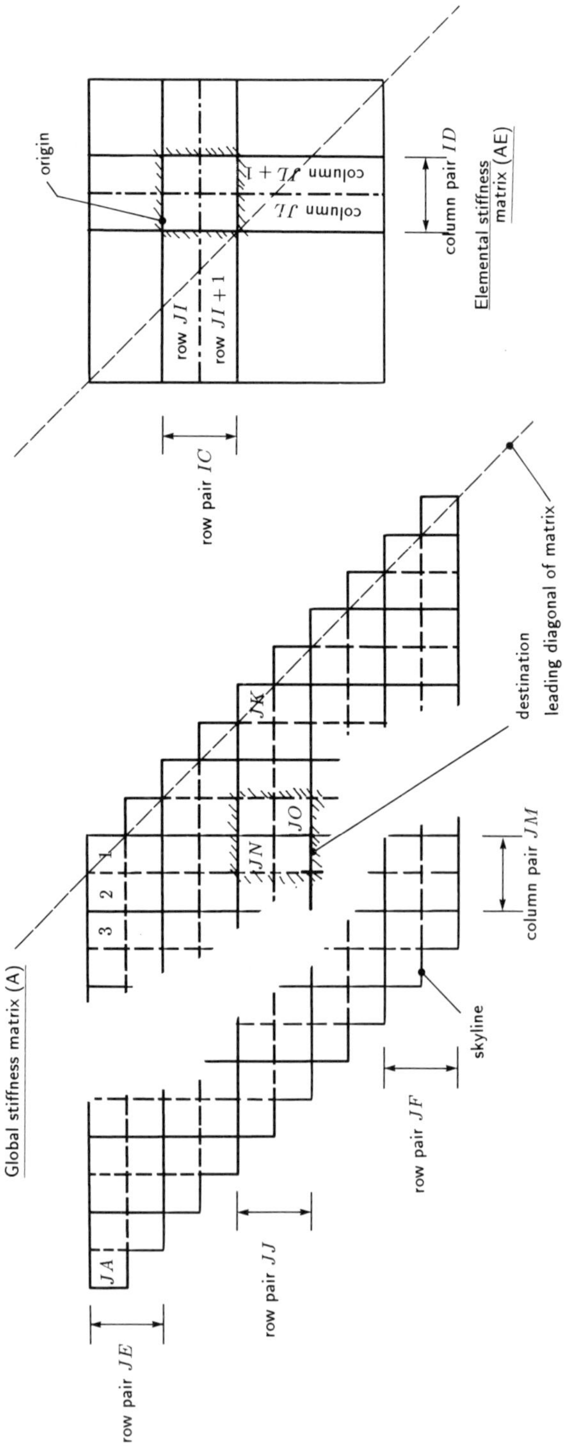
Elemental stiffness matrix (AE)
origin
row JI
row JI + 1
column JL
column JL + 1
column pair ID
row pair IC
Global stiffness matrix (A)
JA
1
2
3
JK
JN
JO
destination
leading diagonal of matrix
column pair JM
skyline
row pair JE
row pair JJ
row pair JF

The subscripts shown are as in the innermost loop of the scatter process, with control variable ID and ending at 9 CONTINUE. Where the coefficients of an elemental stiffness matrix scatter into more than one block, the elemental stiffness matrix is computed more than once. This can be avoided by pre-calculating all the elemental matrices and writing them to a file, but this in turn incurs an input-output overhead. In this and in other respects, a sliding triangle solver (even variable bandwidth) is less efficient than a frontal solver in which global stiffness matrix assembly proceeds simultaneously with bifactorisation.

Before a block is written to the file LUA, nodal constraints are applied. For each displacement to be set to zero, the leading diagonal coefficient multiplying that displacement is set to 1 and all other coefficients of the matrix row containing that coefficient are zeroised. During preprocessing, the subroutine DFIXF (file DOVLC.FOR) sets to zero the corresponding coefficient of the equivalent nodal force vector for each load case. Because only coefficients on and below the leading diagonal of the global stiffness matrix are stored, it is necessary to zeroise coefficients in both rows and columns containing leading diagonal coefficients which are set to 1. The logic of application of constraints becomes more complicated if constraints inclined to the coordinate axes are to be modelled, or if it is to be possible to specify nonzero given displacements.

Finally, note that in the OPEN statement for LUU, the value of RECL is given in *bytes*. On some computer systems, RECL must be specified in different units of storage such as 4 bytes or one word.

DSTIF

This subroutine computes the stiffness matrices of both quadrilateral and triangular elements: KA equals 8 for a quadrilateral, 6 for a triangle. For the 2×2 Gaussian product formula used in this routine the products of Gauss weights are all equal to 1 and so the array GWT is not required; however, it is retained in case it is desired to change the number of Gauss points. Integrations are performed for coefficients of the stiffness matrix on and above the leading diagonal, then coefficients below the leading diagonal are set. The purpose of setting all coefficients of the elemental stiffness matrix instead of those of an upper or lower triangle only is to simplify the logic of the scatter process in subroutine DFORM.

FILE DOVLE.FOR

DMFAC

This subroutine bifactorises the global stiffness matrix by the constant bandwidth sliding triangle technique. The logic of the routine is most easily understood by reference to a diagram showing the method of storage of the sliding triangle of equation coefficients, for a case in which for the purpose of illustration the maximum node numbering difference NDIF equals 3 (impossible for elements with quadratic variation):

variation):

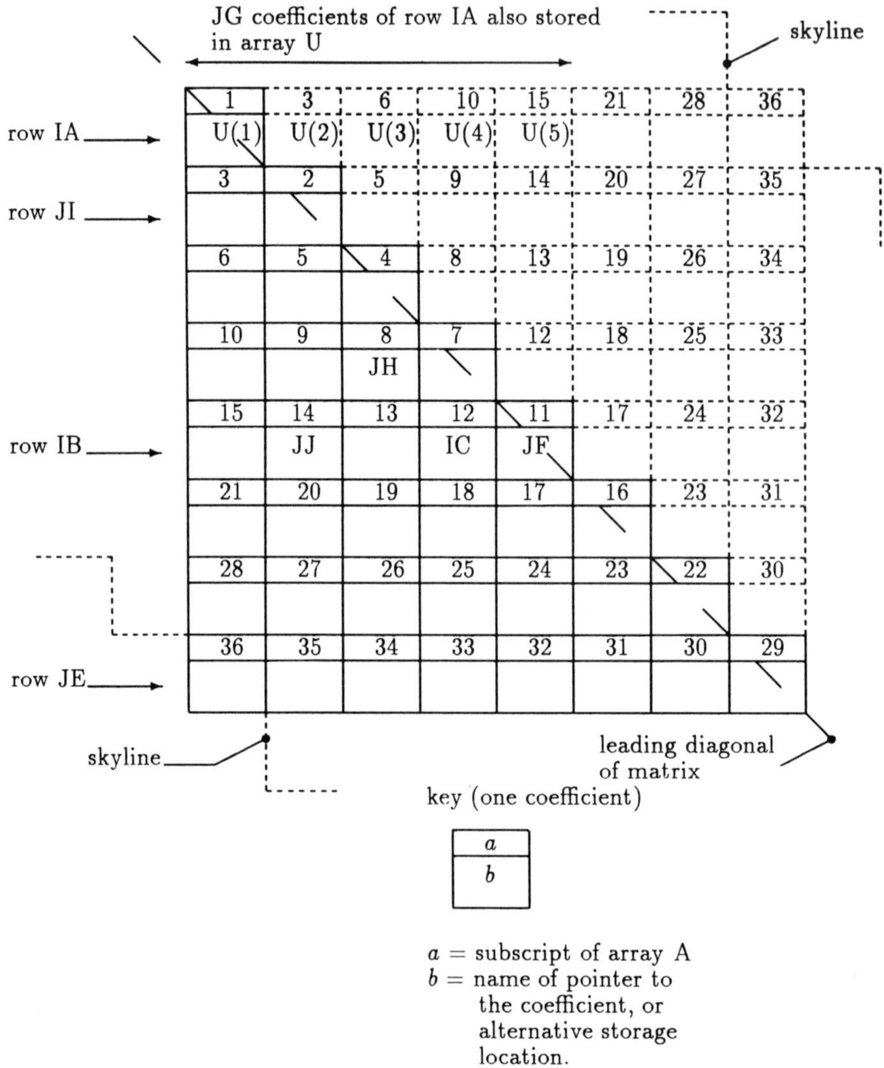

Each square represents one coefficient, and coefficients on and below the leading diagonal constitute the sliding triangle. The situation illustrated is that during the second time round the innermost loop (with control variable IC) when IB = IA+4, before JE attains the value JB. The value of JK is 4, and ELA equals $A(15)/A(1)$. Note that as the coefficients are modified they are moved up one row and to the left one column in the triangle, to allow the next row of the unreduced matrix to be read from file LUA into the last row, after the elimination of unknown IA has been completed.

Since in the back substitution the rows of the upper triangular factor are referred to in reverse order, they are stored in a direct access file. Whereas logical records of a sequential file can be blocked by operating systems into a smaller number of physical records for transfer to or from disc, each WRITE or READ of a direct access file invokes an actual transfer of data to or from the disc. To avoid the potential inefficiency of transferring data as a large number of small records, rows of the upper triangular matrix are copied into the array UB, and the contents of this array written to the file LUU only when there is insufficient room left for another matrix row. The integer ICUE which is equivalenced to the last coefficient of UB is a pointer to the leading diagonal coefficient of the last matrix row held in UB.

DFRBS

This subroutine carries out both forward reduction and back substitution of the load vectors. Computer time could be saved by performing the forward reduction at the same time as bifactorisation of the matrix (i.e. in DMFAC), but this would complicate the implementation of iterative procedures in which the load vectors only are modified at each iteration, such as elastoplastic analysis by the initial stiffness technique. Since the global stiffness matrix is symmetric, the lower triangular factor required for forward reduction of the load vectors is the transpose of the upper triangular factor computed by DMFAC; therefore, DFRBS reads the file LUU sequentially during forward reduction and in reverse order during back substitution. The first subscript of the array ENF denotes equation or degree of freedom number, the second the load case. When back substitution is complete, ENF contains the desired solution vectors which are then written to the file LUS.

```
      PROGRAM DEMON
C     DEMONSTRATION FINITE ELEMENT PROGRAM BY
C     JOHN WATSON, MINING ENGINEERING, UNIVERSITY OF NEW SOUTH WALES
      IMPLICIT DOUBLE PRECISION (A-H,O-Z)
      COMMON/CLOCK/CLO(10)
      COMMON/FORM/FOR(30)
      COMMON/PLOT/PLO(60)
      COMMON/FLAG/IGER,ILER
      COMMON/TITLE/TIT(120)
      COMMON/PROP/PRO(80)
      COMMON/GEOM/GEO(10)
      COMMON/LOAD/ALO(30)
      COMMON/CHANGE/CHA(20000)
      CALL PTIME (0)
      CALL PTICK (1)
C     PREPROCESSING
      CALL PTICK (2)
      CALL DOVLA
      CALL DOVLB
      CALL DOVLC
      CALL PTOCK (2)
      IF (IGER.NE.0.OR.ILER.NE.0) GOTO 1
C     CONSTRUCT STIFFNESS MATRIX
      CALL DOVLD
      IF (IGER.NE.0) GOTO 1
C     EQUATION SOLUTION
      CALL DOVLE
      IF (IGER.NE.0) GOTO 1
C     POSTPROCESSING
      CALL DOVLF
    1 CALL PTOCK (1)
      CALL DSTAT
      END

      SUBROUTINE PTIME (KA)
C     TIMING
      IMPLICIT DOUBLE PRECISION (A-H,O-Z)
      INTEGER*2 NH,NM,NS,ND
      COMMON/CLOCK/SEC(10)
      DATA LSEC/10/
C     INITIALISATION
      CALL PZERO (SEC,LSEC)
      RETURN
C     START CLOCK KA
      ENTRY PTICK (KA)
      CALL GETTIM (NH,NM,NS,ND)
      T=3600*NH+60*NM+NS+0.01D0*ND
      SEC(KA)=SEC(KA)-T
      RETURN
C     STOP CLOCK KA
      ENTRY PTOCK (KA)
      CALL GETTIM (NH,NM,NS,ND)
      T=3600*NH+60*NM+NS+0.01D0*ND
      SEC(KA)=SEC(KA)+T
      RETURN
      END
```

```
      SUBROUTINE DSTAT
C     PRINT TIMES TAKEN BY VARIOUS STAGES OF THE ANALYSIS
      IMPLICIT DOUBLE PRECISION (A-H,O-Z)
      COMMON/CLOCK/SEC(10)
      COMMON/FORM/LUO,LIN,LUI,LUT,KKFO(56)
      COMMON/GEOM/MXA,NXB,MEA,NEB,NDIF,LUA,LUU,KKGE(13)
      COMMON/LOAD/NCC,LUB,LUS,KKLO(57)
C     CLOSE PLOT FILE
      CALL DFINI
      CLOSE (UNIT=LIN,IOSTAT=JA,ERR=2)
    2 CLOSE (UNIT=LUI,IOSTAT=JA,ERR=3)
    3 CLOSE (UNIT=LUT,IOSTAT=JA,ERR=4)
    4 CLOSE (UNIT=LUB,IOSTAT=JA,ERR=5)
    5 CLOSE (UNIT=LUA,IOSTAT=JA,ERR=6)
    6 CLOSE (UNIT=LUU,IOSTAT=JA,ERR=7)
    7 CLOSE (UNIT=LUS,IOSTAT=JA,ERR=8)
    8 WRITE (LUO,100) (SEC(IO),IO=1,6)
  100 FORMAT (
     1 29H1    ELAPSED TIMES IN SECONDS//
     2 23H   TOTAL ALL ROUTINES  ,0PF7.1/
     3 23H   PREPROCESSING       ,0PF7.1/
     4 23H   MATRIX CONSTRUCTION ,0PF7.1/
     5 23H   BIFACTORISATION     ,0PF7.1/
     6 23H   LOAD VECTORS        ,0PF7.1/
     7 23H   POSTPROCESSING      ,0PF7.1)
      CLOSE (UNIT=LUO,IOSTAT=JA,ERR=9)
    9 CONTINUE
      RETURN
      END

      SUBROUTINE DFERR (LUN,IERR)
C     REPORT FILE OPEN OR WRITE ERROR AND TERMINATE RUN
      COMMON/FORM/LUO,KKFO(59)
      IF (IERR.EQ.0) GOTO 1
      WRITE (LUO,200) IERR,LUN
  200 FORMAT (/  14H ERROR NUMBER ,I5,
     1  9H ON FILE ,I2)
      CALL DSTAT
      STOP
    1 CONTINUE
      RETURN
      END

      SUBROUTINE PZERO (A,KA)
C     ZEROISE REAL ARRAY
      IMPLICIT DOUBLE PRECISION (A-H,O-Z)
      DIMENSION A(KA)
      DO 1 IA=1,KA
      A(IA)=0.0D0
    1 CONTINUE
      RETURN
      END
```

```
      SUBROUTINE PZERI (K,KA)
C     ZEROISE INTEGER ARRAY
      DIMENSION K(KA)
      DO 1 IA=1,KA
      K(IA)=0
    1 CONTINUE
      RETURN
      END

      SUBROUTINE PSETF (A,B,KA)
C     COPY REAL ARRAY
      IMPLICIT DOUBLE PRECISION (A-H,O-Z)
      DIMENSION A(KA),B(KA)
      DO 1 IA=1,KA
      B(IA)=A(IA)
    1 CONTINUE
      RETURN
      END

      SUBROUTINE PSETI (K,L,KA)
C     COPY INTEGER ARRAY
      DIMENSION K(KA),L(KA)
      DO 1 IA=1,KA
      L(IA)=K(IA)
    1 CONTINUE
      RETURN
      END

      SUBROUTINE PSRCH (KA,LEX,LEXL,KB)
C     LOOK FOR KA IN LEX - KB=0 IF NOT FOUND
      DIMENSION LEX(LEXL)
      KB=0
      DO 1 IA=1,LEXL
      IF (KA.NE.LEX(IA)) GOTO 1
C     FOUND
      KB=IA
      GOTO 2
    1 CONTINUE
    2 CONTINUE
      RETURN
      END

      SUBROUTINE DSHAD (XJ,FUN,DER,KA)
C     SHAPE FUNCTIONS AND DERIVATIVES W R TO INTRINSIC COORDINATES
      IMPLICIT DOUBLE PRECISION (A-H,O-Z)
      DIMENSION XJ(2),FUN(8),DER(2,8)
      IF (KA.NE.8) GOTO 1
C     QUADRILATERAL
      CALL DSHED (XJ,FUN,DER)
      GOTO 2
C     TRIANGLE
    1 CALL DSHTD (XJ,FUN,DER)
    2 CONTINUE
      RETURN
      END
```

```
      SUBROUTINE DSHED (XJ,FUN,DER)
C     SHAPE FUNCTIONS AND DERIVATIVES W R TO INTRINSIC
C     COORDINATES - QUADRILATERAL
      IMPLICIT DOUBLE PRECISION (A-H,O-Z)
      DIMENSION XJ(2),FUN(8),DER(2,8)
C     INTRINSIC COORDINATES
      XA=XJ(1)
      XB=XJ(2)
      XAP=1.0D0+XA
      XAN=1.0D0-XA
      XBP=1.0D0+XB
      XBN=1.0D0-XB
      XPP=XAP*XBP
      XPN=XAP*XBN
      XNP=XAN*XBP
      XNN=XAN*XBN
      PAP=XPP*XBN
      PAN=XNP*XBN
      PBP=XPP*XAN
      PBN=XPN*XAN
      XAS=XAP*XAN
      XBS=XBP*XBN
C     CORNER NODE FUNCTIONS
      FUN(1)=-0.25D0*XNN*(XA+XBP)
      FUN(2)=0.25D0*XPN*(XA-XBP)
      FUN(3)=0.25D0*XPP*(XA-XBN)
      FUN(4)=-0.25D0*XNP*(XA+XBN)
C     SIDE NODE FUNCTIONS
      FUN(5)=0.5D0*PBN
      FUN(6)=0.5D0*PAP
      FUN(7)=0.5D0*PBP
      FUN(8)=0.5D0*PAN
C     CORNER NODE DERIVATIVES
      SAP=0.5D0*XA+0.25D0*XB
      SAN=0.5D0*XA-0.25D0*XB
      SBP=0.5D0*XB+0.25D0*XA
      SBN=0.5D0*XB-0.25D0*XA
      DER(1,1)=XBN*SAP
      DER(2,1)=XAN*SBP
      DER(1,2)=XBN*SAN
      DER(2,2)=XAP*SBN
      DER(1,3)=XBP*SAP
      DER(2,3)=XAP*SBP
      DER(1,4)=XBP*SAN
      DER(2,4)=XAN*SBN
C     SIDE NODE DERIVATIVES
      HAS=0.5D0*XAS
      HBS=0.5D0*XBS
      DER(1,5)=-XA*XBN
      DER(2,5)=-HAS
      DER(1,6)=HBS
      DER(2,6)=-XB*XAP
      DER(1,7)=-XA*XBP
      DER(2,7)=HAS
      DER(1,8)=-HBS
      DER(2,8)=-XB*XAN
      RETURN
      END
```

```
      SUBROUTINE DSHTD (XJ,FUN,DER)
C     SHAPE FUNCTIONS AND DERIVATIVES W R TO INTRINSIC
C     COORDINATES FOR INTEGRATION - TRIANGLE
      IMPLICIT DOUBLE PRECISION (A-H,O-Z)
      DIMENSION XJ(2),FUN(6),DER(2,6)
C     INTRINSIC COORDINATES
      ELA=XJ(1)
      XB=XJ(2)
      XAB=ELA*XB
C     AREA COORDINATES
      ELB=0.5D0*(XB-ELA-XAB-1.0D0)
      ELC=0.5D0*(XAB-ELA-XB-1.0D0)
      ELAP=1.0D0+ELA
      ELBP=1.0D0+ELB
      ELCP=1.0D0+ELC
      ELAQ=0.5D0+ELA
      ELBQ=0.5D0+ELB
      ELCQ=0.5D0+ELC
      XAN=0.5D0*(1.0D0-ELA)
      XBP=0.5D0*(1.0D0+XB)
      XBN=0.5D0*(1.0D0-XB)
C     CORNER NODE FUNCTIONS
      FUN(1)=0.5D0*ELA*ELAP
      FUN(2)=0.5D0*ELB*ELBP
      FUN(3)=0.5D0*ELC*ELCP
C     SIDE NODE FUNCTIONS
      FUN(4)=ELAP*ELBP
      FUN(5)=ELBP*ELCP
      FUN(6)=ELCP*ELAP
C     CORNER NODE DERIVATIVES
      DER(1,1)=ELAQ
      DER(2,1)=0.0D0
      DER(1,2)=-ELBQ*XBP
      DER(2,2)=ELBQ*XAN
      DER(1,3)=-ELCQ*XBN
      DER(2,3)=-ELCQ*XAN
C     SIDE NODE DERIVATIVES
      DER(1,4)=ELBP-ELAP*XBP
      DER(2,4)=ELAP*XAN
      DER(1,5)=-ELBP*XBN-ELCP*XBP
      DER(2,5)=(ELC-ELB)*XAN
      DER(1,6)=ELCP-ELAP*XBN
      DER(2,6)=-ELAP*XAN
      RETURN
      END


      SUBROUTINE DPICK (X,KEL,KA,XA)
C     COORDINATES OF NODES OF ONE ELEMENT
      IMPLICIT DOUBLE PRECISION (A-H,O-Z)
      DIMENSION X(2,*),KEL(KA),XA(2,KA)
      DO 1 IA=1,KA
      JA=KEL(IA)
      CALL PSETF (X(1,JA),XA(1,IA),2)
    1 CONTINUE
      RETURN
      END
```

```
      SUBROUTINE DAREA (DER,XA,KA,AREA,ADER,XDER,IERR)
C     COMPUTE DETERMINANT AND INVERSE OF JACOBIAN MATRIX
      IMPLICIT DOUBLE PRECISION (A-H,O-Z)
      DIMENSION DER(2,KA),XA(2,KA),ADER(2,2),XDER(2,2)
C     INTRINSIC COORDINATE
      DO 1 IA=1,2
C     CARTESIAN COORDINATE
      DO 2 IB=1,2
      SUM=0.0D0
      DO 3 IC=1,KA
      SUM=SUM+DER(IA,IC)*XA(IB,IC)
    3 CONTINUE
C     JACOBIAN MATRIX COEFFICIENT
      XDER(IB,IA)=SUM
    2 CONTINUE
    1 CONTINUE
C     DETERMINANT
      AREA=XDER(1,1)*XDER(2,2)-XDER(2,1)*XDER(1,2)
      IF (AREA.LT.1.0D-20) GOTO 999
C     INVERSE - FIRST SUBSCRIPT DENOTES INTRINSIC COORDINATE
      ADER(1,1)=XDER(2,2)/AREA
      ADER(2,1)=-XDER(2,1)/AREA
      ADER(1,2)=-XDER(1,2)/AREA
      ADER(2,2)=XDER(1,1)/AREA
      IERR=0
    4 CONTINUE
      RETURN
C     ZERO OR NEGATIVE JACOBIAN
  999 IERR=1
      GOTO 4
      END


      SUBROUTINE DDMAT (DMAT,KA)
C     PLANE STRAIN STRESS-STRAIN RELATIONS FOR MATERIAL KA
C     SUBSCRIPT ORDER IS XX, YY, XY
      IMPLICIT DOUBLE PRECISION (A-H,O-Z)
      DIMENSION DMAT(3,3),EMAT(3,3),
     1 TMAT(3,3),XMAT(3,3),TMTT(3,3)
      COMMON/PROP/MMAT,IMAT(11),
     1 YMX(10),YMY(10),PRX(10),PRY(10),
     2 GEE(10),BET(10),DEN(10),KKPR(8)
C     YOUNG'S MODULI, POISSON'S RATIOS AND SHEAR MODULUS
C     FOR TRANSVERSELY ISOTROPIC MATERIAL
      YTX=YMX(KA)
      YTY=YMY(KA)
      PTX=PRX(KA)
      PTY=PRY(KA)
      GET=GEE(KA)
C     OTHER CONSTANTS
      ELA=YTX/YTY
      ELB=GET/YTY
      PTA=1.0D0+PTX
      PTB=1.0D0-PTX
      PTC=ELA*PTY
      PTD=PTC*PTY
      ELC=PTC*PTA
      ELD=PTA*(PTB-2.0D0*PTD)
      FAC=YTY/ELD
C     LOCAL COMPONENTS OF ELASTICITY MATRIX
      CALL PZERO (EMAT,9)
```

```
      EMAT(1,1)=ELA*(1.0D0-PTD)
      EMAT(2,2)=PTA*PTB
      EMAT(1,2)=ELC
      EMAT(2,1)=ELC
      EMAT(3,3)=ELB*ELD
      DO 1 IA=1,3
      DO 2 IB=1,3
      EMAT(IB,IA)=EMAT(IB,IA)*FAC
    2 CONTINUE
    1 CONTINUE
C     COMPUTE GLOBAL COMPONENTS
      CALL DTMAT (TMAT,TMTT,BET(KA))
      CALL DMATM (TMAT,EMAT,XMAT)
      CALL DMATM (XMAT,TMTT,DMAT)
      RETURN
      END

      SUBROUTINE DTMAT (TMAT,TMTT,BET)
C     TRANSFORMATION MATRIX FOR STRESS PSEUDO-VECTOR, AND
C     ITS TRANSPOSE - SUBSCRIPT ORDER IS XX, YY, XY
      IMPLICIT DOUBLE PRECISION (A-H,O-Z)
      DIMENSION TMAT(3,3),TMTT(3,3)
C     FUNCTIONS OF ANGLE BET
      CBT=DCOS(BET)
      SBT=DSIN(BET)
      CBS=CBT*CBT
      SBS=SBT*SBT
      CSB=CBT*SBT
C     TRANSFORMATION MATRIX
      TMAT(1,1)=CBS
      TMAT(2,2)=CBS
      TMAT(1,2)=SBS
      TMAT(2,1)=SBS
      TMAT(3,3)=CBS-SBS
      TMAT(2,3)=2.0D0*CSB
      TMAT(1,3)=-2.0D0*CSB
      TMAT(3,2)=-CSB
      TMAT(3,1)=CSB
C     TRANSPOSE OF TRANSFORMATION MATRIX
      CALL PSETF (TMAT,TMTT,9)
      TMTT(2,3)=TMAT(3,2)
      TMTT(1,3)=TMAT(3,1)
      TMTT(3,2)=TMAT(2,3)
      TMTT(3,1)=TMAT(1,3)
      RETURN
      END

      SUBROUTINE DMATM (TMAT,EMAT,XMAT)
C     MATRIX PRODUCT TMAT*EMAT
      IMPLICIT DOUBLE PRECISION (A-H,O-Z)
      DIMENSION TMAT(3,3),EMAT(3,3),XMAT(3,3)
      DO 1 IA=1,3
      DO 2 IB=1,3
      SUM=0.0D0
      DO 3 IC=1,3
      SUM=SUM+TMAT(IA,IC)*EMAT(IC,IB)
    3 CONTINUE
      XMAT(IA,IB)=SUM
    2 CONTINUE
    1 CONTINUE
      RETURN
      END
```

```
      SUBROUTINE DBMAT (DER,ADER,BMAT,KA)
C     STRAINS DUE TO NODAL DISPLACEMENTS OF AN ELEMENT
      IMPLICIT DOUBLE PRECISION (A-H,O-Z)
      DIMENSION DER(2,KA),ADER(2,2),BMAT(3,2,KA)
      DO 1 IA=1,KA
C     DERIVATIVES IN CARTESIAN DIRECTIONS OF SHAPE FUNCTION IA
      DMA=DER(1,IA)*ADER(1,1)+DER(2,IA)*ADER(2,1)
      DMB=DER(1,IA)*ADER(1,2)+DER(2,IA)*ADER(2,2)
C     STRAINS XX, YY, XY DUE TO X-DISPLACEMENT AT NODE IA
      BMAT(1,1,IA)=DMA
      BMAT(2,1,IA)=0.0D0
      BMAT(3,1,IA)=DMB
C     STRAINS XX, YY, XY DUE TO Y-DISPLACEMENT AT NODE IA
      BMAT(1,2,IA)=0.0D0
      BMAT(2,2,IA)=DMB
      BMAT(3,2,IA)=DMA
    1 CONTINUE
      RETURN
      END

      SUBROUTINE DSMAT (DMAT,BMAT,SMAT,KA)
C     STRESSES DUE TO NODAL DISPLACEMENTS OF AN ELEMENT
      IMPLICIT DOUBLE PRECISION (A-H,O-Z)
      DIMENSION DMAT(3,3),BMAT(3,2,KA),SMAT(3,2,KA)
      DO 1 IA=1,KA
      DO 2 IB=1,2
C     STRESSES XX, YY, XY
      DO 3 IC=1,3
      SUM=0.0D0
      DO 4 ID=1,3
      SUM=SUM+DMAT(IC,ID)*BMAT(ID,IB,IA)
    4 CONTINUE
      SMAT(IC,IB,IA)=SUM
    3 CONTINUE
    2 CONTINUE
    1 CONTINUE
      RETURN
      END

      SUBROUTINE DUNPK (KA,KB,KC)
C     UNPACKS BITS OF KA INTO KB (NOT VERY EFFICIENTLY)
      DIMENSION KB(KC)
      JA=KA
      DO 1 IA=1,KC
      JB=JA/2
      KB(IA)=JA-2*JB
      JA=JB
    1 CONTINUE
      RETURN
      END
```

```
      SUBROUTINE DPLOT
C     PLOTTING UTILITY ROUTINE
      IMPLICIT DOUBLE PRECISION (A-H,O-Z)
      SAVE JBUF,JZ,JA
      INTEGER*2 IXP,JCHA,JBUF
      CHARACTER IFOR*12
      DIMENSION X(2),LIT(*),IXP(2),JCHA(40),JBUF(500)
      COMMON/FORM/LUO,LIN,LUI,LUT,KKFO(56)
      COMMON/PLOT/ECH,SIZ(2),KKPL(114)
      DATA MINT/32000/
      DATA LPL,MPL/9,500/
      JZ=1
      OPEN (UNIT=LPL,STATUS='UNKNOWN',FILE='DPLOT.PLT',
     1 FORM='UNFORMATTED',IOSTAT=IERR,ERR=304)
  304 CALL DFERR (LPL,IERR)
      RETURN
C     INITIALISATION OF PLOT TYPE KA
      ENTRY DINIT (KA)
      JY=JZ+4
      IF (JY.LE.MPL) GOTO 6
C     EMPTY BUFFER OF PLOT DATA
      JBUF(JZ)=0
      WRITE (LPL,IOSTAT=IERR,ERR=304) JBUF
      JZ=1
    6 JBUF(JZ)=1
C     SCALE FACTOR MM OR DEGREES TO PLOT FILE UNITS
      JBUF(JZ+1)=MINT/DMAX1(SIZ(1),SIZ(2),360.0D0)
      ECH=JBUF(JZ+1)
      JBUF(JZ+2)=ECH*SIZ(1)
      JBUF(JZ+3)=ECH*SIZ(2)
      JZ=JZ+4
      JA=KA
      RETURN
C     CLOSE PLOT FILE
      ENTRY DFINI
      JBUF(JZ)=2
      WRITE (LPL,IOSTAT=IERR,ERR=304) JBUF
      REWIND LPL
      CLOSE (UNIT=LPL,IOSTAT=IERR,ERR=304)
      RETURN
C     MOVE PEN IN STRAIGHT LINE TO THE POINT X (USER COORDINATES)
C     PEN DOWN IF IPEN=0, UP IF IPEN=1
      ENTRY DMOVE (X,IPEN)
      CALL DPCOR (X,IXP,JA)
    1 JY=JZ+3
      IF (JY.LE.MPL) GOTO 7
C     EMPTY BUFFER OF PLOT DATA
      JBUF(JZ)=0
      WRITE (LPL,IOSTAT=IERR,ERR=304) JBUF
      JZ=1
    7 JBUF(JZ)=IPEN+3
      JBUF(JZ+1)=IXP(1)
      JBUF(JZ+2)=IXP(2)
      JZ=JZ+3
      RETURN
C     X IN MM ON PLOTTER
      ENTRY PMOVE (X,IPEN)
      CALL PPCOR (X,IXP)
      GOTO 1
C     WRITE TEXT HT MM HIGH, LEN CHARACTERS AT TH DEGREES TO X AXIS
      ENTRY DTEXT (X,HT,LIT,TH,LEN)
      CALL DPCOR (X,IXP,JA)
```

```
C     COPY DATA IN INTEGER*4 ARRAY TO INTEGER*2 ARRAY
    2 JB=(LEN-1)/4+1
      WRITE (LUT,102,IOSTAT=IERR,ERR=304) (LIT(IO),IO=1,JB)
      REWIND LUT
    5 JC=(LEN-1)/2+1
      READ (LUT,103,END=304) (JCHA(IO),IO=1,JC)
      REWIND LUT
  102 FORMAT (20A4)
  103 FORMAT (40A2)
      JY=JZ+JC+6
      IF (JY.LE.MPL) GOTO 8
C     EMPTY BUFFER OF PLOT DATA
      JBUF(JZ)=0
      WRITE (LPL,IOSTAT=IERR,ERR=304) JBUF
      JZ=1
    8 JBUF(JZ)=5
      JBUF(JZ+1)=IXP(1)
      JBUF(JZ+2)=IXP(2)
      JBUF(JZ+3)=ECH*HT
      JBUF(JZ+4)=ECH*TH
      JBUF(JZ+5)=LEN
      JZ=JZ+6
      DO 9 IA=1,JC
      JBUF(JZ)=JCHA(IA)
      JZ=JZ+1
    9 CONTINUE
      RETURN
C     X IN MM ON PLOTTER
      ENTRY PTEXT (X,HT,LIT,TH,LEN)
      CALL PPCOR (X,IXP)
      GOTO 2
C     INTEGER HT MM HIGH, FIELDWIDTH LEN AT TH DEGREES TO X AXIS
      ENTRY DNUMI (X,HT,KB,TH,LEN)
      CALL DPCOR (X,IXP,JA)
C     RUN-TIME FORMAT FOR WRITING TO LUT
    3 WRITE (IFOR,100,IOSTAT=IERR,ERR=300) LEN
  100 FORMAT (2H(I,I1,4H,1X))
  300 WRITE (LUT,IFOR,IOSTAT=IERR,ERR=5) KB
      REWIND LUT
      GOTO 5
C     X IN MM ON PLOTTER
      ENTRY PNUMI (X,HT,KB,TH,LEN)
      CALL PPCOR (X,IXP)
      GOTO 3
C     WRITE DOUBLE PRECISION NUMBER HT MM HIGH, NDEC DECIMAL PLACES
C     AT TH DEGREES TO X AXIS
      ENTRY DNUMF (X,HT,DP,TH,LEN,NDEC)
      CALL DPCOR (X,IXP,JA)
C     RUN-TIME FORMAT FOR WRITING TO LUT
    4 WRITE (IFOR,101,IOSTAT=IERR,ERR=302) LEN,NDEC
  101 FORMAT (4H(0PF,I2,1H.,I1,4H,1X))
  302 WRITE (LUT,IFOR,IOSTAT=IERR,ERR=5) DP
      REWIND LUT
      GOTO 5
C     X IN MM ON PLOTTER
      ENTRY PNUMF (X,HT,DP,TH,LEN,NDEC)
      CALL PPCOR (X,IXP)
      GOTO 4
      END
```

```
      SUBROUTINE DPCOR (X,IXP,KA)
C     COMPUTE PLOTTER COORDINATES FROM USER COORDINATES
C     KA=1 IS FOR MESH, =2 IS FOR DISPLACEMENT, =3 IS FOR STRESS
      IMPLICIT DOUBLE PRECISION (A-H,O-Z)
      INTEGER*2 IXP
      DIMENSION X(2),IXP(2)
      COMMON/PLOT/ECH,SIZ(2),SCA(3),
     1 CENP(2),CEN(2,3),DCO(4,3),KKPL(68)
      XA=X(1)-CEN(1,KA)
      XB=X(2)-CEN(2,KA)
C     ROTATION
      YA=DCO(1,KA)*XA+DCO(2,KA)*XB
      YB=DCO(3,KA)*XA+DCO(4,KA)*XB
C     SCALING AND CENTREING
      IXP(1)=ECH*(SCA(KA)*YA+CENP(1))
      IXP(2)=ECH*(SCA(KA)*YB+CENP(2))
      RETURN
C     X ARE PLOTTER COORDINATES IN MILLIMETRES
      ENTRY PPCOR (X,IXP)
      IXP(1)=ECH*X(1)
      IXP(2)=ECH*X(2)
      RETURN
      END

      SUBROUTINE DADDC (XPI,X,XA,KA)
C     ADD PLOTTER COORDINATE INCREMENTS XPI TO USER COORDINATES X
C     TO GIVE USER COORDINATES XA, PLOT TYPE KA
      IMPLICIT DOUBLE PRECISION (A-H,O-Z)
      DIMENSION XPI(2),X(2),XA(2)
      COMMON/PLOT/ECH,SIZ(2),SCA(3),
     1 CENP(2),CEN(2,3),DCO(4,3),KKPL(68)
      SCAR=1.0D0/SCA(KA)
      XA(1)=X(1)+(XPI(1)*DCO(1,KA)+XPI(2)*DCO(3,KA))*SCAR
      XA(2)=X(2)+(XPI(1)*DCO(2,KA)+XPI(2)*DCO(4,KA))*SCAR
      RETURN
      END

      SUBROUTINE DCSCA (BIG,HI,SCA)
C     CHOOSE SCALE FACTOR SCA SO THAT GREATEST VALUE HI
C     IS PLOTTED NO LONGER THAN BIG
      IMPLICIT DOUBLE PRECISION (A-H,O-Z)
      DIMENSION SC(4)
      DATA SC/1.0D0,0.5D0,0.4D0,0.2D0/
      SCO=1.0D10
      DO 1 IA=1,20
      DO 2 IB=1,4
      SCA=SCO*SC(IB)
      IF (SCA*HI.LE.BIG) GOTO 3
C     SCALE TOO BIG
    2 CONTINUE
      SCO=0.1D0*SCO
    1 CONTINUE
    3 CONTINUE
      RETURN
      END
```

```
      SUBROUTINE DELEP (X,KEL,KA,IPEN,MID)
C     PLOT ALL EDGES OF AN ELEMENT
C     NOTE THAT FOR PLOTTING OF THE MESH IN PREPROCESSING IT IS
C     PREVIOUSLY CHECKED THAT ALL NODES EXIST, AND MID IS RESET
C     TO ZERO UPON RETURN
      IMPLICIT DOUBLE PRECISION (A-H,O-Z)
      DIMENSION X(2,*),KEL(KA),MID(*),
     1 IEQ(4),IED(21),
     2 XA(2,8),XB(2,3)
      DATA IEQ/0,0,13,1/
      DATA IED/
     1 1,2,5, 2,3,6, 3,4,7, 4,1,8,
     2 1,2,4, 2,3,5, 3,1,6/
C     NODAL COORDINATES OF ELEMENT
      CALL DPICK (X,KEL,KA,XA)
      JA=KA/2
      JB=IEQ(JA)
      DO 1 IA=1,JA
      JD=IED(JB+2)
      JC=KEL(JD)
      IF (MID(JC).NE.0) GOTO 2
C     EDGE IA NOT ALREADY PLOTTED - PICK NODAL COORDINATES
      CALL DPICK (XA,IED(JB),3,XB)
C     DRAW EDGE SOLID OR DASHED ACCORDING TO THE VALUE OF IPEN
      CALL DEDGE (XB,IPEN)
      MID(JC)=1
    2 JB=JB+3
    1 CONTINUE
      RETURN
      END


      SUBROUTINE DEDGE (XA,IPEN)
C     DRAWS ONE EDGE OF AN ELEMENT
C     IPEN=0 FOR SOLID, =1 FOR DASHED
      IMPLICIT DOUBLE PRECISION (A-H,O-Z)
      DIMENSION XA(2,3),IPE(2),XB(2)
      IPE(1)=IPEN
      IPE(2)=0
C     MOVE PEN TO FIRST NODE OF EDGE
      CALL DMOVE (XA,1)
      XI=1.0D0
      DO 1 IA=1,5
      DO 2 IB=1,2
C     DECREMENT INTRINSIC COORDINATE AND COMPUTE SHAPE FUNCTIONS
      XI=XI-0.2D0
      XIA=XI+1.0D0
      XIB=XI-1.0D0
      XIH=0.5D0*XI
      SHA=XIH*XIB
      SHB=XIH*XIA
      SHC=-XIA*XIB
C     CARTESIAN COORDINATES
      DO 3 IC=1,2
      XB(IC)=SHA*XA(IC,1)+SHB*XA(IC,2)+SHC*XA(IC,3)
    3 CONTINUE
C     MOVE PEN TO POINT XI
      CALL DMOVE (XB,IPE(IB))
    2 CONTINUE
    1 CONTINUE
      RETURN
      END
```

```
      SUBROUTINE DPBOX (KA,KB,KC)
C     DRAW BOX AND TITLE
      IMPLICIT DOUBLE PRECISION (A-H,O-Z)
      DIMENSION XT(2),XL(2),XMN(2),XMO(2),
     1 XSM(2),XSN(2),XSO(2),
     2 XSD(2),XSE(2),XSF(2),
     3 LN(4),LS(2),LM(5),LD(6),LE(4),
     4 XY(2)
      COMMON/TITLE/IDENT(20),JDENT(20,5),KKTI(120)
      COMMON/PLOT/ECH,SIZ(2),SCA(3),
     1 CENP(2),CEN(2,3),DCO(4,3),
     2 LOT(3),KKPL(65)
      DATA XT/20.0D0,23.0D0/
      DATA XL/20.0D0,18.0D0/
      DATA XMN/20.0D0,13.0D0/
      DATA XMO/68.0D0,13.0D0/
      DATA XSM/20.0D0,8.0D0/
      DATA XSN/41.0D0,8.0D0/
      DATA XSO/71.0D0,8.0D0/
      DATA XSD/20.0D0,3.0D0/
      DATA XSE/41.0D0,3.0D0/
      DATA XSF/71.0D0,3.0D0/
      DATA LN/4HMATE,4HRIAL,4H NUM,4HBER /
      DATA LS/4H1 MM,4H =  /
      DATA LM/4H UNI,4HTS O,4HF DI,4HSTAN,4HCE  /
      DATA LD/4H UNI,4HTS O,4HF DI,4HSPLA,4HCEME,4HNT  /
      DATA LE/4H UNI,4HTS O,4HF ST,4HRESS/
      DATA TMAR/30.0D0/
      DATA HT/3.0D0/
C     INITIALISE PLOT
      CALL DINIT (KA)
      LOT(KA)=-1
C     DRAW BOX
      XY(1)=0.0D0
      XY(2)=0.0D0
      CALL PMOVE (XY,1)
      XY(1)=SIZ(1)
      CALL PMOVE (XY,0)
      XY(2)=SIZ(2)
      CALL PMOVE (XY,0)
      XY(1)=0.0D0
      CALL PMOVE (XY,0)
      XY(2)=0.0D0
      CALL PMOVE (XY,0)
      XY(2)=TMAR
      CALL PMOVE (XY,1)
      XY(1)=SIZ(1)
      CALL PMOVE (XY,0)
C     MAXIMUM LENGTH OF TITLE THAT CAN BE PLOTTED
      JA=(SIZ(1)-2.0D0*XT(1))/HT
      JA=MIN0(JA,76)
```

```
C     TITLE OF PROBLEM AND MESH SCALE
      CALL PTEXT (XT,HT,IDENT,0.0D0,JA)
      SCAI=1.0D0/SCA(KA)
      CALL PTEXT (XSM,HT,LS,0.0D0,7)
      NDEC=2-DLOG10(SCAI)
      NDEC=MAX0(NDEC,1)
      NDEC=MIN0(NDEC,8)
      CALL PNUMF (XSN,HT,SCAI,0.0D0,10,NDEC)
      CALL PTEXT (XSO,HT,LM,0.0D0,18)
      IF (KA.EQ.1) GOTO 1
C     LOAD CASE TITLE
      CALL PTEXT (XL,HT,JDENT(1,KB),0.0D0,JA)
      IF (KA.EQ.2) GOTO 1
C     MATERIAL NUMBER
      CALL PTEXT (XMN,HT,LN,0.0D0,16)
      CALL PNUMI (XMO,HT,KC,0.0D0,2)
    1 CONTINUE
      RETURN
C     DISPLACEMENT SCALE
      ENTRY DPSCD (SCD)
      SCDI=1.0D0/SCD
      CALL PTEXT (XSD,HT,LS,0.0D0,7)
      NDEC=2-DLOG10(SCDI)
      NDEC=MAX0(NDEC,1)
      NDEC=MIN0(NDEC,8)
      CALL PNUMF (XSE,HT,SCDI,0.0D0,10,NDEC)
      CALL PTEXT (XSF,HT,LD,0.0D0,22)
      RETURN
C     STRESS SCALE
      ENTRY DPSCS (SCS)
      SCSI=1.0D0/SCS
      CALL PTEXT (XSD,HT,LS,0.0D0,7)
      NDEC=2-DLOG10(SCSI)
      NDEC=MAX0(NDEC,1)
      NDEC=MIN0(NDEC,8)
      CALL PNUMF (XSE,HT,SCSI,0.0D0,10,NDEC)
      CALL PTEXT (XSF,HT,LE,0.0D0,16)
      RETURN
      END
```

```
      SUBROUTINE DOVLD
      CALL DFORM
      RETURN
      END

      SUBROUTINE DFORM
C     CONSTRUCT GLOBAL STIFFNESS MATRIX BY BLOCK OF ROWS
C     AND WRITE TO FILE LUA
      IMPLICIT DOUBLE PRECISION (A-H,O-Z)
      DIMENSION JELO(400),JENO(400),JEMO(400),
     1 KELO(8,400),IFXO(1600),
     2 JFX(2)
      COMMON/FORM/LUO,KKFO(59)
      COMMON/FLAG/IGER,ILER
      COMMON/GEOM/MXA,NXB,MEA,NEB,NDIF,LUA,LUU,LUUL,KKGE(12)
      COMMON/CHANGE/X(2,1600),
     1 JEL(400),JEN(400),JEM(400),KEL(8,400),
     2 IFX(1600),A(13000),AE(16,16),KKCH(1088)
      EQUIVALENCE (JELO(1),A(201)),
     1 (JENO(1),A(401)),(JEMO(1),A(601)),
     2 (KELO(1,1),A(1001)),(IFXO(1),A(2601))
      DATA LENU/1000/
      DATA LENA/13000/
      CALL PTICK (3)
      LUUL=LENU
C     COMPRESS /CHANGE/
      JA=8*NEB
      CALL PSETI (JELO,JEL,NEB)
      CALL PSETI (JENO,JEN,NEB)
      CALL PSETI (JEMO,JEM,NEB)
      CALL PSETI (KELO,KEL,JA)
      CALL PSETI (IFXO,IFX,NXB)
C     COEFFICIENTS/MATRIX ROW, NODES/BLOCK, HOW MANY BLOCKS
      JA=2*(NDIF+1)
      JB=(LENA/2)/JA
      JC=(NXB-1)/JB+1
      OPEN (UNIT=LUA,STATUS='SCRATCH',
     1 FORM='UNFORMATTED',IOSTAT=IERR,ERR=300)
  300 CALL DFERR (LUA,IERR)
      OPEN (UNIT=LUU,STATUS='SCRATCH',ACCESS='DIRECT',RECL=8*LUUL,
     1 IOSTAT=IERR,ERR=301)
  301 CALL DFERR (LUU,IERR)
      JD=2*JA
C     FIRST, LAST NODE OF BLOCK
      JE=1
      JF=JB
      DO 1 IA=1,JC
      JF=MIN0(JF,NXB)
C     HOW MANY ROWS THIS BLOCK
      JG=2*(JF-JE+1)
      JH=JG*JA
      CALL PZERO (A,JH)
      DO 2 IB=1,NEB
```

```
C     LOOK FOR NODE OF ELEMENT IB IN RANGE (JE,JF)
      JH=JEN(IB)
      DO 3 IC=1,JH
      JI=KEL(IC,IB)
      IF (JI.GE.JE.AND.JI.LE.JF) GOTO 4
    3 CONTINUE
      GOTO 2
C     FOUND - ELEMENT IB CONTRIBUTES TO THIS BLOCK
    4 CALL DSTIF (X,KEL(1,IB),JH,JEM(IB),AE,IERR)
      IF (IERR.EQ.0) GOTO 6
      WRITE (LUO,200) JEL(IB)
  200 FORMAT (/  24H ZERO JACOBIAN, ELEMENT ,I3)
      CALL PZERI (KEL(1,IB),JH)
      IGER=1
      GOTO 2
C     SCATTER COEFFICIENTS OF ELEMENTAL STIFFNESS MATRIX INTO BLOCK
    6 JI=1
      DO 7 IC=1,JH
      JJ=KEL(IC,IB)
      IF (JJ.LT.JE.OR.JJ.GT.JF) GOTO 8
C     ROW PAIR IS IN THIS BLOCK
      JK=JD*(JJ-JE)+1
      JL=1
      DO 9 ID=1,JH
      JM=KEL(ID,IB)
      IF (JM.GT.JJ) GOTO 10
C     CONTRIBUTION IS ON OR BELOW LEADING DIAGONAL
      JN=JK+2*(JJ-JM)
      JO=JN+JA
C     SUBMATRIX DIAGONAL TERMS
      A(JN)=A(JN)+AE(JI,JL)
      A(JO)=A(JO)+AE(JI+1,JL+1)
C     SUBMATRIX OFFDIAGONAL TERMS
      A(JO+1)=A(JO+1)+AE(JI+1,JL)
      IF (JM.EQ.JJ) GOTO 10
      A(JN-1)=A(JN-1)+AE(JI,JL+1)
   10 JL=JL+2
    9 CONTINUE
    8 JI=JI+2
    7 CONTINUE
    2 CONTINUE
C     APPLY CONSTRAINTS ON DISPLACEMENTS OF NODES
      JH=MAX0(JE-NDIF,1)
C     FIRST, LAST ROW OF BLOCK
      JI=2*JE-1
      JJ=2*JF
```

```
C     FIRST, LAST ROW OF LOWER TRIANGULAR BAND
      JK=2*JH-1
      JL=JK+JA-1
      JM=JA+1
      DO 11 IB=JH,JF
C     UNPACK FIXITY
      CALL DUNPK (IFX(IB),JFX,2)
      DO 12 IC=1,2
      IF (JFX(IC).EQ.0) GOTO 13
C     FOR POINTER TO A
      JN=JA*(JK-JI)+1
      JO=JI-JK+1
C     FIRST, LAST ROWS TO ZEROISE, AND POINTER TO A
      JP=MAX0(JI,JK)
      JQ=MIN0(JJ,JL)
      JR=MAX0(JN,JO)
C     ZEROISE A COLUMN OF THE BLOCK
      JS=JR
      DO 14 ID=JP,JQ
      A(JS)=0.0D0
      JS=JS+JM
   14 CONTINUE
      IF (JK.LT.JI) GOTO 13
C     ZEROISE A ROW AND PUT 1.0 ON LEADING DIAGONAL
      CALL PZERO (A(JR),JA)
      A(JR)=1.0D0
   13 JK=JK+1
      JL=JL+1
   12 CONTINUE
   11 CONTINUE
C     WRITE BLOCK TO LUA BY ROW
      JH=1
      JI=JA
      DO 15 IB=1,JG
      WRITE (LUA,IOSTAT=IERR,ERR=302) (A(IO),IO=JH,JI)
  302 CALL DFERR (LUA,IERR)
      JH=JH+JA
      JI=JI+JA
   15 CONTINUE
      JE=JE+JB
      JF=JF+JB
    1 CONTINUE
      REWIND LUA
      CALL PTOCK (3)
      RETURN
      END
```

```
      SUBROUTINE DSTIF (X,KEL,KA,KB,AE,IERR)
C     COMPUTE STIFFNESS MATRIX OF AN ELEMENT
      IMPLICIT DOUBLE PRECISION (A-H,O-Z)
      DIMENSION X(2,*),KEL(KA),AE(16,16),
     1 GOF(2,4),GWT(4),
     2 DMAT(3,3),XA(2,8),FUN(8),DER(2,8),ADER(2,2),XDER(2,2),
     3 BMAT(3,2,8),SMAT(3,2,8)
      DATA GOF/
     1  0.5773502692D0,  0.5773502692D0,
     2  0.5773502692D0, -0.5773502692D0,
     3 -0.5773502692D0,  0.5773502692D0,
     4 -0.5773502692D0, -0.5773502692D0/
      DATA GWT/4*1.0D0/
      DATA NGP/4/
      CALL PZERO (AE,256)
C     STRESS-STRAIN RELATIONS AND NODAL COORDINATES
      CALL DDMAT (DMAT,KB)
      CALL DPICK (X,KEL,KA,XA)
      DO 1 IA=1,NGP
C     SHAPE FUNCTIONS AND JACOBIAN AT GAUSS POINT IA
      CALL DSHAD (GOF(1,IA),FUN,DER,KA)
      CALL DAREA (DER,XA,KA,AREA,ADER,XDER,IERR)
      IF (IERR.NE.0) GOTO 2
C     JACOBIAN*GAUSS WEIGHT, AND STRAINS AND STRESSES DUE
C     TO UNIT NODAL DISPLACEMENTS
      ELA=AREA*GWT(IA)
      CALL DBMAT (DER,ADER,BMAT,KA)
      CALL DSMAT (DMAT,BMAT,SMAT,KA)
C     COLUMN OF STIFFNESS MATRIX
      JA=1
      DO 3 IB=1,KA
      DO 4 IC=1,2
C     ROW OF STIFFNESS MATRIX - COMPUTE UPPER TRIANGULAR
C     2*2 SUBMATRICES ONLY
      JB=1
      DO 5 ID=1,IB
      DO 6 IE=1,2
C     SUM PRODUCTS OF STRESS AND STRAIN COMPONENTS (XX, YY, XY)
      SUM=0.0D0
      DO 7 IF=1,3
      SUM=SUM+BMAT(IF,IE,ID)*SMAT(IF,IC,IB)
    7 CONTINUE
C     MULTIPLY BY JACOBIAN*GAUSS WEIGHT AND ADD INTO MATRIX
      AE(JB,JA)=AE(JB,JA)+SUM*ELA
      JB=JB+1
    6 CONTINUE
    5 CONTINUE
      JA=JA+1
    4 CONTINUE
    3 CONTINUE
    1 CONTINUE
C     SET LOWER TRIANGULAR COEFFICIENTS
      JA=2*KA
      DO 8 IA=1,JA
      DO 9 IB=IA,JA
      AE(IB,IA)=AE(IA,IB)
    9 CONTINUE
    8 CONTINUE
    2 CONTINUE
      RETURN
      END
```

```
      SUBROUTINE DOVLE
      COMMON/FLAG/IGER,ILER
      CALL DMFAC
      IF (IGER.NE.0) GOTO 1
      CALL DFRBS
    1 CONTINUE
      RETURN
      END

      SUBROUTINE DMFAC
C     BIFACTORISATION OF GLOBAL STIFFNESS MATRIX
C     WRITE UPPER TRIANGULAR FACTOR BY ROWS TO FILE LUU
      IMPLICIT DOUBLE PRECISION (A-H,O-Z)
      COMMON/FORM/LUO,KKFO(59)
      COMMON/FLAG/IGER,ILER
      COMMON/GEOM/MXA,NXB,MEA,NEB,NDIF,LUA,LUU,LUUL,LUUM,KKGE(11)
      COMMON/CHANGE/A(18528),U(192),UB(1000),KKCH(560)
      EQUIVALENCE (ICUE,UB(1000))
      CALL PTICK (4)
      JA=2*(NDIF+1)
C     INITIALISE SLIDING TRIANGLE OF MATRIX COEFFICIENTS
      JB=0
      DO 1 IA=1,JA
      JC=JB+1
      JB=JB+IA
      READ (LUA) (A(IO),IO=JC,JB)
    1 CONTINUE
      JB=2*NXB
      JC=JB-1
      JD=JA-1
      JE=JA
      LA=1
      LUUM=1
      DO 2 IA=1,JC
      IF (ABS(A(1)).LT.1.0D-20) GOTO 999
C     ELIMINATE UNKNOWN IA
      JF=1
      JG=1
      JH=1
C     LEADING DIAGONAL COEFFICIENT OF TRIANGULAR FACTOR
      U(1)=A(1)
      JI=IA+1
      DO 3 IB=JI,JE
```

```
C     ELIMINATE FROM EQUATION IB
C     POINTERS TO COEFFICIENTS ON LEADING DIAGONAL AND IN COLUMN IA
      JF=JF+JG
      JJ=JF+JG
      JG=JG+1
C     COEFFICIENT OF TRIANGULAR FACTOR
      U(JG)=A(JJ)
      ELA=U(JG)/U(1)
      JJ=JJ-1
C     SUBTRACT ELA*EQUATION IA FROM EQUATION IB
C     NOTE RELOCATION OF COEFFICIENTS IN ARRAY A
      JK=JG
      DO 4 IC=JF,JJ
      A(JH)=A(IC)-ELA*U(JK)
      JH=JH+1
      JK=JK-1
    4 CONTINUE
    3 CONTINUE
      IF (LA+JG.LT.LUUL) GOTO 6
C     EMPTY UPPER TRIANGULAR MATRIX BUFFER
      WRITE (LUU,REC=LUUM,IOSTAT=IERR,ERR=300) UB
  300 CALL DFERR (LUU,IERR)
      LUUM=LUUM+1
      LA=1
C     COPY ROW OF UPPER TRIANGULAR MATRIX
    6 CALL PSETF (U,UB(LA),JG)
      ICUE=LA
      LA=LA+JG
      IF (JE.EQ.JB) GOTO 2
C     APPEND A ROW OF THE UNREDUCED MATRIX TO THE SLIDING TRIANGLE
      JI=JH+JD
      READ (LUA) (A(IO),IO=JH,JI)
      JE=JE+1
    2 CONTINUE
      IF (ABS(A(1)).LT.1.0D-20) GOTO 999
C     LAST ROW OF UPPER TRIANGULAR MATRIX
      UB(LA)=A(1)
      ICUE=LA
      WRITE (LUU,REC=LUUM,IOSTAT=IERR,ERR=301) UB
  301 CALL DFERR (LUU,IERR)
    5 REWIND LUA
      CALL PTOCK (4)
      RETURN
  999 WRITE (LUO,200)
  200 FORMAT (/  16H SINGULAR MATRIX)
      IGER=1
      GOTO 5
      END
```

```
      SUBROUTINE DFRBS
C     FORWARD REDUCTION AND BACK SUBSTITUTION FOR EQUIVALENT NODAL
C     FORCE VECTORS
      IMPLICIT DOUBLE PRECISION (A-H,O-Z)
      COMMON/GEOM/MXA,NXB,MEA,NEB,NDIF,LUA,LUU,LUUL,LUUM,KKGE(11)
      COMMON/LOAD/NCC,LUB,LUS,KKLO(57)
      COMMON/CHANGE/ENF(3200,5),U(192),UB(1000),KKCH(5616)
      EQUIVALENCE (ICUE,UB(1000))
      CALL PTICK (5)
      JA=2*(NDIF+1)
      JB=2*NXB
C     READ EQUIVALENT NODAL FORCE VECTORS
      DO 1 IA=1,NCC
      READ (LUB) (ENF(IO,IA),IO=1,JB)
    1 CONTINUE
      REWIND LUB
      JC=JB-1
C     FORWARD REDUCTION
      JD=JA
      LA=LUUL
      LB=0
      DO 2 IA=1,JC
C     ELIMINATE UNKNOWN IA
      JD=MIN0(JD,JB)
      JE=JD-IA+1
      IF (LA+JE.LT.LUUL) GOTO 10
C     REFILL BUFFER
      LB=LB+1
      READ (LUU,REC=LB) UB
      LA=1
C     COLUMN IA OF LOWER TRIANGULAR FACTOR
   10 CALL PSETF (UB(LA),U,JE)
      LA=LA+JE
      JF=IA+1
      DO 3 IB=1,NCC
      ELA=ENF(IA,IB)/U(1)
      JG=2
      DO 4 IC=JF,JD
C     ELIMINATE FROM EQUATION IC LOAD CASE IB
      ENF(IC,IB)=ENF(IC,IB)-ELA*U(JG)
      JG=JG+1
    4 CONTINUE
    3 CONTINUE
      JD=JD+1
    2 CONTINUE
      U(1)=UB(LA)
C     SOLVE LAST EQUATION
      DO 5 IA=1,NCC
      ENF(JB,IA)=ENF(JB,IA)/U(1)
    5 CONTINUE
```

```
C     BACK SUBSTITUTION
      JD=JC
      JE=2
      DO 6 IA=1,JC
C     COMPUTE UNKNOWN JD
      JE=MIN0(JE,JA)
      LA=LA-JE
      IF (LA.GT.0) GOTO 11
C     REFILL BUFFER
      LB=LB-1
      READ (LUU,REC=LB) UB
      LA=ICUE
C     ROW JD OF UPPER TRIANGULAR FACTOR
   11 CALL PSETF (UB(LA),U,JE)
      DO 7 IB=1,NCC
C     SUM OVER UNKNOWNS ALREADY COMPUTED LOAD CASE IB
      SUM=0.0D0
      JF=JD+1
      DO 8 IC=2,JE
      SUM=SUM+ENF(JF,IB)*U(IC)
      JF=JF+1
    8 CONTINUE
      ENF(JD,IB)=(ENF(JD,IB)-SUM)/U(1)
    7 CONTINUE
      JD=JD-1
      JE=JE+1
    6 CONTINUE
      CLOSE (UNIT=LUA,IOSTAT=JA,ERR=300)
  300 CLOSE (UNIT=LUU,IOSTAT=JA,ERR=301)
  301 OPEN (UNIT=LUS,STATUS='SCRATCH',
     1 FORM='UNFORMATTED',IOSTAT=IERR,ERR=302)
  302 CALL DFERR (LUS,IERR)
C     WRITE SOLUTION VECTORS
      DO 9 IA=1,NCC
      WRITE (LUS,IOSTAT=IERR,ERR=303) (ENF(IO,IA),IO=1,JB)
  303 CALL DFERR (LUS,IERR)
    9 CONTINUE
      REWIND LUS
      CALL PTOCK (5)
      RETURN
      END
```

Appendix B
Program BOUND

B1 INTRODUCTION

Program BOUND is a program for the analysis of plane strain by the boundary element method, direct formulation. The program is written for the following purposes.

1. To familiarise intending users of boundary elements with the practical details of using boundary element programs, and with the results that may be expected from boundary element analysis.
2. For those interested in developing their own software, to demonstrate typical programming techniques and provide routines that may be used as a basis for their own work.

The modelling capabilities of BOUND are similar to those of the finite element program DEMON, except in that there may be only one material, which must be isotropic. BOUND in its present form already consists of around 4500 lines of source code: anisotropy would add at least another 500 and subregions to model more than one material over 1000 lines, resulting in a program that would be too long for demonstration purposes. This simply reflects the greater complexity of boundary element methods.

In program BOUND, three-node isoparametric and one-node infinite elements are used. For the isoparametric elements, displacement is taken to vary quadratically with respect to an intrinsic coordinate. The shape functions used to interpolate displacement are also used to interpolate cartesian coordinates, so if the intermediate node does not lie on a straight line joining the end nodes then the element is curved. This allows curved boundaries to be modelled accurately using relatively few elements, and so improves the efficiency of the analysis. For the infinite elements which are used to model the boundary of a semi-infinite region, displacement is taken to tend asymptotically to the displacement corresponding to the given state of stress at infinity. Program BOUND computes displacements and stresses at the boundary nodes, and also at any interior points of the region specified by the user.

Sign convention for stress

In the sign convention of program BOUND, tensile stress is regarded as positive.

Problem size limits

The following limits may not be exceeded.

- Number of boundary nodes 250
- Number of interior nodes 300
- Highest node number 1500
- Number of elements 120
- Highest element number 600
- Number of load cases 5

Note that the user is allowed to leave gaps in the node and element numbering, so that for example the highest node number is not necessarily equal to the number of nodes. Elements and nodes may be numbered in any order.

A load case is some distribution of applied load, for a given structural geometry and set of constraints on displacement. Generally, it takes little more computer time to analyse several load cases than it does to analyse a single load case.

Modes of execution of the program

If the input data ends with the keyword `GO`, then the program carries out the requested analysis if it has found no syntactical errors in the input data. If errors were found then the program stops after reading and checking as fully as possible all the input data but before embarking on any computation requiring significant amounts of computer time. If the input data ends with the keyword `NOGO`, then the program stops after reading and checking the input data, regardless of whether or not syntactical errors were found.

It is advisable to carry out a `NOGO` run before attempting a full analysis and indeed to obtain a plot of the mesh, since errors in specification of the problem to be solved do not necessarily manifest themselves as syntactical errors detectable by the preprocessing routines. Note that program BOUND will always read the input data to the end and provide as full an account as possible of errors, no matter how many errors exist in the data.

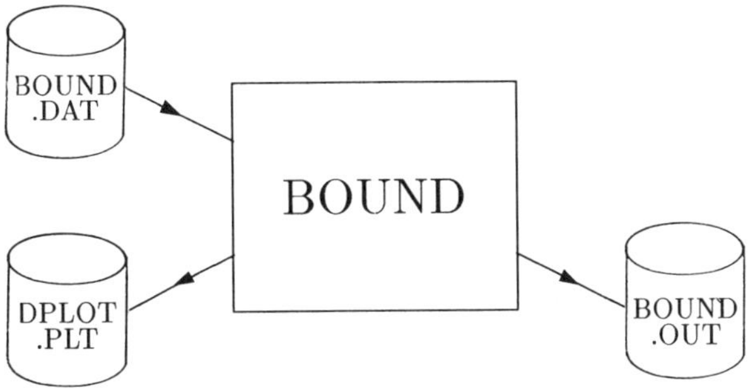

Input and output files

Program BOUND reads input data from the file BOUND.DAT, writes printable output (including any error messages) to the file BOUND.OUT, and writes plot data if requested to the file DPLOT.PLT:

B2 INPUT DATA (FILE BOUND.DAT)

The first four characters of each line of input data are reserved for a keyword which identifies the type of data on the rest of the line. Keywords are from one to four characters in length, and must always begin at the first character of the line. The remaining 76 characters may be used for numeric data or, for problem and load case titles, for text. Items of numeric data must be separated from one another by at least one space. Data must be presented in the following order.

(A) Problem title, plot control data, material properties, symmetries, definition of zero displacement
(B) Coordinates of nodes
(C) Nodes of elements and element orientation
(D) Constraints on displacement
(E) Load case title } repeat for each load case
(F) Applied loads } repeat for each load case
(G) `GO` or `NOGO`

The use of keywords to identify data allows the user some flexibility in the order in which lines of data are presented within each of the groups above; indeed, the order is only of any significance in Group B. The specification of data within each group is as follows.

(A) PROBLEM TITLE, MATERIAL PROPERTIES, PLOT AND OTHER OPTIONS

Problem title

`T` *text*

text = title of problem to appear on printout and plots if any.

Plot of mesh

`MESH` xlo xhi ylo yhi

xlo = minimum x coordinate xhi = maximum x coordinate ylo = minimum y coordinate yhi = maximum y coordinate	all elements which lie in the window defined by these limits are plotted during preprocessing to enable geometry to be checked.

Nodes are represented by triangles. The program chooses the largest convenient scale that will allow the elements within the window to be plotted in the plot area defined by the keyword `SIZE` (see below).

Plot of displacement

`DISP` xlo xhi ylo yhi

Numeric data is as for `MESH`. A plot of the deformed mesh with suitably magnified displacements, superimposed on the undeformed mesh, is produced for each load case.

Plot of principal stresses

`PRST` xlo xhi ylo yhi

Numeric data is as for `MESH`. A plot is produced for each load case, showing the magnitudes and orientations of principal stresses at nodes of elements and interior nodes which lie inside the window defined by the numeric data.

Size of plot

`SIZE` a b

a = first dimension of plot in millimetres

b = second dimension of plot in millimetres

Default values are $a = 290$, $b = 205$ (just smaller than A4).

Annotation of plot

`NUMS`

Node and element numbers are drawn on the plot specified by `MESH`. By default, these numbers are not plotted.

Definition of material type

`MAT` E ν w

E = Young's modulus

ν = Poisson's ratio

w = unit weight (operative in conjunction with load type `GRAV`).

Symmetry

`SYMX` symmetry with respect to x.

`SYMY` symmetry with respect to y.

If there is both structural symmetry and symmetry of loading with respect to x or y, then the amount of data to prepare and computer time required for solution may be reduced by using the `SYMX` or `SYMY` option. If symmetry with respect to x (or y) is specified, then only the boundary of that part of the region which lies in the halfspace $x > 0$ (or $y > 0$) need be modelled; no elements are defined on the plane of symmetry $x = 0$ (or $y = 0$). If symmetry is specified with respect to both x and y, then it is only necessary to model that part of the region for which both x and y are positive. By default, there is taken to be no symmetry.

Definition of zero displaclement

`PRE`

Displacements with respect to the initially strained or pre-mining state as defined by load types `GRAV` and `INF` are computed. By default, the computed displacements are with respect to the unstrained state.

(B) COORDINATES OF NODES

The positions of nodes may be defined by reference to either the cartesian system (x, y) or the associated polar system (r, θ) where $x = r\cos\theta$, $y = r\sin\theta$. Furthermore, the common origin of these systems may be moved from its initial position, by means of the `ORIG` keyword. Prior to reading of an `ORIG` line of data, $x_0 = y_0 = 0$. For details of coordinate systems, see Appendix A. No distinction is made between nodes of elements and interior nodes: the program automatically takes nodes which are not referred to in the elements data to be interior nodes. However, so that the limit on element numbers is not exceeded, the numbers of nodes of elements should not be greater than 600.

Relocation of the origin

`ORIG` x_0 y_0

$\left.\begin{array}{l} x_0 \\ y_0 \end{array}\right\}$ see Appendix A

Cartesian coordinates of single node

`CAR` i x_i y_i

$i =$ node number $1 \leq i \leq 1500$

$x_i = x$ coordinate of node i

$y_i = y$ coordinate of node i

Polar coordinates of single node

`POL` $i \quad r_i \quad \theta_i$

$i =$ node number $1 \leq i \leq 1500$

$r_i = r$ coordinate of node i

$\theta_i = \theta$ coordinate of node i in degrees

Linear interpolation of cartesian coordinates of series of nodes

`CARS` $i \quad j \quad k$

$\left.\begin{matrix} i \\ j \end{matrix}\right\}$ numbers of nodes the coordinates of which have already been specified ($1 \leq i \leq 1500$, $1 \leq j \leq 1500$)

$k =$ node number increment

The program generates the nodes $i+k,\ i+2k,\ \ldots,\ j-k$ such that nodes $i,\ i+k,\ i+2k,\ \ldots,\ j-k,\ j$ are equally spaced along a straight line.

Linear interpolation of polar coordinates of series of nodes

`POLS` $i \quad j \quad k$

$\left.\begin{matrix} i \\ j \end{matrix}\right\}$ numbers of nodes the coordinates of which have already been specified ($1 \leq i \leq 1500$, $1 \leq j \leq 1500$)

$k =$ node number increment

The program generates the nodes $i+k,\ i+2k,\ \ldots,\ j-k$ such that nodes $i,\ i+k,\ i+2k,\ \ldots,\ j-k,\ j$ are spaced at equal intervals of r and θ. Note that the θ coordinates of nodes i and j may not be the same, and that the program always takes the increment of θ to be positive.

Linear interpolation of cartesian coordinates of grid of nodes

`CARG` $i \quad j \quad k \quad l \quad m \quad n$

$\left.\begin{matrix} i \\ j \\ k \\ l \end{matrix}\right\}$ numbers of nodes the coordinates of which have already been specified (in range 1 to 1500)

$m =$ node number increment in direction i to j or l to k.

$n =$ node number increment in direction j to k or i to l.

$i,\ j,\ k$ and l are the nodes at the corners of the grid, in either clockwise or anticlockwise order. The total number of nodes in the grid is

$$\left(\frac{j-i}{m}+1\right)\left(\frac{l-i}{n}+1\right)$$

including the four nodes already defined.

Linear interpolation of polar coordinates of grid of nodes

$\left.\begin{matrix} i \\ j \\ k \\ l \end{matrix}\right\}$ numbers of nodes the coordinates of which have already been specified (in range 1 to 1500)

m = node number increment in direction i to j or l to k.
n = node number increment in direction j to k or i to l.
i, j, k and l are as for CARG. Note that the θ coordinates of nodes i and j, and of k and l, may not be the same, and that the program always takes the increment of θ to be positive from i to j, and from l to k.

(C) NODES OF ELEMENTS AND ELEMENT ORIENTATION

The nodes of an isoparametric element are numbered as shown below.

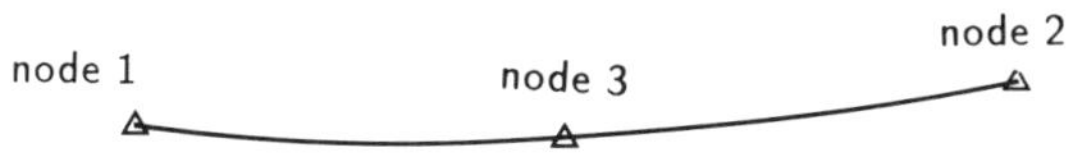

The number of the element is taken to be the number of the third node. Depending upon whether the third node n_3 lies on a straight line joining the nodes n_1 and n_2, the element may be straight or curved. It is desirable to place the node n_3 as nearly as possible equidistant from nodes n_1 and n_2, to avoid localised extreme distortion of the element. Isoparametric elements may be specified individually or in series of elements, the numbers of which are taken to be the same as those of their third nodes.

Infinite elements have only one node, and are straight:

The number of an infinite element is taken to be that of its node. An infinite element is taken by the program to be coincident with the tangent to the adjacent isoparametric element at the node n_1.

The program needs to know to which side of each element the elastic material lies. The user is required to specify by means of the CUE keyword to which side of one element of each separate part of the mesh the material lies, and the program automatically determines this information for the other elements. Subject to no errors having been found earlier in the input data, the program prints the cartesian components of the unit outward normal to the region to be analysed, at the node n_3 of each cue element. It is important to check the correctness of this information as an error could for example result in the analysis of an infinite body problem instead of a finite body problem, or vice versa.

Single isoparametric element

EL n_1 n_2 n_3
n_1 to n_3 = numbers of nodes

Series of isoparametric elements

`ELS` i j

i = number of node n_1 of first element
j = number of node n_2 of last element

The program generates $(j-i)/2$ elements numbered $i+1$, $i+3$, ..., $j-1$. The first of these elements has nodes i, $i+2$, $i+1$, the second $i+2$, $i+4$, $i+3$ and so on. If the coordinates of the middle node $(i+j)/2$ are not defined in Data Group B, then the program takes the nodes i, $i+1$, $i+2$, ..., j to be equally spaced along a straight line from node i to node j, and therefore takes the elements to be of equal length. If the coordinates of node $(i+j)/2$ are defined in Data Group B, then the program takes the nodes i, $i+1$, $i+2$, ..., j to lie either on a straight line from node i to node j or on a circle passing through nodes i, $(i+j)/2$ and j, depending upon whether or not the node $(i+j)/2$ lies on a straight line from node i to node j. Cartesian or polar coordinates of intermediate nodes are determined by interpolation using quadratic shape functions, so if node $(i+j)/2$ is not equidistant from nodes i and j then there will be mesh gradation:

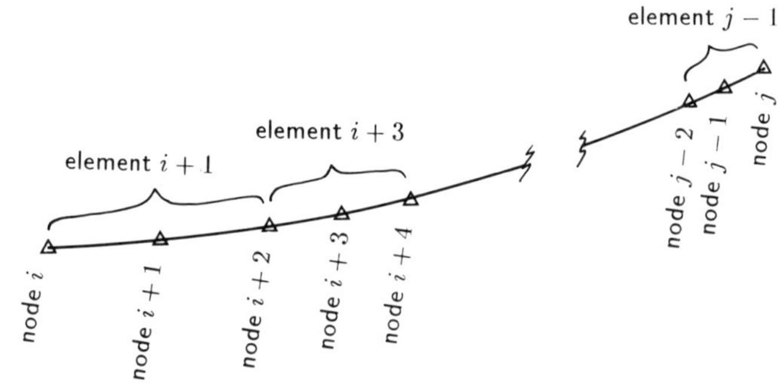

Series of isoparametric elements with wraparound

`ELW` i j k

i = number of node n_1 of first element.
j = number of node n_2 of last element.
k = number of elements to be generated.

The program generates k elements numbered $i+1$, $i+3$, ..., $i+2k-1$. The first of these elements has nodes i, $i+2$, $i+1$, the second $i+2$, $i+4$, $i+3$, and so on until the last element, the nodes of which are $i+2k-2$, j and $i+2k-1$. If the coordinates of the middle node $i+k$ are not defined in Data Group B, then the program takes the nodes i, $i+1$, $i+2$, ..., $i+2k-1$, j to be equally spaced along a straight line from node i to node j, and therefore takes the elements to be of equal length. If the coordinates of node $i+k$ are defined in Data Group B, then the program takes the nodes to lie either on a straight line or on a circle, with or without mesh gradation in the same way as for a series of elements defined by `ELS`.

If $j = i$ and the coordinates of node $i + k$ are defined in Data Group B, then the program takes the nodes $i,\ i+1,\ i+2,\ \ldots,\ i+2k-1$ to be equally spaced on a circle of which the line from node i to node $i+k$ is a diameter, so that the k elements model that circle. Node and element numbers are taken to be increasing in the anticlockwise direction.

Single infinite element

`ELI` i

i = number of node

Cue element

`CUE` e

or

`CUE` $-e$

e = element number. If `CUE` $= e$ then the elastic material is taken to be to the right of the element as viewed from node n_1 to node n_2:

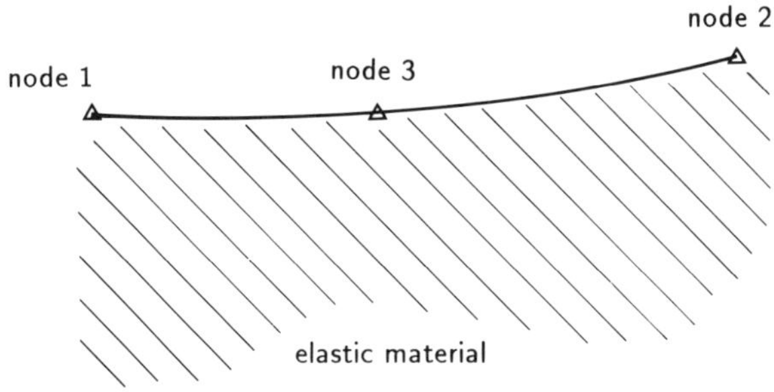

If `CUE` $= -e$ then the elastic material is taken to be to the left of the element as viewed from node n_1 to node n_2:

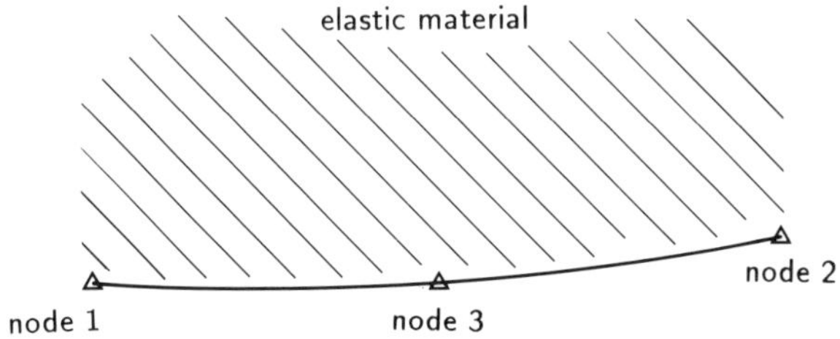

A cue element is required for each separate part of the boundary element mesh. Infinite elements may not be cue elements.

(D) CONSTRAINTS ON DISPLACEMENT

Both elements and nodes may be constrained against displacement. Whereas constraints on elements cause displacement unknowns to be replaced by traction unknowns, constraints on nodes cause degrees of freedom to be eliminated from the system of equations. It is intended that constraints on nodes be specified only where they are required in order to prevent rigid body translation or rotation. Note that, even though the loads to be applied may be self-equilibrating, there must be sufficient constraint on displacement to prevent rigid body motion under any system of loads. Interior nodes may not be constrained.

Constraints may only be specified on isoparametric elements. The program assigns to infinite elements the same constraints as are applied to adjacent isoparametric elements.

Constraint in x direction of single element

`FX` e

e = number of element to be constrained.

Constraint in x direction of series of elements

`FXS` e_i e_j

e_i = number of first element to be constrained

e_j = number of last element to be constrained

The program constrains elements $e_i, e_{i+2}, e_{i+4}, \dots, e_j$

Constraint in y direction of single element

`FY` e

e = number of element to be constrained

Constraint in y direction of series of elements

`FYS` e_i e_j

e_i = number of first element to be constrained

e_j = number of last element to be constrained

The program constrains elements $e_i, e_{i+2}, e_{i+4}, \dots, e_j$

Constraint in x direction of single node

`FNX` n

n = number of node to be constrained

Constraint in y direction of single node

`FNY` n

n = number of node to be constrained

(E) LOAD CASE TITLE

`L` *text*

text = title of load case, to appear on printout and plot if any.

(F) APPLIED LOADS

Within a load case, applied loads may be specified in any order and are additive in effect.

Self weight

`GRAV`

Self weight of w per unit volume (see Data Group A, definition of material properties) is taken to act in the $-y$ direction. Stresses at infinity at $y = 0$ due to `GRAV` are zero.

Initial stress field

`INF` $\sigma_x \quad \sigma_y \quad \tau_{xy} \quad \frac{\partial \sigma_x}{\partial y}$

$\left.\begin{matrix}\sigma_x \\ \sigma_y \\ \tau_{xy}\end{matrix}\right\}$ components of initial stress (Note: σ_y is the value at $y = 0$)

$\frac{\partial \sigma_x}{\partial y}$ = rate of change of σ_x with respect to y.

Traction acting on single element

`TR` $t_x \quad t_y \quad e$

$\left.\begin{matrix}t_x \\ t_y\end{matrix}\right\}$ Components of the traction vector

e = number of element

Traction acting on series of elements

`TRS` $t_x \quad t_y \quad e_i \quad e_j$

$\left.\begin{matrix}t_x \\ t_y\end{matrix}\right\}$ components of the traction vector

e_i = number of first element

e_j = number of last element

The program loads elements $e_i, \ e_{i+2}, \ e_{i+4}, \ \ldots, \ e_j$

Pressure acting on single element

`PR` $p \quad e$

p = magnitude of pressure (compression positive)

e = number of element

Pressure acting on series of elements

`PRS` $p \quad e_i \quad e_j$

p = magnitude of pressure (compression positive)

e_i = number of first element

e_j = number of last element

The program loads elements $e_i, \ e_{i+2}, \ e_{i+4}, \ \ldots, \ e_j$

(G) `GO` OR `NOGO`

The last line of input data is one of the following keywords.

Keyword	Action taken by program
GO	Analysis carried out if no syntactical errors found.
NOGO	Data check only carried out, including any plot requested by keyword MESH.

Examples of input data

The following input data files are on the disc obtainable by reply card.

```
T   CANTILEVER - COARSE MESH
MESH    0.0   5.0     0.0   1.0
DISP    0.0   5.0     0.0   1.0
NUMS
MAT       10000   0.0   1.0
CAR      1     0.0   0.0
CAR      3     0.0   1.0
CAR     13     5.0   1.0
CAR     15     5.0   0.0
ELS     1   3
ELS     3  13
ELS    13  15
ELW    15   1   5
CUE     2
FX     2
FY     2
L   SELF WEIGHT
GRAV
L   UNIT LOAD AT END
TR      0.0  -1.0  14
GO
```

```
T   CANTILEVER - FINE MESH
MESH    0.0   5.0     0.0   1.0
DISP    0.0   5.0     0.0   1.0
NUMS
MAT       10000   0.0   1.0
CAR      1     0.0   0.0
CAR      5     0.0   1.0
CAR     25     5.0   1.0
CAR     29     5.0   0.0
ELS     1   5
ELS     5  25
ELS    25  29
ELW    29   1  10
CUE     2
FXS    2   4
FYS    2   4
L   SELF WEIGHT
GRAV
L   UNIT LOAD AT END
TRS     0.0  -1.0  26   28
GO
```

```
T    HORSESHOE SECTION TUNNEL
PRST    0.0     3.0     -3.0     3.0
PRE
SYMX
MAT     10000    0.2    0.0
POL    1     2.0    90.0
POL    3     2.0    45.0
POL    5     2.0     0.0
CAR    9     2.0    -2.0
CAR   13     0.0    -2.0
POL   101       2.5    90.0
POL   109       2.5     0.0
CAR   117       2.5    -2.5
CAR   125       0.0    -2.5
POL   201       3.0    90.0
POL   209       3.0     0.0
CAR   217       3.0    -3.0
CAR   225       0.0    -3.0
POLG   209  201  101  109    -1  -50
CARG   217  209  109  117    -1  -50
CARG   225  217  117  125    -1  -50
ELS      1   5
ELS      5   9
ELS      9  13
CUE     -2
L    HYDROSTATIC FIELD STRESS
INF     -1.0    -1.0     0.0     0.0
L    HORIZONTAL STRESS = 0.5 TIMES VERTICAL
INF     -0.5    -1.0     0.0     0.0
GO
```

```
T    BEAM WITH HOLE
SYMX
MAT   10000    0.25    0.0
POL      1     125    90
POL      5     125     0
POL      9     125   -90
CAR   101        0   250
CAR   105      300   250
CAR   109     1000   250
CAR   113     1000  -250
CAR   117      300  -250
CAR   121        0  -250
CAR   201        0   150
CAR   204        0   225
CAR   211        0  -150
CAR   214        0  -225
CARS  201   204   1
CARS  211   214   1
ELS      1     9
ELS    101   109
ELS    109   113
ELS    113   121
CUE     -2
CUE    102
FNY    111
L    UNIFORMLY DISTRIBUTED LOAD
PRS    1.0   102   108
TRS    0.0    2.0   110   112
GO
```

B3 PRINTABLE OUTPUT (FILE BOUND.OUT)

BOUND.OUT file records are up to 120 characters in length. Data is written to the file in the following order.

(a) The input data read from the file BOUND.DAT. Note that a blank line is inserted after every five records to improve appearance and each record of input data is preceded by three asterisks so that a blank line of input data can readily be identified as such.
(b) Title of problem, plot control data, material properties and symmetry options.
(c) Coordinates of nodes as specified in Group B of the input data. The coordinates tabulated are cartesian coordinates with respect to the initial position of the origin, before the reading of any `ORIG` data.
(d) Nodes of finite and infinite boundary elements.
(e) A complete table of coordinates of nodes, including any nodes generated by `ELS` and `ELW` data.
(f) Cue elements, signed according to orientation.
(g) The unit outward normal at node 3 of each of the cue elements.
(h) Constraints on elements, and constraints on nodes.
(i) Applied loads.
(j) Computed displacements of, and stresses at, nodes. The following results are written to the file for each node.

  (i) Cartesian components of displacement.
  (ii) Cartesian components of the stress tensor. For a boundary node shared by two elements, the values written to file are the averages of the values computed in the adjacent elements.
  (iii) Principal stresses and angles measured anticlockwise from the x axis to the directions in which they act, computed from the printed cartesian components.

(k) Time taken for execution of the program and for various parts of the program.

B4 PLOT FILE (DPLOT.PLT)

The data on DPLOT.PLT defines the plots requested by keywords `MESH`, `DISP` and `PRST`. For details see Appendix A.

B5 PROGRAM BOUND SOURCE CODE

On the disc obtainable by reply card there are the files BROOT.FOR, BOVLA.FOR, BOVLB.FOR, BOVLC.FOR, BOVLD.FOR, BOVLE.FOR and BOVLF.FOR, which comprise the source code of program BOUND. BROOT.FOR contains the main program and a number of general purpose subroutines, which form the root segment if the program is overlain. The functions of the subroutines on the other files are as follows.

BOVLA.FOR	preprocessing
BOVLB.FOR	preprocessing
BOVLC.FOR	preprocessing
BOVLD.FOR	matrix and load vector construction
BOVLE.FOR	simultaneous equation solution
BOVLF.FOR	postprocessing, including computation of results at interior points

Techniques of programming are the same as for program DEMON. Again there are eight small COMMON blocks and one large block named /CHANGE/, and unit numbers of files are held as variables in the small COMMON blocks. The contents of these files are shown against the variable names in the following table.

File	Contents	Organisation
LIN	the file BOUND.DAT (q.v.)	formatted, sequential
LUO	the file BOUND.OUT (q.v.)	formatted, sequential
LPL	the file DPLOT.PLT (q.v.)	unformatted, sequential
LUI	a copy of the file LIN	formatted, sequential
LUT	a single record of the file LIN, less the keyword	formatted, sequential
LUB	given boundary data and basic geometric data	unformatted, sequential
LUA	matrix of equation coefficients / upper triangular factor	unformatted, direct access
LUL	lower triangular factor	unformatted, sequential
LUG	load vectors and most commonly used Gaussian quadrature formulae	unformatted, sequential
LUS	computed displacements and stresses at nodes of elements and interior nodes	unformatted, sequential

Data (apart from error messages which are always written to LUO) are written to and read from files as shown on p. 444.

Description of files in COMMON blocks

Block	Variable	Description
/CLOCK/		see Appendix A
/FORM/		see Appendix A

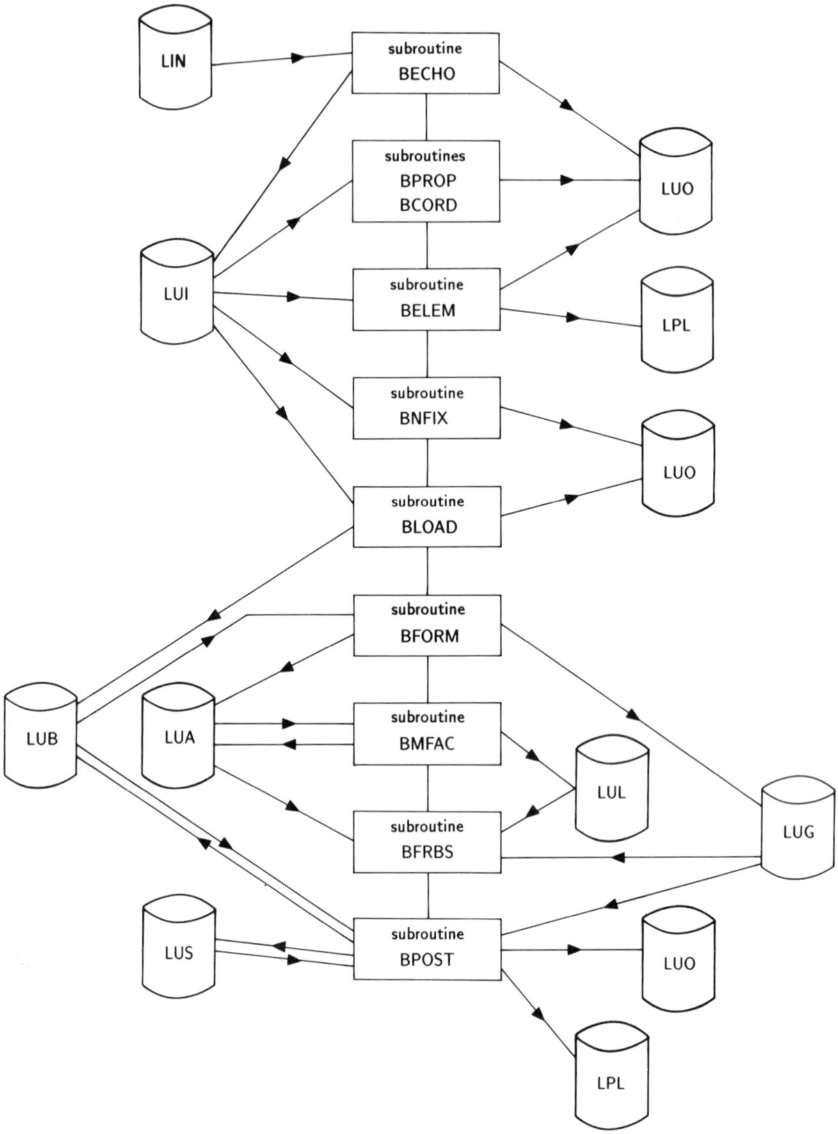
LIN
subroutine
BECHO
subroutines
BPROP
BCORD
LUO
LUI
subroutine
BELEM
LPL
subroutine
BNFIX
LUO
subroutine
BLOAD
subroutine
BFORM
LUB
LUA
subroutine
BMFAC
LUL
LUG
subroutine
BFRBS
LUS
subroutine
BPOST
LUO
LPL

Block	Variable	Description
/PLOT/		see Appendix A
/FLAG/		see Appendix A
/TITLE/		see Appendix A
/PROP/	YM	Young's modulus
	PR	Poisson's ratio
	DEN	unit weight of material
	DAG	unit of distance used in construction of the system of simultaneous equations
	IDIS	=0 if computed displacements are to be with respect to the unstrained state, =1 if they are to be with respect to a pre-mining or other initially strained state
/GEOM/	ISYM(2)	ISYM(1)=1 if there is symmetry with respect to x, zero otherwise. ISYM(2)=1 if there is symmetry with respect to y, zero otherwise
	MXA	maximum allowable node number
	NXB	number of nodes of elements excluding ghost nodes of infinite elements
	NXE	number of nodes of elements including ghost nodes of infinite elements
	MEA	maximum allowable element number
	NEB	number of finite (isoparametric) boundary elements
	NEE	number of elements including infinite elements
	NXY	number of interior nodes
	LUA	unit number of file of matrix of equation coefficients
	LUL	unit number of file of lower triangular factor coefficients
	LUG	unit number of file of load vectors and most commonly used Gaussian quadrature formulae
	ASO(2)	coordinates of reference point for computation of shape functions over infinite elements
	NLIB	number of unknowns = 2*NXB
	LUAL	file LUA record length
	LUAN	how many equations per LUA record
	NBLS	number of fast blocks per slow block
	MUA	number of LUA file records used to hold upper triangular factor
/LOAD/	NCC	number of load cases
	LUB	unit number of file of given boundary data and basic geometric data
	LUS	unit number of file of computed displacements and stresses

Block	Variable	Description
/LOAD/ (cont'd)	NGR(5)	NGR(I)=1 if `GRAV` specified for load case I, zero otherwise
	NIN(5)	NIN(I)=1 if `INF` specified for load case I, zero otherwise
	TIN(3,5)	for each load case, the initial stresses at $y=0$
	TID(5)	for each load case, the rate of change of σ_x with respect to y
/CHANGE/ in subroutine BELEM before call of BPGEO	X(2,1500)	(X(1,I), X(2,I))= coordinates of node with external number I
	IXY(1500)	IXY(I)=1 if node with external number I is defined, zero otherwise
	IEL(600)	IEL(I)=1 if element with external number I is defined, zero otherwise
	JEN(600)	JEN(I)=3 if element with external number I is an isoparametric element, 1 if it is an infinite element
	IPT(600)	IPT(I)=$i+2j+4k$ where $i=1$ if element with external number I appears in `MESH` plot, zero otherwise $j=1$ if element with external number I appears in `DISP` plot, zero otherwise $k=1$ if element with external number I appears in `PRST` plot, zero otherwise
	KEL(3,600)	KEL(I,J)=external number of node I of element with external number J Note: external node and element numbers are those referred to in the input data and understood by the user of the program. In subsequent documentation of /CHANGE/, 1. node and element numbers are internal node and element numbers unless otherwise stated 2. data in subscripted variables are indexed on internal node and element numbers, unless otherwise stated.
/CHANGE/ in subroutine BLOAD	X(2,300)	coordinates of nodes of elements
	IXY(1500)	internal node numbers indexed on external node numbers (nodes of elements only)
	JXY(300)	external node numbers (nodes of elements only)
	IEL(600)	internal element numbers indexed on external element numbers
	JEL(120)	external element numbers
	JEN(120)	JEN(I)=3 if element I is an isoparametric element, 1 if it is an infinite element
	IPT(120)	`IPT(I)`=$i+2j+4k$ as described previously but for element with internal number I

Block	Variable	Description
/CHANGE/ in subroutine BLOAD (cont'd)	KEL(3,120)	KEL(I,J)= number of node I of element J
	Y(2,300)	coordinates of interior nodes
	JYY(300)	external numbers of interior nodes
	IEO(120)	IEO(I) = 1 if element I has positive orientation, -1 if it has negative orientation
	IFN(300)	IFN(I)=$i+2j+4k+8l$ where (nodes of elements only) $i=1$ if the first unknown at node I is t_x, 0 if it is u_x $j=1$ if the second unknown at node I is t_y, 0 if it is u_y $k=1$ if the first unknown at node I is eliminated by symmetry or `FNX` constraint, zero otherwise $l=1$ if the second unknown at node I is eliminated by symmetry or `FNY` constraint, zero otherwise
	IFE(120)	IFE(I) = $i+2j$ where $i=1$ if u_x is given on element I, zero if t_x is given $j=1$ if u_y is given on element I, zero if t_y is given
	BD(2,3,120,5)	(BD(1,I,J,K), BD(2,I,J,K)) = cartesian components of given displacement at node I of element J, load case K
	BT(2,3,120,5)	(BT(1,I,J,K), BT(2,I,J,K)) = cartesian components of given traction at node I of element J, load case K Note: before the call of BPART, these values are of the sum of particular integral and complementary function; after that call, they are of the complementary function only
/CHANGE/ in subroutine BFORM	X(2,300) JXY(300) JEL(120) JEN(120) KEL(3,120) IEO(120) IFN(300) IFE(120)	see /CHANGE/ in BLOAD
	XHW(2,300)	(XHW(1,I), XHW(2,I)) = coordinates of halfway node for construction of auxiliary contour for implicit integration, node I
	XIO(3,3,120)	XIO(I,1,J)=intrinsic coordinate of Gauss point I for integration over element J XIO(I,2,J)=intrinsic coordinate of Gauss point I for integration over the part $1 \geq \xi \geq 0$ of element J XIO(I,3,J)=intrinsic coordinate of Gauss point I for integration over the part $0 \geq \xi \geq -1$ of element J

Block	Variable	Description
/CHANGE/ in subroutine BFORM (cont'd)	XCO(2,3,3,120)	(XCO(1,I,1,J), XCO(2,I,1,J))= cartesian coordinates of Gauss point I for integration over element J
		(XCO(1,I,2,J), XCO(2,I,2,J))= as above but for the part $1 \geq \xi \geq 0$ of element J
		(XCO(1,I,3,J), XCO(2,I,3,J))= as above but for the part $0 \geq \xi \geq -1$ of element J
	UNO(2,3,3,120)	(UNO(1,I,1,J), UNO(2,I,1,J)) = cartesian components of unit outward normal at Gauss point I for integration over element J
		(UNO(1,I,2,J), UNO(2,I,2,J)) = as above but for the part $1 \geq \xi \geq 0$ of element J
		(UNO(1,I,3,J), UNO(2,I,3,J)) = as above but for the part $0 \geq \xi \geq -1$ of element J
	XJO(3,3,120)	XJO(I,1,J) = product of jacobian and Gauss weight for Gauss point I of element J.
		XJO(I,2,J) = as above but for the part $1 \geq \xi \geq 0$ of element J
		XJO(I,3,J) = as above but for the part $0 \geq \xi \geq -1$ of element J
	A(4000)	block of equation coefficients
	B(2500)	load vectors
	TTR(2,2,31)	sums of integrals of the kernel **T**, for adjustment of matrix coefficients in the case of a finite body problem
/CHANGE/ in subroutines BSURF and BINOD called by BPOST	X(2,300) JXY(300) JEL(120) JEN(120) IPT(120) KEL(3,120) IEO(120) IFN(300) IFE(120) Y(2,300) JYY(300)	see /CHANGE/ in BLOAD
	DN(2,300,5)	(DN(1,I,J), DN(2,I,J)) = computed displacements at node I, load case J
	SN(3,300,5)	(SN(1,I,J), SN(2,I,J), SN(3,I,J)) = computed stresses at node I, load case J
		Note: in BSURF these are results at nodes of elements, in BINOD they are results at interior nodes

B6 SELECTED LISTINGS

The complete listing of about 4500 lines of source code may be obtained by printing the files BROOT.FOR etc, referred to earlier. The listings reproduced here are of the routines held on files BROOT.FOR, BOVLD.FOR and BOVLE.FOR, excluding routines already listed in Appendix A. The following notes are to be read in conjunction with the listings.

File BROOT.FOR

BOUND

Except for /FLAG/, COMMON blocks are declared to contain subscripted variables dimensioned to at least the greatest length required at any stage of execution. Generally there is trailing unused space which may be used in future program development. The names of the arrays have no significance elsewhere in the program.

BFERR

The error number printed by BFERR is as specified in the users' manual for the FORTRAN compiler. Normally the only reason for termination of execution in BFERR is insufficient disc space for scratch files.

BCFOR

Program BOUND performs all integrations using three point Gaussian formulae. In subroutine BCFOR the length of each interval over which a three point formula is applied is taken to be the chord length of the element divided by the number of intervals, and the ratio of interval length to minimum distance from the first argument of the kernels is limited to 0.62 for matrix construction and 0.50 for computation of results at interior nodes. The minimum distance from the first argument of the kernels is determined by Newton–Raphson iteration but note that if the ratio of chord length to distance to node 3 of the element is such that it is improbable that more than one interval is required then to save computer time the iteration is bypassed and a single interval specified. Based on experience the computed number of intervals is doubled for an infinite element. Also, because the accurate computation of coefficients of leading diagonal submatrices is felt to be of critical importance and the cost of their computation is only a linear function of the number of elements, the computed number of intervals is doubled for integration over the auxiliary contour Ω, referred to in source code comments as the spiral.

BGFOL

Note that, depending upon whether BGFOL is computing the formula for the first (i.e. infinite) interval of an infinite element or it is computing the formula for an interval adjacent to the first argument of the kernels, the fourth formal parameter KC either is the number of intervals or it identifies the interval for which the formula is required.

File BOVLD.FOR

BFORM

The matrix is constructed as JI blocks, each (except possibly the last) consisting of LUAN pairs of rows. Each block of equation coefficients (except possibly the last) is written to two records of the direct access file LUA. In the array A, equation coefficients are stored by row. Load vectors are stored in the array B, in load case major order. Integrations of kernel-shape function products are carried out by as many applications of the three point Gauss–Legendre formula and the three point Gauss formula with log weight function as are required to satisfy the criterion programmed in subroutine BCFOR. To save computer time, the intrinsic coordinates, cartesian coordinates, unit outward normals and jacobians at the sampling points of the most commonly used Gauss-Legendre formulae are computed and stored in the arrays XIO, XCO, UNO and XJO by subroutine BPREP, prior to construction of the matrix and load vectors.

For each block of the matrix, off-diagonal contributions to matrix and load vector coefficients are computed by element (loop with control variable IB ending at 4 CONTINUE), and for each element by `SYMX` and `SYMY` reflection if any (loops with control variables IC and ID ending at 6 CONTINUE and 7 CONTINUE). Note the use of subroutine PREFL to set up the array XB of reflected nodal coordinates and the arrays XCR and UNR of reflected cartesian coordinates and unit outward normals of the most commonly used Gauss points. The loop with control variable IE ending at 9 CONTINUE is over first arguments of the kernels, each corresponding to a pair of equations. For each first argument, subroutine BCFOR determines the number of intervals of integration JQ required. Integrals of kernel-shape function products are then summed into the arrays TTL and UTL. Symmetry factors are applied to TTL and UTL to account for the fact that they multiply reflected nodal values of displacement and traction. The reflected TTL and UTL are then scattered into the array A, and products of TTL and UTL and given boundary data added to B, by subroutine BMINC. Note that BMINC also increments TTR, the sum of integrals of the kernel **T** for each first argument, which is used to adjust matrix coefficients if the elastic body is of finite extent (see subroutine BDTTR). Integration over elements of which the first argument is a node, and implicit integration, are performed by subroutine BSING which is called immediately before 4 CONTINUE. Note that in the OPEN statement for LUA, the value of RECL is given in *bytes*.

BPREP

In addition to computing the most commonly used Gaussian quadrature formulae, BPREP determines the unit of distance for construction of the system of equations (DAG), the coordinates of the reference point for computation of shape functions over infinite elements (ASO), and the coordinates of the auxiliary contour halfway nodes (XHW). Halfway nodes are located as follows:

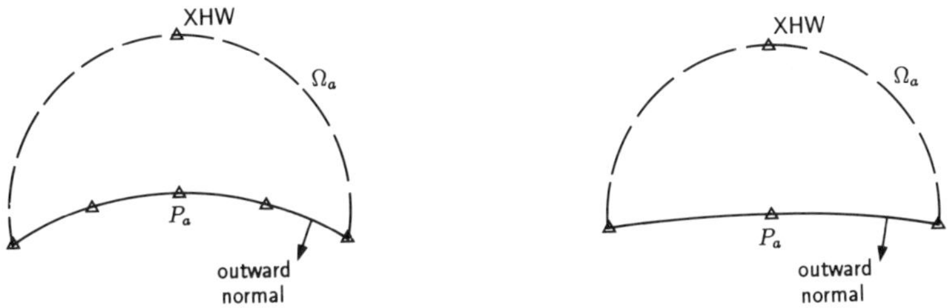

BSING

Subroutine BSING evaluates integrals over elements of which the first argument P_a is a node, including integrals required for implicit evaluation of coefficients of displacement at P_a. Integrals over the auxiliary contour Ω_a are evaluated by subroutine BSPIN (called by BSING), and integrals over the part of an infinite element beyond the end of the auxiliary contour are computed by subroutine BADJI (also called by BSING). The element is divided into two integration intervals if P_a is at node 1 or node 2, or four intervals if P_a is at node 3, and three point Gaussian formulae applied over each interval. Probably a smaller number of intervals would give sufficient accuracy of integration, but since diagonal integration is believed to be of critical importance and the computing cost of such integration is only a linear function of the number of elements, there is every reason to err on the safe side. Integrals of kernel-shape function products are summed into arrays TTL and UTL and scattered into A and B as before, and now the integral of the kernel **T** required for implicit integration is summed into the array AI.

BSPIN

At each call, subroutine BSPIN integrates the kernel **T** over half the auxiliary contour if P_a is at node 1 or node 2 of an element, or over the entire auxiliary contour if P_a is at node 3. One half of the auxiliary contour is represented by as many three node isoparametric elements as necessary for the angle turned by each element to be less than one radian:

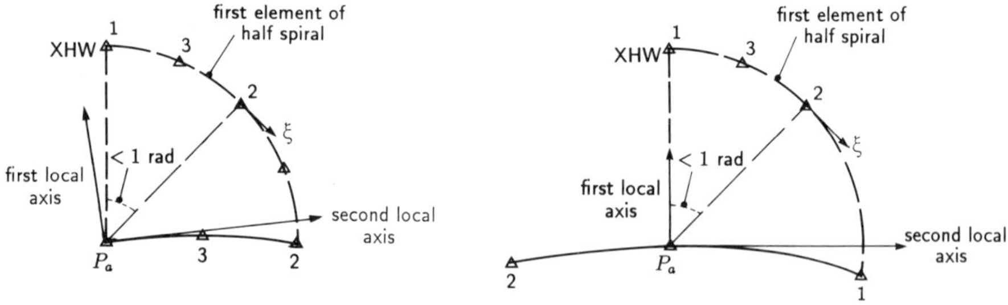

The coordinates of nodes of elements of the half spiral are computed from polar coordinates associated with the local axes shown above, the increments of these polar coordinates being constant from one element of the half spiral to the next.

BMINC

In subroutine BMINC, A and B are shown as multidimensional arrays. The significance of subscripts is as follows.

First subscript of A:	which of a pair of columns	} specification of matrix column
Second subscript of A:	which of NXB pairs of columns	
Third subscript of A:	which of a pair of equations	} specification of equation within block
Fourth subscript of A:	which pair of equations within block	
First subscript of B:	which of a pair of equations	} specification of equation
Second subscript of B:	which of NXB pairs of equations	
Third subscript of B:	which load case	

BDTTR

Again A is shown as a multidimensional array. For a finite elastic body the integrals of the kernel **T** for each first argument should sum to zero, but due to the use of an auxiliary contour for implicit integration this condition is not exactly satisfied by the equation coefficients computed by BFORM and BSING. To avoid local inaccuracy where a finite elastic body is only minimally constrained against displacement, the matrix coefficients are adjusted to make the integrals of **T** sum to exactly zero. Since in theory the integral of **T** is the 2×2 unit matrix for the entire elastic space less finite cavities, BDTTR takes the elastic body to be finite if it has no infinite elements and the greatest coefficient of the array TTR is less than 0.5.

File BOVLE.FOR

BMFAC

Subroutine BMFAC bifactorises the matrix of equation coefficients by the block solution method. Subscripted variables are as follows.

A(14000): slow block
AU(2000): slow block output buffer
FI(2000): fast block
FO(2000): fast block output buffer
AL(119): lower triangular matrix output buffer

The number NBLS of LUA file records per slow block is 7; given that the number LUAN of unreduced equations per LUA file record is at least 4, the blocking factor for eliminations is at least 28.

The overall sequence of operations is governed by the variables

JA= first equation of slow block
JE= number of matrix coefficients per equation (equals NLIB for first slow block)
JF= equations per LUA record (equals LUAN for first slow block)
JG= last equation of slow block

Bifactorisation proceeds as indicated by the following flow chart.

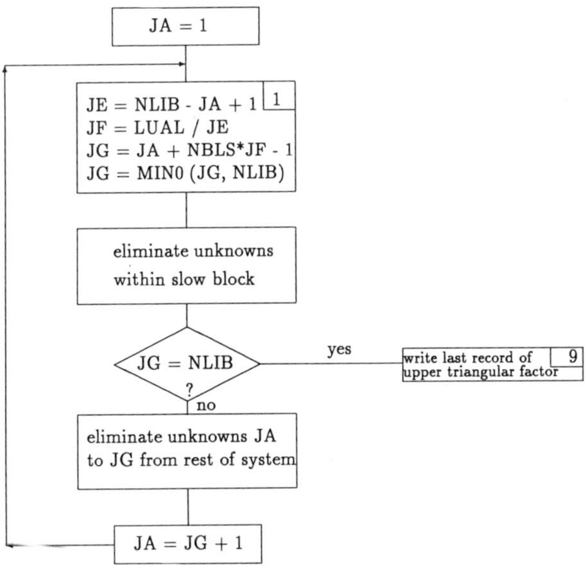

In the diagram below are shown the meanings of pointers and loop control variables during elimination of unknowns within the slow block, as unknown IA is being eliminated from equation IA + IB:

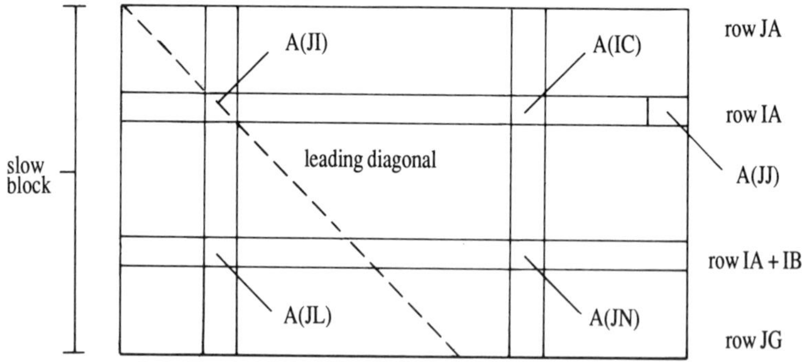

The meanings of various pointers and loop control variables during elimination of unknowns JA to JG from the rest of the system are indicated by the following diagram, in which is shown the situation during execution of the loop with control variable ID ending at 13 CONTINUE. In this loop, unknown IB of the system of equations is being eliminated from equation IC of fast block IA (i.e. equation JP + IC−1 of the system).

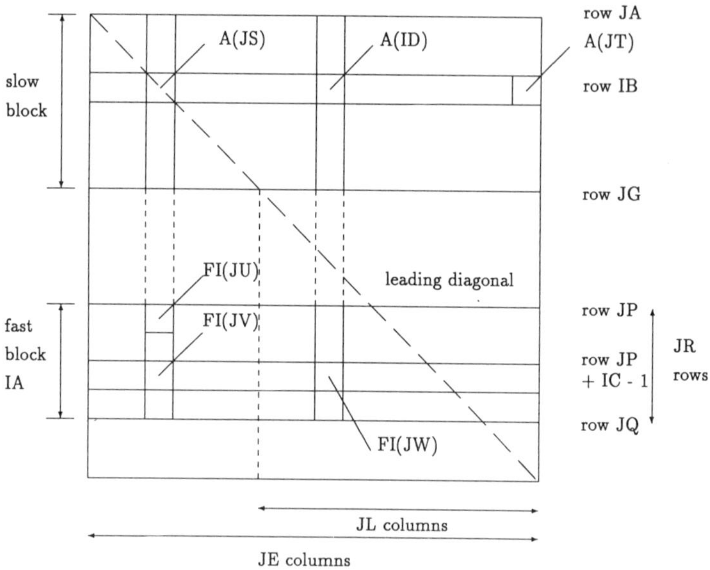

Immediately after the elimination of unknowns within a slow block, rows of the upper triangular factor are copied from A to AU (loop ending at

7 CONTINUE) and the contents of AU written back to the direct access file LUA as necessary with the pointer KUE to the last leading diagonal coefficient of a record set for reference during back substitution. Likewise, after the elimination of unknowns JA to JG from a fast block, partially reduced equations are copied from FI to FO (loop ending at 14 CONTINUE), and the contents of FO written to LUA as necessary but without KUE being set. Meanwhile, throughout eliminations within the slow blocks and in fast blocks, coefficients of the lower triangular factor are copied to AL which is then written to the sequential file LUL. These data are not arranged on LUL by row or column, but in the order in which they are computed during the eliminations.

BFRBS

Subroutine BFRBS carries out both forward reduction and back substitution of the load vectors. Computer time could be saved and the file LUL eliminated by performing the forward reduction at the same time as bifactorisation of the matrix, but this would complicate the implementation of iterative procedures in which the load vectors are modified at each iteration. Load vectors in the array B are held in load case major order. Because the data on LUL must be read in the same order as that in which it is written, in the forward reduction BFRBS retraces the steps taken by BMFAC during bifactorisation: hence the references in source code comments to slow and fast blocks. Back substitution proceeds in the same way as in program DEMON, except in that the upper triangular factor is fully populated. When back substitution is complete, the array B contains the solution vectors.

```
      PROGRAM BOUND
C     DEMONSTRATION BOUNDARY ELEMENT PROGRAM BY
C     JOHN WATSON, MINING ENGINEERING, UNIVERSITY OF NEW SOUTH WALES
      IMPLICIT DOUBLE PRECISION (A-H,O-Z)
      COMMON/CLOCK/CLO(10)
      COMMON/FORM/FOR(30)
      COMMON/PLOT/PLO(60)
      COMMON/FLAG/IGER,ILER
      COMMON/TITLE/TIT(120)
      COMMON/PROP/PRO(10)
      COMMON/GEOM/GEO(20)
      COMMON/LOAD/ALO(30)
      COMMON/CHANGE/CHA(20000)
      CALL PTIME (0)
      CALL PTICK (1)
C     PREPROCESSING
      CALL PTICK (2)
      CALL BOVLA
      CALL BOVLB
      CALL BOVLC
      CALL PTOCK (2)
      IF (IGER.NE.0.OR.ILER.NE.0) GOTO 1
C     CONSTRUCT MATRIX AND LOAD VECTORS
      CALL BOVLD
      IF (IGER.NE.0) GOTO 1
C     EQUATION SOLUTION
      CALL BOVLE
      IF (IGER.NE.0) GOTO 1
C     POSTPROCESSING
      CALL BOVLF
    1 CALL PTOCK (1)
      CALL BSTAT
      END


      SUBROUTINE BSTAT
C     PRINT TIMES TAKEN BY VARIOUS STAGES OF THE ANALYSIS
      IMPLICIT DOUBLE PRECISION (A-H,O-Z)
      COMMON/CLOCK/SEC(10)
      COMMON/FORM/LUO,LIN,LUI,LUT,KKFO(56)
      COMMON/GEOM/ISYM(2),MXA,NXB,NXE,MEA,NEB,NEE,NXY,
     1 LUA,LUL,LUG,KKGE(28)
      COMMON/LOAD/NCC,LUB,LUS,KKLO(57)
C     CLOSE PLOT FILE
      CALL DFINI
      CLOSE (UNIT=LIN,IOSTAT=JA,ERR=2)
    2 CLOSE (UNIT=LUI,IOSTAT=JA,ERR=3)
    3 CLOSE (UNIT=LUT,IOSTAT=JA,ERR=4)
    4 CLOSE (UNIT=LUB,IOSTAT=JA,ERR=5)
    5 CLOSE (UNIT=LUG,IOSTAT=JA,ERR=6)
    6 CLOSE (UNIT=LUA,IOSTAT=JA,ERR=7)
    7 CLOSE (UNIT=LUL,IOSTAT=JA,ERR=8)
    8 CLOSE (UNIT=LUS,IOSTAT=JA,ERR=9)
    9 WRITE (LUO,100) (SEC(IO),IO=1,6)
  100 FORMAT (
     1 28H1     ELAPSED TIME IN SECONDS//
     2 23H    TOTAL ALL ROUTINES  ,0PF7.1/
     3 23H    PREPROCESSING       ,0PF7.1/
     4 23H    MATRIX CONSTRUCTION ,0PF7.1/
     5 23H    BIFACTORISATION     ,0PF7.1/
     6 23H    LOAD VECTORS        ,0PF7.1/
     7 23H    POSTPROCESSING      ,0PF7.1)
```

```
       CLOSE (UNIT=LUO,IOSTAT=JA,ERR=10)
   10 CONTINUE
      RETURN
      END

      SUBROUTINE BFERR (LUN,IERR)
C     REPORT FILE OPEN OR WRITE ERROR AND TERMINATE RUN
      COMMON/FORM/LUO,KKFO(59)
      ENTRY DFERR (LUN,IERR)
      IF (IERR.EQ.0) GOTO 1
      WRITE (LUO,200) IERR,LUN
  200 FORMAT (/  14H ERROR NUMBER ,I5,
     1  9H ON FILE ,I2)
      CALL BSTAT
      STOP
    1 CONTINUE
      RETURN
      END

      SUBROUTINE BSHAP (XI,FUN,KA)
C     QUADRATIC SHAPE FUNCTIONS
      IMPLICIT DOUBLE PRECISION (A-H,O-Z)
      DIMENSION XI(KA),FUN(3,KA)
      DO 1 IA=1,KA
      XH=0.5D0*XI(IA)
      XP=XI(IA)+1.0D0
      XN=XI(IA)-1.0D0
      FUN(1,IA)=XH*XN
      FUN(2,IA)=XH*XP
      FUN(3,IA)=-XP*XN
    1 CONTINUE
      RETURN
      END

      SUBROUTINE BSHAD (XI,FUN,DER,KA)
C     QUADRATIC SHAPE FUNCTIONS AND DERIVATIVES
      IMPLICIT DOUBLE PRECISION (A-H,O-Z)
      DIMENSION XI(KA),FUN(3,KA),DER(3,KA)
      DO 1 IA=1,KA
      XJ=XI(IA)
      XH=0.5D0*XJ
      XP=XJ+1.0D0
      XN=XJ-1.0D0
C     FUNCTIONS
      FUN(1,IA)=XH*XN
      FUN(2,IA)=XH*XP
      FUN(3,IA)=-XP*XN
C     DERIVATIVES
      DER(1,IA)=XJ-0.5D0
      DER(2,IA)=XJ+0.5D0
      DER(3,IA)=-2.0D0*XJ
    1 CONTINUE
      RETURN
      END
```

```
      SUBROUTINE BSHAI (XA,XB,ASO,FUU,FUT,KA,IERR)
C     ASYMPTOTIC SHAPE FUNCTIONS FOR INFINITE ELEMENT
      IMPLICIT DOUBLE PRECISION (A-H,O-Z)
      DIMENSION XA(2,KA),XB(2),ASO(2),FUU(3,KA),FUT(3,KA),XR(2)
C     DISTANCE FROM REFERENCE POINT TO END OF ELEMENT
      CALL BDIST (XB,ASO,XR,RO)
      DO 1 IA=1,KA
      CALL BDIST (XA(1,IA),ASO,XR,RA)
      IF (RA.LT.1.0D-10) GOTO 999
C     FUNCTIONS FOR DISPLACEMENT AND TRACTION
      FUU(1,IA)=RO/RA
      FUT(1,IA)=FUU(1,IA)*FUU(1,IA)
    1 CONTINUE
      IERR=0
    2 CONTINUE
      RETURN
  999 IERR=1
      GOTO 2
      END

      SUBROUTINE BDIST (XA,XB,XR,RAB)
C     CARTESIAN COMPONENTS AND LENGTH OF VECTOR XA-XB
      IMPLICIT DOUBLE PRECISION (A-H,O-Z)
      DIMENSION XA(2),XB(2),XR(2)
      ELA=XA(1)-XB(1)
      ELB=XA(2)-XB(2)
      XR(1)=ELA
      XR(2)=ELB
      RAB=DSQRT(ELA*ELA+ELB*ELB)
      RETURN
      END

      SUBROUTINE PNORM (XA,XB,IERR)
C     NORMALISE VECTOR XA
      IMPLICIT DOUBLE PRECISION (A-H,O-Z)
      DIMENSION XA(2),XB(2)
      ELA=XA(1)
      ELB=XA(2)
      RAD=DSQRT(ELA*ELA+ELB*ELB)
      IF (RAD.LT.1.0D-20) GOTO 999
C     DIVIDE BY LENGTH
      XB(1)=ELA/RAD
      XB(2)=ELB/RAD
      IERR=0
    1 CONTINUE
      RETURN
C     LENGTH IS ZERO
  999 IERR=1
      GOTO 1
      END

      SUBROUTINE BTRAN (XA,DCO,XB)
C     CALCULATE COMPONENTS OF VECTOR XA IN SYSTEM
C     DEFINED BY DIRECTION COSINES DCO
      IMPLICIT DOUBLE PRECISION (A-H,O-Z)
      DIMENSION XA(2),DCO(2,2),XB(2)
      XAA=XA(1)
      XAB=XA(2)
      XB(1)=DCO(1,1)*XAA+DCO(1,2)*XAB
      XB(2)=DCO(2,1)*XAA+DCO(2,2)*XAB
      RETURN
      END
```

```
      SUBROUTINE PREFL (XA,XB,XC,KA)
C     MULTIPLY VECTORS XA BY FACTORS XC
      IMPLICIT DOUBLE PRECISION (A-H,O-Z)
      DIMENSION XA(2,KA),XB(2,KA),XC(2)
      DO 1 IA=1,2
      ELT=XC(IA)
      DO 2 IB=1,KA
      XB(IA,IB)=XA(IA,IB)*ELT
    2 CONTINUE
    1 CONTINUE
      RETURN
      END

      SUBROUTINE BGEOM (KA,XA,XI,KB,XC,XS,UN,EJ,IERR)
C     CARTESIAN COORDINATES, TANGENTS,
C     UNIT OUTWARD NORMALS AND JACOBIANS AT POINTS OF ELEMENT (MAX 3)
      IMPLICIT DOUBLE PRECISION (A-H,O-Z)
      DIMENSION XA(2,3),XI(KB),
     1 XC(2,KB),XS(2,KB),UN(2,KB),EJ(KB),
     2 FUN(3,3),DER(3,3)
      IERR=0
      CALL BSHAD (XI,FUN,DER,KB)
      DO 1 IA=1,KB
C     CARTESIAN DIRECTION
      DO 2 IB=1,2
      SMA=0.0D0
      SMB=0.0D0
      DO 3 IC=1,3
      SMA=SMA+FUN(IC,IA)*XA(IB,IC)
      SMB=SMB+DER(IC,IA)*XA(IB,IC)
    3 CONTINUE
C     COORDINATE AND COMPONENT OF TANGENT AT POINT IA
      XC(IB,IA)=SMA
      XS(IB,IA)=SMB
    2 CONTINUE
C     JACOBIAN
      ELA=XS(1,IA)
      ELB=XS(2,IA)
      EJ(IA)=DSQRT(ELA*ELA+ELB*ELB)
      IF (EJ(IA).GE.1.0D-10) GOTO 4
C     ZERO JACOBIAN
      IERR=1
      GOTO 1
    4 ELC=EJ(IA)
      IF (KA.EQ.1) GOTO 5
C     ELEMENT HAS NEGATIVE ORIENTATION
      ELC=-ELC
    5 UN(1,IA)=-ELB/ELC
      UN(2,IA)=ELA/ELC
    1 CONTINUE
      RETURN
      END

      SUBROUTINE BCFOR (KA,XA,XB,KB,KC,IERR)
C     CHOOSE NUMBER OF INTEGRATION INTERVALS FOR ELEMENT WITH
C     COORDINATES XB, FIRST ARGUMENT OF KERNELS AT XA
C     KA=1 MATRIX CONSTRUCTION, KA=2 INTERIOR POINT
      IMPLICIT DOUBLE PRECISION (A-H,O-Z)
      DIMENSION XA(2),XB(2,3),ASP(2),MUL(3),
     1 XR(2),XC(2),XS(2),UN(2)
```

```
       DATA ASP/0.62D0,0.50D0/
      DATA MUL/2,0,1/
      DATA FAC/0.7D0/
      DATA MIT,DAMP/40,0.75D0/
      DATA EPS,TOL/0.01D0,0.01D0/
      JA=1
C     LENGTH OF ELEMENT AND DISTANCE FROM CENTRE NODE TO XA
      CALL BDIST (XB(1,1),XB(1,2),XR,CHO)
      CALL BDIST (XA,XB(1,3),XR,RO)
C     MINIMUM ALLOWABLE DISTANCE TO ELEMENT FROM XA
      EPSO=EPS*CHO
      IF (RO.LT.EPSO) GOTO 999
      ELJ=FAC*CHO
C     ESTIMATED LOWER BOUND MINIMUM DISTANCE TO ELEMENT FROM XA
      RM=RO-ELJ
      IF (RM.LT.1.0D-6) GOTO 1
      RAT=CHO/RM
      IF (RAT.LE.ASP(KA)) GOTO 2
C     SINGLE INTERVAL MAY NOT BE OK - ITERATE FOR MINIMUM DISTANCE
C     TO ELEMENT FROM XA
    1 XI=0.0D0
      CALL BGEOM (1,XB,XI,1,XC,XS,UN,ELJ,JERR)
      IF (JERR.NE.0) GOTO 999
      DXIM=1.0D0
      DO 3 IA=1,MIT
      SUM=XR(1)*XS(1)+XR(2)*XS(2)
C     INTRINSIC COORDINATE INCREMENT
      DXI=SUM/(ELJ*ELJ)
      DXI=DMIN1(DXI,DXIM)
      DXI=DMAX1(DXI,-DXIM)
      XI=XI+DXI
C     STAY INSIDE ELEMENT
      XI=DMIN1(XI,1.0D0)
      XI=DMAX1(XI,-1.0D0)
      CALL BGEOM (1,XB,XI,1,XC,XS,UN,ELJ,JERR)
      IF (JERR.NE.0) GOTO 999
      CALL BDIST (XA,XC,XR,RM)
      IF (RM.LT.EPSO) GOTO 999
C     CHANGE IN DISTANCE
      DR=ABS(RO-RM)
      IF (DR.LT.TOL*RO) GOTO 4
C     CONTINUE ITERATION
      RO=RM
      DXIM=DXIM*DAMP
    3 CONTINUE
      GOTO 999
C     COMPUTE NUMBER OF INTERVALS
    4 RAT=CHO/RM
      JA=RAT/ASP(KA)+1
C     DOUBLE IF INFINITE ELEMENT OR SPIRAL
    2 KC=JA*MUL(KB)
      IERR=0
    5 CONTINUE
      RETURN
  999 IERR=1
      GOTO 5
      END
```

```
      SUBROUTINE BGFOR (KA,XA,KB,KC,KD,XI,XC,UN,XJ,IERR)
C     COMPUTE INTRINSIC COORDINATES, CARTESIAN COORDINATES,
C     NORMALS AND JACOBIAN*GAUSS WEIGHTS AT GAUSS POINTS OF
C     INTERVAL KD OF ELEMENT SPLIT INTO KC INTERVALS
      IMPLICIT DOUBLE PRECISION (A-H,O-Z)
      DIMENSION XA(2,3),XI(3),XC(2,3),UN(2,3),XJ(3),
     1 GOF(3),GWT(3),TJ(3),XS(2,3),XT(3)
      DATA GOF/
     1 0.7745966692D0, 0.0D0, -0.7745966692D0/
      DATA GWT/
     1 0.5555555555D0, 0.8888888889D0, 0.5555555555D0/
C     INTRINSIC COORDINATE OF MIDPOINT OF INTERVAL
      ELJ=1.0D0/KC
      ORG=1.0D0-ELJ*(2*KD-1)
      DO 1 IA=1,3
C     INTRINSIC COORDINATE AND JACOBIAN*GAUSS WEIGHT FOR INTRINSIC
C     COORDINATE TRANSFORM
      XIA=ORG+ELJ*GOF(IA)
      XIJ=ELJ*GWT(IA)
      XI(IA)=XIA
      IF (KB.NE.1) GOTO 2
C     INFINITE ELEMENT - NONLINEAR TRANSFORM OF INTRINSIC COORDINATE
      ELT=1.0D0-XIA
      XIA=(3.0D0*XIA+1.0D0)/ELT
      XIJ=XIJ*4.0D0/(ELT*ELT)
    2 XT(IA)=XIA
      TJ(IA)=XIJ
    1 CONTINUE
      CALL BGEOM (KA,XA,XT,3,XC,XS,UN,XJ,JERR)
      IF (JERR.NE.0) GOTO 999
C     MULTIPLY JACOBIAN FOR TRANSFORM ARC LENGTH TO INTRINSIC
C     COORDINATE BY THAT FOR INTRINSIC COORDINATE TRANSFORM
      DO 3 IA=1,3
      XJ(IA)=XJ(IA)*TJ(IA)
    3 CONTINUE
      IERR=0
    4 CONTINUE
      RETURN
  999 IERR=1
      GOTO 4
      END
```

```
      SUBROUTINE BGFOL (KA,XA,KB,KC,XI,XC,UN,XJ,ORG,ELJ,IERR)
C     COMPUTE INTRINSIC COORDINATES, CARTESIAN COORDINATES,
C     NORMALS AND JACOBIAN*GAUSS WEIGHTS AT LOG
C     WEIGHT FUNCTION GAUSS POINTS
      IMPLICIT DOUBLE PRECISION (A-H,O-Z)
      DIMENSION XA(2,3),XI(3),XC(2,3),UN(2,3),XJ(3),
     1 GOF(3),GWT(3),ELO(4),ORO(4),
     2 TJ(3),XS(2,3),XT(3)
      DATA GOF/
     1 0.0638907931D0, 0.3689970637D0, 0.7668803039D0/
      DATA GWT/
     1 0.5134045522D0, 0.3919800412D0, 0.0946154066D0/
      DATA ELO/-1.0D0,0.5D0,-0.5D0,1.0D0/
      DATA ORO/1.0D0,0.0D0,0.0D0,-1.0D0/
      IF (KB.NE.1) GOTO 1
C     FIRST OF KC INTERVALS (INFINITE ELEMENT)
      ELJ=-2.0D0/KC
      ORG=1.0D0
      GOTO 2
C     SUBELEMENT ADJACENT TO FIRST ARGUMENT OF KERNEL
    1 ELJ=ELO(KC)
      ORG=ORO(KC)
    2 ELA=ABS(ELJ)
      DO 3 IA=1,3
C     INTRINSIC COORDINATE AND JACOBIAN*GAUSS WEIGHT FOR INTRINSIC
C     COORDINATE TRANSFORM
      XIA=ORG+ELJ*GOF(IA)
      XIJ=ELA*GWT(IA)
      XI(IA)=XIA
      IF (KB.NE.1) GOTO 4
C     FIRST INTERVAL OF INFINITE ELEMENT - NONLINEAR TRANSFORM
      ELT=1.0D0-XIA
      XIA=(3.0D0*XIA+1.0D0)/ELT
      XIJ=XIJ*4.0D0/(ELT*ELT)
    4 XT(IA)=XIA
      TJ(IA)=XIJ
    3 CONTINUE
      CALL BGEOM (KA,XA,XT,3,XC,XS,UN,XJ,JERR)
      IF (JERR.NE.0) GOTO 999
C     MULTIPLY JACOBIAN FOR TRANSFORM ARC LENGTH TO INTRINSIC
C     COORDINATE BY THAT FOR INTRINSIC COORDINATE TRANSFORM
C     AND APPLY LOG FACTOR
      DO 5 IA=1,3
      XJ(IA)=-XJ(IA)*TJ(IA)/DLOG(GOF(IA))
    5 CONTINUE
      IERR=0
    6 CONTINUE
      RETURN
  999 IERR=1
      GOTO 6
      END
```

```
      SUBROUTINE BGRAV (XC,DPI,SPI)
C     COMPUTE PARTICULAR INTEGRAL DUE TO SELF WEIGHT
      IMPLICIT DOUBLE PRECISION (A-H,O-Z)
      DIMENSION XC(2),DPI(2,*),SPI(2,2,*)
      COMMON/PROP/YM,PR,DEN,DAG,IDIS,KKPR(11)
      COMMON/LOAD/NCC,LUB,LUS,NGR(5),KKLO(52)
      SB=DEN*XC(2)
      ELT=SB/(1.0D0-PR)
      SA=PR*ELT
      DB=0.5D0*(1.0D0+PR)*(1.0D0-2.0D0*PR)*ELT*XC(2)/YM
      DO 1 IA=1,NCC
      IF (NGR(IA).EQ.0) GOTO 1
C     SELF WEIGHT LOAD CASE IA
      IF (IDIS.NE.0) GOTO 2
C     COMPUTED DISPLACEMENTS ARE W R TO UNSTRAINED STATE
      DPI(2,IA)=DPI(2,IA)+DB
    2 SPI(1,1,IA)=SPI(1,1,IA)+SA
      SPI(2,2,IA)=SPI(2,2,IA)+SB
    1 CONTINUE
      RETURN
      END

      SUBROUTINE BINFI (XC,DPI,SPI)
C     COMPUTE PARTICULAR INTEGRAL DUE TO STRESS AT INFINITY
      IMPLICIT DOUBLE PRECISION (A-H,O-Z)
      DIMENSION XC(2),DPI(2,*),SPI(2,2,*)
      COMMON/PROP/YM,PR,DEN,DAG,IDIS,KKPR(11)
      COMMON/LOAD/NCC,LUB,LUS,NGR(5),NIN(6),TIN(3,5),TID(5),KKLO(6)
      GG=(1.0D0+PR)/YM
      GF=GG*(1.0D0-PR)
      GR=GG*PR
      DO 1 IA=1,NCC
      IF (NIN(IA).EQ.0) GOTO 1
C     STRESS AT INFINITY LOAD CASE IA
      ELT=TID(IA)*XC(2)
      IF (IDIS.NE.0) GOTO 2
C     COMPUTED DISPLACEMENTS ARE W R TO UNSTRAINED STATE
      EXX=GF*(TIN(1,IA)+ELT)-GR*TIN(2,IA)
      EYY=GF*TIN(2,IA)-GR*(TIN(1,IA)+0.5D0*ELT)
      EXY=GG*TIN(3,IA)
      DPI(1,IA)=DPI(1,IA)+EXX*XC(1)+EXY*XC(2)
      DPI(2,IA)=DPI(2,IA)+EYY*XC(2)+EXY*XC(1)
     1 -0.5D0*GF*TID(IA)*XC(1)*XC(1)
C     STRESSES
    2 SPI(1,1,IA)=SPI(1,1,IA)+TIN(1,IA)+ELT
      SPI(2,2,IA)=SPI(2,2,IA)+TIN(2,IA)
      SPI(1,2,IA)=SPI(1,2,IA)+TIN(3,IA)
      SPI(2,1,IA)=SPI(2,1,IA)+TIN(3,IA)
    1 CONTINUE
      RETURN
      END

      SUBROUTINE BLKER (KA,XI,ORG,ELJ,UT,DIV)
C     COMPUTE LOGARITHMIC COMPONENT OF KERNEL U AT
C     GAUSS POINTS OF INTERVAL, MULTIPLIED BY KA
C     DIV=DAG FOR MATRIX CONSTRUCTION, =YM FOR POSTPROCESSING
      IMPLICIT DOUBLE PRECISION (A-H,O-Z)
      DIMENSION XI(3),UT(2,2,3)
      COMMON/PROP/YM,PR,KKPR(16)
      DATA QPI/12.5663706144D0/
```

```
C     ELASTIC CONSTANT
      ELT=QPI*(1.0D0-PR)*DIV
      EU=(1.0D0+PR)/ELT
      EL=EU*(3.0D0-4.0D0*PR)
      FAC=-KA*EL
      DO 1 IA=1,3
C     INTRINSIC COORDINATE FOR LOG FORMULA INTEGRATION
C     (EQUALS ZERO AT SINGULARITY, 1.0 AT OTHER END OF INTERVAL)
      ETA=(XI(IA)-ORG)/ELJ
      ELL=FAC*DLOG(ETA)
      UT(1,1,IA)=ELL
      UT(1,2,IA)=0.0D0
      UT(2,1,IA)=0.0D0
      UT(2,2,IA)=ELL
    1 CONTINUE
      RETURN
      END

      SUBROUTINE BMKSF (XK,FUN,XJ,XKL,KA)
C     INCREMENT KERNEL-SHAPE FUNCTION-JACOBIAN-WEIGHT PRODUCTS
      IMPLICIT DOUBLE PRECISION (A-H,O-Z)
      DIMENSION XK(2,2,3),FUN(3,3),XJ(3),XKL(2,2,KA)
      DO 1 IA=1,KA
C     SHAPE FUNCTION IA, GAUSS POINT IB
      DO 2 IB=1,3
C     JACOBIAN*GAUSS WEIGHT*SHAPE FUNCTION
      ELB=XJ(IB)*FUN(IA,IB)
C     TRANSPOSE TENSOR COMPONENTS TO ENABLE SYMMETRY FACTORING BY PREFL
      XKL(1,1,IA)=XKL(1,1,IA)+ELB*XK(1,1,IB)
      XKL(2,1,IA)=XKL(2,1,IA)+ELB*XK(1,2,IB)
      XKL(1,2,IA)=XKL(1,2,IA)+ELB*XK(2,1,IB)
      XKL(2,2,IA)=XKL(2,2,IA)+ELB*XK(2,2,IB)
    2 CONTINUE
    1 CONTINUE
      RETURN
      END

      SUBROUTINE BPBOX (KA,KB)
C     DRAW BOX AND TITLE
      IMPLICIT DOUBLE PRECISION (A-H,O-Z)
      DIMENSION XT(2),XL(2),
     1 XSM(2),XSN(2),XSO(2),
     2 XSD(2),XSE(2),XSF(2),
     3 LS(2),LM(5),LD(6),LE(4),
     4 XY(2)
      COMMON/TITLE/IDENT(20),JDENT(20,5),KKTI(120)
      COMMON/PLOT/ECH,SIZ(2),SCA(3),
     1 CENP(2),CEN(2,3),DCO(4,3),
     2 LOT(3),KKPL(65)
      DATA XT/20.0D0,23.0D0/
      DATA XL/20.0D0,18.0D0/
      DATA XSM/20.0D0,8.0D0/
      DATA XSN/41.0D0,8.0D0/
      DATA XSO/71.0D0,8.0D0/
      DATA XSD/20.0D0,3.0D0/
      DATA XSE/41.0D0,3.0D0/
      DATA XSF/71.0D0,3.0D0/
      DATA LS/4H1 MM,4H =  /
      DATA LM/4H UNI,4HTS O,4HF DI,4HSTAN,4HCE  /
      DATA LD/4H UNI,4HTS O,4HF DI,4HSPLA,4HCEME,4HNT  /
      DATA LE/4H UNI,4HTS O,4HF ST,4HRESS/
      DATA TMAR/30.0D0/
      DATA HT/3.0D0/
```

```
C     INITIALISE PLOT
      CALL DINIT (KA)
      LOT(KA)=-1
C     DRAW BOX
      XY(1)=0.0D0
      XY(2)=0.0D0
      CALL PMOVE (XY,1)
      XY(1)=SIZ(1)
      CALL PMOVE (XY,0)
      XY(2)=SIZ(2)
      CALL PMOVE (XY,0)
      XY(1)=0.0D0
      CALL PMOVE (XY,0)
      XY(2)=0.0D0
      CALL PMOVE (XY,0)
      XY(2)=TMAR
      CALL PMOVE (XY,1)
      XY(1)=SIZ(1)
      CALL PMOVE (XY,0)
C     MAXIMUM LENGTH OF TITLE THAT CAN BE PLOTTED
      JA=(SIZ(1)-2.0D0*XT(1))/HT
      JA=MIN0(JA,76)
C     TITLE OF PROBLEM AND MESH SCALE
      CALL PTEXT (XT,HT,IDENT,0.0D0,JA)
      SCAI=1.0D0/SCA(KA)
      CALL PTEXT (XSM,HT,LS,0.0D0,7)
      NDEC=2-DLOG10(SCAI)
      NDEC=MAX0(NDEC,1)
      NDEC=MIN0(NDEC,8)
      CALL PNUMF (XSN,HT,SCAI,0.0D0,10,NDEC)
      CALL PTEXT (XSO,HT,LM,0.0D0,18)
      IF (KA.EQ.1) GOTO 1
C     LOAD CASE TITLE
      CALL PTEXT (XL,HT,JDENT(1,KB),0.0D0,JA)
    1 CONTINUE
      RETURN
C     DISPLACEMENT SCALE
      ENTRY BPSCD (SCD)
      SCDI=1.0D0/SCD
      CALL PTEXT (XSD,HT,LS,0.0D0,7)
      NDEC=2-DLOG10(SCDI)
      NDEC=MAX0(NDEC,1)
      NDEC=MIN0(NDEC,8)
      CALL PNUMF (XSE,HT,SCDI,0.0D0,10,NDEC)
      CALL PTEXT (XSF,HT,LD,0.0D0,22)
      RETURN
C     STRESS SCALE
      ENTRY BPSCS (SCS)
      SCSI=1.0D0/SCS
      CALL PTEXT (XSD,HT,LS,0.0D0,7)
      NDEC=2-DLOG10(SCSI)
      NDEC=MAX0(NDEC,1)
      NDEC=MIN0(NDEC,8)
      CALL PNUMF (XSE,HT,SCSI,0.0D0,10,NDEC)
      CALL PTEXT (XSF,HT,LE,0.0D0,16)
      RETURN
      END
```

```
      SUBROUTINE BOVLD
      CALL BFORM
      RETURN
      END

      SUBROUTINE BFORM
C     CONSTRUCT MATRIX AND LOAD VECTORS
      IMPLICIT DOUBLE PRECISION (A-H,O-Z)
      DIMENSION VO(2),INF(2,2),
     1 XA(2,3),XB(2,3),XR(2),VA(2),VB(2),
     2 BD(2,3,5),BT(2,3,5),JFE(2),JFN(2,3),
     3 XCR(2,3,3),UNR(2,3,3),
     4 XI(3),XC(2,3),UN(2,3),XJ(3),XIL(3),XCL(2,3),UNL(2,3),XJL(3),
     5 TT(2,2,3),UT(2,2,3),FUU(3,3),FUT(3,3),
     6 TTL(2,2,3),UTL(2,2,3)
      COMMON/PROP/YM,PR,DEN,DAG,KKPR(12)
      COMMON/FORM/LUO,KKFO(59)
      COMMON/FLAG/IGER,ILER
      COMMON/GEOM/ISYM(2),MXA,NXB,NXE,MEA,NEB,NEE,NXY,
     1 LUA,LUL,LUG,ASO(2),NLIB,LUAL,LUAN,KKGE(21)
      COMMON/LOAD/NCC,LUB,KKLO(58)
      COMMON/CHANGE/X(2,300),JXY(300),
     1 JEL(120),JEN(120),KEL(3,120),
     2 IEO(120),IFN(300),IFE(120),XHW(2,300),
     3 XIO(3,3,120),XCO(2,3,3,120),UNO(2,3,3,120),XJO(3,3,120),
     4 A(4000),B(2500),TTR(2,2,31),KKCH(9952)
      DATA VO/2*1.0D0/
      DATA INF/1,0,2,3/
      DATA LALO/2000/
      CALL PTICK (3)
      CALL BPREP
      IF (IGER.NE.0) GOTO 1
      JA=ISYM(1)+1
      JB=ISYM(2)+1
      NLIB=2*NXB
C     INITIALISE LOAD VECTORS
      JC=NCC*NLIB
      CALL PZERO (B,JC)
      LUAL=LALO
C     EQUATIONS/LUA RECORD (=NODES/INTEGRATION BLOCK)
C     AND VARIOUS POINTERS AND INCREMENTS
      LUAN=LUAL/NLIB
      JD=2*NLIB
      JF=LUAN*NLIB
      JH=1
C     HOW MANY INTEGRATION BLOCKS
      JI=(NXB-1)/LUAN+1
      OPEN (UNIT=LUA,STATUS='SCRATCH',ACCESS='DIRECT',RECL=8*LUAL,
     1 IOSTAT=IERR,ERR=300)
  300 CALL BFERR (LUA,IERR)
      OPEN (UNIT=LUL,STATUS='SCRATCH',
     1 FORM='UNFORMATTED',IOSTAT=IERR,ERR=301)
  301 CALL BFERR (LUL,IERR)
```

```
C     FIRST, LAST NODE OF BLOCK
      JJ=1
      JK=LUAN
      DO 3 IA=1,JI
      JK=MIN0(JK,NXB)
      JL=JK-JJ+1
C     INITIALISE BLOCK OF MATRIX
      JM=JL*JD
      JN=4*JL
      CALL PZERO (A,JM)
      CALL PZERO (TTR,JN)
      DO 4 IB=1,NEE
C     COORDINATES AND GIVEN BOUNDARY DATA ELEMENT IB
      CALL DPICK (X,KEL(1,IB),3,XA)
      READ (LUB)
     1 (((BD(KO,JO,IO),KO=1,2),JO=1,3),IO=1,NCC),
     2 (((BT(KO,JO,IO),KO=1,2),JO=1,3),IO=1,NCC)
      JM=JEN(IB)
C     ELEMENT AND NODAL CONSTRAINTS
      CALL DUNPK (IFE(IB),JFE,2)
      DO 5 IC=1,JM
      JN=KEL(IC,IB)
      CALL DUNPK (IFN(JN),JFN(1,IC),2)
    5 CONTINUE
      JN=IEO(IB)
      CALL PSETF (VO,VA,2)
C     SYMMETRY W R TO X
      DO 6 IC=1,JA
      JO=JN
      CALL PSETF (VA,VB,2)
C     SYMMETRY W R TO Y
      DO 7 ID=1,JB
      CALL PREFL (XA,XB,VB,3)
      CALL PREFL (XCO(1,1,1,IB),XCR,VB,9)
      CALL PREFL (UNO(1,1,1,IB),UNR,VB,9)
      DO 9 IE=JJ,JK
      CALL PSRCH (IE,KEL(1,IB),JM,JP)
      IF (JP.EQ.0) GOTO 10
C     IE IS NODE OF ELEMENT IB - SKIP IF IT COINCIDES WITH
C     ITS REFLECTION IN SYMMETRY PLANES
      CALL BDIST (XA(1,JP),XB(1,JP),XR,RAD)
      IF (RAD.LT.1.0D-6) GOTO 9
C     CHOOSE INTEGRATION SCHEME - JQ=NUMBER OF INTERVALS
   10 CALL BCFOR (1,X(1,IE),XB,JM,JQ,IERR)
      IF (IERR.EQ.0) GOTO 12
      WRITE (LUO,200) JEL(IB),JXY(IE)
  200 FORMAT (/  38H INTEGRATION SCHEME ERROR FOR ELEMENT ,I3,
     1 39H AND FIRST ARGUMENT OF KERNELS AT NODE ,I4)
      IGER=1
      GOTO 9
   12 CALL PZERO (TTL,12)
      CALL PZERO (UTL,12)
      DO 13 IF=1,JQ
      IF (JQ.GT.2) GOTO 14
C     INTEGRATION FORMULA WAS COMPUTED BY BPREP
      JR=INF(IF,JQ)
      CALL PSETF (XIO(1,JR,IB),XI,3)
      CALL PSETF (XCR(1,1,JR),XC,6)
      CALL PSETF (UNR(1,1,JR),UN,6)
      CALL PSETF (XJO(1,JR,IB),XJ,3)
      GOTO 15
```

```
C     COMPUTE DATA FOR GAUSS POINTS OF INTERVAL IF
   14 CALL BGFOR (JO,XB,JM,JQ,IF,XI,XC,UN,XJ,IERR)
      IF (IERR.EQ.0) GOTO 15
  999 WRITE (LUO,201) JEL(IB)
  201 FORMAT (/  24H ZERO JACOBIAN, ELEMENT ,I3)
      IGER=1
      GOTO 9
C     COMPUTE KERNELS OF THE INTEGRAL EQUATION
   15 CALL BMKER (X(1,IE),XC,UN,TT,UT,IERR)
      IF (IERR.EQ.0) GOTO 16
      WRITE (LUO,202) JEL(IB),JXY(IE)
  202 FORMAT (/  39H KERNEL ARGUMENTS COINCIDE FOR ELEMENT ,I3,
     1 28H AND FIRST ARGUMENT AT NODE ,I4)
      IGER=1
      GOTO 9
   16 CONTINUE
      IF (JM.EQ.1) GOTO 17
C     FINITE BOUNDARY ELEMENT - COMPUTE SHAPE FUNCTIONS AND INCREMENT
C     PRODUCTS OF KERNELS, SHAPE FUNCTIONS AND JACOBIAN*GAUSS WEIGHT
      CALL BSHAP (XI,FUU,3)
      CALL BMKSF (TT,FUU,XJ,TTL,3)
      CALL BMKSF (UT,FUU,XJ,UTL,3)
      GOTO 13
C     INFINITE ELEMENT - COMPUTE SHAPE FUNCTIONS
   17 CALL BSHAI (XC,XB,ASO,FUU,FUT,3,IERR)
      IF (IERR.EQ.0) GOTO 18
  998 WRITE (LUO,203) JEL(IB)
  203 FORMAT (/
     1 52H CANNOT COMPUTE ASYMPTOTIC SHAPE FUNCTIONS, ELEMENT ,I3)
      IGER=1
      GOTO 9
C     INCREMENT PRODUCTS OF KERNELS, SHAPE FUNCTIONS
C     AND JACOBIAN*GAUSS WEIGHT
   18 CALL BMKSF (TT,FUU,XJ,TTL,1)
      CALL BMKSF (UT,FUT,XJ,UTL,1)
      IF (IF.NE.1) GOTO 13
C     FIRST INTERVAL - LOG WEIGHT FUNCTION GAUSS FORMULA
      CALL BGFOL (JO,XB,1,JQ,XIL,XCL,UNL,XJL,ORG,ELJ,IERR)
      IF (IERR.NE.0) GOTO 999
C     SUBTRACT LOG COMPONENT COMPUTED BY GAUSS-LEGENDRE FORMULA
      CALL BLKER (1,XI,ORG,ELJ,UT,DAG)
      CALL BMKSF (UT,FUT,XJ,UTL,1)
C     ADD LOG COMPONENT COMPUTED BY LOG FORMULA
      CALL BLKER (-1,XIL,ORG,ELJ,UT,DAG)
      CALL BSHAI (XCL,XB,ASO,FUU,FUT,3,IERR)
      IF (IERR.NE.0) GOTO 998
      CALL BMKSF (UT,FUT,XJL,UTL,1)
   13 CONTINUE
C     APPLY SYMMETRY FACTORS
      CALL PREFL (TTL,TTL,VB,6)
      CALL PREFL (UTL,UTL,VB,6)
C     SCATTER INTEGRALS TO MATRIX AND LOAD VECTORS
      CALL BMINC (TTL,UTL,BD,BT,A,B,TTR,JJ,NXB,NCC,
     1 KEL(1,IB),IE,JFE,JFN,JM)
    9 CONTINUE
      JO=-JO
      VB(2)=-VB(2)
    7 CONTINUE
      JN=-JN
      VA(1)=-VA(1)
    6 CONTINUE
```

```
C     INTEGRATION OVER ELEMENTS OF WHICH FIRST ARGUMENT OF KERNELS
C     IS A NODE, AND IMPLICIT INTEGRATION
      CALL BSING (IB,JJ,JK,XA,BD,BT,JFE,JFN,JM)
    4 CONTINUE
      REWIND LUB
C     ADJUST INTEGRALS T
      CALL BDTTR (TTR,A,IFN,ISYM,NXB,JL,NEB,NEE)
C     PERFORM ELIMINATIONS AND WRITE BLOCK OF MATRIX TO FILE LUA
      CALL BELIM (A,B,NXB,NCC,IFN,JJ,JK)
      JM=(2*JL-1)/LUAN+1
      JN=1
      JO=LUAL
      DO 20 IB=1,JM
      WRITE (LUA,REC=JH,IOSTAT=IERR,ERR=302) (A(IO),IO=JN,JO)
  302 CALL BFERR (LUA,IERR)
      JH=JH+1
      JN=JN+JF
      JO=JO+JF
   20 CONTINUE
      JJ=JJ+LUAN
      JK=JK+LUAN
    3 CONTINUE
      OPEN (UNIT=LUG,STATUS='SCRATCH',
     1 FORM='UNFORMATTED',IOSTAT=IERR,ERR=303)
  303 CALL BFERR (LUG,IERR)
C     WRITE LOAD VECTORS AND INTEGRATION FORMULAE COMPUTED BY BPREP
      WRITE (LUG,IOSTAT=IERR,ERR=304) (B(IO),IO=1,JC)
  304 CALL BFERR (LUG,IERR)
      DO 21 IA=1,NEE
      WRITE (LUG,IOSTAT=IERR,ERR=305)
     1 ((XIO(JO,IO,IA),JO=1,3),IO=1,3),
     2 (((XCO(KO,JO,IO,IA),KO=1,2),JO=1,3),IO=1,3),
     3 (((UNO(KO,JO,IO,IA),KO=1,2),JO=1,3),IO=1,3),
     4 ((XJO(JO,IO,IA),JO=1,3),IO=1,3)
  305 CALL BFERR (LUG,IERR)
   21 CONTINUE
      REWIND LUG
    1 CALL PTOCK (3)
      RETURN
      END


      SUBROUTINE BPREP
C     PREPARATION FOR MATRIX AND LOAD VECTOR CONSTRUCTION
      IMPLICIT DOUBLE PRECISION (A-H,O-Z)
      DIMENSION NI(3),NJ(3),
     1 JOY(300),JOL(120),JON(120),KOL(3,120),
     2 IOO(120),ION(300),IOE(120),ICH(6360),
     3 XA(2,3,2),XHI(2),XLO(2),JEO(2),
     4 UNA(2),XC(2),XS(2),UN(2)
      COMMON/FORM/LUO,KKFO(59)
      COMMON/FLAG/IGER,ILER
      COMMON/GEOM/ISYM(2),MXA,NXB,NXE,MEA,NEB,NEE,NXY,
     1 LUA,LUL,LUG,ASO(2),KKGE(24)
      COMMON/PROP/YM,PR,DEN,DAG,KKPR(12)
      COMMON/CHANGE/X(2,300),JXY(300),
     1 JEL(120),JEN(120),KEL(3,120),
     2 IEO(120),IFN(300),IFE(120),XHW(2,300),
     3 XIO(3,3,120),XCO(2,3,3,120),UNO(2,3,3,120),XJO(3,3,120),
     4 KKCH(23200)
```

```
       EQUIVALENCE (ICH(1),X(1,1)),
     1 (JOY(1),ICH(2701)),(JOL(1),ICH(3601)),
     2 (JON(1),ICH(3721)),(KOL(1,1),ICH(3961)),
     3 (IOO(1),ICH(5821)),(ION(1),ICH(5941)),
     4 (IOE(1),ICH(6241))
      DATA NI/1,2,2/
      DATA NJ/1,1,2/
C     COMPRESS /CHANGE/
      JA=3*NEE
      CALL PSETI (JOY,JXY,NXB)
      CALL PSETI (JOL,JEL,NEE)
      CALL PSETI (JON,JEN,NEE)
      CALL PSETI (KOL,KEL,JA)
      CALL PSETI (IOO,IEO,NEE)
      CALL PSETI (ION,IFN,NXB)
      CALL PSETI (IOE,IFE,NEE)
C     COMPUTE DATA FOR GAUSS POINTS OF ONE INTERVAL AND TWO INTERVAL
C     INTEGRATION SCHEMES
      DO 1 IA=1,NEE
      CALL DPICK (X,KEL(1,IA),3,XA)
      DO 2 IB=1,3
      CALL BGFOR (IEO(IA),XA,JEN(IA),NI(IB),NJ(IB),
     1 XIO(1,IB,IA),XCO(1,1,IB,IA),UNO(1,1,IB,IA),XJO(1,IB,IA),IERR)
      IF (IERR.EQ.0) GOTO 2
      WRITE (LUO,200) JEL(IA)
  200 FORMAT (/ 24H ZERO JACOBIAN, ELEMENT ,I3)
      IGER=1
    2 CONTINUE
    1 CONTINUE
C     COMPUTE COORDINATE EXTREMA AND COORDINATES OF SPIRAL
C     HALFWAY NODES FOR IMPLICIT INTEGRATION
      DO 3 IA=1,2
      XHI(IA)=-1.0D20
      XLO(IA)=1.0D20
    3 CONTINUE
      DO 4 IA=1,NXB
C     UPDATE COORDINATE EXTREMA
      DO 5 IB=1,2
      XHI(IB)=DMAX1(XHI(IB),X(IB,IA))
      XLO(IB)=DMIN1(XLO(IB),X(IB,IA))
    5 CONTINUE
C     FIND ELEMENTS WITH NODE IA
      JA=0
      DO 6 IB=1,NEE
      CALL PSRCH (IA,KEL(1,IB),JEN(IB),JB)
      IF (JB.EQ.0) GOTO 6
C     IA IS NODE JB OF ELEMENT IB
      JC=JB
      JA=JA+1
      JEO(JA)=IEO(IB)
      CALL DPICK (X,KEL(1,IB),3,XA(1,1,JA))
    6 CONTINUE
```

```
      IF (JA.EQ.2) GOTO 7
C     IA IS NODE OF ONLY ONE ELEMENT - SEE IF MIDNODE
      IF (JC.EQ.3) GOTO 7
C     IS IN PLANE OF SYMMETRY - CONSTRUCT REFLECTED ELEMENT
      CALL PSETF (XA(1,1,1),XA(1,1,2),6)
      DO 8 IB=1,2
      IF (ISYM(IB).EQ.0.OR.X(IB,IA).GT.1.0D-6) GOTO 8
C     ON SYMMETRY IB
      DO 9 IC=1,3
      XA(IB,IC,2)=-XA(IB,IC,2)
    9 CONTINUE
    8 CONTINUE
C     ORIENTATION OF REFLECTED ELEMENT
      JEO(2)=-JEO(1)
      JA=2
C     ADD NORMALS AND JACOBIANS AT MIDNODES OF ELEMENTS
C     OF WHICH IA IS A NODE
    7 CALL PZERO (UNA,2)
      ELA=0.0D0
      DO 10 IB=1,JA
      CALL BGEOM (JEO(IB),XA(1,1,IB),0.0D0,1,
     1 XC,XS,UN,ELT,IERR)
      IF (IERR.EQ.0) GOTO 11
      WRITE (LUO,201) JXY(IA)
  201 FORMAT (/
     1 41H ZERO JACOBIAN, ELEMENT ADJACENT TO NODE ,I4)
      IGER=1
      GOTO 4
   11 UNA(1)=UNA(1)+UN(1)
      UNA(2)=UNA(2)+UN(2)
      ELA=ELA+ELT
   10 CONTINUE
      CALL PNORM (UNA,UNA,IERR)
      IF (IERR.EQ.0) GOTO 12
      WRITE (LUO,202) JXY(IA)
  202 FORMAT (/
     1 37H UNABLE TO CONSTRUCT SPIRAL FOR NODE ,I4)
      IGER=1
      GOTO 4
C     COORDINATES OF HALFWAY NODE
   12 XHW(1,IA)=X(1,IA)-ELA*UNA(1)
      XHW(2,IA)=X(2,IA)-ELA*UNA(2)
    4 CONTINUE
C     COMPUTE COORDINATES OF REFERENCE POINT FOR INFINITE ELEMENTS,
C     AND DISTANCE SCALE FACTOR
      DO 13 IA=1,2
      ASO(IA)=0.5D0*(XHI(IA)+XLO(IA))
      XC(IA)=XHI(IA)-XLO(IA)
      IF (ISYM(IA).EQ.0) GOTO 13
C     PUT REFERENCE POINT IN PLANE OF SYMMETRY
      ASO(IA)=0.0D0
      XC(IA)=2.0D0*XC(IA)
   13 CONTINUE
      DAG=DSQRT(XC(1)*XC(1)+XC(2)*XC(2))
      RETURN
      END
```

```
      SUBROUTINE BSING (KA,KB,KC,XA,BD,BT,JFE,JFN,KD)
C     INTEGRATE EXPLICITLY AND IMPLICITLY COMPUTED COMPONENTS OVER
C     ELEMENTS OF WHICH FIRST ARGUMENT IS A NODE, AND IMPLICITLY
C     COMPUTED COMPONENT OVER SPIRAL, ELEMENT KA NODES KB TO KC
      IMPLICIT DOUBLE PRECISION (A-H,O-Z)
      DIMENSION XA(2,3),BD(2,3,5),BT(2,3,5),
     1 JFE(2),JFN(2,3),
     2 NIN(3),NIL(4,3),
     3 XI(3),XC(2,3),UN(2,3),XJ(3),
     4 XIL(3),XCL(2,3),UNL(2,3),XJL(3),
     5 TT(2,2,3),UT(2,2,3),FUU(3,3),FUT(3,3),FUI(3,3),
     6 TTL(2,2,3),UTL(2,2,3),AI(2,2)
      COMMON/PROP/YM,PR,DEN,DAG,KKPR(12)
      COMMON/FORM/LUO,KKFO(59)
      COMMON/FLAG/IGER,ILER
      COMMON/GEOM/ISYM(2),MXA,NXB,NXE,MEA,NEB,NEE,NXY,
     1 LUA,LUL,LUG,ASO(2),NLIB,KKGE(23)
      COMMON/LOAD/NCC,KKLO(59)
      COMMON/CHANGE/X(2,300),JXY(300),
     1 JEL(120),JEN(120),KEL(3,120),
     2 IEO(120),IFN(300),IFE(120),XHW(2,300),
     3 XIO(3,3,120),XCO(2,3,3,120),UNO(2,3,3,120),XJO(3,3,120),
     4 A(4000),B(2500),TTR(2,2,31),KKCH(9952)
      DATA NIN/2,2,4/
      DATA NIL/
     1 0,4,0,0,
     2 1,0,0,0,
     3 0,2,3,0/
      DO 2 IA=KB,KC
      CALL PSRCH (IA,KEL(1,KA),KD,JB)
      IF (JB.EQ.0) GOTO 2
C     IA IS NODE JB OF ELEMENT KA
      JC=NIN(JB)
      CALL PZERO (TTL,12)
      CALL PZERO (UTL,12)
      CALL PZERO (AI,4)
      DO 4 IB=1,JC
C     COMPUTE DATA FOR GAUSS POINTS OF INTERVAL IB
      CALL BGFOR (IEO(KA),XA,3,JC,IB,XI,XC,UN,XJ,IERR)
      IF (IERR.EQ.0) GOTO 5
  999 WRITE (LUO,200) JEL(KA)
  200 FORMAT (/  24H ZERO JACOBIAN, ELEMENT ,I3)
      IGER=1
      GOTO 2
C     COMPUTE KERNELS OF THE INTEGRAL EQUATION
    5 CALL BMKER (X(1,IA),XC,UN,TT,UT,IERR)
      IF (IERR.EQ.0) GOTO 6
      WRITE (LUO,201) JEL(KA),JXY(IA)
  201 FORMAT (/  39H KERNEL ARGUMENTS COINCIDE FOR ELEMENT ,I3,
     1 28H AND FIRST ARGUMENT AT NODE ,I4)
      IGER=1
      GOTO 2
    6 CONTINUE
      IF (KD.EQ.1) GOTO 7
C     FINITE BOUNDARY ELEMENT - COMPUTE SHAPE FUNCTIONS
      CALL BSHAP (XI,FUU,3)
      CALL PSETF (FUU,FUT,9)
      GOTO 8
C     INFINITE ELEMENT - COMPUTE SHAPE FUNCTIONS
    7 CALL BSHAI (XC,XA,ASO,FUU,FUT,3,IERR)
      IF (IERR.EQ.0) GOTO 8
```

```
  998 WRITE (LUO,202) JEL(KA)
  202 FORMAT (/
     1 52H CANNOT COMPUTE ASYMPTOTIC SHAPE FUNCTIONS, ELEMENT ,I3)
      IGER=1
      GOTO 2
C     COMPUTE IMPLICIT INTEGRATION REMAINDER AND INCREMENT PRODUCTS
    8 DO 9 IC=1,3
      FUI(1,IC)=1.0D0-FUU(JB,IC)
      FUU(JB,IC)=0.0D0
    9 CONTINUE
C     EXPLICITLY COMPUTED INTEGRALS
      CALL BMKSF (TT,FUU,XJ,TTL,KD)
      CALL BMKSF (UT,FUT,XJ,UTL,KD)
C     IMPLICITLY COMPUTED INTEGRAL
      CALL BMKSF (TT,FUI,XJ,AI,1)
      JD=NIL(IB,JB)
      IF (JD.EQ.0) GOTO 4
C     INTERVAL WITH FIRST ARGUMENT OF KERNEL AT ONE END
C     LOG WEIGHT FUNCTION GAUSS FORMULA
      CALL BGFOL (IEO(IB),XA,0,JD,XIL,XCL,UNL,XJL,ORG,ELJ,IERR)
      IF (IERR.NE.0) GOTO 999
C     SUBTRACT LOG COMPONENT COMPUTED BY GAUSS-LEGENDRE FORMULA
      CALL BLKER (-1,XI,ORG,ELJ,UT,DAG)
      CALL BMKSF (UT,FUT,XJ,UTL,KD)
C     ADD LOG COMPONENT COMPUTED BY LOG FORMULA
      CALL BLKER (1,XIL,ORG,ELJ,UT,DAG)
      IF (KD.EQ.1) GOTO 10
      CALL BSHAP (XIL,FUT,3)
      GOTO 11
   10 CALL BSHAI (XCL,XA,ASO,FUU,FUT,3,IERR)
      IF (IERR.NE.0) GOTO 998
   11 CALL BMKSF (UT,FUT,XJL,UTL,KD)
    4 CONTINUE
C     INTEGRATE OVER SPIRAL
      CALL BSPIN (IEO(KA),XA,JB,XHW(1,IA),AI,JEL(KA),JXY(IA))
      IF (KD.NE.1) GOTO 12
C     INTEGRATE OVER INFINITE ELEMENT BEYOND END OF SPIRAL
      CALL BADJI (IEO(KA),XA,TTL,UTL,JEL(KA),JXY(IA),ASO)
C     ADD IMPLICITLY COMPUTED COMPONENT OF BOUNDARY INTEGRAL
   12 DO 13 IB=1,2
      DO 14 IC=1,2
      TTL(IB,IC,JB)=TTL(IB,IC,JB)-AI(IB,IC)
   14 CONTINUE
   13 CONTINUE
      DO 15 IB=1,2
      IF (ISYM(IB).EQ.0.OR.X(IB,IA).GT.1.0D-6) GOTO 15
C     APPLY SYMMETRY FACTOR
      DO 16 IC=1,2
      DO 17 ID=1,2
      DO 18 IE=1,KD
      TTL(IC,ID,IE)=2.0D0*TTL(IC,ID,IE)
      UTL(IC,ID,IE)=2.0D0*UTL(IC,ID,IE)
   18 CONTINUE
   17 CONTINUE
   16 CONTINUE
   15 CONTINUE
      CALL BMINC (TTL,UTL,BD,BT,A,B,TTR,KB,NXB,NCC,
     1 KEL(1,KA),IA,JFE,JFN,KD)
    2 CONTINUE
      RETURN
      END
```

```
      SUBROUTINE BSPIN (KA,XA,KB,XHW,AI,KC,KD)
C     INTEGRATE KERNEL T OVER SPIRAL, ELEMENT KC NODE KD
      IMPLICIT DOUBLE PRECISION (A-H,O-Z)
      DIMENSION XA(2,3),XHW(2),AI(2,2),
     1 NSP(3),ISO(2,3),XIN(3),FUI(3,3),IFN(2,3),
     2 XC(2),XS(2),UN(2),
     3 DCO(2,2),DCI(2,2),XRA(2),XRB(2),
     4 R(3),TH(3),XT(2),XB(2,3),
     5 XI(3),XG(2,3),UG(2,3),XJ(3),
     6 TT(2,2,3),UT(2,2,3)
      COMMON/FORM/LUO,KKFO(59)
      COMMON/FLAG/IGER,ILER
      DATA NSP/1,1,2/
      DATA XIN/-1.0D0,1.0D0,0.0D0/
      DATA IFN/2,0, 1,0, 1,2/
      DATA ISO/1,0, -1,0, -1,1/
      DATA FUI/9*1.0D0/
      DATA THM/1.0D0/
      CALL BGEOM (KA,XA,XIN(KB),1,XC,XS,UN,ELJ,IERR)
      IF (IERR.EQ.0) GOTO 1
  998 WRITE (LUO,200) KD,KC
  200 FORMAT (/  37H UNABLE TO CONSTRUCT SPIRAL FOR NODE ,I4,
     1 13H AND ELEMENT ,I3)
      IGER=1
      GOTO 2
C     FIRST LOCAL AXIS
    1 DCO(1,1)=-UN(1)
      DCO(1,2)=-UN(2)
      JA=NSP(KB)
      DO 3 IA=1,JA
C     SECOND LOCAL AXIS FOR HALF-SPIRAL
      ELT=ISO(IA,KB)/ELJ
      DCO(2,1)=ELT*XS(1)
      DCO(2,2)=ELT*XS(2)
C     DIRECTION COSINES FOR TRANSFORM LOCAL TO GLOBAL COORDINATES
      DCI(1,1)=DCO(1,1)
      DCI(1,2)=DCO(2,1)
      DCI(2,1)=DCO(1,2)
      DCI(2,2)=DCO(2,2)
      JB=IFN(IA,KB)
C     DISTANCES TO HALFWAY NODE AND FAR NODE OF ELEMENT FROM
C     FIRST ARGUMENT OF KERNELS
      CALL BDIST (XHW,XC,XRA,RA)
      CALL BDIST (XA(1,JB),XC,XRB,RB)
      IF (RA.LT.1.0D-6.OR.RB.LT.1.0D-6) GOTO 998
      CALL BTRAN (XRA,DCO,XRA)
      CALL BTRAN (XRB,DCO,XRB)
C     ANGLES IN LOCAL POLAR COORDINATES TO HALFWAY NODE AND FAR NODE
      THA=ATAN2(XRA(2),XRA(1))
      THB=ATAN2(XRB(2),XRB(1))
      THD=THB-THA
      IF (THD.LT.0.001D0) GOTO 998
C     HOW MANY ELEMENTS TO REPRESENT THE HALF-SPIRAL
      JB=THD/THM+1.0D0
      RI=(RB-RA)/JB
      THI=THD/JB
```

```
C     ORIENTATION OF SPIRAL ELEMENTS AND
C     LOCAL POLAR COORDINATES OF NODES OF FIRST ELEMENT
      JC=-ISO(IA,KB)*KA
      R(1)=RA
      R(3)=RA+0.5D0*RI
      R(2)=RA+RI
      TH(1)=THA
      TH(3)=THA+0.5D0*THI
      TH(2)=THA+THI
      DO 4 IB=1,JB
      DO 5 IC=1,3
C     LOCAL CARTESIAN COORDINATES OF NODE OF SPIRAL ELEMENT
      XT(1)=R(IC)*DCOS(TH(IC))
      XT(2)=R(IC)*DSIN(TH(IC))
      CALL BTRAN (XT,DCI,XT)
C     GLOBAL COORDINATES
      XB(1,IC)=XT(1)+XC(1)
      XB(2,IC)=XT(2)+XC(2)
C     INCREMENT LOCAL POLAR COORDINATES
      R(IC)=R(IC)+RI
      TH(IC)=TH(IC)+THI
    5 CONTINUE
C     CHOOSE INTEGRATION SCHEME - JD=NUMBER OF INTERVALS
      CALL BCFOR (1,XC,XB,1,JD,IERR)
      IF (IERR.EQ.0) GOTO 6
  999 WRITE (LUO,201) KD,KC
  201 FORMAT (/  42H UNABLE TO INTEGRATE OVER SPIRAL FOR NODE ,I4,
     1 13H AND ELEMENT ,I3)
      IGER=1
      GOTO 2
    6 DO 7 IC=1,JD
C     COMPUTE DATA FOR GAUSS POINTS OF INTERVAL IC
      CALL BGFOR (JC,XB,3,JD,IC,XI,XG,UG,XJ,IERR)
      IF (IERR.NE.0) GOTO 999
C     COMPUTE KERNELS
      CALL BMKER (XC,XG,UG,TT,UT,IERR)
      IF (IERR.NE.0) GOTO 999
C     INCREMENT PRODUCT OF KERNEL AND JACOBIAN*GAUSS WEIGHT
      CALL BMKSF (TT,FUI,XJ,AI,1)
    7 CONTINUE
    4 CONTINUE
    3 CONTINUE
    2 CONTINUE
      RETURN
      END
```

```
      SUBROUTINE BADJI (KA,XA,TTL,UTL,KB,KC,ASO)
C     EXPLICIT INTEGRATION OVER INFINITE ELEMENT OF WHICH THE FIRST
C     ARGUMENT OF KERNELS IS A NODE, BEYOND END OF SPIRAL
      IMPLICIT DOUBLE PRECISION (A-H,O-Z)
      DIMENSION XA(2,3),TTL(2,2,3),UTL(2,2,3),ASO(2),
     1 XI(3),XC(2,3),UN(2,3),XJ(3),
     2 XIL(3),XCL(2,3),UNL(2,3),XJL(3),
     3 TT(2,2,3),UT(2,2,3),
     4 FUU(3,3),FUT(3,3)
      COMMON/PROP/YM,PR,DEN,DAG,KKPR(12)
      COMMON/FORM/LUO,KKFO(59)
      COMMON/FLAG/IGER,ILER
      DATA NIN/2/
      JB=2*NIN
      DO 1 IA=1,NIN
C     COMPUTE DATA FOR GAUSS POINTS INTERVAL IA
      CALL BGFOR (KA,XA,1,JB,IA,XI,XC,UN,XJ,IERR)
      IF (IERR.EQ.0) GOTO 2
  999 WRITE (LUO,200) KB
  200 FORMAT (/  24H ZERO JACOBIAN, ELEMENT ,I3)
      IGER=1
      GOTO 1
C     COMPUTE KERNELS OF THE INTEGRAL EQUATION
    2 CALL BMKER (XA,XC,UN,TT,UT,IERR)
      IF (IERR.EQ.0) GOTO 3
      WRITE (LUO,201) KB,KC
  201 FORMAT (/  39H KERNEL ARGUMENTS COINCIDE FOR ELEMENT ,I3,
     1 28H AND FIRST ARGUMENT AT NODE ,I4)
      IGER=1
      GOTO 1
C     COMPUTE SHAPE FUNCTIONS
    3 CALL BSHAI (XC,XA,ASO,FUU,FUT,3,IERR)
      IF (IERR.EQ.0) GOTO 4
  998 WRITE (LUO,202) KB
  202 FORMAT (/
     1 52H CANNOT COMPUTE ASYMPTOTIC SHAPE FUNCTIONS, ELEMENT ,I3)
      IGER=1
      GOTO 1
C     INCREMENT PRODUCTS OF KERNELS, SHAPE FUNCTIONS
C     AND JACOBIAN*GAUSS WEIGHT
    4 CALL BMKSF (TT,FUU,XJ,TTL,1)
      CALL BMKSF (UT,FUT,XJ,UTL,1)
      IF (IA.NE.1) GOTO 1
C     FIRST INTERVAL - LOG WEIGHT FUNCTION GAUSS FORMULA
      CALL BGFOL (KA,XA,1,JB,XIL,XCL,UNL,XJL,ORG,ELJ,IERR)
      IF (IERR.NE.0) GOTO 999
C     SUBTRACT LOG COMPONENT COMPUTED BY GAUSS-LEGENDRE FORMULA
      CALL BLKER (1,XI,ORG,ELJ,UT,DAG)
      CALL BMKSF (UT,FUT,XJ,UTL,1)
C     ADD LOG COMPONENT COMPUTED BY LOG FORMULA
      CALL BLKER (-1,XIL,ORG,ELJ,UT,DAG)
      CALL BSHAI (XCL,XA,ASO,FUU,FUT,3,IERR)
      IF (IERR.NE.0) GOTO 998
      CALL BMKSF (UT,FUT,XJL,UTL,1)
    1 CONTINUE
      RETURN
      END
```

```
      SUBROUTINE BMKER (XA,XC,UN,TT,UT,IERR)
C     COMPUTE KERNELS T AND U AT GAUSS POINTS OF INTERVAL
      IMPLICIT DOUBLE PRECISION (A-H,O-Z)
      DIMENSION XA(2),XC(2,3),UN(2,3),TT(2,2,3),UT(2,2,3)
      COMMON/PROP/YM,PR,DEN,DAG,KKPR(12)
      DATA QPI/12.5663706144D0/
C     DIMENSIONLESS ELASTIC CONSTANTS, DIVIDED BY DAG BECAUSE
C     BGFOR DOES NOT SCALE THE JACOBIAN
      ELT=QPI*(1.0D0-PR)*DAG
      EU=(1.0D0+PR)/ELT
      EL=EU*(3.0D0-4.0D0*PR)
      ET=(1.0D0-2.0D0*PR)/ELT
      ES=2.0D0/ELT
      RDAG=1.0D0/DAG
      DO 1 IA=1,3
      CA=(XA(1)-XC(1,IA))*RDAG
      CB=(XA(2)-XC(2,IA))*RDAG
      SUM=CA*CA+CB*CB
      IF (SUM.LT.1.0D-20) GOTO 999
C     1/R AND LOG(1/R)
      RAD=1.0D0/DSQRT(SUM)
      RAL=DLOG(RAD)
C     DIRECTION COSINES OF VECTOR FROM GAUSS POINT TO XA
      CA=CA*RAD
      CB=CB*RAD
      CAA=CA*CA
      CBB=CB*CB
      CAB=CA*CB
C     NORMAL AT GAUSS POINT
      UNA=UN(1,IA)
      UNB=UN(2,IA)
      UNCO=RAD*(UNA*CA+UNB*CB)
      UNCA=RAD*(UNA*CB-UNB*CA)
C     LOGARITHMIC AND BOUNDED COMPONENTS OF KERNEL U
      ELL=EL*RAL
      EUAA=EU*CAA
      EUBB=EU*CBB
      EUAB=EU*CAB
C     SYMMETRIC AND ANTISYMMETRIC COMPONENTS OF KERNEL T
      TAA=UNCO*(ES*CAA+ET)
      TBB=UNCO*(ES*CBB+ET)
      TAB=UNCO*ES*CAB
      TAN=UNCA*ET
C     KERNEL U
      UT(1,1,IA)=EUAA+ELL
      UT(1,2,IA)=EUAB
      UT(2,1,IA)=EUAB
      UT(2,2,IA)=EUBB+ELL
C     KERNEL T
      TT(1,1,IA)=TAA
      TT(1,2,IA)=TAB+TAN
      TT(2,1,IA)=TAB-TAN
      TT(2,2,IA)=TBB
    1 CONTINUE
      IERR=0
    2 CONTINUE
      RETURN
  999 IERR=1
      GOTO 2
      END
```

```
      SUBROUTINE BMINC (TTL,UTL,BD,BT,A,B,TTR,KC,NXB,NCC,
     1 KEL,KA,JFE,JFN,KB)
C     INCREMENT MATRIX AND LOAD VECTORS
      IMPLICIT DOUBLE PRECISION (A-H,O-Z)
      DIMENSION TTL(2,2,KB),UTL(2,2,KB),BD(2,3,NCC),BT(2,3,NCC),
     1 A(2,NXB,2,*),B(2,NXB,NCC),TTR(2,2,*),KEL(KB),
     2 JFE(2),JFN(2,KB)
      COMMON/PROP/YM,PR,DEN,DAG,KKPR(12)
      JA=KA-KC+1
      DO 1 IA=1,2
C     ELEMENT CONSTRAINT IN DIRECTION IA
      JB=JFE(IA)
      DO 2 IB=1,KB
      JC=KEL(IB)
C     INTEGRAL T FOR POSSIBLE ADJUSTMENT
      TTR(IA,1,JA)=TTR(IA,1,JA)+TTL(IA,1,IB)
      TTR(IA,2,JA)=TTR(IA,2,JA)+TTL(IA,2,IB)
      IF (JB.EQ.0) GOTO 3
C     ELEMENT IS CONSTRAINED
      A(IA,JC,1,JA)=A(IA,JC,1,JA)-UTL(IA,1,IB)
      A(IA,JC,2,JA)=A(IA,JC,2,JA)-UTL(IA,2,IB)
      GOTO 4
    3 CONTINUE
      IF (JFN(IA,IB).NE.0) GOTO 4
C     ELEMENT AND NODE ARE FREE
      A(IA,JC,1,JA)=A(IA,JC,1,JA)+TTL(IA,1,IB)
      A(IA,JC,2,JA)=A(IA,JC,2,JA)+TTL(IA,2,IB)
C     INCREMENT LOAD VECTORS
    4 DO 5 IC=1,NCC
      DO 6 ID=1,2
      B(ID,KA,IC)=B(ID,KA,IC)
     1 +UTL(IA,ID,IB)*BT(IA,IB,IC)/YM
     2 -TTL(IA,ID,IB)*BD(IA,IB,IC)/DAG
    6 CONTINUE
    5 CONTINUE
    2 CONTINUE
    1 CONTINUE
      RETURN
      END
```

```
      SUBROUTINE BDTTR (TTR,A,IFN,ISYM,NXB,KA,NEB,NEE)
C     ADJUST MATRIX COEFFICIENTS TO MAKE INTEGRAL T EXACTLY ZERO
      IMPLICIT DOUBLE PRECISION (A-H,O-Z)
      DIMENSION TTR(2,2,KA),A(2,NXB,2,KA),IFN(NXB),ISYM(2),JFN(2)
      IF (NEB.NE.NEE) GOTO 1
C     EITHER FINITE BODY OR INFINITE WITH NO INFINITE ELEMENTS
      FAC=1.0D0/NXB
      TTRS=0.0D0
      DO 2 IA=1,2
      IF (ISYM(IA).NE.0) GOTO 2
      DO 3 IB=1,KA
      DO 4 IC=1,2
C     GREATEST ADJUSTMENT AND ADJUSTMENT PER COLUMN
      TTRS=DMAX1(TTRS,TTR(IA,IC,IB))
      TTR(IA,IC,IB)=TTR(IA,IC,IB)*FAC
    4 CONTINUE
    3 CONTINUE
    2 CONTINUE
      IF (TTRS.GE.0.5D0) GOTO 1
C     FINITE BODY
      DO 5 IA=1,2
      IF (ISYM(IA).NE.0) GOTO 5
C     NO SYMMETRY - ADJUST INTEGRALS OF T
      DO 6 IB=1,NXB
      CALL DUNPK (IFN(IB),JFN,2)
      IF (JFN(IA).NE.0) GOTO 6
C     DISPLACEMENT UNKNOWN - ADJUST MATRIX COLUMN
      DO 7 IC=1,KA
      DO 8 ID=1,2
      A(IA,IB,ID,IC)=A(IA,IB,ID,IC)-TTR(IA,ID,IC)
    8 CONTINUE
    7 CONTINUE
    6 CONTINUE
    5 CONTINUE
    1 CONTINUE
      RETURN
      END

      SUBROUTINE BELIM (A,B,NXB,NCC,IFN,KA,KB)
C     ELIMINATIONS AT NODES KA-KB
      IMPLICIT DOUBLE PRECISION (A-H,O-Z)
      DIMENSION A(2,NXB,2,*),B(2,NXB,NCC),IFN(NXB),KFN(4)
      JA=2*NXB
      DO 1 IA=KA,KB
C     ELIMINATIONS NODE IA
      CALL DUNPK (IFN(IA),KFN,4)
      DO 2 IB=1,2
      IF (KFN(IB+2).EQ.0) GOTO 2
C     ELIMINATE UNKNOWN IB AT NODE IA - MODIFY MATRIX
      JC=IA-KA+1
      CALL PZERO (A(1,1,IB,JC),JA)
      A(IB,IA,IB,JC)=1.0D0
C     MODIFY LOAD VECTORS
      DO 4 IC=1,NCC
      B(IB,IA,IC)=0.0D0
    4 CONTINUE
    2 CONTINUE
    1 CONTINUE
      RETURN
      END
```

```
      SUBROUTINE BOVLE
      COMMON/FLAG/IGER,ILER
      CALL BMFAC
      IF (IGER.NE.0) GOTO 1
      CALL BFRBS
    1 CONTINUE
      RETURN
      END

      SUBROUTINE BMFAC
C     BIFACTORISATION OF MATRIX OF EQUATION COEFFICIENTS ON FILE LUA
C     L MATRIX WRITTEN TO FILE LUL, U MATRIX BACK TO LUA
      IMPLICIT DOUBLE PRECISION (A-H,O-Z)
      DIMENSION AL(119)
      COMMON/FORM/LUO,KKFO(59)
      COMMON/FLAG/IGER,ILER
      COMMON/GEOM/ISYM(2),MXA,NXB,NXE,MEA,NEB,NEE,NXY,
     1 LUA,LUL,LUG,ASO(2),NLIB,LUAL,LUAN,NBLS,MUA,KKGE(19)
      COMMON/CHANGE/A(14000),AU(2000),
     1 FI(2000),FO(2000)
      EQUIVALENCE (KUE,AU(2000))
      DATA NBLO/7/
      CALL PTICK (4)
      NBLS=NBLO
C     FIRST EQUATION OF SLOW BLOCK AND SLOW BLOCK RECORD NUMBER
      JA=1
      JB=1
C     U MATRIX RECORD NUMBER AND POINTER
      JC=1
      JD=1
C     COEFFICIENTS/EQUATION AND EQUATIONS/RECORD
    1 JE=NLIB-JA+1
      JF=LUAL/JE
C     LAST EQUATION OF SLOW BLOCK AND NUMBER OF RECORDS TO READ
      JG=JA+NBLS*JF-1
      JG=MIN0(JG,NLIB)
      JH=(JG-JA)/JF+1
C     READ SLOW BLOCK
      JI=JF*JE
      JJ=1
      JK=LUAL
      DO 2 IA=1,JH
      READ (LUA,REC=JB) (A(IO),IO=JJ,JK)
      JB=JB+1
      JJ=JJ+JI
      JK=JK+JI
    2 CONTINUE
C     LEADING DIAGONAL INCREMENT AND POINTER
      JH=JE+1
      JI=1
      IF (JG.EQ.JA) GOTO 3
C     ELIMINATIONS WITHIN SLOW BLOCK
      JJ=JE
      JK=JG-1
      DO 4 IA=JA,JK
      IF (ABS(A(JI)).LT.1.0D-20) GOTO 999
```

```
C     ELIMINATE UNKNOWN IA
      JL=JI+JE
      JM=JG-IA
      DO 5 IB=1,JM
C     FROM EQUATION IA+IB
      ELT=A(JL)/A(JI)
      AL(IB)=ELT
      JN=JL
      DO 6 IC=JI,JJ
      A(JN)=A(JN)-ELT*A(IC)
      JN=JN+1
    6 CONTINUE
      JL=JL+JE
    5 CONTINUE
C     L MATRIX COEFFICIENTS
      WRITE (LUL,IOSTAT=IERR,ERR=300) (AL(IO),IO=1,JM)
  300 CALL BFERR (LUL,IERR)
      JI=JI+JH
      JJ=JJ+JE
    4 CONTINUE
    3 CONTINUE
      IF (ABS(A(JI)).LT.1.0D-20) GOTO 999
C     COPY U MATRIX ROWS TO OUTPUT BUFFER AND WRITE TO FILE
      JI=1
      JJ=JE
      DO 7 IA=JA,JG
      JK=JD+JJ
      IF (JK.LE.LUAL) GOTO 8
C     NOT ROOM FOR ROW IA - SET POINTER TO LAST LEADING DIAGONAL
C     COEFFICIENT OF RECORD AND WRITE TO LUA
      KUE=JD-JJ-1
      WRITE (LUA,REC=JC,IOSTAT=IERR,ERR=301) AU
  301 CALL BFERR (LUA,IERR)
      JC=JC+1
      JD=1
C     COPY ROW
    8 CALL PSETF (A(JI),AU(JD),JJ)
      JD=JD+JJ
      JI=JI+JH
      JJ=JJ-1
    7 CONTINUE
      IF (JG.EQ.NLIB) GOTO 9
C     ELIMINATE UNKNOWNS JA TO JG FROM REST OF SYSTEM
C     FAST INPUT RECORD NUMBER, AND OUTPUT RECORD NUMBER AND POINTER
      JI=JB
      JJ=JB
      JK=1
C     COEFFICIENTS/EQUATION AND EQUATIONS/RECORD AFTER ELIMINATIONS
      JL=NLIB-JG
      JM=LUAL/JL
      JN=1
C     RECORDS TO READ, FIRST AND LAST EQUATIONS OF RECORD
      JO=(JL-1)/JF+1
      JP=JG+1
      JQ=JG+JF
      DO 10 IA=1,JO
      JQ=MIN0(JQ,NLIB)
      JR=JQ-JP+1
      READ (LUA,REC=JI) FI
```

```
       JS=1
      JT=JE
      JU=1
      DO 11 IB=JA,JG
C     ELIMINATE UNKNOWN IB
      JV=JU
      DO 12 IC=1,JR
C     FROM EQUATION IC OF FAST BLOCK
      ELT=FI(JV)/A(JS)
      AL(IC)=ELT
      JW=JV
      DO 13 ID=JS,JT
      FI(JW)=FI(JW)-ELT*A(ID)
      JW=JW+1
   13 CONTINUE
      JV=JV+JE
   12 CONTINUE
C     L MATRIX COEFFICIENTS
      WRITE (LUL,IOSTAT=IERR,ERR=302) (AL(IO),IO=1,JR)
  302 CALL BFERR (LUL,IERR)
      JS=JS+JH
      JT=JT+JE
      JU=JU+1
   11 CONTINUE
C     COPY PARTIALLY REDUCED EQUATIONS TO OUTPUT BUFFER
C     AND WRITE TO FILE AS NECESSARY
      DO 14 IB=1,JR
      IF (JN.LE.JM) GOTO 15
      WRITE (LUA,REC=JJ,IOSTAT=IERR,ERR=303) FO
  303 CALL BFERR (LUA,IERR)
      JJ=JJ+1
      JK=1
      JN=1
C     COPY EQUATION
   15 CALL PSETF (FI(JU),FO(JK),JL)
      JK=JK+JL
      JU=JU+JE
      JN=JN+1
   14 CONTINUE
      JI=JI+1
      JP=JP+JF
      JQ=JQ+JF
   10 CONTINUE
C     LAST RECORD OF PARTIALLY REDUCED EQUATIONS
      WRITE (LUA,REC=JJ,IOSTAT=IERR,ERR=304) FO
  304 CALL BFERR (LUA,IERR)
      JA=JG+1
      GOTO 1
C     LAST RECORD OF U MATRIX ROWS
    9 KUE=JD-1
      WRITE (LUA,REC=JC,IOSTAT=IERR,ERR=305) AU
  305 CALL BFERR (LUA,IERR)
      MUA=JC
      REWIND LUL
   16 CALL PTOCK (4)
      RETURN
  999 WRITE (LUO,200)
  200 FORMAT (/  16H SINGULAR MATRIX)
      IGER=1
      GOTO 16
      END
```

```
      SUBROUTINE BFRBS
C     FORWARD REDUCTION AND BACK SUBSTITUTION OF LOAD VECTORS
      IMPLICIT DOUBLE PRECISION (A-H,O-Z)
      COMMON/GEOM/ISYM(2),MXA,NXB,NXE,MEA,NEB,NEE,NXY,
     1 LUA,LUL,LUG,ASO(2),NLIB,LUAL,LUAN,NBLS,MUA,KKGE(19)
      COMMON/LOAD/NCC,KKLO(59)
      COMMON/CHANGE/B(2500),A(2000),AL(119),KKCH(30762)
      EQUIVALENCE (KUE,A(2000))
      CALL PTICK (5)
C     READ LOAD VECTORS
      JA=NLIB*NCC
      READ (LUG) (B(IO),IO=1,JA)
      REWIND LUG
C     FORWARD REDUCTION - FIRST EQUATION OF SLOW BLOCK
      JA=1
    1 JB=NLIB-JA+1
      JC=LUAL/JB
C     LAST EQUATION OF SLOW BLOCK
      JD=JA+NBLS*JC-1
      JD=MIN0(JD,NLIB)
      IF (JD.EQ.JA) GOTO 2
C     ELIMINATIONS WITHIN SLOW BLOCK
      JE=JD-1
      DO 3 IA=JA,JE
C     ELIMINATE UNKNOWN IA - READ L MATRIX COEFFICIENTS
      JF=JD-IA
      READ (LUL) (AL(IO),IO=1,JF)
      JG=IA
      DO 4 IB=1,NCC
C     LOAD CASE IB
      ELT=B(JG)
      JH=JG+1
      DO 5 IC=1,JF
      B(JH)=B(JH)-AL(IC)*ELT
      JH=JH+1
    5 CONTINUE
      JG=JG+NLIB
    4 CONTINUE
    3 CONTINUE
    2 CONTINUE
      IF (JD.EQ.NLIB) GOTO 6
C     ELIMINATE UNKNOWNS JA TO JD FROM REST OF SYSTEM
      JE=NLIB-JD
      JF=(JE-1)/JC+1
C     FIRST AND LAST EQUATIONS OF FAST BLOCK
      JG=JD+1
      JH=JD+JC
      DO 7 IA=1,JF
      JH=MIN0(JH,NLIB)
      JI=JH-JG+1
      DO 8 IB=JA,JD
C     ELIMINATE UNKNOWN IB - READ L MATRIX COEFFICIENTS
      READ (LUL) (AL(IO),IO=1,JI)
      JJ=IB
      JK=JG
      DO 9 IC=1,NCC
```

```
C     LOAD CASE IC
      ELT=B(JJ)
      JL=JK
      DO 10 ID=1,JI
      B(JL)=B(JL)-AL(ID)*ELT
      JL=JL+1
   10 CONTINUE
      JJ=JJ+NLIB
      JK=JK+NLIB
    9 CONTINUE
    8 CONTINUE
      JG=JG+JC
      JH=JH+JC
    7 CONTINUE
      JA=JD+1
      GOTO 1
    6 REWIND LUL
      JA=NLIB
C     U MATRIX ROWS FOR BACK SUBSTITUTION
      JB=MUA
      READ (LUA,REC=JB) A
      JC=KUE
C     COMPUTE LAST UNKNOWN
      JD=JA
      DO 11 IA=1,NCC
      B(JD)=B(JD)/A(JC)
      JD=JD+NLIB
   11 CONTINUE
      DO 12 IA=2,NLIB
      JA=JA-1
      JC=JC-IA
      IF (JC.GT.0) GOTO 13
C     U MATRIX ROWS
      JB=JB-1
      READ (LUA,REC=JB) A
      JC=KUE
   13 JD=JC+1
      JE=JC+IA-1
      JF=JA
      DO 14 IB=1,NCC
C     LOAD CASE IB - SUM OVER UNKNOWNS ALREADY CALCULATED
      SUM=0.0D0
      JG=JF+1
      DO 15 IC=JD,JE
      SUM=SUM+A(IC)*B(JG)
      JG=JG+1
   15 CONTINUE
C     COMPUTE UNKNOWN JA
      B(JF)=(B(JF)-SUM)/A(JC)
      JF=JF+NLIB
   14 CONTINUE
   12 CONTINUE
      CLOSE (UNIT=LUA,IOSTAT=JA,ERR=300)
  300 CLOSE (UNIT=LUL,IOSTAT=JA,ERR=301)
  301 CALL PTOCK (5)
      RETURN
      END
```

Appendix C
Fundamental Solutions

The solutions given here are for point sources and displacement discontinuities in an infinite domain and give the solution for the displacement or potential at a field or receiving point with coordinates x, y, $[z]$ due to the influence of the source which is located at x_0, y_0, $[z_0]$. Some derivatives of the fundamental solutions such as strain and stress are also presented. The following are referred to in the solutions (see diagram below).

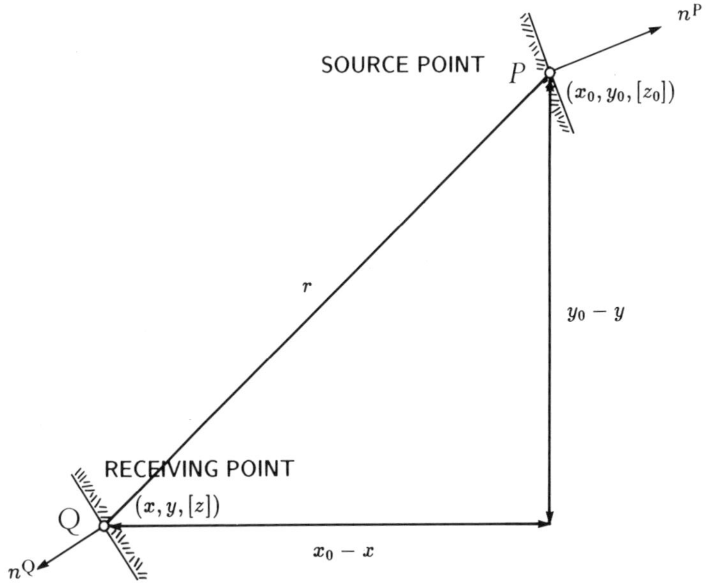

1. Outward normal to boundary at the source $(x_0, y_0, [z_0])$

$$\mathbf{n}^P = \begin{Bmatrix} n_x \\ n_y \\ [n_z] \end{Bmatrix}^P$$

2. Outward normal to boundary at the receiving point $(x, y, [z])$

$$\mathbf{n}^Q = \begin{Bmatrix} n_x \\ n_y \\ [n_z] \end{Bmatrix}^Q$$

The distance between the source and field points is given by

$$r = \sqrt{(x_0 - x)^2 + (y_0 - y)^2 + [(z_0 - z)^2]}$$

We define further

$$r_x = \frac{x_0 - x}{r}$$

$$r_y = \frac{y_0 - y}{r}$$

$$\left[r_z = \frac{z_0 - z}{r}\right]$$

$$\cos\theta = -r_x n_x^P - r_y n_y^P \left[-r_z n_z^P\right]$$

$$\cos\psi = r_x n_x^Q + r_y n_y^Q \left[+r_z n_z^Q\right]$$

$$\cos\beta = n_x^P n_x^Q + n_y^P n_y^Q \left[+n_z^P n_z^Q\right]$$

Note that values in square brackets are for the 3-D case only.
The values of n and $C, C_1, \ldots, C_8$ are given by the table below.

	Plane strain	Plane stress	3-D
n	1	1	2
C	$1/(8\pi G(1-\nu))$	$(1+\nu)/(8\pi G)$	$1/(16\pi G(1-\nu))$
C_1	$3-4\nu$	$3-4\frac{\nu}{1+\nu}$	$3-4\nu$
C_2	$1/(4\pi(1-\nu))$	$(1+\nu)/(4\pi)$	$1/(8\pi(1-\nu))$
C_3	$1-2\nu$	$(1-\nu)/(1+\nu)$	$1-2\nu$
C_4	2	2	3
C_5	$G/(2\pi(1-\nu))$	$(1+\nu)G/(2\pi)$	$G/(4\pi(1-\nu))$
C_6	4	4	5
C_7	$1-4\nu$	$(1-3\nu)/(1+\nu)$	$1-4\nu$
C_8	8	8	15

In the table, G is the shear modulus $\left(G = \frac{E}{2(1+\nu)}\right)$ and ν is Poisson's ratio.

C1 LAPLACE EQUATION—ISOTROPIC

$$\nabla^2 U = 0$$

Solution for potential:

$$U(P,Q) = \frac{1}{2\pi}\ln\left(\frac{1}{r}\right) \qquad \text{2-D}$$

$$U(P,Q) = \frac{1}{4\pi r} \qquad \text{3-D}$$

Solution for gradient:

$$T(P,Q) = \frac{\partial U}{\partial \mathbf{n}^Q} = \frac{\cos\theta}{2\pi r^n}$$

where

$$n = 1 \qquad \text{for 2-D}$$

$$n = 2 \qquad \text{for 3-D}$$

C2 ELASTICITY—ISOTROPIC POINT LOAD (KELVIN) SOLUTION

Solution for displacement:

$$U_{kk}(P,Q) = C\left[C_1 \ln\left(\frac{1}{r}\right) + r_k^2\right] \qquad \text{2-D}$$

$$U_{kk}(P,Q) = \frac{C}{r}\left[C_1 + r_k^2\right] \qquad \text{3-D}$$

$$U_{kl}(P,Q) = \frac{C}{r^{n-1}} r_k r_l \qquad l \neq k$$

Solution for strain:

$$\mathbf{E} = \mathbf{BU}$$

where

$$\mathbf{E} = \begin{bmatrix} E_{xxx} & E_{xxy} & [E_{xxz}] \\ E_{yyx} & E_{yyy} & [E_{yyz}] \\ [E_{zzx} & E_{zzy} & E_{zzz}] \\ E_{xyx} & E_{xyy} & [E_{xyz}] \\ [E_{yzx} & E_{yzy} & E_{yzz}] \\ [E_{xzx} & E_{xzy} & E_{xzz}] \end{bmatrix}$$

$$E_{kkk} = \frac{C}{r^n}\left[C_3 2r_k - r_k + C_4 r_k^3\right]$$

$$E_{kkl} = \frac{C}{r^n}\left[-r_k + C_4 r_k^2 r_l\right] \qquad l \neq k$$

$$E_{klk} = E_{lkk} = \frac{C}{r^n}\left[C_3 r_l + C_4 r_k^2 r_l\right] \qquad l \neq k$$

$$E_{ilk} = \frac{C}{r^n}\left[C_4 r_i r_l r_k\right] \qquad i \neq l \neq k$$

Solution for stress:

$$\mathbf{S} = \mathbf{DE}$$

where

$$\mathbf{S} = \begin{bmatrix} S_{xxx} & S_{xxy} & [S_{xxz}] \\ & etc & \\ & & \end{bmatrix}$$

$$S_{kkk} = \frac{C_2}{r^n}\left[C_3 r_k + C_4 r_k^3\right]$$

$$S_{kkl} = \frac{C_2}{r^n}\left[-C_3 r_k + C_4 r_k^2 r_l\right] \quad l \neq k$$

$$S_{klk} = S_{lkk} = \frac{C_2}{r^n}\left[C_3 r_l + C_4 r_k^2 r_l\right] \quad l \neq k$$

$$S_{ilk} = \frac{C_2}{r^n}\left[C_4 r_i r_l r_k\right] \quad i \neq l \neq k$$

Solution for traction at receiving point $(\mathbf{n}^Q)$:

$$T_{kk}(P,Q) = \frac{C_2}{r^n}\left(C_3 + C_4 r_k^2\right)\cos\psi$$

$$T_{kl}(P,Q) = \frac{C_2}{r^n}\left[C_4 r_k r_l \cos\psi - C_3\left(n_k^Q r_l - n_l^Q r_k\right)\right] \quad l \neq k$$

Derivative of traction solution (for internal point calculation):

$$\mathbf{R} = \mathbf{DBT}$$

where

$$\mathbf{R} = \begin{bmatrix} R_{xxx} & R_{xxy} & [R_{xxz}] \\ & etc & \\ & & \end{bmatrix}$$

$$R_{kkk} = \frac{C_5}{r^{n+1}} \left[C_4 \cos\psi \left(C_3 r_k + \nu 2 r_k - C_6 r_k^3\right) + 2n_k \left(C_4 \nu r_k^2 + C_3\right) \right.$$
$$\left. + n_k \left(C_4 C_3 r_k^2 - C_7\right)\right]$$

$$R_{kkl} = \frac{C_5}{r^{n+1}} \left[C_4 \cos\psi \left(C_3 r_l - C_6 r_k^2 r_l\right) \right.$$
$$\left. + n_k 2 C_4 \nu r_k r_l + n_l \left(C_4 C_3 r_k^2 - C_7\right)\right] \qquad l \neq k$$

$$R_{klk} = R_{lkk} = \frac{C_5}{r^{n+1}} \left[C_4 \cos\psi \left(\nu r_l - C_6 r_k^2 r_l\right) \right.$$
$$\left. + n_k C_4 \nu r_k r_l + n_l \left(C_4 \nu r_k^2 + C_3\right) + n_k C_4 C_3 r_k r_l\right] \qquad l \neq k$$

$$R_{ilk} = \frac{C_5}{r^{n+1}} \left[C_4 \cos\psi \left(-C_6 r_i r_l r_k\right) \right.$$
$$\left. + n_i C_4 \nu r_l r_k + n_l C_4 \nu r_i r_k + n_k C_4 C_3 r_i r_l\right] \qquad i \neq l \neq k$$

where k, l, i is either x, y, $[z]$.

C3 ELASTICITY–ISOTROPIC DISPLACEMENT DISCONTINUITY SOLUTION

Solution for displacement:

$$T_{kk}(P,Q) = \frac{C_2}{r^n} \left(C_3 + C_4 r_k^2\right) \cos\theta$$

$$T_{kl}(P,Q) = \frac{C_2}{r^n} \left[C_4 r_k r_l \cos\theta - C_3 \left(n_k^P r_l - n_l^P r_k\right)\right]$$

Solution for traction:

$$Z_{kk}(P,Q) = \frac{C_5}{r^n} \left\{\left[C_3 \ \cos\beta - C_4 \nu \cos\theta \cos\psi\right] \right.$$
$$+ \left[-C_4 C_3 \cos\theta + C_4 \nu \cos\psi n_k^P \right.$$
$$\left. + C_4 \nu \cos\beta r_k + C_8 \cos\theta \cos\psi r_k \right] r_l$$
$$+ \left[-(1-4\nu) n_k^Q + C_4 C_3 \cos\psi r_k\right] n_l^P$$
$$\left. + \left[C_3 n_k^P - C_4 \nu \cos\theta r_k\right] n_l^Q \right\}$$

where k, l is either $x, y, [z]$

$$
\begin{aligned}
Z_{kl}(P,Q) = \frac{C_5}{r^n} \Big\{ & [-C_4\, C_3 \cos\theta + C_4\nu \cos\psi n_k^P \\
& + C_4\nu \cos\beta r_k + C_8 \cos\theta \cos\psi r_k\,] r_l \\
& + [-(1-4\nu) n_k^Q + C_4 C_3 \cos\psi r_k] n_l^P \\
& + \left[C_3 n_k^P - C_4\nu \cos\theta r_k\right] n_l^Q \Big\}
\end{aligned}
$$

where $k \neq l$.

Appendix D Green's Symmetric Identity and the Boundary Integral Equation of the Direct Method for Laplace's Equation

According to the divergence theorem[8],

$$\int_V \nabla \mathbf{a} \, \mathrm{d}V = \int_S \mathbf{a} \cdot \mathbf{n} \, \mathrm{d}S \tag{D.1}$$

where $\mathbf{a}$ is any continuously differentiable vector field, and $\mathbf{n}$ is the unit outward normal to the surface S of the domain V. If the cartesian components of $\mathbf{a}$ are (a_x, a_y, a_z) then

$$\nabla \mathbf{a} = \frac{\partial a_x}{\partial x} + \frac{\partial a_y}{\partial y} + \frac{\partial a_z}{\partial z} \tag{D.2}$$

and

$$\mathbf{a} \cdot \mathbf{n} = a_x n_x + a_y n_y + a_z n_z \tag{D.3}$$

To see why $\mathbf{a}$ must be continuously differentiable, let us consider the case in which $\mathbf{a}$ is independent of y and z, and the domain V is a prism of unit cross sectional area as shown in Figure D1. Then $\nabla \mathbf{a} = \mathrm{d}a_x/\mathrm{d}x$, and on the faces $x = x_1$ and $x = x_2$ the dot product $\mathbf{a} \cdot \mathbf{n}$ equals $-a_x$ and a_x, respectively. Over the rest of S, $\mathbf{a} \cdot \mathbf{n} = 0$, and equation (D.1) reduces to the fundamental theorem of integral calculus:

$$\int_{x_1}^{x_2} \frac{\mathrm{d}a_x}{\mathrm{d}x} \, \mathrm{d}x = a_x(x_2) - a_x(x_1) \tag{D.4}$$

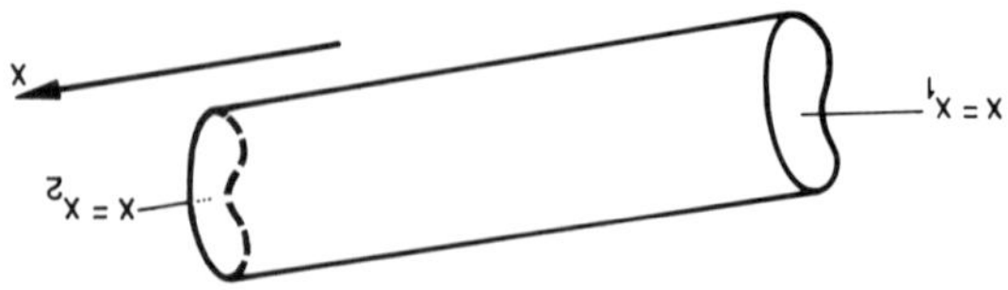

Figure D1 Prismatic domain of unit cross sectional area

In Figure D2 is shown a function a_x which is not continuously differentiable. Clearly equation (D.4) cannot apply for this function, as $a_x(x_2) - a_x(x_1)$ is negative but $\mathrm{d}a_x/\mathrm{d}x$ is positive in the interval $[x_1, x_2]$. An example of a function which is not continuously differentiable is $x/|x|$ in the interval $[-1, 1]$. As will be seen later, care must be taken to satisfy the differentiability condition when deriving boundary integral equations.

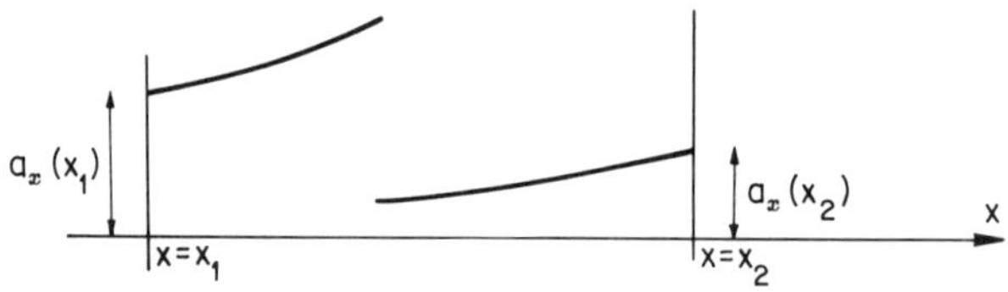

Figure D2 Function a_x not continuously differentiable

The formula leading to the boundary integral equation for Laplace's equation is derived by making certain substitutions in the divergence theorem. Let us replace $\mathbf{a}$ by the product $\phi\boldsymbol{\nabla}\psi$ where ϕ and ψ are arbitrary scalar fields, and according to the usual vector notation

$$\boldsymbol{\nabla}\psi = \left(\frac{\partial\psi}{\partial x}, \frac{\partial\psi}{\partial y}, \frac{\partial\psi}{\partial z}\right) \tag{D.5}$$

We then have

$$\phi\boldsymbol{\nabla}\psi = \left(\phi\frac{\partial\psi}{\partial x}, \phi\frac{\partial\psi}{\partial y}, \phi\frac{\partial\psi}{\partial z}\right) \tag{D.6}$$

and so equation (D.1) becomes

$$\int_V \left[\frac{\partial}{\partial x}\left(\phi\frac{\partial\psi}{\partial x}\right) + \frac{\partial}{\partial y}\left(\phi\frac{\partial\psi}{\partial y}\right) + \frac{\partial}{\partial z}\left(\phi\frac{\partial\psi}{\partial z}\right)\right] \mathrm{d}V$$

$$= \int_S \left[\phi\frac{\partial\psi}{\partial x}n_x + \phi\frac{\partial\psi}{\partial y}n_y + \phi\frac{\partial\psi}{\partial z}n_z\right] \mathrm{d}S \tag{D.7}$$

By differentiating the products on the left side of this equation and rearranging, we obtain the result

$$\int_V \left\{ \phi \left[\frac{\partial^2\psi}{\partial x^2} + \frac{\partial^2\psi}{\partial y^2} + \frac{\partial^2\psi}{\partial z^2} \right] + \left[\frac{\partial\phi}{\partial x}\frac{\partial\psi}{\partial x} + \frac{\partial\phi}{\partial y}\frac{\partial\psi}{\partial y} + \frac{\partial\phi}{\partial z}\frac{\partial\psi}{\partial z} \right] \right\} \mathrm{d}V$$
$$= \int_S \phi \left[n_x \frac{\partial\psi}{\partial x} + n_y \frac{\partial\psi}{\partial y} + n_z \frac{\partial\psi}{\partial z} \right] \mathrm{d}S \quad \text{(D.8)}$$

By noting that

$$n_x \frac{\partial\psi}{\partial x} + n_y \frac{\partial\psi}{\partial y} + n_z \frac{\partial\psi}{\partial z} = \frac{\partial\psi}{\partial \mathbf{n}} \quad \text{(D.9)}$$

and using vector notation, we may rewrite equation (D.8) as follows.

$$\int_V \left[\phi\nabla^2\psi + \boldsymbol{\nabla}\phi \cdot \boldsymbol{\nabla}\psi \right] \mathrm{d}V = \int_S \phi \frac{\partial\psi}{\partial \mathbf{n}} \mathrm{d}S \quad \text{(D.10)}$$

This result is known as Green's first identity. In our investigation of boundary value problems for Laplace's equation we have some interest in formulae with the operator ∇^2 under the integral sign as this may result in volume integrals vanishing; however, we do not wish to retain the term $\boldsymbol{\nabla}\phi \cdot \boldsymbol{\nabla}\psi$. Guided by this principle we rewrite equation (D.10) with ϕ and ψ interchanged:

$$\int_V \left[\psi\nabla^2\phi + \boldsymbol{\nabla}\psi \cdot \boldsymbol{\nabla}\phi \right] \mathrm{d}V = \int_S \psi \frac{\partial\phi}{\partial \mathbf{n}} \mathrm{d}S \quad \text{(D.11)}$$

By subtracting equation (D.11) from equation (D.10), we obtain the result known as Green's symmetric identity:

$$\int_V \left[\phi\nabla^2\psi - \psi\nabla^2\phi \right] \mathrm{d}V = \int_S \left[\phi \frac{\partial\psi}{\partial \mathbf{n}} - \psi \frac{\partial\phi}{\partial \mathbf{n}} \right] \mathrm{d}S \quad \text{(D.12)}$$

For this identity to be true, both ϕ and ψ must be twice continuously differentiable.

Let us take ϕ to be the solution u of the boundary value problem for Laplace's equation. Then, since $\nabla^2 u = 0$ in V,

$$\int_V u(Q)\nabla^2\psi(Q)\, \mathrm{d}V_Q = \int_S \left[u(Q) \frac{\partial\psi}{\partial \mathbf{n}}(Q) - \psi(Q) \frac{\partial u}{\partial \mathbf{n}}(Q) \right] \mathrm{d}S_Q \quad \text{(D.13)}$$

Now let us suppose that we know both u and $\partial u/\partial \mathbf{n}$ at all points of S, and that we wish to obtain by suitable choice of the second arbitrary function ψ a formula for u at some point P inside V. If ψ is chosen such that $\nabla^2\psi$ is zero everywhere except within a small neighbourhood $v(P,\epsilon)$ of P, then

$$u(P)\int_{v(P,\epsilon)} \nabla^2\psi(Q)\,\mathrm{d}V_Q \simeq \int_S \left[u(Q)\frac{\partial\psi}{\partial\mathbf{n}}(Q) - \psi(Q)\frac{\partial u}{\partial\mathbf{n}}(Q)\right]\mathrm{d}S_Q \quad \text{(D.14)}$$

The smaller the neighbourhood $v(P,\epsilon)$, the smaller the error in supposing that $u(Q) = u(P)$ in that neighbourhood and so the more accurately is $u(P)$ determined by equation (D.14). It is apparent that the ideal might be to take $\psi(Q)$ to be the fundamental solution $U(P,Q)$, so that $\nabla^2\psi$ (the differentiations being with respect to the coordinates of Q) would be zero everywhere except at P itself; there would then be no variation of $u(Q)$ over $v(P,\epsilon)$ and equation (D.14) would be exact. However, $U(P,Q)$ is not even once continuously differentiable at P. To overcome this difficulty, we write Green's symmetric identity not for the region V bounded by the surface S, but for a region $V - v(P,\epsilon)$ with surfaces S and $s(P,\epsilon)$ as shown in Figure D3. The neighbourhood $v(P,\epsilon)$ with surface $s(P,\epsilon)$ is arbitrarily taken to be a sphere of radius ϵ centred at P, and since P is inside V it is always possible to set ϵ small enough for $v(P,\epsilon)$ to lie entirely within V. Since $\nabla^2\psi$ equals zero everywhere in $V - v(P,\epsilon)$, the volume integral in equation (D.13) vanishes and we have

$$0 = \int_{S+s(P,\epsilon)} \left[u(Q)\frac{\partial}{\partial\mathbf{n}}U(P,Q) - U(P,Q)\frac{\partial u}{\partial\mathbf{n}}(Q)\right]\mathrm{d}S_Q \quad \text{(D.15)}$$

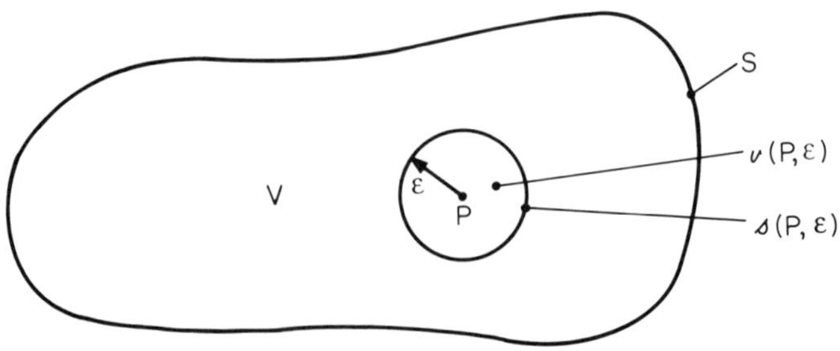

Figure D3 The region V–$v(P,\epsilon)$ with surface $S + s\,(P,\epsilon)$

In this equation, the differentiations are in the outward normal direction at Q. In the notation of equation (6.3), we have

$$0 = \int_S \left[T(P,Q)u(Q) - U(P,Q)\frac{\partial u}{\partial \mathbf{n}}(Q)\right] \mathrm{d}S_Q + \int_{s(P,\epsilon)} T(P,Q)u(Q)\,\mathrm{d}S_Q - \int_{s(P,\epsilon)} U(P,Q)\frac{\partial u}{\partial \mathbf{n}}(Q)\,\mathrm{d}S_Q \tag{D.16}$$

We now let ϵ tend to zero. The area of $s(P,\epsilon)$ is $4\pi\epsilon^2$, and in the last integral of equation (D.16), $U(P,Q) = \frac{1}{4\pi\epsilon}$. Therefore, if $\left|\frac{\partial u}{\partial \mathbf{n}}\right| \leq K$,

$$\left|\int_{s(P,\epsilon)} U(P,Q)\frac{\partial u}{\partial \mathbf{n}}\,\mathrm{d}S_Q\right| \leq 4\pi\epsilon^2 \cdot \frac{1}{4\pi\epsilon} \cdot K = K\epsilon \tag{D.17}$$

The value of the last integral therefore tends to zero as ϵ tends to zero. The second integral of equation (D.16) may be written as the sum of two parts:

$$\int_{s(P,\epsilon)} T(P,Q)u(Q)\,\mathrm{d}S_Q = u(P)\int_{s(P,\epsilon)} T(P,Q)\,\mathrm{d}S_Q + \int_{s(P,\epsilon)} T(P,Q) \cdot [u(Q) - u(P)]\,\mathrm{d}S_Q \tag{D.18}$$

The outward normal to the region $V - v(P,\epsilon)$ at a point Q on $s(P,\epsilon)$ is directed towards P, and so the angle θ of equation (6.3) is zero. Therefore, in equation (D.18), $T(P,Q) = \frac{1}{4\pi\epsilon^2}$. Noting also that $|[u(Q) - u(P)]| \leq K\epsilon$ (i.e. u is Lipschitz smooth),

$$\left|\int_{s(P,\epsilon)} T(P,Q)\,[u(Q) - u(P)]\,\mathrm{d}S_Q\right| \leq 4\pi\epsilon^2 \cdot \frac{1}{4\pi\epsilon^2} \cdot K\epsilon = K\epsilon \tag{D.19}$$

The value of the last integral of equation (D.18) therefore tends to zero as ϵ tends to zero. Finally,

$$\int_{s(P,\epsilon)} T(P,Q)\,\mathrm{d}S_Q = 4\pi\epsilon^2 \cdot \frac{1}{4\pi\epsilon^2} = 1 \tag{D.20}$$

Therefore, in the limit as ϵ tends to zero, equation (D.16) becomes

$$u(P) = \int_S \left[U(P,Q)\frac{\partial u}{\partial \mathbf{n}}(Q) - T(P,Q)u(Q)\right] \mathrm{d}S_Q \tag{D.21}$$

This result is known as the Poisson formula.

In the boundary value problems under consideration only one of the functions $u(Q)$ and $\frac{\partial u}{\partial \mathbf{n}}(Q)$ is given at each point of S and so the Poisson formula cannot be applied directly to compute results at interior points of V. The equation which enables the unknown boundary data to be computed from the given data is obtained by writing Green's symmetric identity with $\phi(Q) = u(Q)$ and $\psi(Q) = U(P,Q)$, but with P on S rather than inside V. Again the continuity requirements for validity of the identity must be respected, so the identity is written for the domain $V - v(P,\epsilon)$ with surface $S - S(P,\epsilon) + s(P,\epsilon)$ shown in Figure D4; $v(P,\epsilon)$ is now that part of V which lies inside a sphere of radius ϵ centred at P, and $s(P,\epsilon)$ is that part of the surface of the sphere which lies inside V. The region of exclusion $S(P,\epsilon)$ is that part of S which lies within the sphere, as shown in Figure D4. We have in place of equation (D.16)

$$0 = \int_{S-S(P,\epsilon)} \left[T(P,Q)u(Q) - U(P,Q)\frac{\partial u}{\partial \mathbf{n}}(Q) \right] \mathrm{d}S_Q + \int_{s(P,\epsilon)} T(P,Q)u(Q)\,\mathrm{d}S_Q - \int_{s(P,\epsilon)} U(P,Q)\frac{\partial u}{\partial \mathbf{n}}(Q)\,\mathrm{d}S_Q \qquad \text{(D.22)}$$

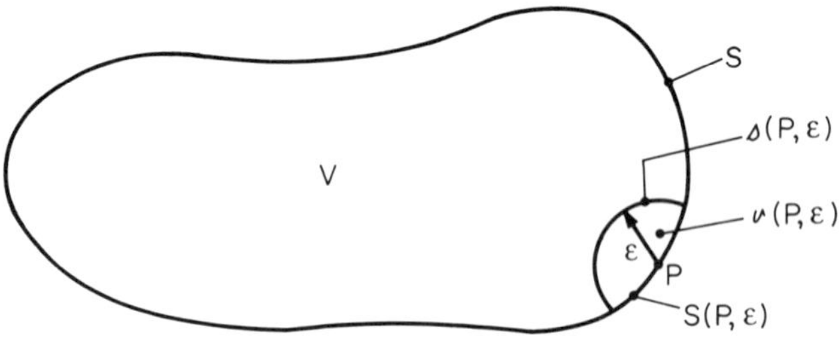

Figure D4 The domain V–$v(P,\epsilon)$ with surface S–$S(P,\epsilon) + s(P,\epsilon)$

We now let ϵ tend to zero. Equations (D.17) and (D.19) apply as before, so

$$\lim_{\epsilon\to 0}\left\{ \int_{s(P,\epsilon)} T(P,Q)u(Q)\,\mathrm{d}S_Q - \int_{s(P,\epsilon)} U(P,Q)\frac{\partial u}{\partial \mathbf{n}}(Q)\,\mathrm{d}S_Q \right\} = u(P)\lim_{\epsilon\to 0}\int_{s(P,\epsilon)} T(P,Q)\,\mathrm{d}S_Q \qquad \text{(D.23)}$$

The analysis now differs from that for an interior point P because $s(P,\epsilon)$ is no longer the entire surface of a sphere. Since $T(P,Q) = \frac{1}{4\pi\epsilon^2}$ on $s(P,\epsilon)$,

$$C(P) = \lim_{\epsilon\to 0}\int_{s(P,\epsilon)} T(P,Q)\,\mathrm{d}S_Q = \frac{\gamma(P)}{4\pi} \qquad \text{(D.24)}$$

where $\gamma(P)$ is the solid angle subtended at P by $s(P,\epsilon)$ in the limit as ϵ tends to zero. If P is on a smooth part of the surface, $\gamma(P) = 2\pi$ and so $C(P) = 1/2$. On a right-angled external edge, $\gamma(P) = \pi$ and so $C(P) = 1/4$. Meanwhile, the limiting value as ϵ tends to zero of the first integral of equation (D.22) is the Cauchy principal value (see Section 6.5). In the limit as ϵ tends to zero, therefore,

$$C(P)u(P) + \int_S T(P,Q)u(Q)\,\mathrm{d}S_Q = \int_S U(P,Q)\frac{\partial u}{\partial \mathbf{n}}(Q)\,\mathrm{d}S_Q \tag{D.25}$$

where $C(P)$ is given by equation (D.24). In practice it is not necessary to compute a Cauchy principal value because, due to the presence of the factor $\cos\theta$ (see equation (6.3)), the kernel $T(P,Q)$ is only weakly singular at P: that is, it tends to infinity as $1/r$, not $1/r^2$, as Q approaches P.

The analysis leading to the Poisson formula for a two-dimensional domain is similar to that for the three-dimensional case, except in that $s(P,\epsilon)$ is taken to be the boundary of a disc centred at P rather than a sphere. The Poisson formula and integral equation are the same, except in that $T(P,Q)$ and $U(P,Q)$ are now given by equations (6.6) and (6.4), and

$$C(P) = \lim_{\epsilon\to 0}\int_{s(P,\epsilon)} T(P,Q)\,\mathrm{d}S_Q = \frac{\gamma(P)}{2\pi} \tag{D.26}$$

where $\gamma(P)$ is the angle measured through the domain between the limiting tangents to S to the left and to the right of P. If P is on a smooth part of the surface, then $\gamma(P) = \pi$ and so $C(P) = 1/2$. If P is at a corner, then $\gamma(P)$ is the angle shown in Figure D5.

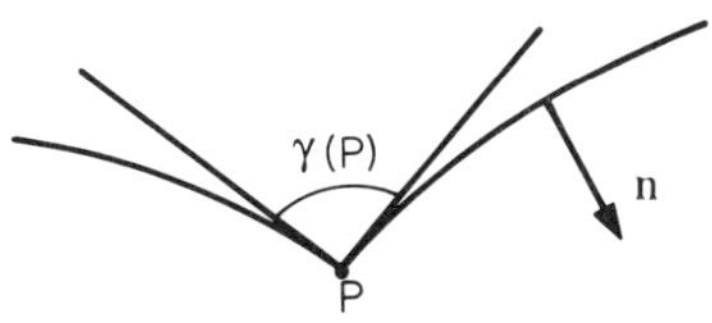

Figure D5 The angle $\gamma(P)$ for a two-dimensional domain

Appendix E
Betti's Theorem and the Boundary Integral Equation of the Direct Method for Elasticity

In the theory of elasticity, the roles played by stress and surface traction are the same as those played by gradient and normal derivative of the potential in the theory of Laplace's equation. With this parallel in mind, we consider the arbitrary displacement fields $\mathbf{u}^{(1)}$ and $\mathbf{u}^{(2)}$, to which correspond the stress tensors $[\boldsymbol{\sigma}]^{(1)}$ and $[\boldsymbol{\sigma}]^{(2)}$. The stress tensor is the two dimensional array

$$[\boldsymbol{\sigma}] = \begin{bmatrix} \sigma_x & \tau_{xy} & \tau_{xz} \\ \tau_{yx} & \sigma_y & \tau_{yz} \\ \tau_{zx} & \tau_{zy} & \sigma_z \end{bmatrix} \tag{E.1}$$

The formula from which the boundary integral equation for elasticity is obtained is derived by substituting for $\mathbf{a}$ in the divergence theorem, equation (D.1), the vectors $[\boldsymbol{\sigma}]^{(1)}\mathbf{u}^{(2)}$ and $[\boldsymbol{\sigma}]^{(2)}\mathbf{u}^{(1)}$. Taking $\mathbf{a} = [\boldsymbol{\sigma}]^{(1)}\mathbf{u}^{(2)}$ in the divergence theorem,

$$\begin{aligned}
&\int_V \left\{ \frac{\partial}{\partial x}\left[\sigma_x^{(1)}u_x^{(2)} + \tau_{xy}^{(1)}u_y^{(2)} + \tau_{xz}^{(1)}u_z^{(2)}\right] + \frac{\partial}{\partial y}\left[\tau_{yx}^{(1)}u_x^{(2)} + \sigma_y^{(1)}u_y^{(2)} + \tau_{yz}^{(1)}u_z^{(2)}\right] \right. \\
&\qquad\qquad \left. + \frac{\partial}{\partial z}\left[\tau_{zx}^{(1)}u_x^{(2)} + \tau_{zy}^{(1)}u_y^{(2)} + \sigma_z^{(1)}u_z^{(2)}\right]\right\} \mathrm{d}V \\
&= \int_S \left\{ \left[\sigma_x^{(1)}u_x^{(2)} + \tau_{xy}^{(1)}u_y^{(2)} + \tau_{xz}^{(1)}u_z^{(2)}\right] n_x + \left[\tau_{yx}^{(1)}u_x^{(2)} + \sigma_y^{(1)}u_y^{(2)} + \tau_{yz}^{(1)}u_z^{(2)}\right] n_y \right. \\
&\qquad \left. + \left[\tau_{zx}^{(1)}u_x^{(2)} + \tau_{zy}^{(1)}u_y^{(2)} + \sigma_z^{(1)}u_z^{(2)}\right] n_z \right\} \mathrm{d}S
\end{aligned} \tag{E.2}$$

We now differentiate the products on the left side of the equation and rearrange terms on both sides, to give

$$\int_V \left\{ \left[\frac{\partial \sigma_x^{(1)}}{\partial x} + \frac{\partial \tau_{yx}^{(1)}}{\partial y} + \frac{\partial \tau_{zx}^{(1)}}{\partial z} \right] u_x^{(2)} + \left[\frac{\partial \tau_{xy}^{(1)}}{\partial x} + \frac{\partial \sigma_y^{(1)}}{\partial y} + \frac{\partial \tau_{zy}^{(1)}}{\partial z} \right] u_y^{(2)} + \left[\frac{\partial \tau_{xz}^{(1)}}{\partial x} + \frac{\partial \tau_{yz}^{(1)}}{\partial y} + \frac{\partial \sigma_z^{(1)}}{\partial z} \right] u_z^{(2)} \right\} \mathrm{d}V$$
$$+ \int_V \left\{ \sigma_x^{(1)} \frac{\partial u_x^{(2)}}{\partial x} + \tau_{xy}^{(1)} \frac{\partial u_y^{(2)}}{\partial x} + \tau_{xz}^{(1)} \frac{\partial u_z^{(2)}}{\partial x} + \tau_{yx}^{(1)} \frac{\partial u_x^{(2)}}{\partial y} + \sigma_y^{(1)} \frac{\partial u_y^{(2)}}{\partial y} + \tau_{yz}^{(1)} \frac{\partial u_z^{(2)}}{\partial y} + \tau_{zx}^{(1)} \frac{\partial u_x^{(2)}}{\partial z} + \tau_{zy}^{(1)} \frac{\partial u_y^{(2)}}{\partial z} + \sigma_z^{(1)} \frac{\partial u_z^{(2)}}{\partial z} \right\} \mathrm{d}V$$
$$= \int_S \left\{ u_x^{(2)} \left[n_x \sigma_x^{(1)} + n_y \tau_{yx}^{(1)} + n_z \tau_{zx}^{(1)} \right] + u_y^{(2)} \left[n_x \tau_{xy}^{(1)} + n_y \sigma_y^{(1)} + n_z \tau_{zy}^{(1)} \right] + u_z^{(2)} \left[n_x \tau_{xz}^{(1)} + n_y \tau_{yz}^{(1)} + n_z \sigma_z^{(1)} \right] \right\} \mathrm{d}S \tag{E.3}$$

The terms in the square brackets in the first volume integral appear in the differential equations of equilibrium in terms of stresses, so for example

$$\frac{\partial \sigma_x^{(1)}}{\partial x} + \frac{\partial \tau_{yx}^{(1)}}{\partial y} + \frac{\partial \tau_{zx}^{(1)}}{\partial z} = -b_x^{(1)} \tag{E.4}$$

where $b_x^{(1)}$ is a component of the body force vector corresponding to the displacement field $\mathbf{u}^{(1)}$. In the surface integral, the terms in the square brackets are components of the traction vector corresponding to the displacement field $\mathbf{u}^{(1)}$, so for example

$$n_x \sigma_x^{(1)} + n_y \tau_{yx}^{(1)} + n_z \tau_{zx}^{(1)} = t_x^{(1)} \tag{E.5}$$

where $t_x^{(1)}$ is a component of the traction vector. Taking account of the symmetry of the stress tensor, the second integral may be rewritten in terms of the pseudo-vector $\boldsymbol{\sigma}$ and differential operator $\mathbf{B}$ as follows

$$\int_V \left\{ \sigma_x^{(1)} \frac{\partial u_x^{(2)}}{\partial x} + \tau_{xy}^{(1)} \frac{\partial u_y^{(2)}}{\partial x} + \tau_{xz}^{(1)} \frac{\partial u_z^{(2)}}{\partial x} + \tau_{yx}^{(1)} \frac{\partial u_x^{(2)}}{\partial y} + \sigma_y^{(1)} \frac{\partial u_y^{(2)}}{\partial y} + \tau_{yz}^{(1)} \frac{\partial u_z^{(2)}}{\partial y} + \tau_{zx}^{(1)} \frac{\partial u_x^{(2)}}{\partial z} + \tau_{zy}^{(1)} \frac{\partial u_y^{(2)}}{\partial z} + \sigma_z^{(1)} \frac{\partial u_z^{(2)}}{\partial z} \right\} \mathrm{d}V$$
$$= \int_V \boldsymbol{\sigma}^{(1)\mathrm{T}} \mathbf{B} \mathbf{u}^{(2)} \, \mathrm{d}V \tag{E.6}$$

Equation (E.3) may therefore be rewritten as follows:

$$\int_V -\mathbf{b}^{(1)} \cdot \mathbf{u}^{(2)}\, \mathrm{d}V + \int_V \boldsymbol{\sigma}^{(1)\mathrm{T}}\mathbf{B}\mathbf{u}^{(2)}\, \mathrm{d}V = \int_S \mathbf{t}^{(1)} \cdot \mathbf{u}^{(2)}\, \mathrm{d}S \tag{E.7}$$

where $\mathbf{b}^{(1)}$, $\mathbf{t}^{(1)}$ and $\boldsymbol{\sigma}^{(1)}$ are the body force vector, traction vector and stress pseudo-vector corresponding to the displacement field $\mathbf{u}^{(1)}$. Equation (E.7), obtained by substituting $\mathbf{a} = [\boldsymbol{\sigma}]^{(1)}\mathbf{u}^{(2)}$ in the divergence theorem, is analogous to Green's first identity, equation (D.10), for Laplace's equation.

Taking $\mathbf{a} = [\boldsymbol{\sigma}]^{(2)}\mathbf{u}^{(1)}$ in the divergence theorem and repeating the analysis leading to equation (E.7), we obtain the result

$$\int_V -\mathbf{b}^{(2)} \cdot \mathbf{u}^{(1)}\, \mathrm{d}V + \int_V \boldsymbol{\sigma}^{(2)\mathrm{T}}\mathbf{B}\mathbf{u}^{(1)}\, \mathrm{d}V = \int_S \mathbf{t}^{(2)} \cdot \mathbf{u}^{(1)}\, \mathrm{d}S \tag{E.8}$$

where $\mathbf{b}^{(2)}$, $\mathbf{t}^{(2)}$ and $\boldsymbol{\sigma}^{(2)}$ are the body force vector, traction vector and stress pseudo-vector corresponding to the displacement field $\mathbf{u}^{(2)}$. Let us consider the second volume integrals of equations (E.7) and (E.8). In equation (E.7), $\boldsymbol{\sigma}^{(1)\mathrm{T}} = \boldsymbol{\epsilon}^{(1)\mathrm{T}}\mathbf{D}^{\mathrm{T}}$ where $\boldsymbol{\epsilon}^{(1)}$ is the strain pseudo-vector corresponding to the displacement field $\mathbf{u}^{(1)}$, and $\mathbf{B}\mathbf{u}^{(2)} = \boldsymbol{\epsilon}^{(2)}$ where $\boldsymbol{\epsilon}^{(2)}$ is the strain pseudo-vector corresponding to the displacement field $\mathbf{u}^{(2)}$. Therefore,

$$\begin{aligned}
\boldsymbol{\sigma}^{(1)\mathrm{T}}\mathbf{B}\mathbf{u}^{(2)} &= \boldsymbol{\epsilon}^{(1)\mathrm{T}}\mathbf{D}^{\mathrm{T}}\boldsymbol{\epsilon}^{(2)} \\
&= \epsilon_x^{(1)}\epsilon_x^{(2)}D_{11} + \epsilon_y^{(1)}\epsilon_x^{(2)}D_{12} + \ldots + \gamma_{zx}^{(1)}\epsilon_x^{(2)}D_{16} \\
&\quad + \epsilon_x^{(1)}\epsilon_y^{(2)}D_{21} + \epsilon_y^{(1)}\epsilon_y^{(2)}D_{22} + \ldots + \gamma_{zx}^{(1)}\epsilon_y^{(2)}D_{26} \\
&\quad . \\
&\quad . \\
&\quad . \\
&\quad + \epsilon_x^{(1)}\gamma_{zx}^{(2)}D_{61} + \epsilon_y^{(1)}\gamma_{zx}^{(2)}D_{62} + \ldots + \gamma_{zx}^{(1)}\gamma_{zx}^{(2)}D_{66}
\end{aligned} \tag{E.9}$$

Similarly, in equation (E.8),

$$\begin{aligned}
\boldsymbol{\sigma}^{(2)\mathrm{T}}\mathbf{B}\mathbf{u}^{(1)} &= \boldsymbol{\epsilon}^{(2)\mathrm{T}}\mathbf{D}^{\mathrm{T}}\boldsymbol{\epsilon}^{(1)} \\
&= \epsilon_x^{(2)}\epsilon_x^{(1)}D_{11} + \epsilon_y^{(2)}\epsilon_x^{(1)}D_{12} + \ldots + \gamma_{zx}^{(2)}\epsilon_x^{(1)}D_{16} \\
&\quad + \epsilon_x^{(2)}\epsilon_y^{(1)}D_{21} + \epsilon_y^{(2)}\epsilon_y^{(1)}D_{22} + \ldots + \gamma_{zx}^{(2)}\epsilon_y^{(1)}D_{26} \\
&\quad . \\
&\quad . \\
&\quad . \\
&\quad + \epsilon_x^{(2)}\gamma_{zx}^{(1)}D_{61} + \epsilon_y^{(2)}\gamma_{zx}^{(1)}D_{62} + \ldots + \gamma_{zx}^{(2)}\gamma_{zx}^{(1)}D_{66}
\end{aligned} \tag{E.10}$$

Since $\mathbf{D}$ is symmetric, it can be seen from equations (E.9) and (E.10) that

$$\boldsymbol{\sigma}^{(2)\mathrm{T}}\mathbf{B}\mathbf{u}^{(1)} = \boldsymbol{\sigma}^{(1)\mathrm{T}}\mathbf{B}\mathbf{u}^{(2)} \tag{E.11}$$

and so the second volume integrals of equations (E.7) and (E.8) are equal. By subtracting equation (E.8) from equation (E.7) we obtain the result analogous to Green's symmetric identity:

$$\int_V \left[-\mathbf{b}^{(1)} \cdot \mathbf{u}^{(2)} + \mathbf{b}^{(2)}\mathbf{u}^{(1)}\right] \,\mathrm{d}V = \int_S \left[\mathbf{t}^{(1)} \cdot \mathbf{u}^{(2)} - \mathbf{t}^{(2)} \cdot \mathbf{u}^{(1)}\right] \,\mathrm{d}S \tag{E.12}$$

Equation (E.12) is called Betti's theorem. We take $\mathbf{u}^{(1)}$ to be the solution $\mathbf{u}$ of the boundary value problem of elasticity under consideration, this being a problem in which there is no body force. Since $\mathbf{b}^{(1)} = \mathbf{0}$, this yields the result

$$\int_V \mathbf{b}^{(2)}(Q)\mathbf{u}(Q) \,\mathrm{d}V_Q = \int_S \left[\mathbf{u}^{(2)}(Q)\mathbf{t}(Q) - \mathbf{t}^{(2)}(Q)\mathbf{u}(Q)\right] \,\mathrm{d}S_Q \tag{E.13}$$

where $\mathbf{t}$ is traction due to the displacement field $\mathbf{u}$.

By analogy with the procedure for Laplace's equation, we take $\mathbf{u}^{(2)}(Q)$ to be the fundamental solution $\mathbf{U}(P,Q)$ defined by equation (6.8), where P is an interior point of V. Since $\mathbf{U}(P,Q)$ is not twice continuously differentiable at P, we write Betti's theorem for the domain $V - v(P,\epsilon)$ with surfaces S and $s(P,\epsilon)$ shown in Figure D3. Since $\mathbf{b}^{(2)}(Q)$ equals zero everywhere in $V - v(P,\epsilon)$, the volume integral in equation (E.13) equals zero and we have

$$\begin{aligned}\mathbf{0} = \int_S [\mathbf{U}(P,Q)\mathbf{t}(Q) - \mathbf{T}(P,Q)\mathbf{u}(Q)] \,\mathrm{d}S_Q + \int_{s(P,\epsilon)} \mathbf{U}(P,Q)\mathbf{t}(Q) \,\mathrm{d}S_Q \\ - \int_{s(P,\epsilon)} \mathbf{T}(P,Q)\mathbf{u}(Q) \,\mathrm{d}S_Q\end{aligned} \tag{E.14}$$

where $\mathbf{T}(P,Q)$ is defined by equation (6.15). We now let ϵ tend to zero. In the same way as for Laplace's equation, it can be shown that the integral of $\mathbf{U}(P,Q)\mathbf{t}(Q)$ over $s(P,\epsilon)$ tends to zero, whereas that of $\mathbf{T}(P,Q)\mathbf{u}(Q)$ tends to the value $\mathbf{u}(P)$. Equation (E.14) therefore becomes a formula for the displacement vector at an arbitrary interior point of V:

$$\mathbf{u}(P) = \int_S [\mathbf{U}(P,Q)\mathbf{t}(Q) - \mathbf{T}(P,Q)\mathbf{u}(Q)] \,\mathrm{d}S_Q \tag{E.15}$$

This result is known as Somigliana's identity, and is analogous to the Poisson formula, equation (D.21), for Laplace's equation.

In the boundary value problems under consideration not all of the boundary data $\mathbf{t}(Q)$ and $\mathbf{u}(Q)$ are given. The integral equation which enables the unknown boundary data to be computed from the given data is obtained by writing Betti's theorem with $\mathbf{u}^{(1)}(Q) = \mathbf{u}(Q)$ and $\mathbf{u}^{(2)}(Q) = \mathbf{U}(P,Q)$, but with P on S rather than inside V, for the domain $V - v(P,\epsilon)$ with surface $S - S(P,\epsilon) + s(P,\epsilon)$ shown in Figure D4. We have in place of equation (E.14)

$$\mathbf{0} = \int_{S-S(P,\epsilon)} [\mathbf{U}(P,Q)\mathbf{t}(Q) - \mathbf{T}(P,Q)\mathbf{u}(Q)]\ \mathrm{d}S_Q + \int_{s(P,\epsilon)} \mathbf{U}(P,Q)\mathbf{t}(Q)\ \mathrm{d}S_Q$$
$$- \int_{s(P,\epsilon)} \mathbf{T}(P,Q)\mathbf{u}(Q)\ \mathrm{d}S_Q \qquad \text{(E.16)}$$

We now let ϵ tend to zero. It is found that

$$\lim_{\epsilon\to 0}\left(\int_{s(P,\epsilon)} \mathbf{U}(P,Q)\mathbf{t}(Q)\ \mathrm{d}S_Q - \int_{s(P,\epsilon)} \mathbf{T}(P,Q)\mathbf{u}(Q)\ \mathrm{d}S_Q\right) = -\mathbf{C}(P)\mathbf{u}(P) \qquad \text{(E.17)}$$

where

$$\mathbf{C}(P) = \lim_{\epsilon\to 0}\int_{s(P,\epsilon)} \mathbf{T}(P,Q)\ \mathrm{d}S_Q \qquad \text{(E.18)}$$

The limiting value defined by equation (E.18) exists whether P is on a smooth part of the surface or at an edge or corner; if P is on a smooth part of S then $\mathbf{C}(P) = \mathbf{I}/2$ where $\mathbf{I}$ is the unit matrix. The limiting value as ϵ tends to zero of the first integral of equation (E.16) also exists because the sum of that integral and those over $s(P,\epsilon)$ equals zero at every stage of the limiting process, and is equal to the Cauchy principal value as defined in Section 6.5. In the limit as ϵ tends to zero, therefore,

$$\mathbf{C}(P)\mathbf{u}(P) + \int_S \mathbf{T}(P,Q)\mathbf{u}(Q)\ \mathrm{d}S_Q = \int_S \mathbf{U}(P,Q)\mathbf{t}(Q)\ \mathrm{d}S_Q \qquad \text{(E.19)}$$

where $\mathbf{C}(P)$ is given by equation (E.18) and the Cauchy principal value of the integral on the left side is taken. The kernel $\mathbf{U}(P,Q)$ is only weakly singular and so it is not necessary to take the Cauchy principal value of the integral on the right side of the equation.

The analysis for plane strain is similar to that for the three-dimensional case. Somigliana's identity and the integral equation are the same, except in that $\mathbf{T}(P,Q)$ and $\mathbf{U}(P,Q)$ are given by equations (6.20) to (6.23), and in equation (E.18), $s(P,\epsilon)$ is that part of a circle of radius ϵ centred at P which lies inside V. If P is on a smooth part of the boundary then $\mathbf{C}(P) = \mathbf{I}/2$ as before. Note that for elasticity, $\mathbf{C}(P)$ cannot in general be expressed as a multiple of the angle $\gamma(P)$ referred to in Appendix D, for either the three-dimensional case or for plane strain.

Index